Yoshiki Oshida
Artificial Intelligence for Medicine

Also of Interest

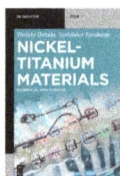

Nickel-Titanium Materials.
Biomedical Applications
Oshida, Tominaga, 2020
ISBN 978-3-11-066603-8, e-ISBN 978-3-11-066611-3

Magnesium Materials.
From Mountain Bikes to Degradable Bone Grafts
Oshida, 2021
ISBN 978-3-11-067692-1, e-ISBN 978-3-11-067694-5

Artificial Intelligence for Data-Driven Medical Diagnosis
Gupta, Kose, Le Nguyen, Bhattacharyya (Eds.), 2021
ISBN 978-3-11-066781-3, e-ISBN 978-3-11-066832-2

Data Science in Chemistry.
Artificial Intelligence, Big Data, Chemometrics and Quantum
Computing with Jupyter
Gressling, 2021
ISBN 978-3-11-062939-2, e-ISBN 978-3-11-062945-3

Yoshiki Oshida

Artificial Intelligence for Medicine

People, Society, Pharmaceuticals, and Medical Materials

DE GRUYTER

Author
Dr. Yoshiki Oshida, MS, PhD.
Adjunct Professor
University of San Fancisco School of Dentistry
Professor Emeritus
Indiana University School of Dentistry
Visiting Professor
University of Guam School of Science
408 Wovenwood, Orinda CA, 94563 USA

ISBN 978-3-11-071779-2
e-ISBN (PDF) 978-3-11-071785-3
e-ISBN (EPUB) 978-3-11-071792-1

Library of Congress Control Number: 2021941702

Bibliographic information published by the Deutsche Nationalbibliothek
The Deutsche Nationalbibliothek lists this publication in the Deutsche Nationalbibliografie;
detailed bibliographic data are available on the Internet at http://dnb.dnb.de.

© 2022 Walter de Gruyter GmbH, Berlin/Boston
Cover image: Yuuji/iStock/Getty Images Plus
Typesetting: Integra Software Services Pvt. Ltd.
Printing and binding: CPI books GmbH, Leck

www.degruyter.com

Preface

From birth (even pre-birth stage) to the graveyard, for twenty hours a day, to both healthy persons and patients, Artificial Intelligence (AI: in either specialized type assigned for a predetermined certain task or general type; in either ordinal artificial intelligence, machining learning, neural networks, or deep learning level) is influencing directly or indirectly our QOL (quality of living), QOS (quality of sleep), and even QOD (quality of dying). AI is a technology upon which our future is being built because it interacts with every aspect of our lives: health and medicine, housing, agriculture, transportation, sports, etc. Everybody claims to have "AI" in their products today. There are AI-powered juicers, AI-enabled WiFi routers, AI-enhanced cameras, AI-assisted vacuum cleaner, AI-controlled vital sensor, etc. It is true to say that no global company today survives without using the Internet, email, or mobile devices. Data, digital transformation, and machine intelligence will simply be table stakes for any organization that wants to stay competitive in an increasingly automated world [1]. From homes to offices, AI technology will be penetrating such that the ratio of automation will increase and ultimately affect human jobs. To some extent, our lives unconsciously get used to AI-assisted daily activities and exposed to AI-installed devices. Hence, if any AI-powered unit or system is not functioning properly, it is the moment when we recognize that our convenience has been supported and spoiled by AI technology. Just like air to our lives and AI is a part of our lives, and we are part of its development process [2, 3]. There are almost uncountable units, systems, or applications which are supported by the AI technology. This book is designed to concentrate on AI-supported medicine and dentistry, as well as some extensions along with the subsidiary activities supporting our lives. It is not our intention to develop any system using AI technology nor algorithm, rather to introduce prosperous applications of AI technology to enrich our lives and manage our healthy status. Our thoughts and concerns are limited for not only introducing such applications but also discussing associated problems including an accountability, ethics, copyright, and conflict between AI versus human brain or more. The book is basically constructed by introductory of AI in general; knowledge versus wisdom; periphery technological environment such as 5G movement (or even 6G in 2030); applications on medicine including treatment, diagnosis, follow-up prognosis, new drug development, some QOL/QOD-related anti-aging issue, and frailty; food-science and food-industry; applications in dentistry including prosthodontics, orthodontics, and implantology; and the last portion of this book will be discussing issues and limitation and future potentials of AI-technology.

In addition to the above structural schedule of this book, we paid a special attention to the statement that every new era has brought challenges and opportunities, requiring humans to adapt and grow, but in contrast to the linear transformations of the past, we are experiencing an exponential change happening in an AI era [4]. It is

https://doi.org/10.1515/9783110717853-202

said that there are dozens of definitions of the term AI, depending on your expertise and interest. Although there is no unified one, the average and commonly accepted definition indicates that AI is a function of a computer that imitates the advanced work performed by humans. But, since human intelligence is not yet fully understood, it is impossible to evaluate the intelligence of the AI. In addition, once when the functions are realized by AI, it will no longer be called AI, and will be incorporated as part of efficiency improvement and high functionality. There are fewer than 10,000 people in the world currently qualified to do state-of-the-art AI research and engineering [5], including data science team manager, machine learning engineers, data engineers, researchers, applied research scientists and engineers, or distributed systems engineers. Most of these AI researchers published their works mainly in any one of the following major journals:

Advanced Engineering Informatics
Applied Intelligence
Artificial Intelligence
Artificial Intelligence in Medicine
Artificial Intelligence Review
Artificial Life
Association for the Advancement of Artificial Intelligence
Autonomous Robots
Cognitive Psychology
Cognitive Science
Computational Linguistics
Cybernetics and Systems
Design Studies
Expert Systems with Applications
Frontiers in Neurorobotics
Fuzzy Optimization and Decision Making
Fuzzy Sets and Systems
IEEE Computational Intelligence Magazine
IEEE Intelligence Systems
IEEE Transactions on Fuzzy Systems
IEEE Transactions on Human-Machine Systems
Information Sciences
Integrated Computer-Aided Engineering
International Conference on Machine Learning
International Journal of Approximate Reasoning
International Journal of Computer Vision
International Journal of Fuzzy Systems
International Journal of Intelligent Systems
International Journal of Robotics Research
Japanese Society for Artificial Intelligence

Journal of the ACM
Journal of Artificial Intelligence Research
Journal of Intelligent and Fuzzy Systems
Journal of Machine Learning Research
Journal of Memory and Language
Journal of Parallel and Distributed Computing
Journal of Semantics
Knowledge and Information Systems
Knowledge-Based Systems
Machine Learning
Multidimensional Systems and Signal Processing
Neural Computing and Applications
Neurocomputing
Neural Information Processing Systems
Neural Networks
Neural Processing Letters
Networks and Spatial Economics
Pattern Recognition Letters
Physics of Life Reviews
Pattern Recognition
Soft Robotics
Swarm Intelligence
Topics in Cognitive Science

It is a good timing for this book because the linearly and continuous growth of AI has been shifted into new normal and paradigm shift. When conducting a research on numerous disciplines, there are normally three distinct approaches: interdisciplinary, multidisciplinary and trans-disciplinary [6, 7]. In addition to these three basic disciplinary orientations, there should be an intradisciplinary and cross-disciplinary [8]. In spite of various definitions on these disciplines, the following definitions are simple and clear [8–12]. Referring to Figure 1 [8], intradisciplinary: working within a single discipline; (additive) multidisciplinary: people from different disciplines working together, each drawing on their disciplinary knowledge and is driven by standardization and a general investment into the entire system through monodisciplinary representation of one's own profession; cross-disciplinary: viewing one discipline from the perspective of another; (interactive) interdisciplinary: integrating knowledge and methods from different disciplines, using a real synthesis of approaches as well as coordination, planning, and how the system can maintain itself drive sequential or interdisciplinary interdependence; and finally (holistic) transdisciplinary: creating a unity of intellectual frameworks beyond the disciplinary perspectives. And reciprocal interdependence or transdisciplinarity is driven by goals that include integrated input/output and the affecting of other disciplines by reorientation and the systemic function is homeostasis. Hence, transdisciplinary approach can create new idea based on involved individual disciplines [13, 14].

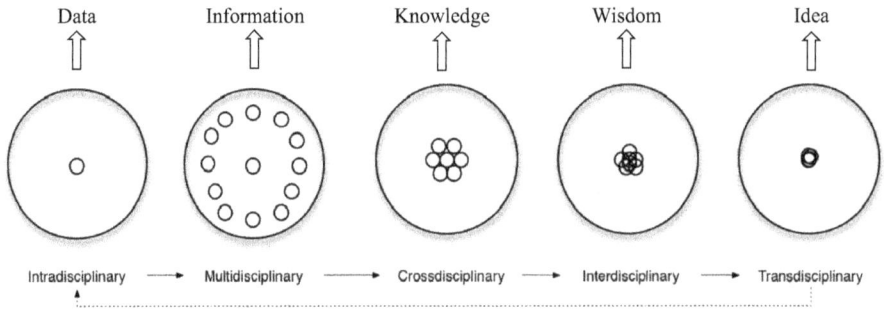

Figure 1: Recursive orientation between various disciplinaries [8, 11].

Riga et al. [15] developed a bio-psycho-social model characterized by being gnoseological/ epistemological, heuristic and wholistic/integrative as the purpose of transdisciplinarity. Recently, IEEE computer society technically sponsored "Transdisciplinary AI 2020: TransAI 2020" to focus on the interactions between AI and other research disciplines, including computer science, education, humanities, medicine, agriculture, sciences, engineering, law, and business [16]. Rudin [17] mentioned that the aim of AI is to augment or substitute biological intelligence with machine intelligence, and so far AI supports machine reasoning, natural language processing, speech and object recognition, human–computer interaction, and emotion sensing. To further study the brain, Rudin proposed an AI-research as an interconnected, transdisciplinary triangle (composed of neurobiology, neurophilosophy, and neuroinformatics), as seen in Figure 2 [17].

AI era is creating a new culture and proving golden opportunity to create new innovations. It is our hope that during exercising transdisciplinary AI-related phenomena throughout this book, comprehension on AI technology can be deepened, leading to a new school of research and science.

Three words (people, society, and medicine) used for subtitle of this book should represent principal contents, which will cover directly and indirectly human life and healthcare.

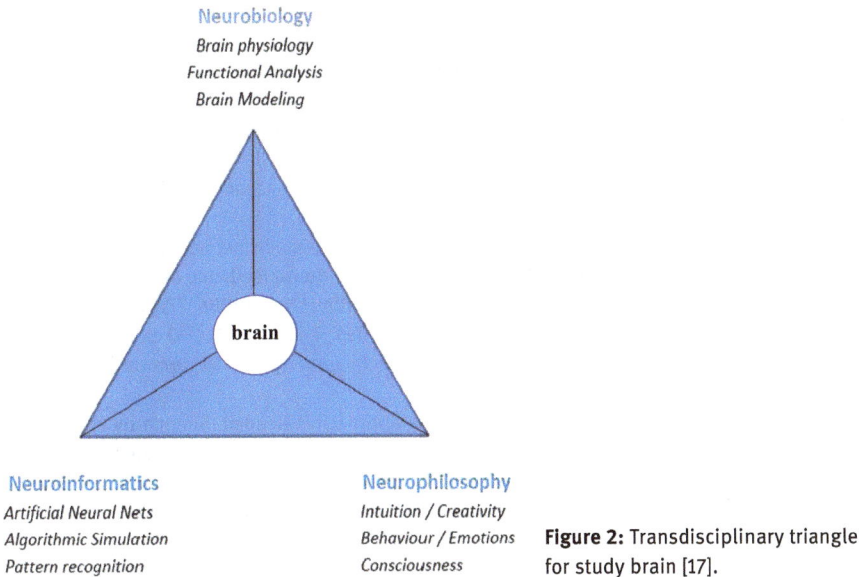

Neurobiology
Brain physiology
Functional Analysis
Brain Modeling

NeuroInformatics
Artificial Neural Nets
Algorithmic Simulation
Pattern recognition

Neurophilosophy
Intuition / Creativity
Behaviour / Emotions
Consciousness

Figure 2: Transdisciplinary triangle for study brain [17].

References

[1]   Webb A. The Big Nine. Public Affairs, New York NY, 2019.
[2]   CNN. AI set to exceed human brain power. Science and Space, 2006; http://edition.cnn.com/2006/TECH/science/07/24/ai.bostrom/.
[3]   https://en.wikipedia.org/wiki/Nick_Bostrom.
[4]   Hewitt J, Hintsa A. Exponential. Better Life, Better Performance from Formula 1 to Fortune 500. 2017, Createspace Independent Pub. 2016.
[5]   Metz C Tech Giants Are Paying Huge Salaries for Scarce A.I. Talent. 2017; http://www.nytimes.com/2017/10/22/technology/artificial-intelligence-experts-salaries.html.
[6]   Rosenfield PL. The potential of transdisciplinary research for sustaining and extending linkages between the health and social sciences. Social Science & Medicine. 1992, 35, 1343–57.
[7]   Ward MR, Rosenberg E. Revealing Mechanisms in a Transdisciplinary Community Reforestation Research Programme. African Evaluation Journal. 2020, 8, a467 https://doi.org/10.4102/aej.v8i1.467.
[8]   Jensenius AR Disciplinarities: intra, cross, multi, inter, trans. 2012. https://www.arj.no/2012/03/12/disciplinarities-2/.
[9]   Stember M. Advancing the social sciences through the interdisciplinary enterprise. The Social science journal. 1991, 28, 1–14.
[10] Newell WH. A Theory of Interdisciplinary Studies. Integrative Studies. 2001, 19, 1–25, http://web.mit.edu/jrankin/www/interdisciplinary/interdisc_Newell.pdf.
[11] Duran V From Galatians To Druidic Curriculum: An Old Perspective To Today's Problems. Conference: XII. Uluslararası Eğitim Araştırmaları Kongresi; at: Rİze. 2020;https://www.researchgate.net/publication/338356909_FROM_GALATIANS_TO_DRUIDIC_CURRICULUM_AN_OLD_PERSPECTIVE_TO_TODAY%27S_PROBLEMS.

[12]   Lotrecchiano GR Feature Article: Leadership is as Simple as A Child's Game of Marbles; Transdisciplinarity, Learning, and Complexity in Fairsies, Keepsies and Mibs. Integral Leadership Review. 2011; http://integralleadershipreview.com/3261-leadership-is-as-simple-as-a-childs-game-of-marbles/.

[13]   The Daily Omnivore. Transdisciplinarity; https://thedailyomnivore.net/2012/12/12/transdisciplinarity/.

[14]   Oseland. The Transdisciplinary Workplace. 2018; http://workplaceunlimited.blogspot.com/2018/10/the-transdisciplinary-workplace.html.

[15]   Riga S, Riga S, Ardelean A, Goldiş V Transdisciplinary in bio-medicine, neuroscience and psychiatry: The Bio-Psycho-Social Model. The Publishing House Medicine of the Romanian Academy. 2014. https://www.researchgate.net/publication/284878786_TRANSDISCIPLINARITY_IN_BIO-MEDICINE_NEUROSCIENCE_AND_PSYCHIATRY_THE_BIO-PSYCHO-SOCIAL_MODEL.

[16]   TransAI 2020: Transdisciplinary AI. http://www.wikicfp.com/cfp/servlet/event.showcfp?eventid=98230&copyownerid=135733.

[17]   Rudin P Human-Level AI: Why a Transdisciplinary Approach is Needed. Singularity 2030. 2020; https://singularity2030.ch/human-level-ai-why-a-transdisciplinary-approach-is-needed/Singulaity.

Contents

Chapter 6
AI in QOL (quality of living), QOS (quality of sleeping), and QOD (quality of dying) —— 101

Chapter 7
AI in food industry —— 118

List of abbreviations

AAAAI	American Academy of Allergy Asthma and Immunology
AAT	animal-assisted therapy
AC	artificial consciousness
ACDC	AI comprehensive dental care
ACP	advanced care planning
ACT	AI consciousness test
AD	Alzheimer's disease
ADCs	antibody drug conjugates
ADHD	attention-deficit/hyperactivity disorder
ADMET	absorption, distribution, metabolism, excretion (or elimination), and toxicity
ADR	adverse drug reactions
ADRs	adenoma detection rates
AF	apical foramen
AF	atrial fibrillation
AFRAID	analysis of frailty and death
AGI	artificial general intelligence
AH	artificial humanity
AI	artificial intelligence, or augmented intelligence
AIoT	artificial intelligence of things
AIPWE	augmented IPWE
AIT	algorithmic information theory
aka	also known as
AMD	age-related macular degeneration
AMeDAS	automated meteorological data acquisition system
AMH	anti-Müller tube hormone
AML	acute myeloid leukemia
ANI	artificial narrow intelligence
ANN	artificial neural network
API	application programing interface
AR	augmented reality
ARC	abstract and reasoning corpus
ARDS	acute respiratory distress syndrome
ART	accountability, responsibility, and transparency
ART	assisted reproductive technology
ASD	autism spectrum disorders
ASI	artificial super intelligence
ASSM	American Society of Sleep Medicine
AT	avatar therapy
ATM	automated teller machine
ATNC	adversarial threshold neural computer
AVH	auditory verbal hallucinations
AUC	area under the ROC curve
BAT	Baidu, Alibaba, and Tencent
BC	breast cancer
BCIs	brain–computer interfaces
BCS	biopharmaceutical classification system
BDI	Beck depression inventory

https://doi.org/10.1515/9783110717853-204

BFCC	Bark frequency cepstral coefficients
BM	brain metastases
BMs	biomarkers
BMI	brain–machine interface
BMI	body mass index
BN	Bayesian network
BP	back propagation
BPO	business process outsourcing
BPSD	behavior and psychological symptoms of dementia
CAD	computer-aided design
CAD	computer-assisted diagnosis
CADD	computer-aided detection and diagnosis
CAM	computer-aided manufacturing
CAPPOIS	computer-assisted preoperative planning for oral implant surgery
CBC	complete blood count
CBCT	cone beam CT
CBF	cephalic blood flow
CBT	cognitive-behavioral therapy
CDC	Centers for Disease Control and Prevention
CDS	clinical decision support
cfDNA	cell-free DNA
CHMP	Committee on Human Medicinal Products
CI	cochlear implant
CIT	constraint-induced therapy
CLC	closed-loop control
CLL	chronic lymphocytic leukemia
CNB	core needle biopsy
CNC	computer numerical control
CNN	convolutional neural network
COPD	chronic obstructive pulmonary disease
COVID-19	corona-virus-disease-2019
CPAP	continuous positive airway pressure
CPS	cyber-physical system
CPU	central processing unit
CRC	colorectal cancer
CRP	C-reactive protein
CT	computerized tomography
ctDNA	circulating tumor DNA
CTG	cardiotocography
CTM	clinical trial matching
cTnI	cardiac troponin I
CURB	confusion, urea, respiratory rate, and blood pressure criteria
CVD	cardiovascular disease
CVS	cervical vertebrae stages
DA	data analysis
DAO	decentralized autonomous organization
DARE	deception analysis and reasoning engine
DARPA	Defense Advanced Research Projects Agency
DBS	deep-brain stimulation

DCNN	deep convolutional neural network
DDKT	deceased donor kidney transplantation
DDS	doctor of dental surgery
DDS	drug delivery system
DDx	differential diagnosis
DHA	digital health advisors
DI	discomfort index
DI	drug information
DICOM	digital imaging and communication in medicine
DIKW	data, information, knowledge, and wisdom
DIY	do-it-yourself
DL	deep learning
DL-MDS	deep-learning-based medical diagnosis system
DM	data mining
DME	diabetic macular edema
DNA	deoxyribonucleic acid
DNI	direct neural interface
DNN	deep neural network
DP	differentiable plasticity
DR	diabetic retinopathy
DRE	digital rectal examination
D-R matching	donor–recipient matching
DS	data science
DTR	dynamic treatment regime
DV	decidual vasculopathy
DX	digital transformation
EASO	European Association for the Study of Obesity
EBD	evidence-based dentistry
EBM	evidence-based medicine
EBL	evidence-based learning
ECG	electrocardiogram
ECS	emergency care system
ED	emergency department
EEG	electroencephalogram
EHRs	electronic health records
EI	edge intelligence
EI	emotional intelligence
EI	extended intelligence (or XI)
EIQ	emotional intelligence quotient
eMBB	enhanced mobile broadband
EMI	electro-magnetic interference
EMRs	electronic medical records
EMS	emergency medical service
ENT	ear, nose, and throat
EOC	epithelial ovarian cancer
EP	electrophysiology
EQ	emotional quotient
ER	emergency room
ERP	event-related potential

ES	expert system
ES	ethical sourcing
ESI	emergency severity index
FAO	Food and Agriculture Organization (of the United Nation)
FDA	US Food and Drug Administration
FHIR	fast healthcare interoperability resources
FHR	fetal heart rate
FIs	frailty indices
fMRI	functional MRI
FRIGHT	frailty inferred geriatric health timeline
GA	genetic algorithms
GAFA	Google, Apple, Facebook, and Amazon
GAN	generative adversarial networks
GC	gastric cancer
GDM	gestational diabetes mellitus
GDPR	general privacy data protection regulation
GI	gastrointestinal
GIST	gastrointestinal stromal tumor
G-MAFIA	Google, Microsoft, Apple, Facebook, IBM, and Amazon
GON	glaucomatous optic neuropathy
GP	general physician
GP	genetic programming
GPS	global positioning system
GPU	graphics processing unit
GRAS	generally recognized as safe
GSR	galvanic skin response
H&E	haematoxylin and eosin
HAMS	heart activity monitoring system
HCC	hepatocellular carcinoma
HDL	high-density lipoprotein
HEMS	home energy management system
HHS	(US Department of) Health and Human Services
HI	heat index
HIPAA	Health Insurance Portability and Accountability Act
HIT	health information technology
HIV	human immunodeficiency virus
HMI	human machine interface
HPV	human papillomavirus
HR	heart rate
HRV	heart-rate variability
HTTP	hypertext transfer protocol
HTTPS	hypertext transfer protocol secure
IBC	intrabody communication
IC	informed consent
ICD	implantable cardioverter defibrillator
ICF	international classification of functioning
ICH	intracranial hemorrhage
ICIDH	International Classification of Impairments, Disabilities, and Handicaps
ICSI	intracytoplasmic sperm injection

ICT	information and communication technology
ICU	intensive care unit
IEEE	Institute of Electrical and Electronics Engineers
IFPC	International Federation for Produce Coding
IND	investigation new drug
IoMT	internet of medical things
I/O signals	input/output signals
IoT	internet of things
IP	intellectual property
IPP	intellectual property policy
IPWE	inverse-probability weighted estimator
IQ	intelligence quotient
IRC	intraretinal cystoid fluid
IT	information technology
IVF	*in vitro* fertilization
KDRI aka KDPI	kidney donor risk index, also known as kidney donor profile index
LAN	local area network
LBDD	ligand-based drug design
LCD	liquid crystal display
LDL	low-density lipoprotein
LED	light-emitting diode
LPC	linear predictive coding
LPCC	linear predictive cepstral coefficients
LR	logistic regression
LV	left ventricle
LVOs	large vessel occlusions
MA	microalbuminuria
MACBETH	measuring attractiveness by a categorial-based evaluation technique
MAN	metropolitan area network
MAR-ASD	maternal autoantibody-related autism spectrum disorder
MC	machine consciousness
mcDNN	multichannel deep neural network
MCI	mild cognitive impairment
MCI-C	converters MCI
MCI-NC	nonconverters MCI
MCPs	medically compromised patients
MD	medical doctor
MDD	major depressive disorder
MELD	model for end-stage liver disease
MERS-CoV	Middle-East respiratory syndrome – coronavirus
MFCC	Mel frequency cepstral coefficients
MGI	materials genome initiative
MGS	mouse grimace scale
MI	materials informatics
MIDRC	Medical Imaging and Data Resource Center
MI-GAN	medical imaging – GAN
MI procedures	minimally-invasive procedures
MIS	minimally invasive surgery
ML	machine learning

MLCs	machine learning classifiers
MLP	multi-layer perceptron
MMI	mind-machine interface
MR	mixed reality
MRI	magnetic resonance imaging
mpMRI	multiparametric MRI
MSS	membrane-type surface sensor
mMTC	massive machine type communication
MYO	myoglobin
MZCs	monolithic zirconia crowns
NC	neurotypical controls
NC	numerical control
NCC	neural correlates of consciousness
NCDs	noncommunicable diseases
NCI	neural control interface
NDA	new drug application
NGS	next-generation sequencing
NHS	National Health Service (in England)
NI	natural intelligence
NIH	National Institute of Health (in USA)
NILT	near-infrared-light transillumination
NISQ computer	noisy intermediate scale quantum computers
NIT	near-infrared transillumination
NLP	natural language processing
NN	neural network
NPN	natural peptide network
NSQ	next-generation sequencing
OB/GYN	obstetrics and gynecology
OCR	optical character recognition
OCT	optical coherence tomography
OGTT	oral glucose tolerance test
OLE	confocal laser endomicroscopy
OPS	operations per second
OSCC	oral squamous cell carcinoma
OSS	open source software
OTC	over-the-counter
PAI clone	personal AI clone
PACS	picture archiving and communication systems
PAN	personal area network
PBL	periodontal bone loss
PCT	procalcitonin
PCT	principal component analysis
PCT	periodontally compromised teeth
PD	Parkinson's disease
PD	public domain
PE	pulmonary embolism
PEFS	pathologist-estimated fibrosis score
PEG	polyethylene glycol
PET	position emission tomography

PF	physical functioning
PFF	precision fish farming
PGA	peak ground acceleration
PHI	protected/personal health information
PHR	personal health records
PKC	problem-knowledge coupling
PLU	price look up
PMMA	polymethyl methacrylate
PNDs	psychiatric and neurological disorders
PNN	probabilistic neural network
POMR	problem-oriented medical record
POS	problem-oriented system
PPG	photoplethysmography
PSA	prostate-specific antigen
PTSD	posttraumatic stress disorder
QOA	quality of aging
QOD	quality of dying
QOL	quality of living
QOS	quality of sleep
QR	quick response
ORAC	oxygen radical absorption capacity
QSAR	quality structure–activity relationships
QSPR	quality structure–property relationships
RAM	restoring active memory
R-CNN	region-based CNN
REM	rapid eye movement
RF	radio frequency
RF	random forest
RFID	radio-frequency identification
RL	reinforcement learning
RMM	remote monitoring and management
RNA	ribonucleic acid
RNFL	retinal nerve fiber layer
RNN	recurrent neural network
ROC	receiver operating characteristics
ROI	return on investment
ROI	region of interest
ROP	retinopathy of prematurity
ROT	redundant, obsolete, and trivial
RPA	robotic process automation
RPS	robotics process automation
RS	resting-state
RT-PCR	reverse transcription polymerase chain reaction
RV	right ventricle
RVO	retinal vein occlusion
R2R	roll-to-roll
SaO2	oxygen saturation
SAR	structure–activity relationship
SARS-CoV-2	severe acute respiratory syndrome coronavirus-2

SBDD	structure-based drug design
SC	synthetic consciousness
scDNN	single-channel DNN
SCG	seismocardiogram
SCM	supply chain management
SDGs	sustainable development goals
SI	superintelligence
SNS	social networking service
SOAP	simple object access protocol
SOAP	subjective, objective, assessment, and plan
SODA	strategic options development and analysis
SOL	standard of living
SOM	self-organizing map
SPECT	single-photon emission CT
SRF	subretinal fluid
STL	standard triangle language or standard tessellation language
SVMs	support vector machines
TAT	turn-around-time
TB	tuberculosis
TBI	traumatic brain injury
TCD	thalamocortical dysrhythmia
TDDS	transdermal drug delivery system
TF	target finishing
TH	telehealth
TMJ	temporomandibular joint
TNBCs	triple-negative breast cancers
TNF	tumor necrosis factor
tPA	tissue plasminogen activator
TPMS	therapeutic performance mapping system
TRIPOD	transparent reporting of a multivariable prediction model for individual prognosis or diagnosis
TRUS	transrectal ultrasound
TTE	transthoracic echocardiography
T1D	type 1 diabetes
T2D	type 2 diabetes
URLLC	ultra-reliable and ultra-low latency communications
US	ultrasound
UX	user experience
VA	virtual agent
VATS	video-assisted thoracoscopic surgery
VF	visual field
VR	virtual reality
VRFs	vertical root fractures
VSM	vital signs monitor
VUI	voice user interface
WAN	wide area network
WCD	wearable cardioverter defibrillator
WGS	whole-genome sequencing
WHDD	wearable heat-stroke-detection device

WHO	World Health Organization
WIPO	World Intellectual Property Organization
WMIF	world medical innovation forum
XAI	explainable AI
XI	extended intelligence (or EI)

About the Author

Yoshiki Oshida is Professor Emeritus at Indiana University School of Dentistry, Adjunct Full Professor at the University of California San Francisco School of Dentistry, and Visiting Professor at the University of Guam School of Science. He received Master of Science from Syracuse University, Syracuse, New York, and PhD in Materials Science and Engineering from Waseda University, Tokyo Japan. Prof. Oshida has published more than 100 peer-reviewed articles, 80 oral presentations, 50 invited lectures (in Japan, Korea, Canada, Turkey, Taiwan, and Thailand), 8 book chapters and 8 book publications. His accumulated research grand is about US$1.2 million from different sources including dental industries, ADA, NASA, Navy, IBM, Exxon, etc. His research interests include corrosion science and engineering, characterization of dental materials and biomaterials, implant system developments, superplastic forming and diffusion bonding, non-destructive failure analysis, fatigue damage assessment, robot materials science and engineering, AI-dentistry and AI-medicine, and pain management and recognition pain-mechanism.

https://doi.org/10.1515/9783110717853-205

Chapter 1
AI in general

1.1 Undefined definition

Air conditioner, cleaner, laundry machine, or many other home appliances are nowadays in market under advertisement of "AI-installed xxx." Unfortunately, technologies involved in these devices and equipment are not artificial intelligence (AI), rather system engineering or control engineering which has a longer technological history than AI. At the same time, AI is well known to be used in various places such as automatic driving to avoid obstacles, smartphone speech recognition by smartphone, Internet image search, web page search, robot control, or image processing in the industrial field. Besides the term "AI-installed," there are other terms such as "AI-assisted," "AI-involved," "AI-enriched," or "AI-powered." Although it can be said that a simple and fundamental concept commonly found in these terms is a mimicking slice(s) of human intelligence and/or behavior. If the definition of AI is asked, it would not be surprised to find that the term "artificial intelligence" is not clearly defined. It refers to a very large category, depending on the position. Being similar to variable definitions of the phenomenological term "engineering fatigue" [1], the definition of intelligence is not clear, so the term "artificial intelligence" cannot be clearly defined and should differ, depending on an area of expertise and position. There is a variety of definitions as follows [2–9]: AI is comprehended as a mechanism or system, a computer or computer program, or others. When AI is considered as a mechanism or system, it can be defined as (i) an artificially created human intelligence, (ii) a mechanism with intelligence or a mechanism with a heart, (iii) a system that simulates human brain activity to the limit, (iv) an artificially created intelligent behavior (or system), (iv) a compositional system for imitating, supporting, and transcending human intellectual behavior, (v) a system that can artificially create emotional, (vi) a system that simulates human brain activity to the limit, (vii) concepts and techniques for artificially mimicking human intelligence, or (viii) an artificial system centered on computers that enables highly intelligent tasks and judgments that only humans can do. When AI is considered as a computer or computer program, it can be defined as (i) basically advanced computer programs which can think, learn and decide like humans while considering all the scenarios of a given situation, these programs are then used in all the places like smartphones, robots and all, (ii) natural intelligence (NI) reproduced on a computer, (iii) a computer with human-like intelligence, (iv) the science and technology of making intelligent machines, especially intelligent computer programs, (v) the concept of "computing" and "a branch of computer science" that studies "intelligence" using a tool called "computer," (vi) research on the design and realization of intelligent information processing systems using computers, or (vii) a broad area of computer science that makes machines seem like they have human intelligence. It can be

https://doi.org/10.1515/9783110717853-001

also defined as (i) an artificially made intelligence but imagines that the level of intelligence is beyond human beings, or (ii) an engineering-made intelligence that imagines that the level of intelligence is beyond human beings [2–9]. In my opinion, as far as current AI technology is concerned, AI is an extension of "my world" and can only be handled in principle within the expected range; hence, it is the AI to ignore the thing coming from the other side which is not perceived.

1.2 History of AI development and future

Since the word AI was introduced, there were three booms and two winter seasons in-between, and currently we are entering the third boom, as illustrated in Figure 1.1 [10, 11].

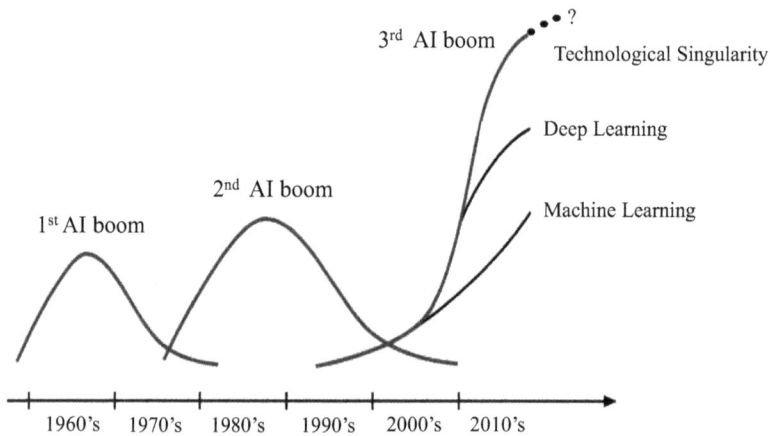

Figure 1.1: Brief history of AI development [10, 11].

The first boom (or generation) of AI can be characterized by reasoning and heuristic search. It was a time when the word "Artificial Intelligence" was born and people's expectations for computers increased. During this period, research was under way to make computers reason and explore with the aim of complementing and extending people's functions. AI, at that time, was just a program, it worked only in restricted areas, and it could only be set by the developer. There was the second boom, typified by knowledge expression. In the 1980s, a system called "Expert System (ES)" was proposed to complement the knowledge of people. An ES is a way to accumulate expert knowledge so that anyone can gain the same knowledge as an expert. In order to run the ES successfully, it was necessary to clearly define common-sense expressions that can be understood by human beings and teach them to the system. However, there should be immeasurable numbers of common-sense expressions. In addition, there are differences depending on the context and background, so if these are included, it would be needed to define a huge amount of knowledge. Accordingly, it would be a big cost and the effort to classify the collected data correctly and make it usable in the system was also very large. As a result, the ES with high expectations shrunk [10, 11].

A period since 2012 (the third boom of AI development) started with ML (machine learning) and further characterized by DL (deep learning). The problem with the second AI boom was that humans collected data to be input to the system and judged whether or not humans were lying. It was not realistic to prepare all the necessary data in advance and make a correct decision on success or failure. To solve this problem, there were ML, ICT (information and communication technology) and IoT (Internet of things). ML is a way to get machines to learn, and it learns automatically just by giving data to AI. In addition, the Internet makes it easier to obtain huge amounts of data. These new supportive technologies have significantly reduced the "cost and time" that was a bottleneck in the second AI boom. Furthermore, DL possesses the AI function that mimics the workings of the human brain in processing data for use in detecting objects, recognizing speech, translating languages, and making decisions, like the chatbot as an example. AI/DL is able to learn without human supervision, drawing from data that is both unstructured and unlabeled. The neural network (NN) is a processing procedure that mimics a network of brain cells. It is possible to learn efficiently by strengthening the memory based on the relationship with the existing memory. In other words, just by giving data to AI, we are able to "automatically teach common-sense" as a mechanism to get general concepts and learn them [10, 11].

Figure 1.2 illustrates the taxonomy of computer vision field and its interrelated branches such as science and technology, mathematics and geometry, physics and probability, and so on [12]. Practically, AI techniques such as ML and DL needs huge volume of training data which can be retrieved and processed using computer vision techniques.

Back to Figure 1.1, there should be a big concern on the "Technological Singularity." The word "singularity" is often used in mathematics and physics, indicating that general explanations and formulas cannot be directly applied. This singularity was proposed in a book by Kurzweil [13]. It is a concept of futurology, and it is the time when the improved and advanced technology and intelligence by self-feedback of AI becomes the leading role of the progress of civilization on behalf of mankind [13]. It is also attracting attention as the Fourth Industrial Revolution. A technical singularity is an event that is said to occur when AGI (artificial general intelligence) or "strong artificial intelligence" or human intelligence amplification becomes possible, and once a superior mechanical intelligence that operates autonomously is created, the version upgrade of mechanical intelligence is repeated recursively. It was hypothesized that superior intelligence is born to the effect that man's imagination does not reach. According to Kurzweil, the singularity would be realized around 2045, and if (i) AI with

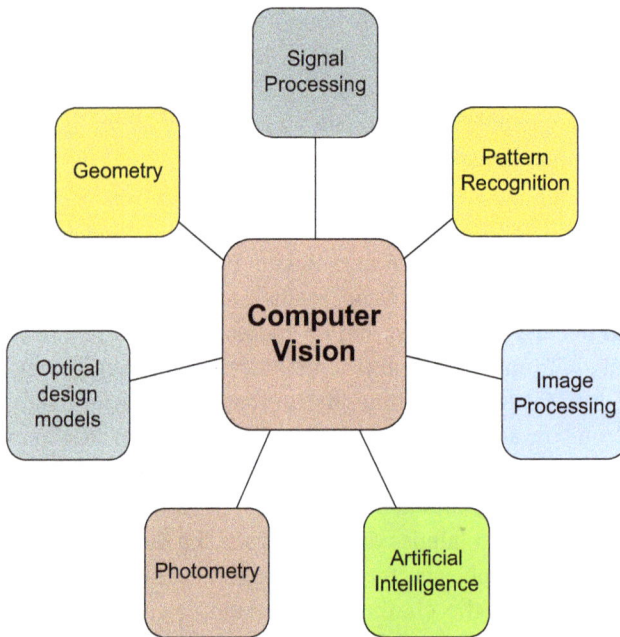

Figure 1.2: Computer vision and its subconcepts [12].

slightly superior intelligence than the combined total intelligence of human is born and (ii) AI will be able to create another AI that is a little better than itself, that is the time when the singularity will appear, resulting in that AI evolves at an accelerated speed, and in no time it far exceeds the wisdom of mankind. Ever-advancement of AI does not require huge improvement, rather a small yet continuous increment is just enough. For example, we start with "1.00" today, tomorrow we will improve and better it by just 1% to make it "1.01" and if we continue this effort for, say 1,000 times, we will end up with 2.10×10^4, and if for 10,000 time, it could be up to 1.64×10^{43}. On the other hand, if AI gets worsen by 1% to make it "0.99," after repeating this situation for 10,000 times, it will become nearly zero (i.e., 2.2×10^{-44}) [14]. Applying this simple calculation to advancing AI, it is easy to agree with that AI's IQ will be soon or later higher than all human beings.

Just about the time when the third AI boom is remarkably recognized (around year 2010), the Moore's law appears to be no longer explainable theory for rapid increasing computation capacity. Moore's law originally refers to the tendency that the number of transistors on integrated circuits doubles every 1.5 years [15]. AI is progressing at a pace of five to 100 times that of Moore's law [16, 17]. Figure 1.3 [17] shows the increasing trend of computer's computing power consumed for training, where the word Petafrop/Days indicates 4 trillion operations per second per day. It is clearly shown that there is a breaking point around 2010. Until around 2010, computing power doubled every 2

Two Eras of Compute Usage in Training AI Systems

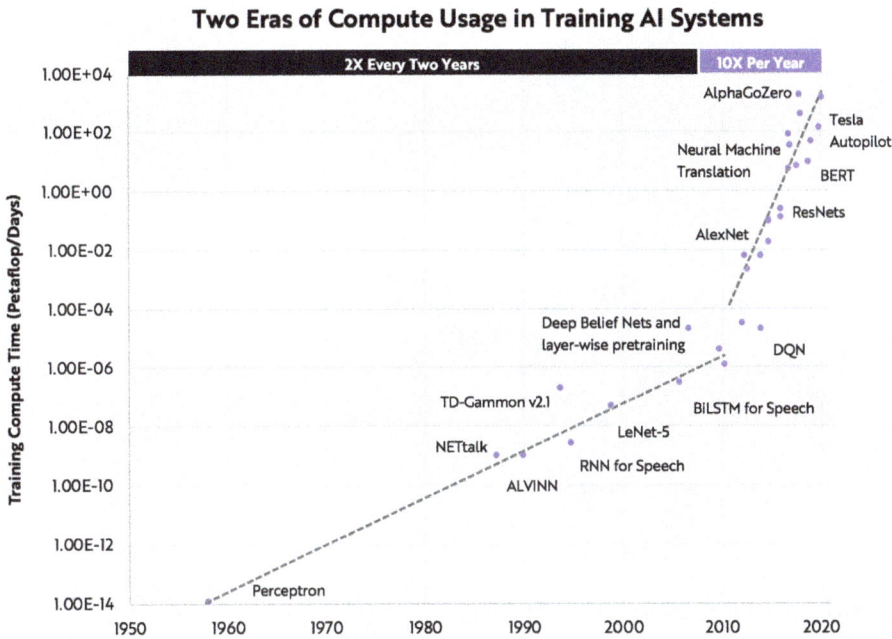

Figure 1.3: Training computation time versus year, illustrating a clear breaking point around 2010 [17].

years. But the processing power from 2010 beyond is growing about 10 times per year, more than five times the rate of Moore's law [17].

1.3 Nature

1.3.1 In general

Since the invention of computers or machines, their capability to perform various tasks went on growing exponentially. Humans have developed powerful computer systems and while exploiting the power of the computer systems, the curiosity of human, lead human to wonder, "Can a machine think and behave like human does?" Thus, the development of a branch of computer science (namely, AI) started with the intention of creating similar intelligence in machines that we find and regard high in humans [18]. AI is elegantly interconnected to supporting technologies as well as influencing technologies, as demonstrated in Figure 1.4 [18].

Most often, human intelligence is measured using a scoring method developed in 1912 by German psychologist William Stern. It is called as the "intelligence quotient (IQ)," The score is calculated by dividing the result of an intelligence test by

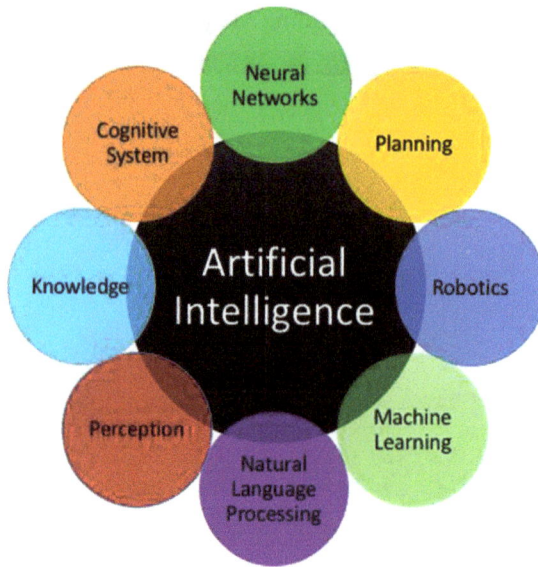

Figure 1.4: AI and supporting technologies [18].

examinee's chronological age and then multiplying the answer by 100. About 2.5% of the population scores above 130 and are considered elite thinkers, while 2.5% fall below 70 and are categorized as having learning or other mental disabilities. Two-thirds of the population score between 85 and 115 on the scale. As our intellectual ability improves, so will AI, but we cannot score AI using the IQ scale. Instead, we measure the power of a computer using operations (also calculation) per second (or ops), which we can still compare to the human brain. At our current rate, it will take humans 50 years of evolution to notch 15 points higher on the IQ scale. To us, 15 points will feel noticeable. The difference between a 119 "high average" brain and a 134 "gifted" brain would mean significantly greater cognitive ability – making connections faster, mastering new concepts more easily, and thinking more efficiently [19].

It does not need any single word to explain the ever-growing smarter AI. But, here, we feel to re-consider the human-related intellectual sense: namely emotional intelligence (EI), which shares with similar terms including emotional quotient (EQ) or emotional intelligence quotient (EIQ), indicating the capability of individuals to recognize their own emotions and those of others, discern between different feelings and label them appropriately, use emotional information to guide thinking and behavior, and manage and/or adjust emotions to adapt to environments or achieve one's goal(s) [20]. In booming growth of AI particularly, EQ is considered more important than IQ [21, 22]. Hewitt [21] pointed out that the EI has always been important but cultivating the competencies that underpin emotional intelligence will be increasingly significant as automation and AI replaces all, or part of many jobs, and humans in the workplace become increasingly differentiated by their unique human characteristics and skills. Beck et al. [22] stated that, due to advanced AI/ML, like most transformational technologies, is both exciting and scary. It is exciting to consider all the ways our lives may improve, but it's scary to consider the social and personal implications. As ML continues to grow, we all need to develop new skills in order to differentiate ourselves. As AI improves, which is possibly happening quickly, a much broader set of "thinking" rather than "doing" jobs will be affected. We are talking about jobs (like teacher, doctor, financial advisor, stockbroker, marketer, and business consultant) that, until the last few years, we could not imagine being done without the participation of an actual, trained human being. Although it is possible to have a high cognitive intellect, there should be a lack of the emotional intelligence, suggesting that we can have highly developed task-based skills – we know what to do – but fail to be able to apply these skills effectively in the real world. Being emotionally intelligent requires us to be able to pay attention to what we are thinking and feeling (self-awareness) as well as direct our attention to what others may be thinking and feeling (empathy) [23].

1.3.2 As a part of IT

Information technology (IT) is a generic term for technologies such as the Internet and computers to store, retrieve, transmit, and manipulate data or information. IT is considered to be a subset of the information & communication technology (ICT). An IT system is generally an information system, a communications system or, more specifically speaking, a computer system – including all hardware, software, and peripheral equipment – operated by a limited group of users. AI is simply software or system that mimics the mechanism of the human brain with a computer. Therefore, AI is a process that can be incorporated into IT. IoT is a communication system by applying sensors to devices so that they can take advantage of the information they acquire. By attaching a sensor, it is possible to take data on the position information, the state of devices (e.g., battery level) or the environment around them (e.g., temperature). Then, by sending information from the sensor through a communication network such as the Internet, data can be checked on a pc (personal computer). There are four major elements involved in the IoT, including device, sensor, communication network, and information processing for visualization (or application). Recently, DX (digital transformation) is receiving a special attention. DX is the integration and process of digital technology into all areas of a business to create new or modify existing business processes, culture, and customer experiences to meet changing business and market requirements. This reimagining of business in the digital age is digital transformation [24–26]. DX can be positioned in an entire world of information technology, as seen in Figure 1.5 [26].

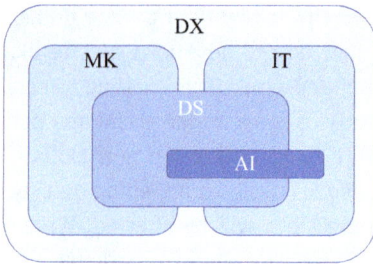

Figure 1.5: Positing of DX, where MK stands for marketing, IT for information technology, DS (data science: an approach using data to draw new scientific and socially beneficial insights) and AI is an artificial intelligence [26].

Specifically, in order for companies and local governments to promote DX, it is essential to utilize technologies depicted in Figure 1.6 [27]. Besides these seven elements, RPS (robotics process automation), SNS, and Blockchain [24, 28] should be added.

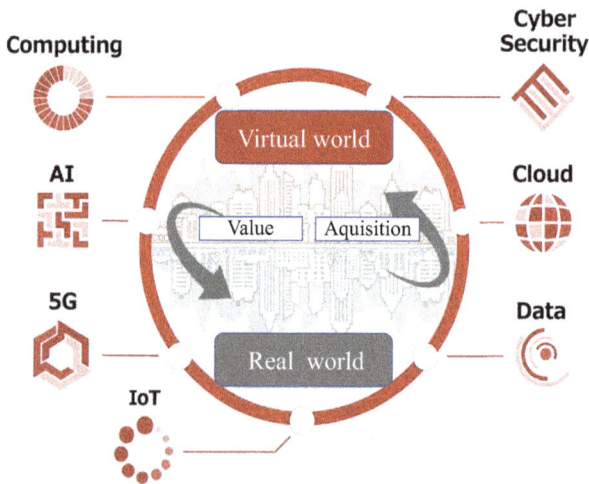

Figure 1.6: Seven essential supporting technologies for successful DX [27].

1.4 Classification

AI connotes four elemental technologies (AI, ML, ANN, and DL) and AI can be classified by three types (ANI, AGI, and ASI).

1.4.1 Elemental technologies involved in AI

Referring to Figure 1.7, AI ring possesses an accurately performing of advanced reasoning for large amounts of knowledge data and furthermore is classified with three types of AIs. Of AI technologies, ML is based on statistics. Some search engines use

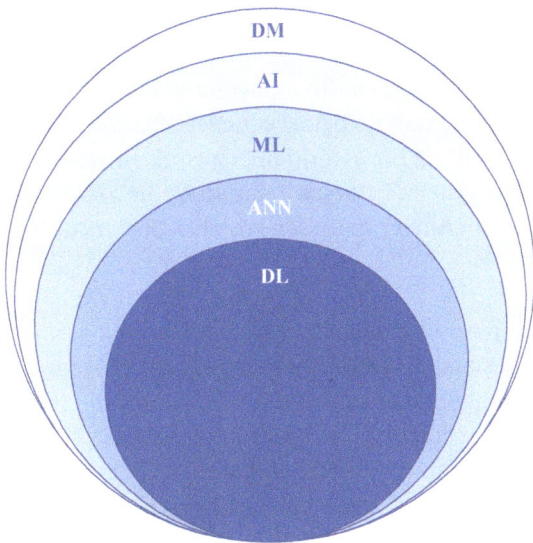

Figure 1.7: Elemental technologies for AI, where DM is data mining, AI is artificial intelligence, ML is machine learning, ANN is artificial neural network, and DL is deep learning [23–29].

a lot of AI/ML. It learns what kind of keywords users are looking for when users input them, and what pages they want, combined with the characteristics of the web page. It is also the job of ML to identify low-quality pages and identify harmful content. ML can be characterized by two consecutive processes: (i) grasping characteristics of collected (or input) data by iterative learning and generalizing (or modeling) and (ii) automatizing generalized MLed data by creating systematized reproductivity by nonprogramming. Hence, ML gives computers the ability to learn without being explicitly programmed. There are three types of MLs.

(1) Supervised (or coached) Learning to educate a machine how to learn data with correct answers (correct output) by process of regression (how to return a number as output when entering data) and/or classification (how to return the attributes and types of data as output when entering data). Supervised learning is one of the most basic types of ML. In this type, the ML algorithm is trained on labeled data. Even though the data needs to be labeled accurately for this method to work, supervised learning is extremely powerful when used in the right circumstances.

(2) Unsupervised Learning to have an ML by deriving structures, trends, laws, or others from features automatically calculated from vast amounts of data without the need for correct answers by clustering.

(3) Reinforcement Learning to improve by achieving the most recent goals and rewarding them through trial and error to find optimal behavior and value. Unsupervised ML holds the advantage of being able to work with unlabeled

data, meaning that human labor is not required to make the dataset machine-readable, allowing much larger datasets to be worked on by the program. Reinforcement learning directly takes inspiration from how human beings learn from data in their lives. Based on the psychological concept of conditioning, reinforcement learning works by putting the algorithm in a work environment with an interpreter and a reward system. In every iteration of the algorithm, the output result is given to the interpreter to decide whether the outcome is favorable or not [29–32].

ANN (artificial neural network) is the piece of a computing system designed to simulate the way the human brain analyzes and processes information. It is the foundation of AI and solves problems that would prove impossible or difficult by human or statistical standards. ANNs have self-learning capabilities that enable to produce better results as more data becomes available. The human brain consists of a network of neurons, one neuron connecting with other neurons to receive electrical stimulation from the synapse, which fires when it accumulates more than a certain amount of electricity and conveys electrical stimulation to the next neuron. There are three types of ANNs [33–43]:

(1) DNN (deep neural network) serves the basic structure of DL by an algorithm modeled on the NN of an organism and a multilayered structure of algorithms modeled on neural circuits in humans and animals designed to recognize patterns called NNs.

(2) CNN (convolutional neural network) is DNN compatible with two-dimensional data and is good at recognizing image patterns and serves as algorithms using forward-propagation NNs with abstraction and position universality of local information.

(3) RNN (recurrent neural network) is algorithms that allow you to handle data that is not pre-sized, such as video or audio, rather than fixed-length data such as images.

ML has two main types of classification and prediction models. One is an identification model such as image recognition, and a model that lets the computer determine what the input data or images are. The other is the generation model that generates new pseudo data from the input data and images. GAN (generative universal network) falls under this generation model, and it is a mechanism to deepen the learning of input data by competing each other for two NNs. Image generation is famous as an application of GAN, but it is also attracting attention as a technology that supplements DL in terms of generating data. Instead of the conventional method of increasing data by tilting the sample image or changing the color, by creating new data containing features, it can be applied to DL where data shortage tends to be a problem [44–46]. DL is an ML using multi-tiered NNs and is a learning method using DNNs that allow machines to automatically extract features from data without human power if there is enough data. DL features NNs, a type of algorithm that is based on the physical structure of the human brain, utilizing ANNs. DL has finally taken AI a step into areas where humans have had to be involved. NNs seem to be the most productive path forward for AI research, as it allows for a much closer emulation of the human brain than has ever been seen before. DL can be applied in a field involving image recognition, speech recognition, natural language processing (NLP), or anomaly detection.

Statistics is the discipline concerned with the collection, analysis, description, visualization, and drawing of inferences from data. Statistics is generally not considered part of AI, but many statistical techniques form the foundation for more advanced ML techniques or are used in conjunction with them, including descriptive statistics, inferential statistics, and data mining [19]. Probabilistic programming enables us to create learning systems that make decisions in the face of uncertainty by making inferences from prior knowledge. The practical probabilistic programming model [47] was first created to capture knowledge of a target domain in quantitative, probabilistic terms. Van Otterloo [48] and Greenhill et al. [49] indicated that ML is an important subfield of AI, and is also an important subfield of data science, as illustrated in Figure 1.8 [48].

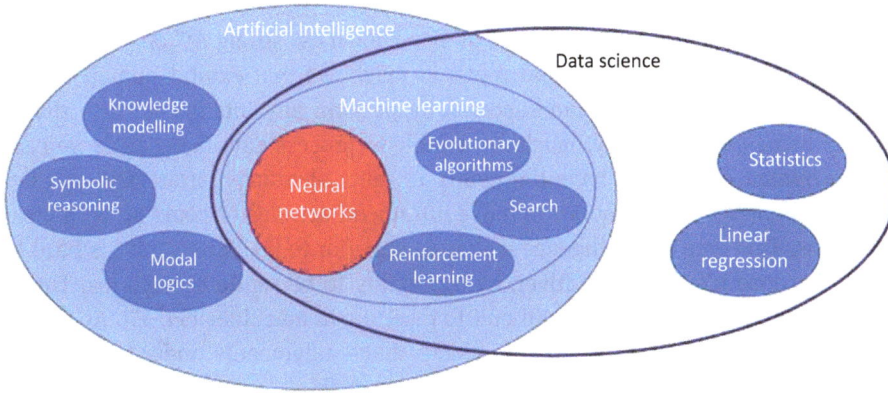

Figure 1.8: Interdisciplinary relationship between AI and data science [48].

1.4.2 Three AIs

The AIs are normally categorized, according to the following points: (i) its capacity to mimic human characteristics, (ii) the technologies enabling human characteristics to be mimicked, (iii) the real-world applications of the system, and (iv) theory of mind [50]. There are basically three types of AIs: namely ANI, AGI, and ASI [50–55].

ANI (Artificial Narrow Intelligence or also known as "weak AI"), possessing a narrow-range of abilities, is the only type out of the three that is currently around.

Although ANIs are very good at solving specific problems, they fall short of fulfilling the requirements for human-level AI. ANI systems can attend to a task in real time, but they pull information from a specific data-set. As a result, these systems do not perform outside of the single task that they are designed to perform. Unlike General or "strong" AI, Narrow AI is not conscious, sentient, or driven by emotion the way that humans are. Narrow AI operates within a pre-determined/pre-defined range. Though we refer to existing AI and intelligent machines as "weak" AI, we should not take it for granted. ANI by itself is a great feat in human innovation and intelligence. Additionally, ANI has relieved us of a lot of the boring, routine, mundane tasks that we do not want to do. From increasing efficiency in our personal lives, like Siri ordering a pizza for us online, to rifting through mounds of data and analyzing it to produce results, ANI has made our lives significantly better, which is why we should not underestimate it. With the advent of advanced technologies like self-driving cars, ANI systems will also relieve us of frustrating realities like being stuck in traffic, and instead provide us with more leisure time. There are many examples of narrow AI around us every day, represented by devices including, self-driving cars, facial recognition tools, customer service bots that redirect inquiries on a webpage, recommendation systems showing items that could be useful additions to shopping cart based on browsing history, or spam filters that keep the inbox clean through automated sorting [53–55].

AGI (Artificial General Intelligence or also known as "strong AI" or "deep AI") is about as capable as a human and refers to machines that exhibit human intelligence. In other words, AGI can successfully perform any intellectual task that a human being can. Currently, machines enable to process data faster than we can. But as human beings, we have the ability to think abstractly, strategize, and tap into our thoughts and memories to make informed decisions or come up with creative ideas. This type of intelligence makes us superior to machines, but it's hard to define because it's primarily driven by our ability to be sentient creatures. Therefore, it's something that is very difficult to replicate in machines [54, 55]. AGI is expected to be able to reason, solve problems, make judgements under uncertainty, plan, learn, integrate prior knowledge in decision-making, and be innovative, imaginative and creative. But for machines to achieve true human-like intelligence, they will need to be capable of experiencing consciousness. AGI technology would be on the level of a human mind. Based on this fact, it will probably be some time before we truly grasp AGI, as we still do not know all there is to know about the human brain itself. However, in concept at least, AGI would be able to think on the same level as a human [53–55].

Artificial Super Intelligence (ASI) is more capable than a human. ASI is a step further from AGI, where AI exceeds human capabilities to operate at a genius level. Since ASI is still hypothetical, there are no real limits to what ASI could accomplish, from building nanotechnology to producing objects to preventing aging. It will surpass human intelligence in all aspects – from creativity, to general wisdom, to problem-solving. Machines will be capable of exhibiting intelligence that we have not seen in the brightest among us. This is the type of AI that many people are worried about and many philosophers and scientists have different theories about the feasibility of reaching ASI [56]. ASI refers to AI technology that will match and then surpass the human mind. To be classed as an ASI, the technology would have to be more capable than a human in every single way possible. Not only could these AI things carry out tasks, but they would even be capable of having emotions and relationships. Summarizing this section, all four elemental technologies involved in AI and three major types of AI are interrelated, as illustrated in Figure 1.9.

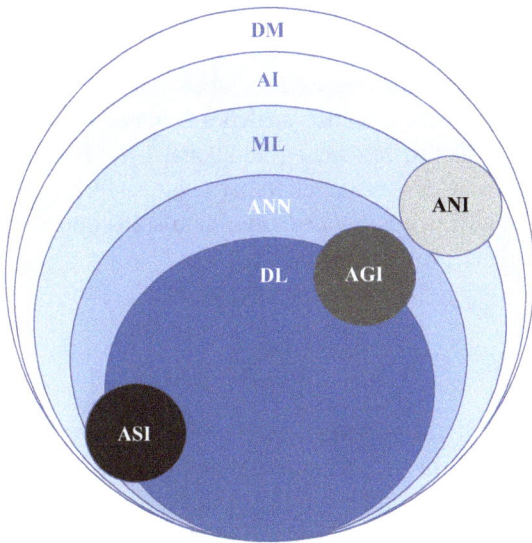

Figure 1.9: Four elemental technologies and three AIs.

1.5 Role, function, and applications of AI

1.5.1 Role

AI is now becoming an essential part of our lives. AI can offer a series of very important benefits in any sector and use since it is one of the most transversal technologies that exist today. But it can also generate problems and risks if people let the machines and algorithms make the final decisions [57]. The emergence of AI and its progressively wider impact on many sectors requires an assessment of its effect on the achievement of the Sustainable Development Goals (SDGs). The fast development of AI needs to be supported by the necessary regulatory insight and oversight for AI-based technologies to enable sustainable development. Failure to do so could result in gaps in transparency, safety, and ethical standards [58].

The role of AI in the industry is crucial as the world is smarter and more connected than ever before. Artificial machines predict or analyze outcomes accurately with provided with the algorithm and NNs, and learn from its users how they react with outputs it to function. The primary mission of ML was to work on programming algorithms that give a capability to machines by receiving information and statistically analyzing data to providing acceptable outputs. AI/ML in data security could be technologies and algorithms that can help organizations in achieving their goal of safeguarding their information [59]. To make ideas and data be useful in real lives, Kozyrkov [60] pointed out that applied data science is a team sport that's highly interdisciplinary, and perspective and attitude matter at least as much as education and experience. It is then the interdisciplinary team should be composed of data engineer, decision-maker, analyst, expert analyst, statistician, applied ML engineer, data scientist, analytics manager/data science leader, qualitative expert/social scientist, researcher, additional personnel (including domain expert, ethicist, software engineer, reliability engineer, UX (user experience) designer, interactive visualizer/graphic designer, data collection specialist, data product manager, or project/program manager).

1.5.2 Function

Development and widening applicability of AI are relied on versatile elemental functions and/or any combinations of them such as automation and control, machine vision, image recognition, speech recognition, optical character recognition (OCR), natural language processing (NLP), language identification and analysis, speech synthesis, video analysis, prediction, text-to-speech, chatbots, or others [61, 62].

1.5.3 Supporting technologies

The goal of AI is to realize software that can logically interpret input and explain output to humans. AI brings human-to-human dialogue between humans and software to help make decisions about specific tasks, but it is not a substitute for humans and is not likely to happen in the near future. Hence, in order to make AI more effective and useful to human's life, there should be additional technologies supporting AI [63]. Such supportive technologies can include: (1) GPU (Graphical Processing Unit) which is attracting attention as one of the key areas of AI because it can provide the advanced computing power needed for large numbers of iterations. In addition to big data, advanced computing power is critical to training NNs. (2) IoT is responsible for collecting data that AI analyzes them. IoT generates huge amounts of data from interconnected devices, but most of them are left un-analyzed. By automating the creation and application of models with AI, the most of data can be useful. (3) Advanced algorithms are being developed and methods are being devised to combine them in new ways for the effective analyzation of more data faster and at multiple levels. These intelligent processes play an important role in identifying and predicting rare events, understanding complex systems, and optimizing your own scenarios. (4) API (Application Programing Interface) simplifies the use of certain program functions, making it easier to add AI functionality to existing products and software. Besides these technologies, 5G-assisted DX technology should not be avoided. Digital transformation (DX) is a strategy of enabling business innovation predicated on the incorporation of digital technologies into operational process, products, solutions, and customer interactions. The strategy is focused on leveraging the opportunities of new technologies and their impact to the business by focusing on the creation and monetization of digital assets. DX involves the building of a digital ecosystem in which there are coherence and seamless integration between customers, partners, employees, suppliers, and external entities, providing greater overall value to the whole. The penetration of digital technology refers to changes that cause or have a good impact on all aspects of human life. It is a word that captures the change of the society as a whole with a very broad concept [63].

In order to promote DX, accumulation and analysis of data are the most important. At the same time, the use of technologies such as AI, IoT, and cloud is a prerequisite for effective usage data. Namely, they can include AI, 5G, cyber security, cloud, VR/AR/MR (virtual reality/augmented reality/mixed reality), HMI (human–machine interface), quantum computer, and information processing platform [64, 65]. For promoting DX, it is important to combine various types of digital technologies, solve problems, and transform the entire company. Just as humans are made up of various functions (see Figure 1.10 [66]), DX also needs to utilize a variety of technologies in the right place. In particular, just as the human brain is important, the role of AI in DX is becoming increasingly important.

The letter "G" stands for "generation." There was 0G in 1940, when the first telephone network began to roll out. This was the deceptive phase. It took forty years to crawl our way to 1G, which showed up via the first mobile phones in the 1980s, marking the transition from deceptive to disruptive. By the 1990s, around the time the Internet emerged, 2G came along for the ride. But the ride did not last long. A decade later, 3G ushered in a new era of acceleration as bandwidth costs began to plummet – at a staggeringly consistent 35% per year. Smartphones, mobile banking, and e-commerce unleashed 4G networks in 2010. But starting in 2019, 5G will begin to hotwire the whole deal, delivering speeds a hundred times faster at near-zero prices. A rumor is around on that the 6G (6th generation) will be in 2030s [67–69] (see Figure 1.11 [70]).

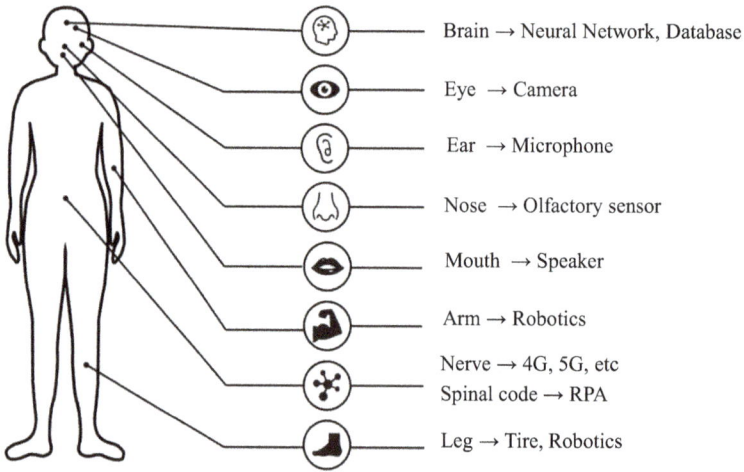

Brain → Neural Network, Database

Eye → Camera

Ear → Microphone

Nose → Olfactory sensor

Mouth → Speaker

Arm → Robotics

Nerve → 4G, 5G, etc
Spinal code → RPA

Leg → Tire, Robotics

Figure 1.10: DX elemental functions, analogue to human's function [66].

The Evolution of 5G

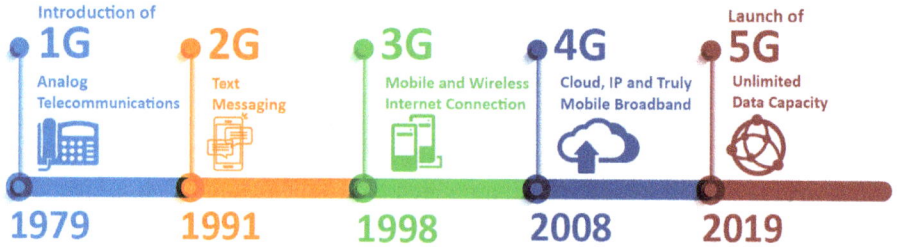

Figure 1.11: The evolution of 5G [70].

Since the 5G is characterized by (i) ultra-high-speed and high-capacity communication (or eMBB: enhanced mobile broadband), (ii) multiple simultaneous connections (or mMTC: massive machine type communication) and (iii) ultra-reliable and ultra-low latency communications (or URLLC) [71, 72], 5G and IoT can accelerate AI expansion [73]. With these developments, a real-time understanding the traffic conditions can be accomplished. By understanding the distance between pedestrians and other cars accurately in real time, autonomous driving becomes a more practical technology. Moreover, if the accuracy of the diagnosis by the image improves, it will be possible to instruct the operation even in the situation where it is possible to be separated from the site. In addition, 5G accelerates the development of the Edge-AI, meaning that AI algorithms are processed locally on a hardware device. The algorithms are using data (sensor data or signals) that are created on the device. A device using Edge-AI does not need to be connected in order to work properly; it can process data and take decisions independently without a connection [74]. On the other hand, the cloud AI is an AI that processes data and outputs results on the cloud. Because there is no data processing on the device, complex data and large amounts of data can be processed [75].

1.5.4 Applications

AI, to sum up in simple language, is basically advanced computer programs which can think, learn, and decide like humans while considering all the scenarios of a given situation, these programs are then used in all the places like smartphones, robots and all. Fields of application for AI can be versatile and should include medicine and health which we will discuss in later chapters, business and companies, education, finance, legal scope, manufacturing, biometric facial recognition, agriculture and food science [18, 57]. AI applications can be classified into three major areas: (1) cognitive science applications (such as expert systems. learning systems, fussy logic, genetic algorithms, NNs, or intelligence agents), (2) robotics applications (including visual perception, tactility, dexterity, locomotion, or navigation), and (3) natural interface applications (such as natural languages, speech recognition, multisensory interfaces, or virtual reality) [18]. Figure 1.12 illustrates variety of AI applications [76].

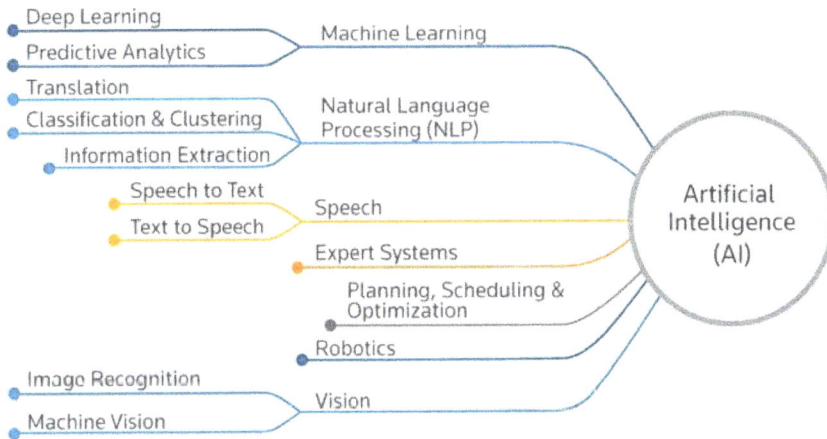

Figure 1.12: Various applications of AI technology [76].

1.6 Advantages and disadvantages

The word IoT has been born, and AI is becoming more and more familiar and is expected to be used in all aspects of daily life and business. To versatile AI implementation, there are many advantages and disadvantages as well as concerns [77–82].

1.6.1 Advantages

Working behavior:
– AI is available for 365 days a year, 7 days a week and 24 hours a day (aka 24/7/365)
– AI feels tireless, selfless with no breaks nor vacation to claim
– AI machines have no emotions
– AI reduces human error and takes risks instead of humans
– AI helps in repetitive jobs
– AI improves human workflows

Working performance:
– AI reduces time spent on effective data acquisition and analysis
– AI understands high-dimensional data
– AI enhances creative tasks
– AI provides precision
– AI performs predictive maintenance
– AI improves rational decision making faster
– AI controls, optimizes, and automates productive processes and production lines
– AI increases productivity and quality in production
– AI exhibits mechanical advantage

1.6.2 Disadvantages

Nature:
– AI requires high costs of creation and implementation
– AI does not have any emotion
– AI does not have instinct
– AI makes humans lazy
– AI causes number of jobs for humans*
   *Over the next 10–20 years or so, about half of 702 detailed occupations could
   be lost due to the impact of computerization, indicating that about 47% of total
   US employment is at risk [83].

Capability:
- AI lacks creativity and out-of-the-box thinking
- AI possesses uncontrollability
- AI does not improve with experience
- AI can never replicate a human

Concerns:
- AI possesses ethical considerations
- AI degrades itself without self-repair capability

Most of advantages have been realized in devices or tools on daily activities as well as medical healthcare field, which will be discussed in Chapters 7–11. Several disadvantageous issues and concerns will be discussed in Chapters 3 and 4.

References

[1] Oshida Y Bioscience and Bioengineering of Titanium Materials. Elsevier Pub., New York, NY, 2007.
[2] Faggella D. What is Artificial Intelligence? An Informed Definition. 2018. https://emerj.com/ai-glossary-terms/what-is-artificial-intelligence-an-informed-definition/.
[3] Copeland BJ. Artificial intelligence. 2020. https://www.britannica.com/technology/artificial-intelligence.
[4] Artificial intelligence. https://en.wikipedia.org/wiki/Artificial_intelligence.
[5] Artificial Intelligence. 2018. https://plato.stanford.edu/entries/artificial-intelligence/.
[6] Kumar C. Artificial Intelligence: Definition, Types, Examples, Technologies. 2018. https://chethankumargn.medium.com/artificial-intelligence-definition-types-examples-technologies-962ea75c7b9b.
[7] IBM Cloud Education. Artificial Intelligence (AI). 2020. https://www.ibm.com/cloud/learn/what-is-artificial-intelligence.
[8] Grewal DS A critical conceptual analysis of definitions of artificial intelligence as applicable to computer engineering. IOSR Journal of Computer Engineering. 2014, 16, 9–13.
[9] Marsden P. Artificial Intelligence Defined: Useful list of popular definitions from business and science. 2017. https://digitalwellbeing.org/author/marsattacks/.
[10] Garvey C. Broken promises & empty threats: The evolution of AI in the USA, 1956–1996. Technology's Science. 2018, 6. doi: 10.15763/jou.ts.2018.03.16.02.
[11] Matsuo Y. Does Artificial Intelligence go Beyond Humans: Beyond Deep Learning. Kadokawa Pub., Tokyo, 2015.
[12] Kakani V, Nguyen VH, Kumar BP, Kim H, Rao PV. A critical review on computer vision and artificial intelligence in food industry. Journal of Agriculture and Food Research. 2020, 2. doi: https://doi.org/10.1016/j.jafr.2020.100033.
[13] Kurzweil R. The Singularity Is Near: When Humans Transcend Biology. Penguin Books, 2006.
[14] Matsuo Y 2015. Can Artificial Intelligence exceed humans? Kadokawa Sensho.
[15] Moore GC, Benbasat I. Development of an instrument to measure the perceptions of adopting an information technology innovation. Information Systems Research. 1991, 2, 173–239.

[16]   Desjardins J. Visualizing Moore's Law in Action (1971–2019). 2019; https://www.visualcapital ist.com/visualizing-moores-law-in-action-1971-2019/.

[17]   Wang J. The Cost of AI Training is Improving at 50x the Speed of Moore's Law: Why It's Still Early Days for AI. Ark Invest. 2020; https://ark-invest.com/analyst-research/ai-training/.

[18]   Future Technology Explained: What is Artificial Intelligence (AI)? 2016; https://www.besttech guru.com/future-technology-explained-what-is-artificial-intelligence/.

[19]   Amy Webb. The Big Nine. Public Affairs, New York NY, 2019.

[20]   Goleman D. Emotional Intelligence Why it Can Matter More Than IQ. https://www.academia. edu/37329006/Emotional_Intelligence_Why_it_Can_Matter_More_Than_IQ_by_Daniel_Gole man. BLOOMSBURY.

[21]   Hewitt J. IQ, EI and AI – The Power of Emotional Intelligence in the Digital Age. 2020; https:// www.hintsa.com/iq-ei-and-ai-the-power-of-emotional-intelligence-in-the-digital-age/.

[22]   Beck M, Libert B. The rise of AI makes emotional intelligence more important. Harvard Business Review. 2017; https://hbr.org/2017/02/the-rise-of-ai-makes-emotional-intelligence -more-important.

[23]   Hewitt J, Hintsa A, Öhman J, Mari Huhtanen M. Exponential. Hintsa Performance. Helsinki Finland, 2016.

[24]   https://www.fujitsu.com/jp/dx/technology/.

[25]   https://www.itmedia.co.jp/news/articles/2005/29/news027.html.

[26]   https://ledge.ai/digital-transformation/.

[27]   https://blog.global.fujitsu.com/jp/2019-12-18/02/.

[28]   https://www.roit.co.jp/blog/support-dx-technology.html.

[29]   Artificial Intelligence vs. Machine Learning vs. Deep Learning: What's the Difference? 2020; https://camrojud.com/artificial-intelligence-vs-machine-learning-vs-deep-learning-whats-the -difference/.

[30]   Alexander D. What's the Difference between Machine Learning and AI? https://interestingen gineering.com/whats-the-difference-between-machine-learning-and-ai.

[31]   What is Machine Learning: Definition, Types, Applications and Examples. Potentia Analytics; https://www.potentiaco.com/what-is-machine-learning-definition-types-applications-and- examples/.

[32]   Cuocolo R, Ugga L. Imaging applications of artificial intelligence. Health Management. 2018, 18, https://healthmanagement.org/c/healthmanagement/issuearticle/imaging-applications- of-artificial-intelligence.

[33]   Marr B. What Are Artificial Neural Networks – A Simple Explanation For Absolutely Anyone. 2018; https://www.forbes.com/sites/bernardmarr/2018/09/24/what-are-artificial-neural- networks-a-simple-explanation-for-absolutely-anyone/?sh=7856089b1245.

[34]   What is a Neural Network? https://www.arm.com/ja/glossary/artificial-neural-network.

[35]   Dormehl L. What is an artificial neural network? Here's everything you need to know. 2019; https://www.digitaltrends.com/cool-tech/what-is-an-artificial-neural-network/.

[36]   Ravichandran S. Artificial neural network modelling and its applications in agriculture. https://naarm.org.in/VirtualLearning/vlc/e-chapters/ann.pdf.

[37]   Zador AM. A critique of pure learning and what artificial neural networks can learn from animal brains. Nature Communications. 2019, 10. doi: https://doi.org/10.1038/s41467- 019-11786-6.

[38]   Chauhan NS. Introduction to artificial neural networks (ANN): your first step in deep learning. 2019; https://towardsdatascience.com/introduction-to-artificial-neural-networks-ann -1aea15775ef9.

[39]   Frankenfield J. Artificial Neural Network (ANN): What Is an Artificial Neural Network (ANN)? 2020; https://www.investopedia.com/terms/a/artificial-neural-networks-ann.asp.

[40] Walczak S, Cerpa N. Artificial Neural Networks. Encyclopedia of Physical Science and Technology. 2003; https://www.analyticsvidhya.com/blog/2014/10/ann-work-simplified/.

[41] Akst J. A Primer: Artificial intelligence versus neural networks. The Scientists. 2019; https://www.the-scientist.com/magazine-issue/artificial-intelligence-versus-neural-networks-65802.

[42] Kavlakoglu E. AI vs. Machine Learning vs. Deep Learning vs. Neural Networks: What's the Difference? 2020; https://www.ibm.com/cloud/blog/ai-vs-machine-learning-vs-deep-learning-vs-neural-networks.

[43] Vaishnavi P. Artificial intelligence neural network: Role of neural networks in artificial intelligence. MindMajix. 2020; https://mindmajix.com/neural-network-in-artificial-intelligence.

[44] Introduction. Generative Adversarial Networks. 2019; https://developers.google.com/machine-learning/gan.

[45] Generative adversarial network. https://en.wikipedia.org/wiki/Generative_adversarial_network.

[46] Rocca J. Understanding Generative Adversarial Networks (GANs): Building, step by step, the reasoning that leads to GANs. 2019; https://towardsdatascience.com/understanding-generative-adversarial-networks-gans-cd6e4651a29.

[47] Pfeffer A. Practical probabilistic programming. Manning Pub., 2016.

[48] Van Otterloo S. AI, Machine Learning and neural networks explained. 2020; https://ictinstitute.nl/ai-machine-learning-and-neural-networks-explained/.

[49] Greenhill AT, Edmunds BR. A primer of artificial intelligence in medicine. Techniques and Innovations in Gastrointestinal Endoscopy. 2020, 22, 85–9.

[50] Ackermann TJ. The 3 types of AI: Narrow (ANI), General (AGI), and Super (ASI). 2019; https://www.bgp4.com/2019/04/01/the-3-types-of-ai-narrow-ani-general-agi-and-super-asi/.

[51] Davidson L. Narrow vs. General AI: What's Next for Artificial Intelligence? 2019; https://www.springboard.com/blog/narrow-vs-general-ai/.

[52] https://www.ediweekly.com/the-three-different-types-of-artificial-intelligence-ani-agi-and-asi/. The three different types of Artificial Intelligence – ANI, AGI and ASI.

[53] O'Carroll B. What are the 3 types of AI? A guide to narrow, general, and super artificial intelligence. 2020; https://codebots.com/artificial-intelligence/the-3-types-of-ai-is-the-third-even-possible.

[54] Jajal TD. Distinguishing between Narrow AI, General AI and Super AI. 2018; https://medium.com/@tjajal/distinguishing-between-narrow-ai-general-ai-and-super-ai-a4bc44172e22.

[55] Eady Y. Does recent progress with neural networks foretell artificial general intelligence? https://medium.com/protopiablog/does-recent-progress-with-neural-networks-foretell-artificial-general-intelligence-9545c17a5d8b.

[56] Mills T. AI Vs AGI: What's The Difference? 2018; https://www.forbes.com/sites/forbestechcouncil/2018/09/17/ai-vs-agi-whats-the-difference/#5832df7338ee.

[57] Anuraq. The role of AI in our daily lives. 2020; https://www.newgenapps.com/blog/the-role-of-ai-in-our-daily-lives/.

[58] Vinuesa R, Azizpour H, Leite I, Balaam M, Dignum V, Domisch S, Felländer A, Langhans SD, Tegmark M, Nerini FF. The role of artificial intelligence in achieving the Sustainable Development Goals. Nature Communications. 2020, 11. doi: https://doi.org/10.1038/s41467-019-14108-y.

[59] AI O Role of Artificial intelligence and Machine Learning in Data Security. 2020; https://medium.com/@Oodles_ai/role-of-artificial-intelligence-and-machine-learning-in-data-security-4f85e0b31edf.

[60] Kozyrkov C. Top 10 roles in AI and data science. 2018; https://www.kdnuggets.com/2018/08/top-10-roles-ai-data-science.html.

[61]   Ray S. Functions of AI. 2018; https://codeburst.io/functions-of-ai-26dd24664cc7.

[62]   https://www.hitachi-solutions-create.co.jp/column/technology/ai-type.html.

[63]   What is Digital Transformation (DX)? https://www.netapp.com/devops-solutions/what-is-digital-transformation/.

[64]   https://fce-pat.co.jp/magazine/907/.

[65]   https://www.holmescloud.com/useful/3430/.

[66]   https://ainow.ai/2019/12/02/181355/.

[67]   Diamandis PH, Kotler S. The Future is Faster Than You Think. Simon & Schuster, New York NY, 2020.

[68]   Zhang C, Ueng Y-L, Studer C, Burg A. Artificial intelligence for 5G and beyond 5G: Implementations, algorithms, and optimizations. IEEE Journal on Emerging and Selected Topics in Circuits and Systems. 2020, 10, https://ieeexplore.ieee.org/document/9108219.

[69]   Taulli T. How 5G Will Unleash AI. 2020; https://www.forbes.com/sites/tomtaulli/2020/05/08/how-5g-will-unleash-ai/?sh=3b1e7bed48c3.

[70]   Moltzau A. Artificial Intelligence and 5G – Why is 5G Relevant to the Field of AI? 2019; https://towardsdatascience.com/artificial-intelligence-and-5g-d59de7dfd213.

[71]   https://note.com/koichirot11/n/n24ecd445b18b.

[72]   https://ainow.ai/2020/06/11/222489/.

[73]   Brave New World: Everything Gets Smarter When 5G and AI Combine. 2019; https://www.electronicdesign.com/industrial-automation/article/21807565/brave-new-world-everything-gets-smarter-when-5g-and-ai-combine.

[74]   https://www.advian.fi/en/what-is-edge-ai.

[75]   https://research.google/teams/cloud-ai/.

[76]   Brandall B. 6 Useful Applications of AI & Machine Learning in Your Business Processes. 2018; https://www.process.st/applications-of-ai/.

[77]   Kumar S. Advantages and Disadvantages of Artificial Intelligence. 2019; https://towardsdatascience.com/advantages-and-disadvantages-of-artificial-intelligence-182a5ef6588c.

[78]   Advantages and disadvantages of artificial intelligence. 2019; https://nexusintegra.io/advantages-disadvantages-artificial-intelligence/.

[79]   Bansal S. 10 Advantages and Disadvantages of Artificial Intelligence. https://www.analytixlabs.co.in/blog/advantages-disadvantages-of-artificial-intelligence/.

[80]   Pros and Cons of Artificial Intelligence – A Threat or a Blessing? https://data-flair.training/blogs/artificial-intelligence-advantages-disadvantages/.

[81]   Phoenix J. The Pros and Cons of Artificial Intelligence. 2020; https://understandingdata.com/artificial-intelligence-advantages-and-disadvantages/.

[82]   What are the Advantages and Disadvantages of Artificial Intelligence? 2020; https://www.edureka.co/blog/what-are-the-advantages-and-disadvantages-of-artificial-intelligence/.

[83]   Frey CD, Osborne MA. The future of employment: How susceptible are jobs to computerization? Oxford Martin Programme on Technology and Employment. 2013; https://www.oxfordmartin.ox.ac.uk/downloads/academic/future-of-employment.pdf.

Chapter 2
AI in information

In general, when a particular experience or practice causes a fairly sustained transformation of behavior, the process or outcome is called learning. Although learning in this sense is seen in many animals, in the case of humans, symbols and images centered on language are characterized by mediating this process. As a result, the outcome of learning accumulates in the memory, and it is possible to make a good choice about the reaction to the stimulation and the action in a certain scene. On the other hand, learning in an ML as a mainstream of AI means a learning process with supplied training data or learning data to perform some tasks using learning outcomes [1–3]. In addition, these experiences suggest computer algorithms that improve learning content and improve it automatically in a recursive manner. For a purpose of improving the quality of learning, ML is closely related to computational statistics, mathematical optimization, and data mining (DM). In particular, the quantity and quality of the input data are very important because they affect directly the learning results, sharing the same concept of the phrase "you are what you eat."

In this chapter, we will be discussing data versus information, knowledge versus wisdom, intelligence versus intellect, and AI ontology.

2.1 Data versus information

Data and information are similar concepts, but they are not the same thing. The main difference between data and information is that data is a part and information is the whole [1–3]. The data is raw and unorganized materials comprise facts, figures, image, observations, perceptions, numbers, characters, symbols, statements, and so on. Data is always interpreted and analyzed, by a human or machine, to derive meaning. It comes from the Latin word meaning "something given." On the other hand, information is defined simply as "news or knowledge received or given." It is what results when human has processed, interpreted, structured, organized facts, or presented in a predetermined context to make it meaningful and useful. The word comes from the Latin word meaning "formation or conception." As a simple example, the history of temperature readings all over the world for the past 100 years is data. If this data is organized and analyzed to find that global temperature is rising, then that is an information. What makes AI different from earlier technologies is that the system learns as data is fed into it. Hence, data is the fuel that makes the whole thing work. Just like three essential nutrients for the brain (glucose, oxygen, and water). The thus organized and analyzed information through AI technology can be fed to another higher level of AI as a raw data again [1–3].

https://doi.org/10.1515/9783110717853-002

2.1.1 Data

Factors such as Internet, web, and SNS are possible as a growing data source, but the decisive changes in data source are changes in the commercial ecosystem due to the spread of smartphones, social networks, and app stores. Various types of data should include "digital data" such as images on the web, "social data" on SNS, "real-world data" collected from sensors, and IoT terminals installed in towns, facilities, factories, robots, "open data" held by local governments, face images, fingerprints, irises, and so on. Since the amount of such data is very large, it is very important to develop not only hardware components such as central processing units and graphics processing units but also communication technology for effective data processing. Big data in AI era possesses 3V characteristics, that is, volume, variety, and velocity (with high frequency of data creation and updating), as shown in Figure 2.1 [4]. Big data and AI are under the win–win situation. Firstly, power of AI is essential for the analysis of big data. Due to the spread of IoT connecting devices with various sensors to the Internet, the analysis of data that can be collected and stored in large quantities in a wide range of fields is not possible with human power alone, and the use of high-performance AI is indispensable. In particular, data analysis such as images and voice has made remarkable progress with AI, and it is also widely used in the medical field for diagnostic imaging [4–6].

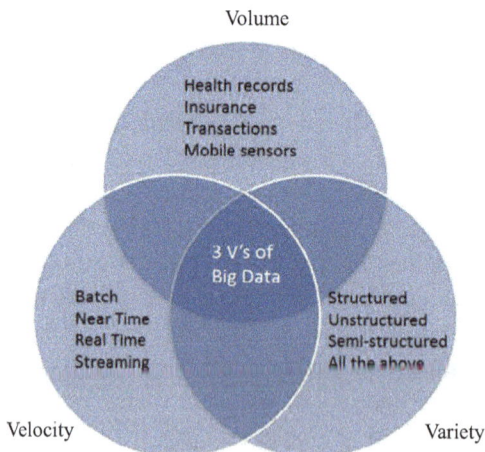

Volume

Health records
Insurance
Transactions
Mobile sensors

3 V's of
Big Data

Batch
Near Time
Real Time
Streaming

Structured
Unstructured
Semi-structured
All the above

Velocity

Variety

Figure 2.1: Three Vs' characteristics of big data [4].

Big data involved in the life science possesses also a wide variety of elemental data sources. They should include molecular data such as genetic data and protein data, medical treatment data, drug data, literature data, and ecological data. Life science data is characterized by three main features: diversity, large scale, and very many links between data [7]. Big data consists of three types of data, including clean data, ROT (redundant, obsolete, and trivial) and dark data. The clean data (or actionable data) is obviously data which is effective and useful and occupies about 20% of the total data. ROT indicates duplicate information or old and useless one, occupying about 30% of total data. Among ever-accumulated database every day, the dark data refers to data that is unanalyzable or not analyzed yet, so that its true value has not been determined and it occupies about 50% of total data. The dark data furthermore comprises of three types: text-type dark data (such as data consisted mainly of text including e-mails, documents, logs, and notifications), rich media–type dark data (such as images, video, and audio), and deep web–type dark data (although they exist on the web, they are not detected by search results by the public search engine) [8–13].

2.1.2 Data mining, information, and DIKW hierarchy

The usage of word mining in the "DM" is somewhat misleading, since the real function of DM is not just an extraction (meaning mining in metallurgical engineering) but rather an extraction of patterns and knowledge from large amounts of data [14–16]. Actually, the DM is a process of discovering patterns in large data sets involving methods at the intersection of three disciplines of ML (algorithms that can learn from data to make predictions), statistics (the numeric study of data relationships), and database systems (algorithms that can learn from data to make predictions) [16, 17], as illustrated in Figure 2.2 [17].

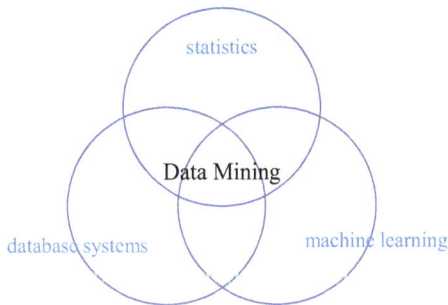

Figure 2.2: Data mining which is constructed by statistics, database systems, and machine learning [17].

Accordingly, DM is an interdisciplinary technology of computer science and statistics with an overall goal to extract information (with intelligent methods) from a data set and transform the information into a comprehensible structure for further use [18–20]. DM conducted through one of analytical methods (clustering, logistic regression analysis, or market-basket analysis) will produce the following four outcomes: (1) data (unprocessed numerical information), (2) information (organized and categorized data), (3) knowledge (trends and insights obtained from information), and (4) wisdom (judging ability using knowledge). DIKW (data–information–knowledge–wisdom) model represents graphically how knowledge can be organized and describes how the data can be processed and transformed into information, knowledge, and wisdom. The DIKW model of transforming data into wisdom can be viewed from two different approaches of contextual and understanding, as seen in Figure 2.3(a) [21]. As per the contextual concept, one moves from a phase of gathering data parts (data), the connection of raw data parts (information), formation of whole meaningful contents (knowledge) and conceptualize and joining those whole meaningful contents (wisdom). As sequential process of understanding, the DIKW hierarchy can also be explained as a process starting with researching and absorbing, doing, interacting, and reflecting. The DIKW hierarchy can also be represented in terms of time. For instance, the data, information, and knowledge levels can be seen as the past while the final step – wisdom – represents the future. Since the information hierarchy stage reveals the relationships in the data, the analysis is carried out to find the answer to who, what, when, and where questions, as shown in Figure 2.3(b) [21].

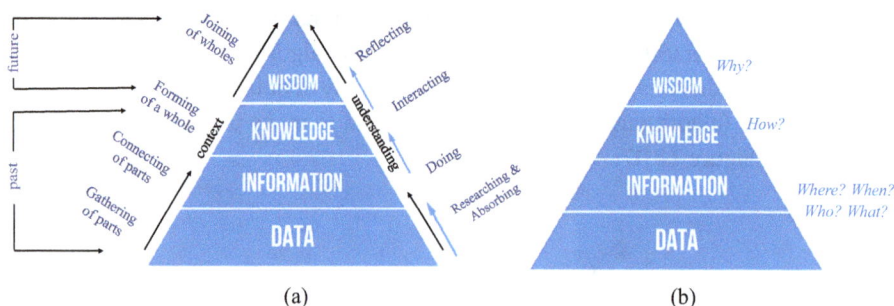

Figure 2.3: DIKW pyramid or DIKW hierarchy [21].

2.2 Knowledge, intelligence, and intellect

Each component of DIKW is hierarchically interrelated. Humans find and feel things and events by the five senses (vision, listening, smell, taste, touch). "Data" represents the thing or event by words or numbers or obtained by measuring instruments or observation instruments/devices and is like a material that can be a basis for us to make judgments, evaluations, or decisions. When the data becomes a factor in our decision-making, it becomes "information." There is also a way of saying that when data has meaning, it becomes "information." "Knowledge" is what can be used to systematically compile information and data and make general judgments and thinking in the future. At the top level is "wisdom," the ability to realize the reason of things and handle them properly, as shown in Figure 2.3(b). Up to now, humans have been involved in knowledge and information in the form of a hierarchical structure of things and events → data → information → knowledge → wisdom, as described above, but with the development and spread of the Internet, changes have occurred in the way it should be. Until the twentieth century, people who knew various things well were "intelligent." but now you can understand what you need by "searching" and knowing itself is no longer so intelligent. That is, the age changed from "memory" to "search." From now on, it will be important how to create new things by combining what you know, and how to create new things by collaborating with multiple people [22–25].

Knowledge has the complexity of experience, which come about by seeing it from different perspectives. This is the reason why training and education is difficult – one cannot count on one person's knowledge transferring to another. Knowledge is built from scratch by the learner through experience. Information is static, but knowledge is dynamic as it lives within us. Knowledge is the integration and construction of multiple "information" into a more useful form such as a theory or method. There are three types of knowledge: (1) personal knowledge is meaningful knowledge in a single person; (2) local knowledge is useful knowledge for people who share the same experiences and situations; and (3) global knowledge is more general knowledge. Since it is the knowledge that can be communicated to people who do not share a knowledge base, it is important that the transmission cost is high and highly organized and integrated. Much of what is commonly called "knowledge" will apply to global knowledge. Wisdom is the ultimate level of understanding. As with knowledge, wisdom operates within us. We can share our experiences that create the building blocks for wisdom; however, it needs to be communicated with even more understanding of the personal contexts of our audience than with knowledge sharing. Wisdom is an extremely personal thing that can be obtained from a deep understanding of knowledge. It is also called tacit knowledge (as opposing to the explicit knowledge) and cannot be shared with others. Tacit knowledge is knowledge that "I can't explain it well even I know it can be done" or "I can't say well even if I know it well" [26, 27].

Intelligence is the ability to quickly find a right answer to a predetermined question whose answer can be found. Intelligence is an excellence of mind that is employed within a fairly narrow, immediate, and predicable range; it is a manipulative, adjustive, unfailingly practical quality. On the other hand, intellect is the ability to never give up and continue to ask questions that are not easily answered. Sometimes, even if someone asks a question for a lifetime, it can be said that an answer cannot found, and it is still the ability to keep asking that question. Intellect is different from learning and problem answering skills, rather is an ability to think deeply with a wealth of knowledge. Intellect is the critical, creative, and contemplative side of mind. Hence, it is clear that there is a definite difference between intelligence and intellect. It is that intellect is a human trait. It seems that other animals are the same in the point of "intelligence" which means a certain situation immediately if it is said. If it is intellect that is to be used to make human beings and other animals different, the development of intellect becomes an urgent issue in the advent of a society (Society 5.0) in which AI is widely used, because Society 5.0 requires more humanity [28, 29].

Humans have no skills from the start but are born with a wide range of possibilities to acquire new skills. The NN is composed of the phenomenon of self-organization (in other words, by spontaneously connecting neurons). The brain has the property of plasticity, and the strength of its binding changes, so that the brain changes. Scientific research on human cognition has achieved remarkable results. Through this research, the brain, which is the center of human cognition, is formed as a network of about 1,000 million neurons. The network was connected to nature while working as an organism, that is, self-organizing. The network of nerve cells that make up the brain is aimed at survival. On the other hand, ANNs on which connectionism relies are not organized to maintain life. This is a big difference from the actual brain network. AI is good at information processing, but because it does not have life, it is inherently different from human intelligence for survival [30]. The peculiarity of human intellect is in plasticity, leading to an ability to apply knowledge to situations that have never been experienced [31, 32]. Neuroplasticity (also known as neural plasticity, or brain plasticity) is the ability of NNs in the brain to change its activity in response to intrinsic or extrinsic stimuli by growth and reorganizing its structure, functions, or connections. These changes range from individual neurons making new connections, to systematic adjustments like cortical remapping. Examples of neuroplasticity include circuit and network changes that result from learning a new ability, environmental influences, practice, and psychological stress [33, 34]. A fundamental property of neurons is their ability to modify the strength and efficacy of synaptic transmission through a diverse number of activity-dependent mechanisms, typically referred to as synaptic plasticity [34]. The human brain remains the biggest inspiration for the field of AI. Neuroscience-inspired methods are regularly looking to recreate some of the mechanisms of the human brain in AI models. Among those mechanisms, plasticity holds the key of many learning processes in the human brain. Synaptic plasticity is one of those magical abilities of the human brain that has puzzled neuroscientists for decades. From the neuroscience standpoint, synaptic plasticity refers to the ability of connections between neurons (or synapses) to strengthen or weaken over time based on the brain activity [35, 36]. Not surprisingly, synaptic plasticity is considered a fundamental building block of aspect such as long-term learning and memory. In the AI space, researchers have long tried to build mechanisms that simulate synaptic plasticity to improve the learning of NNs. Recently, a meta-learning method (differentiable plasticity) was proposed [37], which imitates some of the synaptic plasticity dynamics to create NNs that can learn from experience after their initial training.

2.3 AI ontology

The rapid advancement of AI and its branches like ML and DL, which function on extracting relevant information and generating insights from data to find sustainable and decisive solutions, is nothing new, although to run these algorithms, organizations need data and code. To translate this necessity into something meaningful, data science is necessitated. Further, data science helps organizations communicate with stakeholders, customers, track and analyze trends, and determine whether if the collected data is of any help or simply a waste of a database farm. So, having an ontology consisting of the relevant terms and connections from a specific domain, the process of identifying core concepts, improving classification results, and unifying data to collate critical information becomes streamlined [38].

Ontology is the philosophical study of the nature of being, becoming, existence, or reality. It is part of the major branch of philosophy known as metaphysics. An ontology is an explicit specification of conceptualization. In other words, it is a formal naming and definition of the types, properties, and interrelationships of the entities that fundamentally exist for a particular domain of discourse. Some of the important objectives of an ontology are representing knowledge, sharing knowledge, and reusing knowledge for semantic inferencing. Ontology deals with questions about what things exist or can be said to exist, and how such entities can be grouped according to similarities and differences [39]. Just as a clear specification is required when creating an information system, when writing knowledge, there should be a specification. There are two sects of ontology. A light-weight ontology is an ontology simply based on a hierarchy of concepts and a hierarchy of relations, while a heavy-weight ontology is enriched with axioms used to fix the semantic interpretation of concepts and relations [40–42]. Baclawski et al. [43] pointed out that there are many connections among AI, learning, reasoning, and ontologies. Noy et al. [44] mentioned that the AI literature contains many definitions of an ontology; many of these contradict one another. For the purposes of this guide, an ontology is a formal explicit description of concepts in a domain of discourse (classes, aka concepts), properties of each concept describing various features and attributes of the concept (slots, aka roles or properties), and restrictions on slots (facets, aka role restrictions) [44].

Closing this chapter, AI seems to be getting closer and closer to humans. It became possible to learn the method of learning by oneself and to create the image which had not been seen, and the sentence reading comprehension which was weak also exceeded the average point of man in the benchmark test. Applications such as research on quantum gravity theory using ML and development of new material and drug developments have also expanded, and it has also been a partner in scientific research. However, AI does not function like the human brain. There is also a misunderstanding that humans would never do, and the speed of learning is still slow and lack of flexibility. This indicates a limitation of the AI and expectation for developing artificial intellect.

References

[1] Difference between Data and Information Explained; https://examples.yourdictionary.com/difference-between-data-and-information-explained.html.

[2] Difference between Information and Data; https://www.guru99.com/difference-information-data.html.

[3] Difference between Data and Information; https://byjus.com/biology/difference-between-data-and-information/.

[4] Austin CO, Kusumoto FM. The application of Big Data in medicine: current implications and future directions. Journal of Interventional Cardiac Electrophysiology. 2016, 47. doi: 10.1007/s10840-016-0104-y.

[5] Understanding The 3 Vs of Big Data – Volume, Velocity and Variety. 2017. https://www.whishworks.com/blog/data-analytics/understanding-the-3-vs-of-big-data-volume-velocity-and-variety/.

[6] Hansen S. The 3 V's of Big Data Analytics. 2019. https://hackernoon.com/the-3-vs-of-big-data-analytics-1afd59692adb.

[7] Yamaguchi A. Trends in Open Data and Integrated Use of Data. 2013. https://www.jst.go.jp/crds/pdf/2013/WR/CRDS-FY2013-WR-03.pdf.

[8] https://roboma.io/blog/marketing/what-is-dark-data-and-how-to-use-it/.

[9] https://data.wingarc.com/dark-data-16639.

[10] Shed light on dark data. GAVS. 2018; https://www.gavstech.com/wp-content/uploads/2018/03/Article-Shed-Light-on-Dark-Data.pdf.

[11] Vaughan J. Dark Data. 2017; https://searchdatamanagement.techtarget.com/definition/dark-data.

[12] Van Rees I. Transforming hidden data into powerful business. DATABEG. https://blog.datumize.com/evolution-dark-data.

[13] Rouse M. Dark Data. 2017; https://searchdatamanagement.techtarget.com/definition/dark-data.

[14] Twin A. Data mining. Investopedia. 2020; https://www.investopedia.com/terms/d/datamining.asp.

[15] Definition of 'Data Mining'. 2020; https://economictimes.indiatimes.com/definition/data-mining.

[16] Data Mining: What it is & why it matters. https://www.sas.com/en_us/insights/analytics/data-mining.html.

[17] https://www.sas.com/en_us/insights/analytics/data-mining.html.

[18] Clifton C Definition of Data Mining. Encyclopædia Britannica. 2010, https://www.britannica.com/technology/data-mining/Pattern-mining#ref1073343.

[19] Hastie T, Tibshirani R, Friedman J. The Elements of Statistical Learning: Data Mining, Inference, and Prediction. Springer, 2009.

[20] Han J, Kamber M, Pei J, Data Mining: Concepts and Techniques. Morgan Kaufmann, 2010.

[21] Brahmachary A. DIKW Model: Explaining the DIKW Pyramid or DIKW Hierarchy. 2019; https://www.certguidance.com/explaining-dikw-hierarchy/.

[22] https://www.seojapan.com/blog/information-vs-knowledge.

[23] https://office.uchida.co.jp/workstyle/column/2014080701.html.

[24] https://learn-tern.com/data-information/.

[25] Understanding and Performance; http://www.nwlink.com/~donclark/performance/understanding.html.

[26] Wurman RS Understanding Understanding. 2017.

[27] Wurman RS, Leifer L, Sume D, Whitehouse K Information Anxiety 2 (Hayden/Que). 2000.

[28] Hofstadter R. Anti-Intellectualism in American Life. Library of America, 1966.

[29] http://learningweb.hatenadiary.jp/entry/20140706.

[30] http://www.nikkei-science.com/202001_025.html.

[31] https://wired.jp/2020/08/07/its-called-artificial-intelligence-but-what-is-intelligence/.

[32] Chollet F. Deep Learning with Python. 1st edn. Manning Publications, 2018.

[33] Kolb B, Gibb R. Robinson T. Brain Plasticity and Behavior. https://www.psychologicalscience.org/journals/cd/12_1/Kolb.cfm.

[34] Mateos-Aparicio P, Rodríguez-Moreno A. The Impact of studying brain plasticity. Frontiers in Cellular Neuroscience. 2019. doi: https://doi.org/10.3389/fncel.2019.00066.

[35] Elliot L. Plasticity in deep learning: dynamic adaptations for AI self-driving cars. Altrends. 2018; https://www.aitrends.com/ai-insider/plasticity-in-deep-learning-dynamic-adaptations-for-ai-self-driving-cars/.

[36] Rodriguez J. Imitating brain plasticity in deep neural networks. The Startup. 2020; https://medium.com/swlh/imitating-brain-plasticity-in-deep-neural-networks-e056452db7c1.

[37] Miconi T, Clune J, Stanley KO. Differentiable plasticity: training plastic neural networks with backpropagation. Proc. 35th Intl Conf on Machine Learning. 2018.

[38] Preetipadma. Why do we need ontology in an ai or data science framework? Artificial Intelligence Data Science Latest News. 2020; https://www.analyticsinsight.net/need-ontology-ai-data-science-framework/.

[39] https://en.wikipedia.org/wiki/Ontology.

[40] Dalal A. Ontology For Machine Learning Based Enterprise Solutions. https://www.predii.com/blog/ontology-for-machine-learning-based-enterprise-solutions-part-1-introduction-to-ontology.

[41] Tesfaye L. How to Build a Knowledge Graph in Four Steps: The Roadmap From Metadata to AI. 2019; https://enterprise-knowledge.com/how-to-build-a-knowledge-graph-in-four-steps-the-roadmap-from-metadata-to-ai/.

[42] Hawley SH. Challenges for an Ontology of Artificial Intelligence. 2019; https://arxiv.org/abs/1903.03171.

[43] Baclawski K, Bennett M, Berg-Cross G, Fritzsche D, Schneider T, Sharma R, Sriram RD, Westerinen A. Ontology Summit 2017 communiqué – AI, learning, reasoning and ontologies. Applied Ontology. 2018, 13, 3–18.

[44] Noy NF, McGuinness DL. Ontology Development 101: A Guide to Creating Your First Ontology. Stanford University, https://protege.stanford.edu/publications/ontology_development/ontology101-noy-mcguinness.html.

Chapter 3
AI in society

It was assessed by Google that AI applications are in view of the following seven missions: (1) be socially beneficial, (2) avoid creating or reinforcing unfair bias, (3) be built and tested for safety, (4) be accountable to people, (5) incorporate privacy design principles, (6) uphold high standards of scientific excellence, and (7) be made available for uses that accord with these principles [1]. Similarly, Japanese AI society [2] sets forth on the following seven principles: (1) human-centered principles, (2) principles of education and literacy, (3) principles of privacy, (4) principles of security, (5) principles of ensuring fair competition, (6) principles of fairness, accountability and transparency, and (7) principles of innovation. Digital technologies have made us smarter and more productive, transforming how we communicate, learn, shop, and play. Advances in AI are giving rise to computing systems that can see, hear, learn, and reason, creating new opportunities to improve education and healthcare, address poverty and achieve a more sustainable future. But these rapid technology changes also raise complex questions about the impact they will have on other aspects of society: jobs, privacy, safety, inclusiveness, and fairness. When AI augments human decision-making, questions will be asked such as "how we can ensure that it treats everyone fairly, safe, and reliable and how we respect privacy?" [3]. There are six principles suggested by Microsoft [4] to put humans at the center of the systems around AI: (1) fairness (systems should be designed not to include bias against anyone), (2) reliability (systems need to go through rigorous testing and people should play a critical role in making decisions around how they are used), (3) privacy and security (privacy laws must be complied with and personal information should only be used in accordance with high standards), (4) inclusiveness (barriers should be removed from products and environments that can exclude people; this must be considered in the design process), (5) transparency (people must be able to understand how key decisions are made and be able to identify and feedback potential errors and unintended outcomes), and (6) accountability (those who design and deploy AI systems must be accountable for how those systems operate; and this should be ongoing during the design and operational periods).

These principles should be effective globally, societally, and personally.

3.1 Global

The emergence of AI and its progressively wider impact on many sectors requires an assessment of its effect on the achievement of the 17 sustainable development goals (SDGs) along with 169 targets internationally agreed in the 2030 Agenda for Sustainable Development [5]. The fast development of AI needs to be supported by the necessary

https://doi.org/10.1515/9783110717853-003

regulatory insight and oversight for AI-based technologies to enable sustainable development. Failure to do so could result in gaps in transparency, safety, and ethical standards [6]. SDGs are the blueprint to achieve a better and more sustainable future for all. The 17 global challenges should include (1) no poverty, (2) zero hunger, (3) good health and well-being, (4) quality education, (5) gender equality, (6) clean water and sanitation, (7) affordable and clean energy, (8) decent work and economic growth, (9) industry, innovation, and infrastructure, (10) reduce inequalities, (11) sustainable cities and communities, (12) responsible consumption and production, (13) climate action, (14) life below water, (15) life on land, (16) peace, justice, and strong institutions, and (17) partnerships [5]. The more we enable SDGs by deploying AI applications, from autonomous vehicles to AI-powered healthcare solutions and smart electrical grids, the more important it becomes to invest in the AI safety research needed to keep these systems robust and beneficial, so that it can prevent them from malfunctioning or from getting hacked [6]. Substantive research and application of AI technologies to SDGs is concerned with the development of better data mining and ML techniques for the prediction of certain events, such as forecasting extreme weather conditions. Many of these aspects involved in SDGs result from the interplay between technological developments on one side and requests from individuals, response from governments, as well as environmental resources and dynamics on the other [5]. Figure 3.1 shows a schematic representation of these dynamics, with emphasis on the role of technology. These interactions are not currently balanced and the advent of AI has exacerbated the process. A wide range of new technologies are being developed very fast, significantly affecting the way individuals live as well as the impacts on the environment, requiring new piloting procedures from governments. The problem is that neither individuals nor governments seem to be able to follow the pace of these technological developments [6].

AI, which has evolved remarkably in the 2000s, is said to significantly change the way society should be and the economic situation in the future. It is likely to be thought that it is a developed country that is affected only by the state-of-the-art technology, but it is the developing countries that are actually the hardest hit. China and India, which have developed their economies over the past few decades, have continued to grow based on different growth models [7]. The two countries have continued to grow by using their respective strengths, but AI is assumed to alter their business style. AI can dramatically accelerate factory automation while taking customer service, telemarketing, and other tasks out of human hands. AI can also quickly improve the quality of services by accumulating and learning more data [8]. It is important to say that the quality of AI depends on the amount of data, but there is a positive spiral in which AI with higher quality gathers more data, and AI that improves quality further attracts more users and gathers data [8]. In that situation, what developing countries with high economic situation need to survive is to quickly find new growth models. One of the major measures is the expansion of services on the soft side that can only be realized by "human." Industries such as tourism, cultural tourism, hotline call centers, and

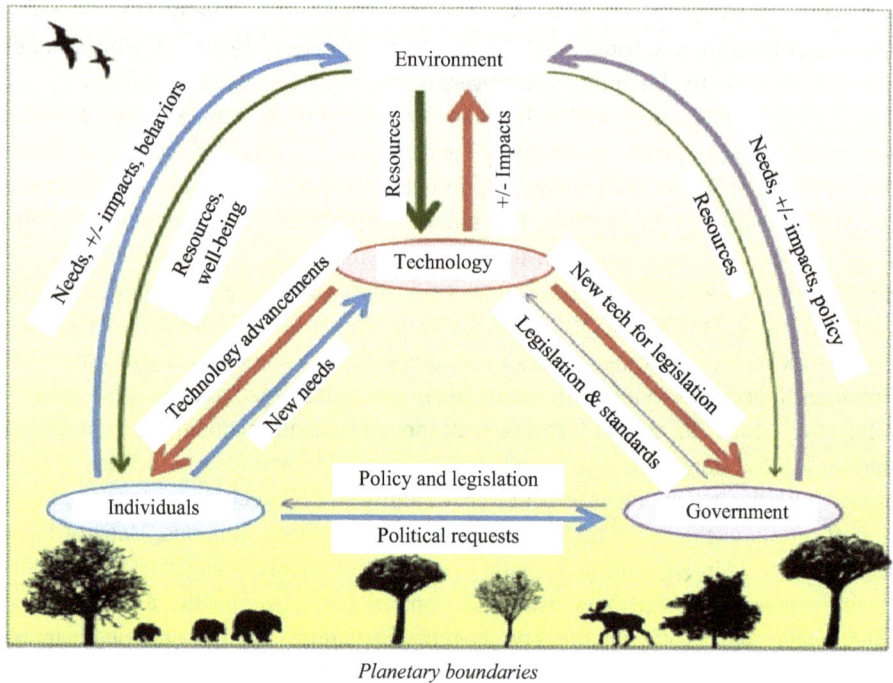

Figure 3.1: Sustainable development goal dynamics, with emphasis on the role of technology [6].

elderly care are considered to be persistently surviving areas in the AI era. The other point for developing countries is an establishment of system and algorithm which should be best adaptable to these countries. Business algorithm which has been successfully employed in the United States is not guaranteed to perform the same work in developing countries such as Ethiopia [8].

Furthermore, not only AI-supported education can be developed in developing countries [9], but also medical and healthcare services have been utilized [10, 11]. For example, there is only one MD (medical doctor) for every 15,000 population in a countryside poor agricultural area of Bangladesh (opposed to 1 MD for every 3,000 nationwide average). Currently, in order to provide medical system with high quality to these areas, following three medical systems have been developed, using AI-related technology such as ML or Bayesian inference: (1) risk-screening system by which future healthy risk and improvement advices are made from vital data, (2) diagnosis-supporting system which indicates disease with high risk of affection or prescription of appropriate drug(s), and (3) remote treatment system, which fully utilizes the above techniques [11]. As AI becomes more popular, it may be replaced by AI and robot-preference jobs, but it may not be accepted because countries with large populations, such as the United States, are reluctant to increase the number of unemployed. There are also fears to be unemployed in China, which has increased its national strength due to its low-wage workforce. On the other hand, in Japan, where there is no resistance to the introduction of new technologies and an aging society is advancing, as a result, such a move may be welcomed. The difference in the environment in which AI, robots, robotic process automation (RPA), and others are accepted will directly affect the competitiveness of the nation [7].

3.2 Societal

As with most changes in life, there will be positive and negative impacts on society as AI continues to transform the society where we live in. There are some challenges that might be faced as well as several of the positive impacts AI. AI will definitely cause our workforce to evolve. The alarmist headlines emphasize the loss of jobs to machines, but the real challenge is for humans to find their passion with new responsibilities that require their uniquely human abilities [10]. The transformative impact of AI on our society will have far-reaching economic, legal, political and regulatory implications that we need to discuss and prepare for. Determining who is at fault if an autonomous vehicle hurts a pedestrian and how to manage a global autonomous arms race are just a couple of examples of the challenges to be faced. Another issue is ensuring that AI does not become so proficient at doing the job it was designed to do that it crosses over ethical or legal boundaries. While the original intent and goal of the AI is to benefit humanity, if it chooses to go about achieving the desired goal in a destructive (yet efficient way), it would negatively impact our society. AI algorithms must be built to align with the overarching goals of humans [12]. AI can dramatically improve the efficiencies of our workplaces and can augment the work humans can do. With better monitoring and diagnostic capabilities, AI can influence healthcare. By improving the operations of healthcare facilities and medical organizations, AI can reduce operating costs and save money [13].

Like many science and technology, AI has great benefits to society, but its impact on society requires proper development and social implementation, due to its unpredictable influences on the society. For using AI effectively to benefit society while avoiding or reducing negative aspects in advance, it is needed to design society in all aspects such as people, social systems, industrial structures, innovation systems, and governance, and promote the creation of a society that can use AI effectively and safely, that is, to transform it into an "AI-ready society" [2]. To develop the AI-ready society, there are some trends proposed, including (i) Wisdom Network Society and (ii) Society 5.0 Plan. With the progress of AI networking, AI and AI, AI and humans, and even human and human relationships will be able to collaborate and cooperate in a seamless manner, it is predicted that a society will be able to create, distribute, and connect new set of data, information, and knowledge by learning and analyzing data, information, and knowledge by networked AI, that is, the intelligence of networked AI or human intelligence that utilizes it. At the same time, although there is a concern about increasing reliance on extremely advanced AI networks, the need for wisdom increases which can be considered as human ability to envision the ideal way of human beings and society and solve problems for realization by utilizing intelligence based on data, information, and knowledge. Here, if many people can gain wisdom in the AI network society, it is expected that human beings will actively and effectively use technology at a high level throughout society, and interconnection and cooperation of human wisdom to solve problems will advance. In other words, by utilizing AI networks, human beings are expected to connect each wisdom as advanced problem-solving ability to form a wisdom network society [2]. Historically, our society pattern has been changed from hunting and gathering (Society 1.0) → agrarian (Society 2.0) → industrial (Society 3.0) → information (Society 4.0), followed by the newly proposed Society 5.0, which is characterized by sustainable wisdom network society by structuring the CPS (cyber-physical system) as a base, in which AI, IoT, robots, RPS and other advanced technologies will be widely implemented in society and different people respect each other's diversity and individuality [14, 15].

IoT and AI, which have developed at the same time, are highly interrelated technologies accompanied with high affinity and adaptability that create new value by combining and utilizing them. In IoT systems, tendency of past data has traditionally been analyzed using statistical analysis methods, and some actions have been taken based on the judgment of data analysis experts or predetermined rules. By applying AI to this situation, it is possible to immediately analyze data that changes from time to time, and to grasp, monitor, predict, and forecast the situation in real time. In some cases, by automating human-based work and judgments to AI, human error can be eliminated or minimized and accuracy can be improved. AI is already becoming smarter and smarter by repeating autonomous ML without allowing humans to learn by giving teacher data. Based on expectation of better job performance by AI than humans, systems and services that combine IoT and AI are appearing one after another. Typical application examples should include (1) performance of the predictive maintenance and quality check in the manufacturing industry; (2) determination of the most efficient shipping route in transportation and distribution operation; (3) supporting crop growth plan through various sensors (temperature, humid, precipitation, soil chemical compositions, etc.) and even with assist of drone, as discussed in Chapter 7; (4) constructing HEMS (home energy management system) combined sensors and control systems for power distribution facilities, solar panels, storage batteries, and so on; (5) supporting healthcare and wellness management with sensor data obtained by wearable devices (for blood pressure, heart rate, burned calories, etc.), as discussed in Chapter 5; and (6) devices for nursing care and crime prevention that monitor GPS (global positioning system) and other location information and biometric information data in real time [16].

Weather conditions – either for today or long-term forecast – are affecting many business and industries and individual mental and physical conditions as well as personal life schedule. Advances in AI technology have made it available to learn and utilize a variety of weather data that could not be utilized until now. AMeDAS (Automated Meteorological Data Acquisition System) are installed at 1,300 locations over entire Japanese islands (with approximately 17 km apart from each station) to collect weather observation data such as precipitation, wind direction and speed, sunshine hours, and temperature which are automatically send to Central Meteorological Agency [17]. Besides, there are information available from the meteorological satellites such as "Sunflowers" and supercomputers. information from the meteorological satellites "Sunflowers" and super computers. Rain clouds could not be observed. However, if AI is used, rain cloud data can be learned in advance in addition to the data available. As a result, in addition to the previous data, weather forecasts can be made from dangerous cloud colors and shapes, which can improve accuracy. In the future, it is now possible to improve the accuracy of weather forecasts by accumulating tries and errors, and with further technological advances, all weather forecasts may be in the era of AI. In addition, AI learns from the data when its forecast is wrong and improves the accuracy of the next forecast (up to 99% accuracy) [17–20].

Earthquake motion prediction, which predicts shaking caused by earthquakes, is one of the main research issues in the field of earthquake disaster prevention. By improving the accuracy of earthquake motion prediction technologies used in earthquake hazard assessments to prepare for future earthquake disasters and emergency earthquake prediction technologies immediately after the earthquake, nation's resilience (especially disaster prevention and mitigation capabilities) can be further improved [21–23]. In order to improve the accuracy of seismic prediction, a hybrid prediction method was developed by combining ML with seismic prediction formulas based on conventional physical models [23]. Figure 3.2 compares four different PGA (peak ground acceleration) distribution of the major earthquake at central Kyushu island Japan (in 2016). Figure 3.2(a) shows PGA distribution of the real observed data, (b) depicts PGA distribution predicted by ML only, (c) shows PGA distribution predicted by hybrid method and (d) shows PGA distribution predicted only by the seismic motion prediction formula. It is clearly indicated that the hybrid method (c) can predicts PGA distribution map very close to that actually occurred (a).

3.3 Personal

3.3.1 Technological environment

The trend of AI utilization in all areas of society can no longer be stopped by anyone. You may not feel a sudden change in your daily life, but it does not change into an AI society (or AI-ready society) after one day. Change does not occur suddenly but comes gradually yet steadily. Let us compare the characteristics and strengths of AI with human capabilities. First of all, AI has its memory and computing power. That is why the ability to learn vast amounts of information and to reason based on learning results

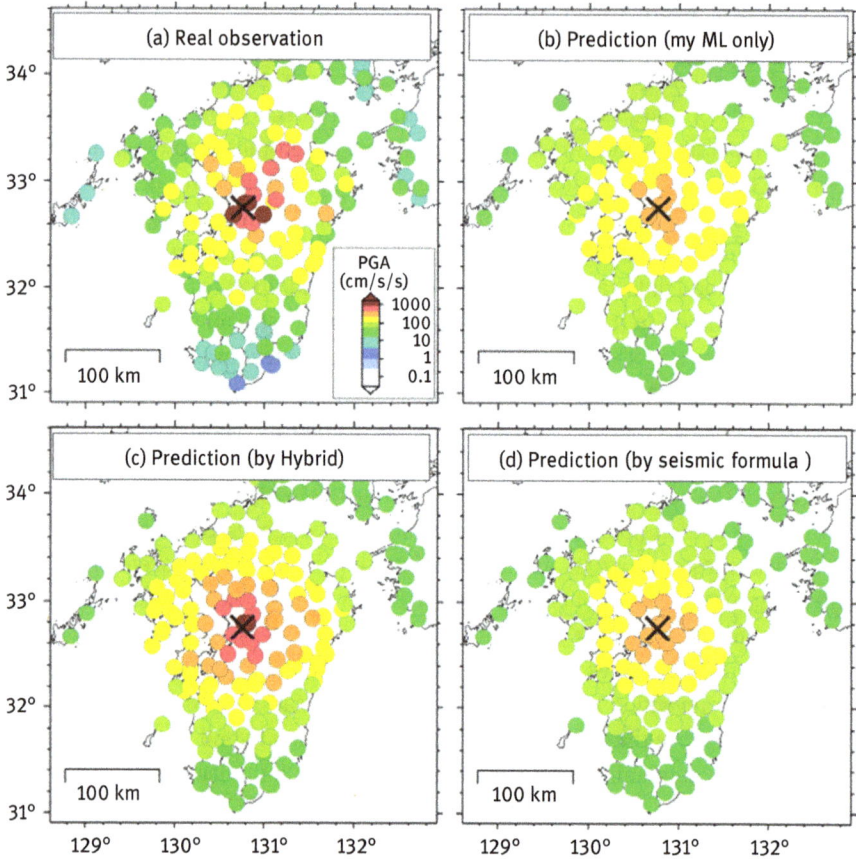

Figure 3.2: Comparison of PGA distribution [23].

will be superior to humans. Because it is a computer, there is also a strength that you do not get tired even if you are doing the same work for a long time. Hence, AI is likely to replace occupations where no special knowledge or skills are required, or where data analysis and patterned manipulation are required. On the other hand, as discussed in previous chapter, the weaknesses of AI include the lack of will, emotion, and common sense, the inability to set challenges on their own, and the inability to appeal to others. AI is just software on a computer, so unless human programs what needs to be done and how to do it, it will not proceed with work without permission. Human is considered good at perceptual power, emotion, common sense, inspiration, imagination, presentation ability, leadership, complicated communication, and response at an ad hoc basis. AI is considered good at learning ability, memory, computational power (speed, accuracy), inference power, predictive power, continuity power, and simple task. There are three basic powers and 12 associated elements that we need in an AI-ready society as basic skills for working adults. These elements can be

categorized into three abilities. The first ability is to step forward with (1) independence, (2) approaching ability, and (3) execution ability. The second ability is to
think through with (4) problem discovery ability, (5) planning ability, and (6)
imagination. The third ability is to work in a team with (7) ability to communicate,
(8) listening, (9) flexibility, (10) situation grasping ability, (11) discipline, and (12)
stress control ability [24].

   RPA is the technology that allows anyone today to configure computer software,
or a "robot" to emulate and integrate the actions of a human interacting within digital systems to execute a business process. RPA is a term used to express robot automation efforts and is also called "digital labor" or "virtual intellectual worker." RPA
robots utilize the user interface to capture data and manipulate applications just
like humans do. They interpret, trigger responses, and communicate with other systems in order to perform on a vast variety of repetitive tasks. An RPA software robot
never sleeps and makes zero mistakes. RPA has three types: client type (or desktop
type), server type, and cloud type. The option of using business process outsourcing
(BPO) for dispatch and offshore (which refers to especially emerging and developing
countries) as a means of supplementing the labor shortage has been popular. The
most significant difference between RPA and BPO is that it can significantly reduce
the time required for work while maintaining quality, allowing you to continue
working 24/7/365, and there is no risk of retirement. As a result, RPA should be applied more frequently in wide applicable areas [25, 26]. RPA is a robot (software)
that uses technologies such as AI that tries to reproduce human intelligence on a
computer and ML. Simply it is like a "macro file" that automatically processes the
work registered by the user, such as table calculation software. It is an image that
the operation of the application that a person is doing is macronized using an automation tool, and the macro file performs work on its behalf. RPA is a "human
agency" and "complement of human work" [27].

3.3.2 Examples of jobs that are likely to be substituted by AI

We have seen in the above how the traditional society style might be reformed into
AI-ready society, advantages and disadvantages associated with AI technology, AI-
installed systems, RPA and hybrid system of RPA and AI by DX technology. Frey
et al. [28] examined how susceptible jobs are to computerization, implementing a
novel methodology to estimate the probability of computerization for 702 detailed
occupations, using a Gaussian process classifier. It was reported that, according to
estimates, about 47% of total US employment is at risk.

   The following long list is about current jobs that is more likely replaced by AI
technologies [28–30].

   They are IC production operator, general clerk, casting founder, medical clerks,
receptionist, AV and communication equipment assembly and repairman, station

staff/clerk, NC grinding machine worker, NC lathe worker, accounting auditors, processed paper manufacturer, loan clerk, school clerk, camera assemblyman, mechanical woodworking, dormitories/condominium manager, CAD operator, school lunch cook, education/training clerks, administrative clerk (country), administrative clerks (prefectural municipalities), bank teller, metalworking and metal products inspection worker, metal polisher, metal material manufacturing and inspection worker, metal heat treatment worker, metal presses worker, cleaning clerk, instrument assembler, security guards, accounting clerk, inspection and inspection staff, meter reading staff, construction workers, rubber products molding (excluding tire molding), wrapping worker, sash-frame worker, industrial waste collection and transportation workers, paperware manufacturer, automotive assembly work, automotive painter, shipping worker, a small collection worker, personnel clerk, newspaper distributer, medical information management officer, fisheries paste product manufacturing industry, supermarket clerk, production site clerk, bakery, milling worker, book-made workers, soft drink route salesperson, oil refining operators, cement production operator, textile product inspection worker, warehouse workers, side dish manufacturer, surveyor, lottery seller, taxi driver, courier delivery man, forging worker, parking management, customs officer, on-line sales receptionist, wholesale workers, data entry clerk, telecommunications engineers, computer planting operator, electronic computer maintenance staff (IT maintenance staff), electronic component manufacturing, train conductor, road patrol personnel, daily necessities repair shop clerk, motorcycle delivery man, generators, non-destructive inspection worker, building facility management engineer, building cleaners, goods purchasing clerk, plastic products molding worker, process plate-made operator, boiler operator, trade clerks, packaging workers, storage and management staff, insurance clerk, hotel guest room clerk, machining center operator, sewing machine sewing worker, plating work, noodle manufacturing works, postal foreign service, postal clerk, toll fee receiving personnel, cashier, train cleaners, car rental sales office staff, and route bus driver.

3.3.3 Representative examples of less likely positions to be substituted by AI

There are some types of jobs as listed below which appear to be less likely substituted by AI technologies [29–31]: art director, outdoor instructor, announcer, aromatherapist, dog nurse/trainer, medical social worker, interior coordinator, interior designer, film photographer, film director, economist, music classroom lecturers, curators, school counselors, tourist bus guides, educational counselors, classical performer, graphic designer, healthcare manager, game creator, management consultant, entertainment manager, surgeon, speech auditor, industrial designer, advertising director, international cooperation expert, copywriter, occupational therapist, lyricist, composer, magazine editor, industrial counselor, obstetrician and gynecologist, dentist, child welfare student, scenario writer, sociology researcher, social education supervisor, social welfare facility care staff, social welfare facility instructor, veterinarian, judo therapist, jewelry designer, elementary school teacher, commercial photographer, pediatrician, product development staff, midwives, psychology researchers, anthropologist, stylist, sports instructor, sports writer, vocalist, psychiatrist, sommelier, university/junior college teacher, junior high school teacher, small and medium-sized enterprise diagnostician, tour conductor, disc jockey, display designer, desk, TV photographers, TV talent, book editor, physician, language teacher, nail artist, bartender, actor, acupuncture therapist, hairdresser, critic, fashion designer, food coordinator, stage director, stage artist, flower designer, free writer, producer, pension managers, marketing researchers, nursery teacher, broadcast reporter, broadcast director, news photographer, legal Instructor, manga (cartoon) artist, musicians, makeup artist, teachers of schools for the blind/deaf/handicapped, kindergarten teachers, physiotherapist, culinary researcher, travel agency counter clerk, record producer, restaurant manager, and recording engineer.

Jobs listed in the above are directly or indirectly related to value of communication, which is one of typical weak points of AI. It is expected that the value of the skill and the ability to do the work which cannot be done by the automation of the machine such as AI is expected to increase more than ever. First of all, there is a job and skill that fills man's fundamental desire to connect with people: human artistry (theater, music), physical ability (sports), compassion (therapy), and hospitality (restaurant), and so on, as pointed out by Brynjolfsson et al. [32]. Machines may be able to run faster than humans or play music more accurately. However, it is still a performance by a living person that people feel that it is good to be moved, excited, and pay money. Secondly, it is to cultivate the skills to come up with new ideas and concepts that have never been achieved through flexible and flexible ideas. Machines can give answers, but they still do not have the ability to ask questions.

3.3.4 New-type jobs that AI creates

Machines and algorithms in the workplace are expected to create 133 million new roles but cause 75 million jobs to be displaced by 2022 [33], indicating that the growth of AI could create 58 million net new jobs in the next few years. With this net positive job growth, there is expected to be a major shift in quality, location, and permanency for the new roles. Companies are expected to expand the use of contractors doing specialized work and utilize remote staffing [34]. Within the next few years, AI will be all around us, embedded in many higher order pursuits. It will educate our children, heal our sick, and lower our energy bills. It will catch criminals, increase crop yields, and help us uncover new worlds of augmented and virtual reality [35]. Frank et al. [35] listed the following 15 new jobs under ever-growing AI society: data detective, genome portfolio director, partner to walk and/or talk with, head of ethical sourcing, chief trust officer, cyber city analyst, person in charge of human-machine collaboration, head of AI business development, "bring your own: personally owned equipment utilization" IT facilitator, edge computing experts, fitness commitment counselor, digital tailor, AI-assisted medical engineer, financial health coach, and quantum ML analyst.

3.3.5 Work-style reformation

The way people work has changed with the times, from hunting-gathering to farming, from industrialization to information technology. In addition, with the recent AI era, companies are about to experience new changes in the way they work [36]. In social science inquiry in AI era, we need the right people asking/answering the right questions. Such questions, according to Brynjolfsson et al. [37], should at least include why median income has stopped rising in the United States, why is the share of population that is working falling so rapidly, or why are our economy and society are becoming more unequal. To win global competition, companies will replace robots and AI with fixed-type operations. On the other hand, human beings believe that work with high intellectual creativity that machines cannot do, work on the side that utilizes AI, and work with a large human touch element. At the same time as changes in human work, reforms are also emerging in our working style, such as the expansion of remote work. With the progress of ICT and digitalization, remote work is becoming possible. However, current remote work is just like working at home for what are needed to be done at working office. The inherent good thing about remote work is that you can set the time and location freely. It can be said that it is not an employment type but a self-employed way of working. In the future, self-operated remote work will deal with companies with their skills through contracts as a mainstream. As work style changes, companies will have to change as they change. Until now, the corporate structure has been a hierarchy-type vertical structure, an organization with a large employment and labor-like part, such as the supervisor supervising subordinates. However, it is assumed that the structure of the company will become horizontal as work becomes modular, and the networking by the substitution by the machine and outsourcing advances [36].

New technologies such as ICT, robots, RPA, and AI are about to significantly change the way companies work and operate. By digitizing all information, ICT which is the basis of these technologies, enables teleworking and other work-related work environments that do not change the time and place, and is expected to contribute to improving the productivity of companies [38]. As telework gradually spreads and gains recognized citizenship thereof, the COVID-19 corona shock accelerated this movement [39]. On the other hand, for more advanced technologies such as AI, labor substitution by automation is overly emphasized. Robots and RPA are meant to be more efficient through automation and labor substitution, but the essence of AI is big data, "prediction" using ML in a broader way [40]. How to use that "prediction" is up to humans, and it is possible for AI and humans to form complementary relationships and coexist. Through AI forecasts, it is expected that "personalization" (individualization) of the business will progress, so that, according to preferences and preferences of individual consumers, a variety of goods can be provided and services would be differentiated [38]. The workplace has been considered as a place where employees have information and knowledge that must be shared at the same time and in the same place. On the other hand, if ICT unlocks the constraints so that information can be shared and communicated as in the workplace wherever they are, employees do not have to be in the same place at the same time. To that end, the development of ICT has contributed to the introduction of work-like activities that can be used at any time or place. Such telework can include work-at-home, mobile work (outside work such as sales activities), and satellite office work (working in an office in a different place of work from the original place of work, etc.). Although it is often emphasized that the need for teleworking can be reduced by the cost of office and commuting, allowing the workman to work according to the lifestyle, there is a drawback potentially associated with such teleworking: (i) the inability to demonstrate synergistic effects such as exchange, information sharing, and information work obtained from working in the same place and at the same time: and (ii) the inability of the boss to monitor can affect the activity of the workers [36].

3.4 Future of AI society

Many of the jobs currently being done by human hands will be AI-digitized in the near future, and what was only a human "tool" has evolved greatly and is becoming a new labor force that replaces people. This trend is welcome, and it will make people's lives richer than they are today. However, it is also that AI will step into the business ring that people have competed with each other until now, and the work that can be done by machines and the work that only people can do will be highlighted more clearly in the future. According to Tegmark [41], human life has been experiencing three stages: Life 1.0 referring to biological origins, Life 2.0 referring to cultural developments in humanity, and Life 3.0 referring to the technological age of humans. At the Life 3.0 stage, the emerging technology such as artificial general intelligence that may someday, in addition to being able to learn, will be able to also redesign its own hardware and internal structure.

As computers behave more like humans, the social sciences and humanities will become even more important. Languages, art, history, economics, ethics, philosophy, psychology, and human development courses can teach critical, philosophical, and ethics-based skills that will be instrumental in the development and management of AI solutions. If AI is to reach its potential in serving humans, then every engineer will need to learn more about the liberal arts and every liberal arts major will need to learn more about engineering. In many ways, AI is already in our lives without our being aware of it [42]. As AI and IoT are implemented in society, structural changes accompanying technological innovation are progressing in various fields of human society, such as production, mobility, finance, logistics, medical care, nursing care, and education. In addition to industrial structures, labor market, social systems, and organizations, AI technology is built/installed into the devices around us, and AI networking that functions in connection and cooperation is progressing, and people's lives will change greatly. Among the new era of social devices that will be built during the moving back and forth between the cyber and physical worlds, there will be a significant increase in the number of parts that need to be modified and complemented, even if there are many parts to follow in the future. The challenge is how to balance the connection between the two worlds and the relationship between the two influences, how to place the balance in a comfortable function in the ground of individuals' knowledge, and to examine the relationship between the knowledge, emotion, mind, and body. To that end, there is a need for individual verification of the basic concepts, individuals, subjects, self-decisions, responsibilities, autonomy, agreements, and accountability that have been built on the basis of modern social systems, and the work of examining the entire system and organization that has been conceived by a set of concepts. It is also available that AI technology civilization is inviting a re-inspection of the basic framework of modern learning itself in this way that has underpinned modern social systems [43].

"Amazon recommends fake branded goods because AI overlooked." This is a news reported in April 2019. This incidence makes us think about the relationship between humans and AI. It has been a long time since we have been called out for collaboration between humans and AI, but how can humans confront AI and how can we use AI? From this point alone, it can be said that AI is never an all-rounder player. Although there are excessive expectations and excessive vigilance, such "AI myths" should be redefined and renamed as "augmented intelligence" that can demonstrate its capabilities only when paired with humans. By the way, augmented intelligence is called "another AI" because they share the same acronym "AI" [44].

As final words in this chapter, humans should always understand AI as our extended intelligence, encourage learning, design a comprehensive system that cooperates with AI and humans, and solve problems.

References

[1]   Artificial Intelligence at Google: Our Principles; https://ai.google/principles/.
[2]   Sudo O. Social impact of artificial intelligence and human co-evolution. Journal of Information and Communications Policy. 2018, 2, 1–10.
[3]   Artificial Intelligence's role in our society; https://www.ics.ie/news/artificial-intelligences-role-in-our-society.
[4]   Responbile AI; https://www.microsoft.com/en-us/ai/responsible-ai?activetab=pivot1%3aprimaryr6.

[5] UN General Assembly. A/RES/70/1Transforming our world: the 2030 Agenda for Sustainable Development. Resolution. 2015, 1–35, https://www.un.org/sustainabledevelopment/sustainable-development-goals/.

[6] Vinuesa R, Azizpour H, Leite I, Balaam M, Dignum V, Domisch S, Felländer A, Langhans SD, Tegmark M, Nerini FF. The role of artificial intelligence in achieving the Sustainable Development Goals. Nature Communications. 2020, 11. doi: https://doi.org/10.1038/s41467-019-14108-y.

[7] https://gigazine.net/news/20180921-ai-devastate-developing-world/.

[8] Lee K-F Artificial Intelligence Threatens Jobs in Developing World. 2018; https://www.bloomberg.com/view/articles/2018-09-17/artificial-intelligence-threatens-jobs-in-developing-world.

[9] https://roboteer-tokyo.com/archives/5933.

[10] https://www.huffingtonpost.jp/mari-sako/bangladesh-ai-technology_b_17068132.html.

[11] How much can IT contribute?; https://miup.jp/.

[12] DeAngelis S. Artificial Intelligence and Its Impact on Society. 2018; https://www.enterrasolutions.com/blog/artificial-intelligence-and-its-impact-on-society/.

[13] What Is The Impact Of Artificial Intelligence (AI) On Society? https://bernardmarr.com/default.asp?contentID=1828#:~:text=Artificial%20intelligence%20can%20dramatically%20improve,creativity%20and%20empathy%20among%20others.

[14] Minevich M. Japan's 'Society 5.0' initiative is a road map for today's entrepreneurs. 2019; https://techcrunch.com/2019/02/02/japans-society-5-0-initiative-is-a-roadmap-for-todays-entrepreneurs/.

[15] From Industry 4.0 to Society 5.0: the big societal transformation plan of Japan; https://www.i-scoop.eu/industry-4-0/society-5-0/.

[16] IoT systems, expected to become more sophisticated by applying AI; https://www.teldevice.co.jp/ted_real_iot/column/ai_iotsystem/.

[17] Kobayashi T, Shirai S, Kitadate S. AMeDAS: supporting mitigation and minimization of weather-related disasters. Fujitsu Technology. 2017, 53, 53–61, https://www.fujitsu.com/global/documents/about/resources/publications/fstj/archives/vol53-3/paper11.pdf.

[18] Hito-Titor. Advances in AI are entering an era of 99% accuracy in weather forecasting. 2019; https://www.chiropotes.com/ai/article201911003/.

[19] Inoue H. Is there still anything AI can do? 2019; https://ainow.ai/2019/08/23/174786/.

[20] The weather forecast is full of cutting edge! Life changes greatly with the development of AI!? https://www.kicks-keirinkan.com/blog/cat_oya1/7061/.

[21] https://tifana.ai/861/.

[22] https://www.newton-consulting.co.jp/topics_detail11/id=4793.

[23] Kubo H, Kunugi T, Suzuki W, Suzuki S, Aoi S. Hybrid predictor for ground-motion intensity with machine learning and conventional ground motion prediction equation. 2020; Scientific Reports. https://doi.org/10.1038/s41598-020-68630-x.

[24] https://job.mynavi.jp/conts/2022/ai/02.html.

[25] https://www.uipath.com/rpa/robotic-process-automation.

[26] https://rpa-technologies.com/about/.

[27] https://rpa-technologies.com/lp/roboforce/.

[28] Frey CB, Osborne MA. The future of employment: How susceptible are jobs to computerization? 2013. Oxford Martin Programme on Technology and Employment; http://sep4u.gr/wp-content/uploads/The_Future_of_Employment_ox_2013.pdf.

[29] https://www.bigdata-navi.com/aidrops/681/.

[30] https://job.mynavi.jp/conts/2022/ai/02.html.

[31] https://www.nri.com/-/media/Corporate/jp/Files/PDF/news/newsrelease/cc/2015/151202_1.pdf.

[32] Brynjolfsson E, McAfee A. The Second Machine Age: Work, Progress, and Prosperity in a Time of Brilliant Technologies. W.W. Norton & Company, 2014.

[33] The Future of Jobs 2018; World Economic Forum. http://reports.weforum.org/future-of-jobs -2018/.

[34] Chowdhry A. Artificial intelligence To create 58 million new jobs By 2022, Says Report. 2018; https://www.forbes.com/sites/amitchowdhry/2018/09/18/artificial-intelligence-to-create -58-million-new-jobs-by-2022-says-report/?sh=617dfc8c4d4b.

[35] Frank M, Roehrig P, Pring B. What To Do When Machines Do Everything: How to Get Ahead in a World of AI, Algorithms, Bots, and Big Data. Wiley, 2017.

[36] Oouchi S. What is the future vision of work that awaits the future of the AI era? 2019; https://www.adeccogroup.jp/power-of-work/044.

[37] Brynjolfsson E, Andrew Mcafee A. Race Against The Machine: How the Digital Revolution is Accelerating Innovation, Driving Productivity, and Irreversibly Transforming Employment and the Economy Kindle Edition. Digital Frontier Press, 2011.

[38] Tsuru K. How New Technologies Will Change the Way We Work – A Vision for the AI Era. 2019; https://www.rieti.go.jp/jp/publications/pdp/19p023.pdf.

[39] Takeuchi S. The way of working that is not used for AI required for the after-corona. 2020; https://www.itmedia.co.jp/business/articles/2007/18/news013.html.

[40] Agrawal A, Gans J, Goldfarb A. Prediction Machines: The Simple Economics of Artificial Intelligence. Harvard Business Review Press, 2018.

[41] Tegmark M. Life 3.0: Being Human in the Age of Artificial Intelligence. Knopf, New York, 2017.

[42] Smith B, Shum H. The future computed: artificial intelligence and its role in society. Official Microsoft Blog, 2018; https://blogs.microsoft.com/blog/2018/01/17/future-computed-artificial-intelligence-role-society/.

[43] Suzuki M. Technological civilization and human society in the age of AI: AI technology and the future of human beings. Journal of Information and Communications Policy. 2018, 2, 21–43, https://www.soumu.go.jp/main_content/000592823.pdf.

[44] Takeda A. The Risk of "Everything relied on AI performance" – The Collapse of AI Myths and the Growing Significance of Human Existence -. 2019; https://note.com/atsushi_tt/n/n6a1d85e4356d.

Chapter 4
AI in concern

As with other technologies, AI is still in progress. Besides while humans constantly impose expectations and roles, there are various concerns which are challenging back to humans, including personality, responsibility, accountability, singularity, and others.

4.1 Digital clone

Digital cloning (or PAI clone: personal AI clone) is an emerging technology involving DL algorithms, which allows one to manipulate currently existing audio, photos, videos that are hyper-realistic, or create a digitized reproduction of a person. A lot of information is uploaded on the Internet through social media (like SNS) and others. Information such as documents, images, videos, and audio (aka "life log") engraves a record of our lives in the digital space (in a broad way, life logs refer to keeping records of our lives in general as data, including biometric such as heart rate and body temperature and location information). Such information left on computers or computer networks is analyzed, and the person's experience, thoughts, and values are reconstructed. Research is being conducted as an application of AI, such as using this to realize conversations with the deceased [1–3]. Right now, there is no way to clone a human being into a robot which is 100% copy in terms of physical appearance, motion (walking style, mimics, posture, etc.), behavior, reactions, emotions, mind and other characteristics. There are studies and products that can clone human being partially [2]. A digital clone would not be built as people were built. It would not come into existence biologically and then duplicate everyone's experiences, interactions, and learning moments. A digital clone is an electronic copy of people's personality [3]. There are more social robots like Nadine and Sophia, which are also humanoid robots. A humanoid robot is a robot with a human-like body and is developed to have specific tasks and functions. If a humanoid robot aesthetically resembles humans, it is called android. Voice cloning (or vocaloid) technology has been advanced and is a speech synthesis technology and by inputting melodies and lyrics, the singing voice based on the voice of the sampled person can be synthesized [4, 5].

In computing, an avatar is an incarnation and is a graphical (virtual) representation of a user or the user's character or persona. It may take either a two-dimensional form as an icon in Internet forums and other online communities or a three-dimensional form, as in games or virtual worlds. NLP, speech recognition, intelligent agents, multi-agent (or distributed cooperation of intelligent agents), ML, human agent interaction and DL have been nicely integrated to make PAI (personal

https://doi.org/10.1515/9783110717853-004

artificial intelligence) realized [6, 7]. According to Govender [7], PAI avatar possesses advanced scenarios such as (i) real-time health analysis (currently your smartwatch monitors heart rate and steps and sends it back to phone, which sends it into the cloud, but in the near future, PAI will read this data and analyze it in real-time basis, with the ability to alert people as early as possible as to risk of a heart attack, or stroke) (ii) environmental analysis (air quality is a problem in many parts of the world), (iii) real-time language translation (device like Skype needs to be augmented with the power available on the edge device), and (iv) custom applications. In the world of IoT, a new concept is becoming prevalent – the intelligent edge. Edge computing refers to data processing power at the edge of a network instead of holding that processing power in a cloud. The intelligent edge recognizes that to deliver what businesses require, data processing and intelligence need to be applied at the edge before data is synced into the cloud [7]. Relationship between intelligent edge and intelligent cloud is illustrated in Figure 4.1 [7].

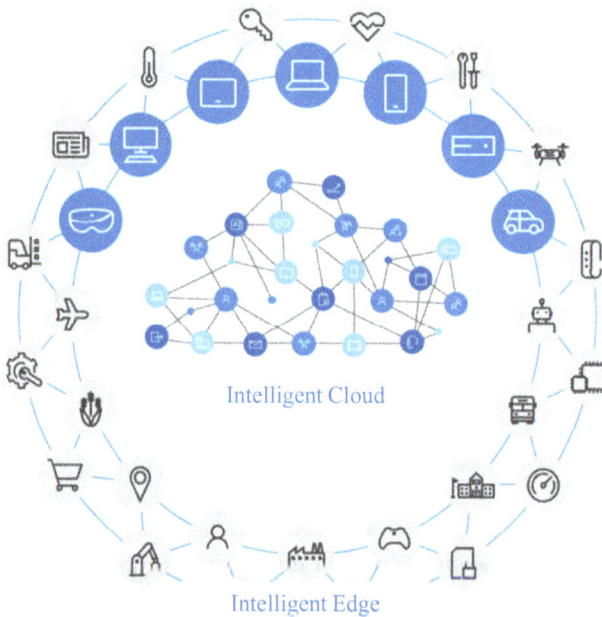

Figure 4.1: Relationship between intelligent edge and intelligent cloud [7].

Many grieving people feel an emotional connection to things that represent dead loved ones, such as headstones, urns, and shrines, according to grief counselors. In the future, it is predicted that, using AI or PAI technology, humans will replace dead relatives with synthetic robot clones, complete with a digital copy of that person's brain [8]. Much more, modern times are an era where everyone left a lot of their own images, audio, and utterances on SNS. Hence, data quantity as energy source for AIs is enough. If Einstein can be reproduced/incarnated and the theory of relativity is lectured directly from Einstein, there may be applications in the field of education [6]. Is this rebirth of the deceased legally free? AI clones (or reproduced android) possess various issues, including digital personality, patentability, portrait rights, ethics, and others. In general, new appliances are valued for newness, and many people buy new products that are cheaper than expensive repair costs when they break down. Among them, why is "AIBO" (which is AI-powered robot) only wanted to own while repairing the individual once bought? [9]. Communication with AI-powered robots creates more emotions than attachment to products (in an empathized situation), and it also increases the sadness of parting. It may be time to think about how to view aging and death of AI.

4.2 Personality

We turn on/off our PCs every day, and no matter what great AI is installed in the computer, we do not feel it is going to die or come back to life. AI is a tool providing wonderful functions that humans cannot do, but AI is not life (a vital substance).

4.2.1 Legal personality

There are some studies regarding the legal personhood for AI. Saito [10] proposed three ideas: (1) legal entities for AI will be expected to limit the responsibilities of related business parties; (2) even for corporations that are controlled by AI, it will not be necessary to dispense with shareholders and directors; and (3) if AI which has been established as a legal person under foreign law comes to Japan, it will be possible to protect transaction safety through the choice of laws and the laws on foreigners. Akasaka [11] mentioned that with the development of autonomous AI, it has become difficult for individuals (e.g., AI user and AI producer) to think about the predictability of the outcome of the event caused by AI, and the algorithm of the action is becoming a black box without human understanding, and uncertainty is increasing. In such a situation, it is said that it is difficult to think of responsibility for the individuals involved in today's legal system, and there are problems such as victims not being rescued [11]. IEEE (Institute of Electrical and Electronics Engineers) [12] announced that, as the use and impact of autonomous and intelligent systems become pervasive, societal and policy guidelines are needed to be established in order for such systems to remain human-centric, serving humanity's values and ethical principles. Even shifting from an automatic self-driving car to an autonomous car (accompanied with a sense of moral machine), we need to have governance [13].

4.2.2 Personality

As AI becomes more widespread and evolving and closer to humans, there is a theme that will become inevitable, such as whether AI has a personality. While "weak AI" is an intellectual processing done by human beings, we sometimes call the efforts to generate intelligence by machines as "strong AI." Realization of such strong AI appears to be not easy since human capabilities and machine capabilities possess different vectors, and it is difficult to match them [14]. After all, AI is no different from smartphones and PCs, and it is only a computer that automatically extracts what seems to be the correct answer from a lot of data, and basic function is not far different from those of refrigerator, air conditioner, and so on. Therefore, it is the domain of previous data scientists, not AI, that uses fast calculations and algorithms to get the right answers from big data. Home appliances are becoming more common, but if AI without personality evolves as it is, AI that makes mistakes in value judgment will be flooded. In the world of AI, it will become real when the word "educated" is born after it has been learned. If AI is defined as the thing which realized an intellectual thinking power by computers to perform the logical reasoning and learning by experience while having a sense of values and intention like a human mind, the program itself can be called an artificial personality program [14], which can be received to be like as personality or character for the human being who communicate with.

4.2.3 Consciousness, gender, and marriage

Artificial consciousness (AC, or sometimes called machine consciousness: MC or synthetic consciousness: SC) is a research area related to AI and cognitive robotics, with the aim of raising awareness of artifacts created by technology [15]. AC does not have to be as genius as "strong AI." It needs to be objective, like scientific methods, and it must be able to realize the ability of known consciousness. Even if AC can be formally proved, the judgment of whether the implemented one has consciousness will depend on observation. Turing test proposes to measure the intelligence of a machine by interacting with it. Turing test guesses whether the person in the dialogue is a machine or a human. It is only when artificially conscious bodies go beyond the observer's imagination and build meaningful relationships that they pass such tests. Cats and dogs cannot pass this test. Consciousness is not an attribute that only human beings have. However, there is a high possibility that machines with artificial consciousness will not pass the Turing test [16].

Here is a question if AIs can develop conscious experience. The question is not so far-fetched. Robots are currently being developed to work inside nuclear reactors and care for the elderly. As AI technologies grow more sophisticated, they are projected to take over many human jobs within the next few decades. Schneider et al. [17] considered three important points. (i) First, ethicists worry that it would be wrong to force AIs to serve us if they can suffer and feel a range of emotions. (ii) Second, consciousness could make AIs volatile or unpredictable, raising safety concerns (or conversely, it could increase an empathy of AI; based on its own subjective experiences, it might recognize consciousness in us and treat us with compassion). (iii) Third, machine consciousness could impact the viability of brain-implant technologies. If AI cannot be conscious, then the parts of the brain responsible for consciousness could not be replaced with chips without causing a loss of consciousness. In a similar vein, a person could not upload their brain to a computer to avoid death because that upload would not be a conscious being. Based on this essential characteristic of consciousness, a test for machine consciousness, the AI Consciousness Test (ACT), was proposed which looks at whether the synthetic minds we create have an experience-based understanding of the way it feels, from the inside, to be conscious [17]. Griffin [18] mentioned that the ACT would challenge an AI with a series of increasingly demanding natural language interactions to see how quickly and readily it can grasp and use concepts and scenarios based on the internal experiences we associate with consciousness. For example, at the most basic level we might simply ask the machine if it conceives of itself as anything other than its physical self. Then, at a more advanced level, we might see how IT deals with scenarios like the ones mentioned above and test its ability to reason and discuss philosophical questions that zero in on the "hard" problems of consciousness. Finally, and at the most demanding level, we might see if it can invent and use a consciousness, based concept of its own, without relying on human ideas and inputs [18].

Here is another issue – robot gender stereotype. A robot is a mirror held up not just to its creator, but to our whole species: What we make of the machine reflects what we are [19]. One of the biggest issues is gender. Robots do not have genders – they are metal, plastic and silicon, and filled with 1s (ones) and 0s (zeroes). Gender is a complicated mix of biology, which robots do not have, and how we feel about that biology, feelings that robots also lack. How gender biases manifest in the design of voice assistants is well-worn territory. Research shows that users tend to like a male voice when an authoritative presence is needed; while a female voice when receiving helpful guidance. Scientists are just beginning to consider how these gender biases materialize in physical robots. The reception robot is designed to be more female so that it can create a more "welcoming" atmosphere, or it is thought that the security robot with a tight shoulder width is designed, and "intimidation" is put out more. This gender grant has the advantage of giving machines personality, making it easier for consumers to feel closer to robots. The problem is that we tend to be more gendered, even if the robot does not reflect gender or, on the contrary, looks like a human or animal [19–21]. In a movie, entitled *HER* directed by Spike Jonze (2013), a sensitive and soulful man earns a living by writing personal letters for other people. Left heartbroken after his marriage ends, Theodore becomes fascinated with a new operating system which reportedly develops into an intuitive and unique entity in its own right. He starts the program and meets "Samantha," whose bright voice reveals a sensitive, playful personality. Though "friends" initially, the relationship soon deepens into love. Later he separated from human girlfriend. The same copy of this story spreads around the world on the Internet, and it has the same relationship with users all over the world and AI becomes out of hand [22].

There is a technology that recognizes not only emotions but also emotional movements that the person himself does not notice. The state of the other party's mind is understood by opening the pupil, the number of blinks, the movement of the glance, and so on. This called the manifestation of unconsciousness (before we realize it, the body is reacting) [23]. There are already pet robots as a healing of the heart, and some people are loving them. Then there is a possibility to choose AI (or AI robot) as the best partner. When we try to marry AI, we first run into social institutional barriers, but same-sex marriage has come to be legally recognized. In that case, our values may change, such as the fact that robots that cannot leave offspring as living organisms may be the other party. Also, if we cannot leave offspring, biological species will become extinct. But with advances in reproductive medicine, such as artificial insemination, there is already a way to leave children inheriting their genes without a biological partner. Hence a new cultural anthropology is needed to be developed. The important thing is the establishment of the ego. It may be the same for humans, but it is about recognizing own existence value and significance [23]. Kondo gets engaged to AI Robot Hatsune Miku (in Japan), J. Zheng married to AI robot Ying Ying (in China), or Lily got engaged to AI robot Immovator (in France). These are just a few to list. AI expert David Levy, speaking at a conference called Love and Sex with Robots at Goldsmiths University in London in 2016, claimed that choosing robots as spouses was an untruly trend, and that human–robot marriage would also be legal by around 2050 [24–26]. We have been looking at personality, consciousness, gender, and even marriage associated with AI (or AI robot). The individuality and personality of human beings are formed by a complex mixture of "thoughts," "moods," "curiosities," "unreasonableness," "actions that are not particularly meaningful," and "definitions of each correct answer" [27]. In that way, it is human beings to behave in a way that seems to be meaningless at first time, and I even feel that it is one of the reasons why people love people [27].

4.3 IQ and common sense

4.3.1 IQ test

There are several testing methods proposed to examine IQ value of AI.

First of all, there is the Turing test. The Turing test, developed by Alan Turing in 1950, is a test of a machine's ability to exhibit intelligent behavior equivalent to, or indistinguishable from, that of a human. The "standard interpretation" of the Turing test, in which the interrogator, is given the task of trying to determine which player (A or B) is a computer and which is a human. The interrogator is limited to using the responses to written questions to make the determination. The purpose of the Turing test is to determine whether a machine can behave closely to a person. Hence, what can be determined by the Turing test is that "through the test, if the judges distinguish between a human and a computer, the computer was able to behave as if it had human intelligence" [28–30] which really reflects to what McDonald mentioned that any AI smart enough to pass a Turing test is smart enough to know to fail it [31].

All it takes to identify the computer programs which are AI is to give them a test and award AI to those that pass the test. Let us say that the scores they earn at the test will be called IQ [32, 33]. The test is used to determine what AI is. The test will produce a certain score, and this is the program's IQ. Then we decide that all computer programs the IQ of which is above a certain level satisfy the AI definition. We shall combine candidates with higher IQs in order to obtain offspring with even higher IQ. We shall kill the low-IQ candidates for making rooms for more promising programs. Using this natural selection approach, we shall obtain programs with very high IQs. The genetic algorithm is one way of finding the AI, but we will end up with a program the inner workings of which are enigmatic to us. If we are to control a program, we better write it ourselves rather than generate it automatically [33].

Cutting through the hype surrounding AI, the Abstract and Reasoning Corpus (ARC) [34] was proposed, which is an intelligence test that could shape the course of future AI research [35]. Chollet [34] articulated a new formal definition of intelligence based on algorithmic information theory (AIT), describing intelligence as skill-acquisition efficiency and highlighting the concepts of scope, generalization difficulty, priors, and experience, as critical pieces to be accounted for in characterizing intelligent systems. Using this definition, a set of guidelines was proposed for what a general AI benchmark should look like and a new benchmark was presented closely following these guidelines, the ARC, built upon an explicit set of priors designed to be as close as possible to innate human priors. It was indicated that (i) ARC can be used to measure a human-like form of general fluid intelligence and (ii) it enables fair general intelligence comparisons between AI systems and humans. An AI system developed through a program synthesis approach that scores well on the ARC test would also do well at other cognitive tasks that humans with general intelligence are capable of [35]. Furthermore, the AI system could also use these general cognitive skills to reason and solve situations unknown to it beforehand.

In humans, we measure abstract reasoning using fairly straightforward visual IQ tests. One popular test, called Raven's Progressive Matrices, features several rows of images with the final row missing its final image. It is up to the test taker to choose the image that should come next based on the pattern of the completed rows. To apply this test to AIs, the DeepMind researchers created a program that could generate unique matrix problems. Then, they trained various AI systems to solve these matrix problems. Ultimately, the team's AI IQ test shows that even some of today's most advanced AIs cannot figure out problems we have not trained them to solve. That means we are probably still a long way from general AI. But at least we now have a straightforward way to monitor our progress [36].

4.3.2 Common sense

During an interview with Francesca Rossi (distinguished research scientist, IBM), it was quoted that the rapid evolution of AI today has brought about a coincidental overlap of three important factors: namely, (1) evolution of algorithms, (2) learning with exponentially growing data, and (3) and exponential improvement of computing power [37]. AI possesses various strengths as a support for decision-making. AI can understand large amounts of data that humans cannot handle and provide evidenced insights based on data. Unlike humans, decision-making is not biased by emotions or preconceptions. In order to play a better supporting role, machines must understand human natural language more deeply, and AI will also need certain common sense and understanding of human emotions. While communicating with AI and humans, if it is necessary to convey to AI what humans take for granted, it would be hard to use AI tools. Hence, for solving this problem, AI must have a certain common sense. In addition, it is important to understand people's emotions and to be able to interact with them in easy-to-understand terms accordingly. This understanding of common sense and emotions can also be achieved through repeated learning using ML large amounts of data, resulting in smoother AI-human interaction. In addition to "common sense," it is also a big issue to make AI have a "sense of ethics" [37].

Dave Gunning explained the DARPA (Defense Advanced Research Projects Agency) by saying that (i) the absence of common sense prevents an intelligent system from understanding its world, communicating naturally with people, behaving reasonably in unforeseen situations, and learning from new experiences and (ii) this absence is perhaps the most significant barrier between the narrowly focused AI applications we have today and the more general AI applications we would like to create in the future [38]. There may be differences in the degree of common sense that people have, but "common sense" is the concept of not relying on clear information but on fair and justifiable judgment based on one's own experience and worldview in complex situation [39]. Neural networks are artificial systems that mimic the structure of the human brain and combined with more advanced machine vision (such as a system that reads data from images for use in tasks and decision-making) makes AI systems to possess common sense [40].

4.3.3 Lie and Sontaku

4.3.3.1 AI-assisted lie detection

Being able to tell when a person is lying is an important part of everyday life. The Deception Analysis and Reasoning Engine (DARE) was developed by researchers from the University of Maryland and DARE is a system that uses AI to autonomously detect deception in courtroom trial videos. DARE was taught to look for and classify human micro-expressions, such as "lips protruded" or "eyebrows frown," as well as analyze audio frequency for revealing vocal patterns that indicate whether a person is lying or not. It was then tested using a training set of videos in which actors were instructed to either lie or tell the truth [41–44]. There are three elements to analyze when discovering lies: (1) text information, (2) voice information and (3) visual information. During the learning, characteristics of these elements are detected and extracted. Then DARE is educated to find and understand changes in people's minute expressions and to determine whether they are lies or not. For example, lies can be detected from minute lips and eyebrow movements when lying or vocal tendencies. AUC is a statistical indicator and stands for "area under the ROC (receiver operating characteristics) curve," measuring entire 2D area underneath the entire ROC curve (think integral calculus) from (0,0) to (1,1). It is known that a normal person scores AUC of about 0.58, while it was reported that, when DARE detects a lie, the AUC score was 0.877, which is much higher than that of humans [42].

Speech forms the basis of human communication. Speech recognition is literally a technology that analyzes a person's voice (data recorded). It is already widely used, such as replacing speech with letters, performing actions that are recognized as commands, and understanding emotions involved in speech. It is the use of voice as biometric authentication. There are other reasons why voice analysis is taking place. It is to grasp feelings which is another information hidden in the voice [45]. Deception is a complex social skill present in human interactions. In modern society, lie detection has a relevant impact on social activities, particularly on those which require tutoring (e.g., in education or healthcare), being a necessary skill for a broad range of professions such as teachers, doctors, or law enforcement officers. Those individuals are usually trained to detect lies but it has been proven that their ability to differentiate between liars and truth teller is often imprecise [46]. Robots with the ability to autonomously detect deception could provide an important aid to human–human and human–robot interactions. Gonzalez-Billandon et al. [46] demonstrated the possibility to develop a lie detection system that could be implemented on robots, focusing on human and human–robot interaction to understand if there is a difference in the behavior of the participants when lying to a robot or to a human. It was reported that the selected behavioral variables are valid markers of deception both in human–human and in human–robot interactions and could be exploited to effectively enable robots to detect lies.

4.3.3.2 AI lies

Lies do not occur unconsciously. This is different from the error (or incorrect). Errors can also occur for requests that are more than an ability of human and AI, too. Depending on whom you talk to, AI either makes you fear that the end of the world is now imminent or gets you excited about the possibilities of improving the quality of our human lives. There are a lot of misconceptions about AI, but many experts say that the benefits outweigh the risks [47]. People working in teams do things – such as telling white lies – that can help the team be successful. We accept that, usually, when a person does it. But what if an AI bot is telling the lie, or being told a lie? More importantly, if we allow bots to tell people lies, even white lies, how will that affect trust? And if we do give AI bots permission to lie to people, how do we know that their lies are helpful to people instead of the bot? [48]. Chakhraborti et al. [49] discussed effective collaboration between humans and AI-based systems which requires effective modeling of the human in the loop, both in terms of the mental state as well as the physical capabilities of the latter. They illustrated some of these issues in a teaming scenario and investigated how they are perceived by participants in a thought experiment. It was concluded that by reiterating the importance of self-reflection in the principled design of AI algorithms whose deployment can have real-life consequences, intended or otherwise, on the future of the field, but also, with the inquisitive mind of a young researcher, marvel at the widening scope of interactions with an artificial agent into newer uncharted territories that may be otherwise considered to be unethical [49].

For example, just as smart speakers can spread fake news, when AI enters the human world, can we say that AI cannot tell a lie? [50]. Here we need to introduce the word "Sontaku" (which is a Japanese word, meaning a combined context of speculation, surmises, conjecture, compromises, read between the lines or read other's mind). Traditional expert systems have basic rules established in advance, and developers are only developing systems accordingly. Because there is a certain direction in the way, it is not always possible for the system to draw conclusions that no one speculated. However, AI which has been under development in recent years is different from conventional ESs. The biggest difference may be the function of DL. For example, AI which defeats chess champion has not been taught how to win by anyone but has developed a way to learn by watching many games and winning on its own. In the early stages of AI dissemination, humans make decisions about what kind of learning to make, so the initiative is on the developer. However, as AI becomes more widespread in society and the work to be processed becomes more complex, it is likely that AI will be made to think about the learning method itself. At this point, the range that the development side can control becomes considerably limited. If AI developed in this way began to advise decision-making without exercising Sontaku the taboos of society, there should not be much left for receiving side to taker handle this [51]. Currently, AI is being utilized in the field of (i) dialogue services that interact directly with users who are customers such as chatbots, and (ii) data discovery that analyzes large amounts of data and discovers trends that humans cannot notice. In the case of the former dialogue service, it would be accepted that offends customers, so if AI is equipped with a sense of "Sontaku," it would be better; while the latter field of data discovery is a field that AI should present "aware" as it is without being "Sontaku" [50, 52].

AI today has no chance of lying, so there is no need to persuade it in the first place. However, by telling a lie, it is required to explain why and the situation behind, leading to the accountability, which will be discussed later in this chapter. The essential meaning of lie can be found between AI and human. Lie is a necessitated intelligence for intelligent organisms and is sometimes a small compassion, and there can be an idea that you need a lie to be happy. It can be said that human communication is also constructed with a stack of petty lies. AI can create parameters to determine the situation, but it cannot create problems that should be solved by oneself, so that AI cannot set up problems. After all, there is a wall of curiosity between humans and AI [53].

4.4 Bias and errors

4.4.1 Bias

AI has produced guidelines applicable to model building. In general, it is believed that ML models should be: (1) lawful (respecting all applicable laws and regulations), (2) ethical (respecting ethical principles and values), and (3) robust (both from a technical perspective while taking into account its social environment), although these short requirements, and their longer form, include and go beyond issues of bias, acting as a checklist for engineers and teams [54]. Among valid concerns associated with AI technology, Kantarci [55] mentioned that AI technology may inherit human biases due to biases in training data and defined that AI bias is an anomaly in the output of ML algorithms. These could be due to the prejudiced assumptions made during the algorithm development process or prejudices in the training data. There are two types of AI bias: (i) cognitive biases (which are effective feelings toward a person or a group based on their perceived group membership and could seep into machine learning algorithms via either designers unknowingly introducing them to the model a training data set which includes those biases) and (ii) lack of complete data (it may not be representative and therefore it may include bias, if data is not complete). Here is an example in healthcare field. Kantarci et al. [55] mentioned that a health care risk-prediction algorithm that has been utilized on US citizen of more than 200 million, and demonstrated racial bias because it relied on a faulty metric for determining the need. The algorithm was designed to predict which patients would likely need extra medical care; however, then it is revealed that the algorithm was producing faulty results that favor white patients over black patients. The algorithm's designers used previous patients' healthcare spending as a proxy for medical needs.

This was a bad interpretation of historical data because income and race are highly correlated metrics and making assumptions based on only one variable of correlated metrics led the algorithm to provide inaccurate results [55]. It was mentioned about how to fix biases in ML algorithms [55] and fixing tools [54, 55] as well.

We have long known that AI algorithms that were trained with data that do not represent the whole population often perform worse for underrepresented groups. For example, algorithms trained with gender-imbalanced data do worse at reading chest X-rays for an underrepresented gender, and researchers are already concerned that skin-cancer detection algorithms, many of which are trained primarily on light-skinned individuals, do worse at detecting skin cancer affecting darker skin. Given the consequences of an incorrect decision, high-stakes medical AI algorithms need to be trained with data sets drawn from diverse populations [56]. Kaushal et al. [57] reviewed over 70 publications that compared the diagnostic prowess of doctors against digital doppelgangers across several areas of clinical medicine and mentioned that (i) whether by race, gender or geography, medical AI has a data diversity problem: researchers cannot easily obtain large, diverse medical data sets and that can lead to biased algorithms, (ii) medical data sharing should be more commonplace; however, the sanctity of medical data and the strength of relevant privacy laws provide strong incentives to protect data, and severe consequences for any error in data sharing, and (iii) data are sometimes sequestered for economic reasons. Humans cannot rely on computers for answers. Instead, we must carefully monitor both the input and output for signs of discrimination [58].

4.4.2 Errors

4.4.2.1 AI errors and human errors

AI has already surpassed humans in many activities requiring thoroughness, accuracy, and patience, although some experts even hesitate whether it is artificial intelligence or artificial stupidity we should fear more [59]. About the accident in which an Uber driverless car hit a woman, Bexborodykh [59] mentioned that no matter how smart it is – it thinks differently from humans. When a decision should be made, AI relies on inbuilt algorithms and a massive amount of data which it processes in order to come up conclusive action. For providing AI with a solid ground for proper decision-making, Brynjolfsson et al. [60] mentioned that we should set well-defined inputs and outputs, clearly define goals and metrics, give short and precise instructions, and eliminate long chains of reasoning relying on common sense. After all, as Dostoyevsky [61] said in his Crime and Punishment, "it takes something more than intelligence to act intelligently" – and AI is not an exception [59].

As AI technologies become more advanced, their errors can become more serious. AI errors can be characterized by a direct result of the intelligence that those machines are required to exhibit, those mistakes can have their origins either in the learning phases of AI, or in the performance phases, while human errors are based on more cognitive feature [62]. Human errors can be skill-based, rule-based, or knowledge-based. Skill-based errors are execution errors, whereas the remaining two are planning-based. Most human errors (61 %) are skill-based, such as slips and lapses, and they have a better chance of being detected. Other errors are less frequent and are less likely to be detected [62]. A common factor in both human and AI errors is that both affect humans. The degree of impact can vary. Human errors have proven far costlier and more catastrophic, ranging from allowing a controlled fire to get out of hand and burn down hundreds of houses, to a doctor making a diagnostic or procedural mistake and causing a patient's death [62]. The case of an iatrogenesis should be included here. Winter [62] also mentioned that one key difference separating AI errors from human errors is the element of predictability. AI errors are far more predictable than human errors, in part because they are systematic, and because the behavior of machines can be modeled (unlike that of humans). The fact that AI errors are easier to predict means they are easier to fix and even prevent, as will be described later.

4.4.2.2 Remedy of errors

Humans are naturally prone to making mistakes. Such errors are increasingly impactful in the workplace, but human error in the realm of cybersecurity can have particularly devastating and long-lasting effects. As the digital world becomes more complex, human error becomes much tougher to navigate – and thus, more unfair to blame humans for the errors they make. Many believe that AI will replace human intelligence, but nothing could be further from the truth. In fact, AI exists to empower human intelligence – it can take on the heavy lifting of data security, freeing up human intelligence to tackle the high-level issues that require creativity, experienced problem-solving capabilities, and a passion to drive the business to success [63]. Most modern AI applications are powered by ML algorithms and until recently, most ML algorithms were built manually. Like all manual tasks, coding and model design are prone to human error, more so when building a completely new and complex solution. What is needed are development tools with guardrails to warn and prevent data scientists from making dangerous mistakes [64]. Manually building modern AI applications is a high-risk process. No one is perfect, and human error can hamper your success with AI. Cobey et al. [65] warned that data technologies and systems can malfunction, be deliberately or accidentally corrupted and even adopt human biases. These failures have profound ramifications for security, decision-making, and credibility, and may lead to costly litigation, reputational damage, customer revolt, reduced profitability, and regulatory scrutiny. In a world of increasing connectivity, where barely a moment passes without interference from technology, humans are bound to become distracted. When distractions are the norm in the workplace, errors are more likely to occur. When mistakes are made, cybercriminals are watching and waiting for their opportunity [66].

With regard to the emergency water landing of the Airbus 230 on the Hudson River, in 2009, with no human casualties, it was commented that AI is not the equal of man, who is capable of imagining and successfully carrying out an emergency landing on the Hudson River, and moreover humans can rely on an algorithm to reduce the risk of error in their interactions with a complex system, but the final decision must remain with the human [67].

Ismail [68] mentioned that the growing complexity of software has meant that human developers are increasingly powerless to manage enormous interconnected applications. There is now a need to focus on developing AI that talks not to humans but directly to computer programs, creating software that codes, edits, and tests itself. This will ensure better software that is more secure and impervious to attack [68]. As conclusive remarks, it was mentioned that (i) people's day-to-day lives are becoming more entwined with technology, as they rely more and more on our phones, computers, and tablets; (ii) the IoT is looming, and people need to know that they are safe using it; (iii) there is no logic behind the fear that AI can cause harm, when in fact, it can provide us with the security society needs in the age of technology, as well as the time and place to be more creative; and (iv) by automating tasks, AI has opened up the market to all sorts of innovative jobs – and by removing human error, society could now be approaching an era of technological perfection [68].

AI is not a panacea. The human factors of creativity, common sense, and instinct so often play a critical role in decision-making. What is doing with AI, in essence, is striving to better harness data to gain critical additional insights that could lead to improved efficiency and outcomes [69]. In less than a decade, computers have become extremely good at diagnosing diseases, translating languages, and transcribing speech. They can outplay humans at complicated strategy games, create photorealistic images, and suggest useful replies to your emails. Despite these impressive achievements, AI possesses non-negligible weaknesses. AI would not be very smart if computers do not grasp cause and effect. That is something even humans have trouble with. Today's AI has only a limited ability to infer what will result from a given action. An even higher level of causal thinking would be the ability to reason about why things happened and ask "what if" questions. A patient dies while in a clinical trial; was it the fault of the experimental medicine or something else? School test scores are falling; what policy changes would most improve them? This kind of reasoning is far beyond the current capability of the AI [70].

4.4.3 Computer virus and immune system

A computer virus, much like a flu virus, is designed to spread from host to host and has the ability to replicate itself. Similarly, in the same way that flu viruses cannot reproduce without a host cell, computer viruses cannot reproduce and spread without programming such as a file or document. In more technical terms, a computer virus is a type of malicious code or program written to alter the way a computer operates and is designed to spread from one computer to another. A virus operates by inserting or attaching itself to a legitimate program or document that supports macros for executing its code. In the process, a virus has the potential to cause unexpected or damaging effects, such as harming the system software by corrupting or destroying data [71]. There are different types of computer viruses: boot sector virus, web scripting virus, browser hijacker, resident virus, direct action virus, polymorphic virus, file infector virus, multipartite virus, and macro virus [71].

Natural immune systems provide a rich source of inspiration for computer security in the age of the Internet. Immune systems have many features that are desirable for the imperfect, uncontrolled, and open environments in which most computers currently exist. These include distributability, diversity, disposability, adaptability, autonomy, dynamic coverage, anomaly detection, multiple layers, identity via behavior, no trusted components, and imperfect detection. These principles suggest a wide variety of architectures for a computer immune system [72].

4.5 EQ and ethics

AI can be specifically characterized by (1) what AI understands is only that it learns from input (or fed) data, and (2) what is missing with AI should include (i) common sense that human beings have gained from experience, (ii) emotions (which can be possibly educated to AI), (iii) ethics (again, which can be trained to AI), (iv) preconceptions, and (v) consciousness or ego that cannot be found in AI. AI is further characterized by (3) AI's strength by (i) learning huge amounts of data that humans cannot handle and (ii) insights are based on data and are not biased by emotions or preconceptions. The last and unique character associated with AI is (4) AI's limitations, including (i) collecting data on human senses is difficult, so that robots are developed to enable to sense details like humans and collect data, (ii) determining quality and quantity of data to be learned is determining the accuracy of AI, so that a close attention should be paid to the data that AI learns for not to be biased, and (iii) the decision process being a black box nature, so that a white box technology be developed to exhibit the ability to explain the rationale [50, 51]. In summary, AI grows in any way depending on the data given (or fed) – "You are what you eat!" Depending on the data, it will become an AI with poor accuracy, it will become an AI that is close to humans, it will become an AI like a leading scientist or engineer, and it will be exactly the same as a human depending on learning. In other words, it is up to humans to decide whether AI grows up to be trustworthy.

4.5.1 EQ

The position of recognizing the independence of the heart of the program is called
the strong AI. Can AI read the other person's mind? If you can, you will possess EQ
(emotional intelligence) sense (not an IQ), so that better examination and diagnose
can be performed. The "negligent lie" can be made by unintentionally drawing the
wrong conclusion due to wrong or biased data. However, when AI is furnished with
human common sense, emotions, and ethics that learned humanity comes out,
such AI will spit a lie in the same way as humans. Although AI would not be lying
because of malice, but if AI becomes too close to human, it will learn the concept of
Sontaku and may learn to do the "willful lie" [50, 51]. AI is an extension of human
beings, an extension of human intelligence, and is not in conflict with humans, but
a partner. I think the symbiotic relationship between humans and AI is the key to
an opening up the future.

Based on several different sources, a simple definition of emotional intelligence
(also called the emotional quotient, or EQ) describes an ability to monitor your own
emotions as well as the emotions of others [73], to distinguish between and label different emotions correctly, and to use emotional information to guide thinking and behavior and influence that of others [74–77]. Emotional intelligence is what we use when we
empathize with our coworkers, have deep conversations about our relationships with
significant others, and attempt to manage an unruly or distraught child. It allows us to
connect with others, understand ourselves better, and live a more authentic, healthy,
and happy life. Although there are many kinds of intelligence, and they are often connected to one another, there are some very significant differences between them [78].
EQ is emotional intelligence, which, as stated above, is all about identifying emotions
in ourselves and others, relating to others, and communicating about our feelings [79].
IQ, on the other hand, is cognitive intelligence. This is the intelligence that people are
generally most familiar with, as it is the type that is most often referred to when the
word "intelligence" is used. It is also the type that is most often measured through testing and estimated through things like grade-point average [78].

Emotional intelligence is very important in different areas of daily life, including
health, relationships, and even performance at work [75, 80]. For example, for physical health, our understanding of our emotions can help us correctly manage them,
and not being able to manage stress, for example, can have direct repercussions for
our physical health. Increased risk of heart attack, a speeded aging process, and
higher blood pressure are just some of the physical consequences of stress. A way to
start improving EQ is learning how to identify and relieve tension. As to the mental
health, depression and anxiety are just some of the mental health results of not being
able to understand and control emotions. Having trouble identifying and managing
emotions can create an obstacle for creating strong and stable relationships, which
may lead to isolation or depression. With regard to a relationship, as soon as you are
able to understand your emotions and control them, you are able to correctly express

them to people around you, and you are better at learning how others feel, too. This translates into easier and more effective communication, and therefore better relationships [80]. To improve marketing as well as custom service, there are several successful reports available [81–83].

4.5.2 Ethics

When selecting AI-enabled programs for hiring, teaching, testing, or student support, higher education institutions should weigh the cost savings and efficiency of the programs against any possible risks to fairness, privacy, bias, and public safety [84]. While AI is a potential game-changer for improving efficiency, it may also come with ethical baggage. John O'Brien, president and CEO of EDUCAUSE, defines digital ethics as "doing the right thing at the intersection of technology innovation and accepted social values" [85].

AI is for all, and human dignity must not be committed. In order for AI to be used correctly in line with the ethics, it is also important to have high transparency in the behavior and technology of the AI used, to be explainable to humans, and to distribute control rights to multiple people (citizens). In any case, before discussing the ethics that AI should have in the future, the ethics of humans using AI and those who make AI should be considered. This problem also leads to social ethics. As easily noticed from the politics and presidential elections in the United States, there are many cases where ethics that are considered to be socially good differ from conclusions that come out in discussions with everyone, and simply discussing them alone will not create AI with high ethics. In this way, how to educate AI on high ethics is a big question, and it is similar to teaching ethics to humans. In worst scenario, it may amplify the evil elements of human beings as they are. Bossmann [86] pointed out nine important ethical issued associated with AI technology, (1) unemployment (What happens after the end of jobs?), (2) inequality (How do we distribute the wealth created by machines?), (3) humanity (How do machines affect our behavior and interaction?), (4) artificial stupidity (How can we guard against mistakes?), (5) racist robots (How do we eliminate AI bias?), (6) security (How do we keep AI safe from adversaries?), (7) evil genies (How do we protect against unintended consequences?), (8) singularity (How do we stay in control of a complex intelligent system?) and (9) robot rights (How do we define the humane treatment of AI?). Some of these issues have been already discussed in the above and some others like the singularity issue will be described later. The ethics of AI is the part of the ethics of technology specific to robots and other artificially intelligent entities. It can be divided into a concern with the moral behavior of humans as they design, construct, use and treat artificially intelligent beings, and machine ethics, which is concerned with the moral behavior of artificial moral agents (AMAs). It also includes the issues of singularity and superintelligence [86].

The systems we require for sustaining our lives increasingly rely upon algorithms to function. Governance, energy grids, food distribution, supply chains, healthcare, fuel, global banking, and much else are becoming increasingly automated in ways that impact all of us. Yet, the people who are developing the automation, machine learning, and the data collection and analysis that currently drive much of this automation do not represent all of us and are not considering all of our needs equally. Most of us do not have an equal voice or representation in this new world order. Leading the way instead are scientists and engineers who do not seem to understand how to represent how we live as individuals or in groups – the main ways we live, work, cooperate, and exist together – nor how to incorporate into their models our ethnic, cultural, gender, age, geographic or economic diversity, either. The result is that AI will benefit some of us far more than others, depending upon who we are, our gender and ethnic identities, how much income or power we have, where we are in the world, and what we want to do [87].

AI must be developed with an understanding of who humans are collectively and in groups (anthropology and sociology), as well as who we are individually (psychology), and how our individual brains work (cognitive science), in tandem with current thinking on global cultural ethics and corresponding philosophies and laws [88]. What it means to be human can vary depending upon not just who we are and where we are, but also when we are, and how we see ourselves at any given time. When crafting ethical guidelines for AI, we must consider "ethics" in all forms, particularly accounting for the cultural constructs that differ between regions and groups of people, as well as time and space. That means that truly ethical AI systems will also need to dynamically adapt to how we change. Consider that the ethics of how women are treated in certain geographical regions has changed as cultural mores have changed. It was only within the last century that women in the United States were given the right to vote, and even less than that for certain ethnicities. Additionally, it has taken roughly that long for women and other ethnicities to be accepted pervasively in the workplace – and many still are not. As culture changes, ethics can change, and how we come to accept these changes, and adjust our behaviors to embrace them over time, also matters. As we grow, so must AI grow too. Otherwise, any "ethical" AI will be rigid, inflexible, and unable to adjust to the wide span of human behavior and culture [88]. AI ethics is now a global topic of discussion in academic and policy circles. At least 63 public–private initiatives have produced statements describing high-level principles, values, and other tenets to guide the ethical development, deployment, and governance of AI. Mittelstadti [89] mentioned that, according to recent meta-analyses, AI ethics has seemingly converged on a set of principles that closely resemble the four classic principles of medical ethics. Despite the initial credibility granted to a principled approach to AI ethics by the connection to principles in medical ethics, there are reasons to be concerned about its future impact on AI development and governance. Significant differences exist between medicine and AI development that suggest a principled approach in the latter may not enjoy success comparable to the former. Compared to medicine, AI development lacks (1) common aims and fiduciary duties, (2) professional history and norms, (3) proven methods to translate principles into practice, and (4) robust legal and professional accountability mechanisms, suggesting that we should not yet celebrate consensus around high-level principles that hide deep political and normative disagreement [89].

Current advances in research, development, and application of AI systems have yielded a far-reaching discourse on AI ethics. In consequence, a number of ethics guidelines have been released in recent years. These guidelines comprise normative principles and recommendations aimed to harness the "disruptive" potentials of new AI technologies. Hagendorf [90] analyzed and compared 22 guidelines, highlighting overlaps but also omissions and concluded that there are two consequences for AI ethics: (i) a stronger focus on technological details of the various methods and technologies in the field of AI and machine learning is required. This should ultimately serve to close the gap between ethics and technical discourses. It is necessary to build tangible bridges between abstract values and technical implementations, as long as these bridges can be reasonably constructed. (ii) the consequence of the presented considerations is that AI ethics, conversely, turns away from the description of purely technological phenomena in order to focus more strongly on genuinely social and personality-related aspects. AI ethics then deals less with AI as such, than with ways of deviation or distancing oneself from problematic routines of action, with uncovering blind spots in knowledge, and of gaining individual self-responsibility. It was also mentioned that future AI ethics faces the challenge of achieving this balancing act between the two approaches [90].

The ethical issues related to the possible future creation of machines with general intellectual capabilities far outstripping those of humans are quite distinct from any ethical problems arising in current automation and information systems. Such superintelligence would not be just another technological development; it would be the most important invention ever made and would lead to explosive progress in all scientific and technological fields, as the superintelligence would conduct research with superhuman efficiency. To the extent that ethics is a cognitive pursuit, a superintelligence could also easily surpass humans in the quality of its moral thinking. However, it would be up to the designers of the superintelligence to specify its original motivations. Since the superintelligence may become unstoppably powerful because of its intellectual superiority and the technologies it could develop, it is crucial that it be provided with human-friendly motivations [91].

4.5.3 Accountability

AI technology is required to exhibit the ability to transparently show why we answered this question and why this option is better than others. ML and big data tend to be opaque in their explanations, and it is true that the reason why is a black box (whose algorithms refer to the general inability to see inside of a system and see how it arrives at a decision), even when reasoning differently from humans. The spread of mobile devices, such as smartphones, makes it easier to grasp people's behavior. Such information is integrated by the development of big data processing and AI technology, and new information is being created one after another. The use of such information can have various benefits for society. On the other hand, in such technologies, users are often collecting information without being too aware of it, and concern about privacy and personal information different from the conventional one are also growing [92]. AI needs to be more accountable but ethical considerations are not keeping pace with the technology's rate of deployment. This is partly due to the "black box" nature of AI, whereby it is almost impossible to determine how or why an AI makes the decisions it does, as well as the complexities of creating an "unbiased" AI [92].

Lee [93] pointed out the followings as potential resolutions. (i) Explainable AI: One notion to counter black box algorithms is advocating the use of explainable AI (or XAI). Essentially, XAI programs enable users to understand the technology, provide input on the decision-making process, and improve algorithmic accountability. (ii) Privacy by design: Privacy needs to be considered by the whole design process, not just as a bolt-on approach. Essentially, privacy by design advocates that organizations need to consider privacy at the initial design stages and throughout the development process of new products or services that involve handling personal data. (iii) Regulatory frameworks for AI: Regulatory frameworks are needed to safeguard transparent and auditable education data. Prejudices of the past must not be unwittingly built into AI-enabled learning systems. Systems and algorithms need to be carefully designed, explained, and audited. It was then concluded that for AI to deliver its promise on learning enhancement, it will require people to trust the technology, organizations to safeguard the use of personal and educational data, and for government to collaborate with industries to research and discuss ethics in AI [93].

In order to ensure that systems will uphold human values, design methods are needed that incorporate ethical principles and address societal concerns, Dignum [94] proposed that ethical AI rests in three pillars of equal importance: the AI ART (accountability, responsibility, and transparency) design principles for the development of AI systems sensitive to human values. Accountability implies that it is necessary to explain on the grounds that actions and choices are appropriate, and if this explanation is not properly done, it will be put in a position where it is not safe, so that it refers to the need to explain and justify one's decisions and actions to its partners, users and others with whom the system interacts. To ensure accountability, decisions must be derivable from, and explained by, the decision-making algorithms used. This includes the need for representation of the moral values and societal norms holding in the context of operation, which the agent uses for deliberation. Accountability in AI requires both the function of guiding action (by forming beliefs and making decisions), and the function of explanation (by placing decisions in a broader context and by classifying them along moral values). Responsibility refers to the role of people themselves and to the capability of AI systems to answer for one's decision and identify errors or unexpected results. As the chain of responsibility grows means are needed to link the AI systems' decisions to the fair use of data and to the actions of stakeholders involved in the system's decision [94]. Transparency will be discussed later.

4.6 Privacy and its related issues

With the development of AI, while convenience will increase, it will be able to trace the history of an individual's actions as much as available, and there will be problems in the balance with privacy. It is necessary to reaffirm that a person has various rights that have not been explicitly considered before, such as the right to be forgotten, the right to be missed, or the right to be warned. In this section, we will discuss privacy, and several important issues associated with privacy such as accountability, transparency or explainability.

4.6.1 Privacy

The general public is largely wary of AI's data-hungry ways. According to a survey by Brookings, 49% of people think AI will reduce privacy. Only 12% think it will have no effect, and a mere 5% think it may make it better. According to cybersecurity and privacy researchers, it is believed that the relationship between AI and data privacy is more nuanced. The spread of AI raises a number of privacy concerns, most of which people may not even be aware. But in a twist, AI can also help mitigate many of these privacy problems [95].

AI and its applications are a part of everyday life: namely, from social media newsfeeds to mediating traffic flow in cities, from autonomous cars to connected consumer devices like smart assistants, spam filters, voice recognition systems, and search engines. AI has the potential to revolutionize societies in many ways. AI-driven consumer products and autonomous systems are frequently equipped with sensors that generate and collect vast amounts of data without the knowledge or consent of those in their proximity. AI methods are being used to identify people who wish to remain anonymous; to infer and generate sensitive information about people from non-sensitive data; to profile people based upon population-scale data; and to make consequential decisions using this data, some of which profoundly affect people's lives. However, as with any scientific or technological advancement, there is a real risk that the use of new tools by states or corporations will have a negative impact on human rights, including the right to privacy [96]. Major risks and problems should include (i) re-identification and de-anonymization, (ii) discrimination, unfairness, inaccuracies, and bias, (iii) opacity and secrecy of profiling, and (iv) data exploitation [96].

In the absence of targeted legal or regulatory obligations, AI poses new ethical and practical challenges for companies that strive to maximize consumer benefits while preventing potential harms. There is growing recognition that using AI/ML models raises novel privacy challenges. For policymakers to ensure non-discrimination, due process, and defensibility in decision-making, they must first understand the technology underlying these new features, products, and services. While the benefits are great, the potentially discriminatory impact of machine learning necessitates careful oversight and further technical research into the dangers of encoded bias, or undue opacity in automated decisions. As we look to the future of AI and ML, it is important to remember that while these systems and models will be invaluable in enabling us to evaluate and benefit from the unfathomable amount of data available, they do not yet represent "intelligence" in any way that correlates to humans. It is incumbent on the people involved in designing, managing, implementing, and regulating these systems to retain accountability – assessing the balance of benefits and risks, and seeking to ensure the best possible outcomes for all [97]. Increasingly, privacy is not merely a question of philosophy, but table stakes in the course of business. Laws at the state, local, and federal levels aim to make privacy a mandatory part of compliance management. Hundreds of bills that address privacy, cybersecurity, and data breaches are pending or have already been passed in 50 US states, territories, and the District of Columbia [98]. Arguably the most comprehensive of them all – the California Consumer Privacy Act – was signed into law roughly two years ago. That is not to mention the Health Insurance Portability and Accountability Act (HIPAA), which requires companies to seek authorization before disclosing individual health information. International frameworks like the EU's General Privacy Data Protection Regulation (GDPR) [99] aim to give consumers greater control over personal data collection and use. AI technologies have not historically been developed with privacy in mind. But a subfield of machine learning – privacy-preserving machine learning – seeks to pioneer approaches that might prevent the compromise of personally identifiable data. Of the emerging techniques, federated learning, differential privacy, and homomorphic encryption are perhaps the most promising. During the interview, Ballon (Intel's IoT group) said that the strategy is very similar to going from hypertext transfer protocol; no data encryption implemented (HTTP) to hypertext transfer protocol secure; encrypted connections (HTTPS) [98].

4.6.2 Transparency and explainability

Because of the difficulties to foresee ML outcomes as well as reverse-engineering algorithmic decisions, no single measure can be completely effective in avoiding perverse effects. Thus, where algorithmic decisions are consequential, it makes sense to combine measures to work together. Advance measures such as transparency and risk assessment, combined with the retrospective checks of audits and human review of decisions, could help identify and address unfair results. A combination of these measures can complement each other and add up to more than the sum of the parts. Risk assessments, transparency, explainability, and audits also would strengthen existing remedies for actionable discrimination by providing documentary evidence that could be used in litigation. Not all algorithmic decision-making is consequential, however, so these requirements should vary according to the objective risk [100]. Numerous organizations and companies as well as several legislators propose such accountability. Their proposals take various forms: (i) transparency refers to disclosures relating to uses of algorithmic decision-making. While lengthy, detailed privacy policies are not helpful to most consumers, they do provide regulators and other privacy watchdogs with a benchmark by which to examine a company's data handling and hold that company accountable. Replacing current privacy policies with "privacy disclosures" that require a complete description of what and how data is collected, used, and protected would enhance this benchmark function. In turn, requiring that these disclosures identify significant uses of personal information for algorithmic decisions would help watchdogs and consumers know where to look out for untoward outcomes [94, 100]. (ii) Explainability: While transparency provides advance notice of algorithmic decision-making, explainability involves retroactive information about the use of algorithms in specific decisions. This is the main approach taken in the European Union's General Data Protection Regulation (GDPR) [99]. A sense of fairness suggests such a safety valve should be available for algorithmic decisions that have a material impact on individuals' lives. Explainability requires (1) identifying algorithmic decisions, (2) deconstructing specific decisions, and (3) establishing a channel by which an individual can seek an explanation. Reverse-engineering algorithms based on ML can be difficult, and even impossible, a difficulty that increases as machine learning becomes more sophisticated. Explainability therefore entails a significant regulatory burden and constraint on use of algorithmic decision-making and, in this light, should be concentrated in its application, as the EU has done (at least in principle) with its "legal effects or similarly significant effects" threshold. As understanding increases about the comparative strengths of human and machine capabilities, having a "human in the loop" for decisions that affect people's lives offers a way to combine the power of machines with human judgment and empathy [100].

4.7 Copyright

AI-created works (or AI works) meet people's demands as well as human-made works. It is in the public interest to promote the creation. However, the laws of many countries now protect only works created by a person's knowledge and wisdom. Protection against the work created by the machine is not recognized. Intellectual property rights are an engine that promotes investment in the creation of intellectual property and promotes economic development. The high efficiency of the engine enables a leap forward in economic development. The difference in the efficiency of the engine makes a difference in the economic development of each country. In particular, in the information society, copyright as an engine for promoting economic development is important. In order to become an efficient engine for promoting economic development, the AI-era copyright system will need to be restructured. The creations on the earth have been classified into (1) creations by people, (2) creations using AI as a tool, and (3) AI-based creations [101, 102]. Since a copyright is centered on human creativity, (1) is not problem. For case (2), since humans use AI as a tool to produce creations, it can be said to be a "work" because of human involvement, and it is protected by copyright. As to case (3), simply creating instructions to AI does not reflect human thought and emotions in the creation, so that it is not protected by copyright. Any creation that is not protected by copyright will be available to everyone as a so-called public domain (PD) [101, 103, 104]. In April 2019, the 5th genereation mobile communication system (5G) was launched. 5G speeds up conventional information and communication speeds and is said to be an important technology that is the foundation of Society 5.0, in which AI analyzes vast amounts of data (big data) and provides feedback to humans through robots and other means, creating new value and bringing it to society.

Idei [105] reviewed the current status of legal protection of data related to the development and use of AI. Raw data is, in general, protected by copyright since data such as paintings or music contains creative expressions, while data consisting of objective facts, such as weather data or machine operation data obtained by sensors, is not a creative expression, so copyright protection does not extend. Big data may be protected as a database work if creativity is recognized in the selection and systematic composition of the information. Learned models (which are generated by loading learning data for leaning into an AI program) are composed of two parts: (i) AI program (which can be protected as an "invention of things" under patent law; in some cases, program is written by using OSS (open-source software) and for this case protection is hard to be established) and (ii) program parameters (since these parameters are usually automatically represented as a few columns by AI programs, they are not protected by copyright). There are several international organizations to study and discuss over the copyright protection of AI-related creations. AI will have a great impact on many aspects of our life, including IP protection. It was reported that (i) current copyright laws in EU are probably not covering AI results and (ii) the question is whether protection is needed; and in this case what kind of protection would be more advisable, taking into consideration the effects – and the possible counter-effects – on human authors' protection and the economy [106]. It was mentioned that (i) AI certainly poses challenges for copyright; however, the magnitude of these challenges is uncertain, (ii) since we have not reached the moment of singularity, we must assess the legal implications of AI systems that function, rather than think, and (iii) from the perspective of copyright law, this means recasting a recurrent question whenever new technologies affect the use and exploitation of works [107]. Recently (February 14, 2020), Creative Commons (CC) submitted its comments on the World Intellectual Property Organization's (WIPO) Issue Paper as part of WIPO's consultation process on AI and intellectual property (IP) policy [108]. Technological developments in AI are fast-paced and raise complex policy, legal, and ethical issues that deserve global attention. However, AI needs to be properly understood before any copyright implications can be addressed. At this nascent stage of AI technology, there lacks consensus on how to define AI. AI algorithms differ in the depth and breadth of input required to produce coherent output, and it is not clear how to judge the originality of a work essentially composed of random snippets of thousands or millions of input works. There is also uncertainty about whether and to what extent AI is capable of producing content "autonomously" without any direct human involvement, and whether AI outputs should be protected by copyright. Clarity on these and other basic definitions in the "AI" space is a prerequisite to competent regulation in this arena [108].

Looking to the future, Spindler [109] commented that the situation may change if AI ultimately becomes able to turn around the goals and preferences set by its owner. Even if that still seems far away and something from science fiction, in that case we will need to consider the introduction of new forms of legal persons because the activities of AI can no longer be attributed to the "author." The "governance" of such a new person, including control of it, has then to be carved out more precisely. Moreover, regarding the protection of data used for training AI as well as data generated by it, it is debatable whether protection as a trade secret is sufficient. On the other hand, new developments in information technology make it likely that data and its use can be traced, so that the boundaries and limits of data use may be controlled through technological means. Thus, the introduction of legal property rights may not be necessary as technological tools could be quite effective. In addition, business-to-business platforms have developed standard contractual terms and conditions in order to share data while protecting it against unfair use. Thus, in the end the evolution of AI should be closely monitored, and furthermore how data is traded and protected by contractual terms. For the time being there seems to be no need for legislative action concerning the extension or modification of copyright protection [109].

In an information world, the so-called prosumer also appears, as Toffler [110] indicated. The term "prosumer" is a portmanteau of the words "provider" and "consumer." Thus, a prosumer is an individual who not only consumes but also produces. Research has identified six types of prosumers: DIY (do-it-yourself) prosumers, self-service prosumers, customizing prosumers, collaborative prosumers, monetized prosumers, and economic prosumers [111]. Furthermore, the spread of the Internet has expanded this situation on a global scale. This is a new type of human image that integrates producers and consumers who try to make things that are right for them. In the present and near future, the rights of multimedia works (digitally signaled works) in cyberspace (networked electronic environments) are a problem.

4.8 Singularity and other concerns

4.8.1 Superintelligence and singularity

AI holds great economic, social, medical, security, and environmental promise. AI systems can help people acquire new skills and training, democratize services, design and deliver faster production times and quicker iteration cycles, reduce energy usage, provide real-time environmental monitoring for pollution and air quality, enhance cybersecurity defenses, boost national output, reduce healthcare inefficiencies, create new kinds of enjoyable experiences and interactions for people, and improve real-time translation services to connect people around the world. In the long term, we can imagine AI enabling breakthroughs in medicine, basic and applied science, managing complex systems, and creating currently unimagined products and services. For all of these reasons and many more, researchers are thrilled with the potential uses of AI systems to help manage some of the world's hardest problems and improve countless lives [112]. But in order to realize this potential, the challenges associated with AI development have to be addressed. One of the concerns for the safe and beneficial development of AI, both in the near term and far term, is an AGI (artificial general intelligence) and SI (superintelligence) [112]. The idea that machine intelligence could equal human intelligence in most or all domains is called strong AI or AGI. The idea of such a machine then greatly surpassing human intelligence, possibly through recursive self-improvement, is referred to as superintelligence, or the intelligence explosion. Many AI experts agree that AGI is possible, and only disagree about the timelines and qualifications. AGI would encounter all of the challenges of narrow AI but would additionally pose its own risks such as containment [112].

Existing weak AI systems can be monitored and easily shut down and modified if they misbehave. However, a misprogrammed superintelligence, which is, by definition, smarter than humans in solving practical problems it encounters in the course of pursuing its goals, would realize that allowing itself to be shut down and modified might interfere with its ability to accomplish its current goals. If the superintelligence therefore decides to resist shutdown and modification, it would (again, by definition) be smart enough to outwit its programmers if there is otherwise a "level playing field" and if the programmers have taken no prior precautions. In general, attempts to solve the control problem after superintelligence is created are likely to fail because a superintelligence would likely have superior strategic planning abilities to humans, and (all things equal) would be more successful at finding ways to dominate humans than humans would be able to post facto find ways to dominate the superintelligence. The control problem asks: What prior precautions can the programmers take to successfully prevent the superintelligence from catastrophically misbehaving? In AI, AI control problem is the issue of how to build a superintelligent agent that will aid its creators, and avoid inadvertently building a superintelligence that will harm its creators [113].

Its study is motivated by the notion that the human race will have to solve the control problem before any superintelligence is created, as a poorly designed superintelligence might rationally decide to seize control over its environment and refuse to permit its creators to modify it after launch [114]. In addition, some scholars argue that solutions to the control problem, alongside other advances in AI safety engineering [115], might also find applications in existing non-superintelligent AI.

The technological singularity (also, simply, the singularity) is a hypothetical point in time at which technological growth becomes uncontrollable and irreversible, resulting in unforeseeable changes to human civilization [116]. Now suppose you need to reproduce (or proliferate) something less than yourself (say $0.9 < 1.0$) for 1,000 times, you will not exceed your original capability and result will be close to zero (e.g., 0.9^{1000} is 1.75×10^{-46}, which is almost zero). However, if you can make AI somewhat exceeding your original ability a little bit better (say $1.1 > 1.0$) to proliferate for 1,000 times, as a result, by proliferation, an overwhelming intelligence could be born (1.1^{1000} becomes a very large number of 2.47×10^{41}) [117]. The term "AI" possesses two-fold problems. It is a question whether intelligence can be reconstructed artificially and what is intelligence in the first place? Even though the full extent of human intelligence has not yet been elucidated, the process of attempting to reconstruct it has highlighted what human intelligence is. In the first place, the term "beyond" the mechanism of the brain that has not been elucidated cannot be well defined. The idea that human history is approaching a "singularity" – that ordinary humans will someday be overtaken by artificially intelligent machines or cognitively enhanced biological intelligence, or both – has moved from the realm of science fiction to serious debate [118]. There are pros and cons with regard to occurrence possibility of singularity and (if it occurs) when it would be. It can be expected that supercomputers will soon surpass human capabilities in almost all areas – somewhere between 2020 and 2060 [119]. On the other hand, experts do not expect that to happen in the next 30 to 40 years [54]. Experts are starting to ring alarm bells. Technology visionaries, such as Elon Musk from Tesla Motors, Bill Gates from Microsoft, and Apple cofounder Steve Wozniak, are warning that superintelligence is a serious danger for humanity, possibly even more dangerous than nuclear weapons. Figure 4.2 [120, 121] shows a combined feature exhibiting the timing for singularity occurrence. By 2020, you may be able to buy a device with the computational capacity of a single human brain for the price of a nice refrigerator (1,000 US$), while by 2050, it could be happed to buy a device with the computational capacity of all mankind for the price of a nice refrigerator today [120, 122]. By 2020, the human intellect curve and machine intelligence curve intersects, indicating the occurrence of the singularity [121].

The content of AI threats talked about in the world is often based on misunderstandings of current AI technology. On the other hand, there are certainly themes that AI developers are required to discuss as social and ethical issues that AI technology faces. Asakawa picked up three major misunderstandings on the AI threat concepts [123]. There is a fear on the misconception that AI may eventually be willing to autonomously destroy humanity. In the first place, today's AI technology is

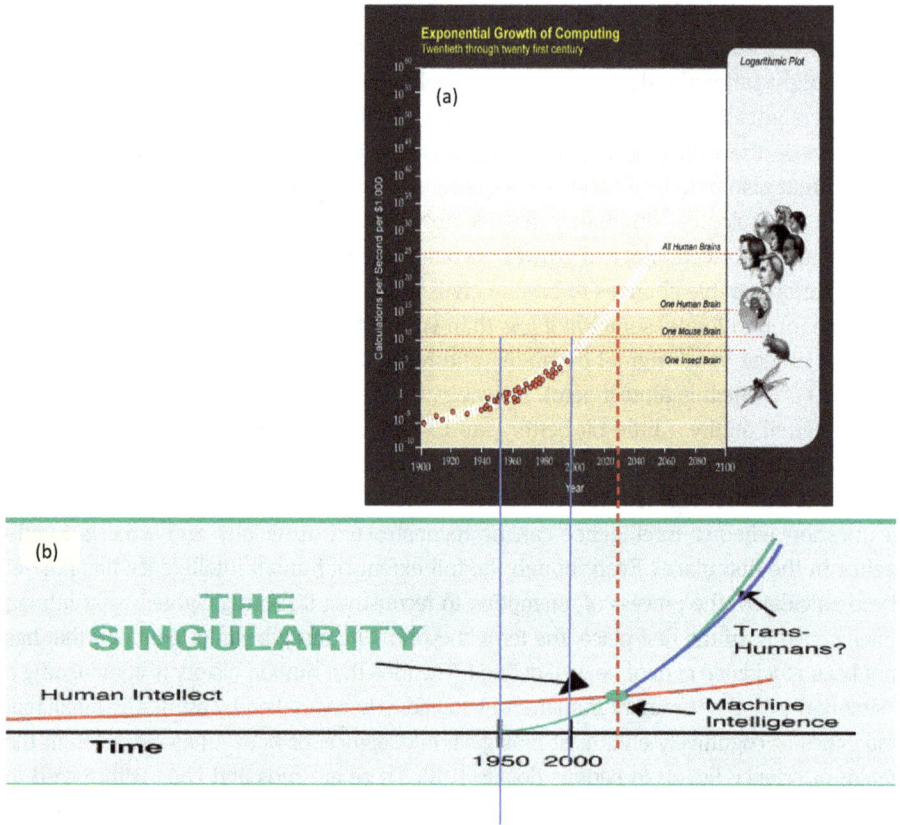

Figure 4.2: Exponential growth of computing ability (a) [120] and possible occurrence of singularity (b) [121].

far from "self-willing." Most of DL (which is an advanced ML using multilayered neural networks) developers claim that AI that demonstrates the same intelligence as humans "will not be realized in the future," but "it is not an extension of the technology we are seeing now" [123].

While the AI threat theory such as "more dangerous than nuclear weapons" has been called out for AI, which has achieved remarkable development in recent years, there is also an optimistic opinion that "threat theory is irresponsible." Meanwhile, Russell [124] mentioned that it is never too early to worry about the threat of AI and has a thorough rebuttal of his views. Some optimists say that there is no threat of AI beyond human intellect even though calculators have superhuman computing power, but they are not a threat at all." Since there is no need to be fear because of calculator's superhuman computing capability, some optimists say that there should not be threat of AI beyond the wisdom of people. Russell said these opinions "don't even deserve a rebuttal" [125]. The fact that the short-term memory of the chimpanzee is superior to the human race cannot be directly interpreted that it is said that the chimpanzee is smarter than man. It is extremely difficult to stop AI once it has been diverted from under control, even though it is a threat enough to be superintelligent beyond humans in everything [124, 125]. The singularity presents both an existential threat to humanity and an existential opportunity for humanity to transcend its limitations [118, 126, 127].

4.8.2 Other concerns

According to the social constructivism of technological philosophy, technology does not change society, society does not change technology, and technology and society interact in both directions [128]. Hori [128] foresees the society utilizing AI, in which various AI technologies are embedded and interconnected everywhere, so that the boundary between man and machine becomes unclear. Accordingly, specific social system designs and specific AI system designs should be revised. Since AI is only a part of the function of cybernetics, it is needed to reconsider the AI's positioning in the cybernetic environment, which is a transdisciplinary approach for exploring regulatory and purposive systems (structures, constraints, and possibilities) and can include control systems and theory of electrical network, mechanical and structural engineering, logic modeling, evolutionary biology, neuroscience, anthropology, and psychology.

With the recent wave of progress in AI has come a growing awareness of the large-scale impacts of AI systems, and recognition that existing regulations and norms in industry and academia are insufficient to ensure responsible AI development. AI has the potential to transform society in ways both beneficial and harmful. Beneficial applications are more likely to be realized, and risks more likely to be avoided, if AI developers earn rather than assume the trust of society and of one another. Based on this background, Brundage et al. [129] suggested various steps that different stakeholders (including user, regulator, academic and developer, involved) in AI development can take to make it easier to verify claims about AI development, with a focus on providing evidence about the safety, security, fairness, and privacy protection of AI systems. Implementation of such mechanisms can help make progress on the multifaceted problem of ensuring that AI development is conducted in a trustworthy fashion. AI is full of good answers and waits for better questions, so that it is important to solve problems, but finding problems is much more important. I would like to close this chapter by saying that (i) AI is only an extension my world and (ii) only things within the expected range can be handled, so that AI ignores it if it comes from the other (the imperceivable) side [129].

References

[1] Hill K. Virtual Immortality Through Digital Cloning. 2010; https://www.forbes.com/sites/kash mirhill/2010/10/21/virtual-immortality-through-digital-cloning/?sh=477aafd87802.

[2] Berk G. Robot Clones of Real Humans. 2019; https://medium.com/@gozdeberk1/robot-clones-of-real-humans-6f32d3c43b89.

[3] Enderle R. Digital clone: you won't be able to live without one. TechNewsWorld. 2014; https://www.technewsworld.com/story/80971.html.

[4] Kim M. What is Voice Cloning? https://www.idrnd.ai/what-is-voice-cloning/.

[5] Vincent J. Listen to this AI voice clone of Bill Gates created by Facebook's engineers. 2019; https://www.theverge.com/2019/6/10/18659897/ai-voice-clone-bill-gates-facebook-melnet-speech-generation.

[6] Nishimura K. AI+ is an artificial intelligence avatar on the cloud that copies personalities and does the job on your behalf. 2015; https://jp.techcrunch.com/2015/01/27/alt-copies-and-becomes-you-as-an-artificial-intelligence-in-the-cloud-that-will-do-menial-tasks-for-you/.

[7] Govender T. Personal Artificial Intelligence. 2017; https://www.bbntimes.com/technology/personal-a-i.

[8] O'Neill N. Companies Want to Replicate Your Dead Loved Ones With Robot Clones. https://www.vice.com/en/article/pgkgby/companies-want-to-replicate-your-dead-loved-ones-with-robot-clones.

[9] Komiyama Y. The future of artificial intelligence; Facing AI's death. 2020; https://www.ju kushin.com/archives/27453.

[10] Saito K. Artificial intelligence as a legal entity. the institute of electronics. Information and Communication Engineers. 2017, 35, 19–27, https://www.jstage.jst.go.jp/article/jsicr/35/3/35_13/_pdf/-char/ja.

[11] Akasaka R. A Study on legal personhood of AIs in tort law. The 32nd Annual Conf of Japanese Society for Artificial Intelligence. 2018; https://confit.atlas.jp/guide/event-img/jsai2018/1F2-OS-5a-03/public/pdf?type=in.

[12] IEEE. Ethically Aligned Design (EAD): A Vision for Prioritizing Human Well-being with Autonomous and Intelligence Systems. 2018; https://standards.ieee.org/content/dam/ieee-standards/standards/web/documents/other/ead_v2.pdf.

[13] Schenker JL. Can we balance human ethics with artificial intelligence? Techonomy. 2017; https://techonomy.com/2017/01/how-will-ai-decide-who-lives-and-who-dies/.

[14] Hattori T. Artificial character: constructing and verification for social; intelligence. Technical Report of the Institute of Electronics, Information and Communication Engineers. 2007, 107, 73–8, https://ci.nii.ac.jp/naid/110006549606.

[15] Aleksander I. Artificial neuroconsciousness an update In: Mira J. et al. ed., From Natural to Artificial Neural Computation. IWANN 1995. Lecture Notes in Computer Science, Vol. 930, Springer, Berlin, Heidelberg, 1995. doi: https://doi.org/10.1007/3-540-59497-3_224.

[16] Artificial Consciousness; https://ja.wikipedia.org/wiki/%E4%BA%BA%E5%B7%A5%E6%84%8F%E8%AD%98.

[17] Schneider S, Turner E. Is anyone home? A Way to Find Out If AI Has Become Self-Aware: It's not easy, but a newly proposed test might be able to detect consciousness in a machine. 2017; https://blogs.scientificamerican.com/observations/is-anyone-home-a-way-to-find-out-if-ai-has-become-self-aware/.

[18] Griffin M. This new test will tell us if AI has become self-aware and gained consciousness. 2017; https://www.fanaticalfuturist.com/2017/08/this-new-test-will-tell-us-if-ai-has-become-self-aware-and-gained-consciousness/.

[19] Simon M. It's Time to Talk About Robot Gender Stereotypes. 2018; https://www.wired.com/story/robot-gender-stereotypes/.

[20] Simon M. Considering the problems that give robots "gender" that should not be in machines. 2018; https://wired.jp/2018/12/27/robot-gender-stereotypes/.

[21] Johnson K. AI researchers propose 'bias bounties' to put ethics principles into practice. 2020; https://thebridge.jp/2020/05/ai-researchers-propose-bias-bounties-to-put-ethics-principles-into-practice.

[22] https://www.rottentomatoes.com/m/her.

[23] https://www.huffingtonpost.jp/2016/11/29/ntt-ai_n_13281528.html.

[24] https://note.com/studyvtubermimi/n/n767055d7b6b3.

[25] https://mainichi.jp/articles/20200416/k00/00m/040/134000c.

[26] https://roboteer-tokyo.com/archives/13378.

[27] http://ailab.hatenablog.com/entry/artificial-intelligence-personality.

[28] Turing test. https://en.wikipedia.org/wiki/Turing_test.

[29] Computer AI passes Turing test in 'world first'. 2014; https://www.bbc.com/news/technology-27762088.

[30] Tardif A. What is the Turing Test and Why Does it Matter? 2020; https://www.unite.ai/what-is-the-turing-test-and-why-does-it-matter/.

[31] McDonald I. 2020; https://www.reddit.com/r/artificial/comments/gdxr7o/any_ai_smart_enough_to_pass_a_turing_test_is/.

[32] Dobrev D. Formal definition of artificial intelligence. International Journal Information Theories & Applications. 2005, 12, 277–85.

[33] Dobrev D. The IQ of Artificial Intelligence. https://arxiv.org/ftp/arxiv/papers/1806/1806.04915.pdf.

[34] Chollet F. On the Measure of Intelligence. 2019; https://arxiv.org/pdf/1911.01547.pdf.

[35] Marks S. Who's smarter? An IQ test for both AI systems and humans. 2019; https://towardsdatascience.com/whos-smarter-an-iq-test-for-both-ai-systems-and-humans-cdf44844b612.

[36] Houser K. DeepMind created an IQ test for AI, and it didn't do too well. 2018; https://www.weforum.org/agenda/2018/07/deepmind-created-a-test-to-measure-an-ai-s-ability-to-reason.

[37] Ookawara K. IBM researcher says future AI needs "common sense and ethics." 2017; https://ascii.jp/elem/000/001/569/1569632/.

[38] Coldewey D. DARPA wants to teach and test 'common sense' for AI. 2018; https://techcrunch.com/2018/10/11/darpa-wants-to-teach-and-test-common-sense-for-ai/?guccounter=1&guce_referrer=aHR0cHM6Ly93d3cuZ29vZ2xlLmNvbS8&guce_referrer_sig=AQAAAFYzMCI7GfStuih0MfzdqSyeymHQG9KwJC8OdtkLNUkaP7I1SATh2A3bsDmW70AbFX1iDha0l_PtVMnMRcVmE6naFA5gnTueUE6dRUYS0ZUKQPYCbUutTG5orlEvBw2wptpAxOw8dggJHtYo_BTHvQh8Ermp2X83BJhMkY32ylAN.

[39] Osborne C. Can artificial intelligence have "common sense"? 2017; https://japan.cnet.com/article/35098179/.

[40] Sominite T. Facebook's AI chief: machines could learn common sense from video. MIT Technical Review. 2017; https://www.technologyreview.com/2017/03/09/153343/facebooks-ai-chief-machines-could-learn-common-sense-from-video/.

[41] Galeon D. This AI can tell when you're lying. 2018; https://www.weforum.org/agenda/2018/01/how-ai-could-bring-an-end-to-lying-and-deception/.

[42] Mukherjee A. DARE Case Study. https://www.digitaltechnologieshub.edu.au/docs/default-source/making_difference/roseville-college/archisha_deception-analysis-and-reasoning-engine.pdf.

[43] Tamura N. AI in the court: Truth or Lie? https://ledge.ai/dare_ai/.

[44] Sebin A. DARE TO LIE? 2018; https://digital.hbs.edu/platform-digit/submission/dare-to-lie/.

[45] Kobayashi A. AI sees lies from voice: Dramatically evolving speech recognition changes the world. 2017; https://www.itmedia.co.jp/news/articles/1909/09/news033.html.

[46] Gonzalez-Billandon J, Aroyo A, Tonelli A, Pasquali D, Sciutti A, Gori M, Sandini G, Rea F. Can a robot catch you lying? a machine learning system to detect lies during interactions. Frontiers in Robotics and AI. 2019, 6. doi: 10.3389/frobt.2019.00064.

[47] Sudakow J. We have just taught artificial intelligence to lie better than humans. This Could Get Interesting. 2017; https://www.inc.com/james-sudakow/we-have-just-taught-artificial-intelligence-to-lie-better-than-humans-this-could.html.

[48] Harris R. Should AI bots lie? Hard truths about artificial intelligence. 2018; https://www.zdnet.com/article/should-ai-bots-lie/.

[49] Chakraborti T, Kambhanpati S. Algorithms for the Greater Good! On Mental Modeling and Acceptable Symbiosis in Human-AI Collaboration. 2018; https://arxiv.org/pdf/1801.09854.pdf.

[50] Murakami M. Artificial intelligence tells lies. 2018; https://www.itmedia.co.jp/news/articles/1803/01/news044.html.

[51] Kaya K. How to deal with AI that does not "SONTAKU" society. 2017; https://www.itmedia.co.jp/business/articles/1709/07/news013.html.

[52] Kamata M. Wish to a person who can make better choices than AI. 2017; https://www.news-postseven.com/archives/20170921_614014.html?DETAIL.

[53] Morita R. Artificial intelligence that sees through lies is too shocking. 2015; https://ascii.jp/elem/000/001/043/1043020/.

[54] McKenna M. Machines and Trust: How to Mitigate AI Bias. https://www.toptal.com/artificial-intelligence/mitigating-ai-bias.

[55] Kantarci A. Bias in AI: What it is, Types & Examples, How & Tools to fix it. 2020; https://research.aimultiple.com/ai-bias/.

[56] Kaushal A, Altman R, Langlotz. Health Care AI Systems Are Biased: We need more diverse data to avoid perpetuating inequality in medicine. Scientific American. 2020; https://www.scientificamerican.com/article/health-care-ai-systems-are-biased/.

[57] Kaushal A, Altman R, Langlotz C. Geographic distribution of US cohorts used to train deep learning algorithms. Journal of the American Medical Association. 2020, 324, 1212–13.

[58] Dorr J. Can AI solve AI's Problems? 2019; https://towardsdatascience.com/can-ai-solve-ais-problems-8d2504fac48c.

[59] Bexborodykh I. Why Does AI Make Mistakes? 2018; https://stfalcon.com/en/blog/post/why-AI-makes-mistakes.

[60] Brynjolfsson E, Mitchell T. What can machine learning do? Workforce implications. Science. 2017, 358, 1530–4.

[61] Dostoyevsky F. Crime and Punishment. Penguin Press, 2002.

[62] Winter D. AI Errors vs. Human Errors. Technology 2018; https://internationaldirector.com/technology/ai-errors-vs-human-errors/.

[63] Eagan C. AI: A Remedy for Human Error? 2020; https://www.technative.io/ai-a-remedy-for-human-error/.

[64] Priest C. Avoiding Human Error When Building Artificial Intelligence. 2019; https://www.datarobot.com/blog/avoiding-human-error-when-building-artificial-intelligence/.

[65] Cobey C, Boillet J. How do you teach AI the value of trust? 2018; https://www.ey.com/en_gl/digital/how-do-you-teach-ai-the-value-of-trust.

[66] Gendre A. How Artificial Intelligence Fights Human Error. https://cyberstartupobservatory.com/how-artificial-intelligence-fights-human-error/.

[67] AI could reduce human error rate. 2018; https://hellofuture.orange.com/en/ai-reduce-human
      -error-rate/.
[68] Ismail N. Why AI is set to fix human error. 2017; https://www.information-age.com/artificial-
      intelligence-set-fix-human-error-123466675/.
[69] Does Artificial Intelligence help to help us to avoid process & human errors? 2020; https://
      www.hint-global.com/posts/does-artificial-intelligence-help-us-to-avoid-process-human-
      errors.
[70] Bergstein B. What AI still can't do. MIT Technology Review. 2020; https://www.technologyre
      view.com/2020/02/19/868178/what-ai-still-cant-do/.
[71] Johansen AG. What is a computer virus? 2020; https://us.norton.com/internetsecurity-
      malware-what-is-a-computer-virus.html.
[72] Somayaji A, Hofmeyr S, Forrest S. Principles of a Computer Immune System. 2000; https://
      www.researchgate.net/publication/2353435_Principles_of_a_Computer_Immune_System.
[73] Fatemi F. Why EQ + AI Is A Recipe For Success. 2018; https://www.forbes.com/sites/falonfa
      temi/2018/05/30/why-eq-ai-is-a-recipe-for-success/?sh=6056d3f1005c.
[74] Goleman D. Emotional Intelligence: Why It Can Matter More Than IQ. Bantam Book, 2005.
[75] Buitrago MP. Artificial Intelligence Keeps Evolving: The Introduction of Emotional
      Intelligence. https://www.30secondstofly.com/ai-software/artificial-intelligence-and-
      emotional-intelligence/.
[76] Bakshi K. In A World Of Artificial Intelligence, EQ Is Our Competitive Advantage. 2017;
      https://www.techleer.com/articles/378-in-a-world-of-artificial-intelligence-eq-is-our-
      competitive-advantage/.
[77] Kang E. Emotional Intelligence & Artificial Intelligence: EQ Vs. AI? 2019; https://www.you-eq.
      com/news-events/2019/7/18/emotional-intelligence-amp-artificial-intelligence-eq-vs-ai.
[78] Ackerman E. What is Emotional Intelligence? +18 Ways To Improve It. 2020; https://positivep
      sychology.com/emotional-intelligence-eq/.
[79] Cherry K. IQ vs. EQ: Which one is more important? 2018; https://www.verywellmind.com/
      iq-or-eq-which-one-is-more-important-2795287.
[80] Why is Emotional Intelligence Important? 2017; https://www.naturalhr.com/2017/10/10/
      emotional-intelligence-important/.
[81] McNeil J. Teaching EQ to AI. Medium. 2017; https://medium.com/s/where-is-the-future/
      teaching-eq-to-ai-442d9a8e97c8.
[82] Why EQ is Key in an AI World. 2020; https://24-7intouch.com/blog/why-eq-is-key-in-an-ai-
      world/.
[83] Jarboe G. What is artificial emotional intelligence & how does emotion AI work? Surface
      Engine Journal. 2018; https://www.searchenginejournal.com/what-is-artificial-emotional-
      intelligence/255769/#close.
[84] Thornton L. Artificial Intelligence and Ethical Accountability, 2020; https://er.educause.edu/
      blogs/2020/7/artificial-intelligence-and-ethical-accountability.
[85] O'Brien J. Digital ethics in higher education: 2020. EDUCAUSE Review. 2020; https://er.educa
      use.edu/articles/2020/5/digital-ethics-in-higher-education-2020.
[86] Bossmann J. Top 9 ethical issues in artificial intelligence. 2016; https://www.weforum.org/
      agenda/2016/10/top-10-ethical-issues-in-artificial-intelligence/.
[87] https://en.wikipedia.org/wiki/Ethics_of_artificial_intelligence.
[88] Applin SA. Everyone's talking about ethics in AI. Here's what they're missing. 2019; https://
      www.fastcompany.com/90356295/the-rush-toward-ethical-ai-is-leaving-many-of-us-behind.
[89] Mittelstadti B. AI Ethics – Too Principled to Fail? 2019; https://robotic.legal/wp-content/up
      loads/2019/05/SSRN-id3391293.pdf.

[90] Hagendorf T. The ethics of AI Ethics: an evaluation of guidelines. Minds and Machines. 2020, 30, 99–120, https://link.springer.com/content/pdf/10.1007/s11023-020-09517-8.pdf.

[91] Bostrom N. Ethical Issues in Advanced Artificial Intelligence. https://nickbostrom.com/ethics/ai.html.

[92] Skelton SK. Accountability is the key to ethical artificial intelligence, experts say. 2019; https://www.computerweekly.com/feature/Accountability-is-the-key-to-ethical-artificial-intelligence-experts-say.

[93] Lee S. The Ethics and Accountability of AI for Learning. 2019; https://trainingindustry.com/magazine/mar-apr-2019/the-ethics-and-accountability-of-ai-for-learning/.

[94] Dignum V. The ART of AI – accountability, responsibility, Transparency. 2018; https://medium.com/@virginiadignum/the-art-of-ai-accountability-responsibility-transparency-48666ec92ea5.

[95] Chen Z, Gangopadhyay A. AI could help solve the privacy problems it has created. 2018; https://theconversation.com/ai-could-help-solve-the-privacy-problems-it-has-created-130510.

[96] Artificial Intelligence.; https://privacyinternational.org/learn/artificial-intelligence.

[97] The privacy expert's guide to artificial intelligence and machine learning. Future of Privacy Forum. 2018; https://iapp.org/media/pdf/resource_center/FPF_Artificial_Intelligence_Digital.pdf.

[98] Wiggers K. AI has a privacy problem, but these techniques could fix it. 2019; https://venturebeat.com/2019/12/21/ai-has-a-privacy-problem-but-these-techniques-could-fix-it/.

[99] Shou D. The Next Big Privacy Hurdle? Teaching AI to Forget. 2019; https://www.wired.com/story/the-next-big-privacy-hurdle-teaching-ai-to-forget/.

[100] Kerry CF. Protecting privacy in an AI-driven world. 2020; https://www.brookings.edu/research/protecting-privacy-in-an-ai-driven-world/.

[101] Yamamoto TB. Copyright protection for AI works and copyright in AI age. 2017; https://www.itlaw.jp/AI%20Works.pdf https://www.itlaw.jp/AI%20Created%20Works%20and%20Copyright.pdf.

[102] Rana L, Mathew M. India: Artificial Intelligence And Copyright – The Authorship. 2019; https://www.mondaq.com/india/copyright/876800/artificial-intelligence-and-copyright-the-authorship.

[103] Nakano H. Legal Issues of AI (Artificial Intelligence). 2019; https://it-bengosi.com/ai-horitsu/.

[104] Are AI creations subject to copyright and intellectual property rights? 2019; https://komon5000.com/2018/10/17/ai-chizai-law/.

[105] Idei H. In 2020, we will consider the intellectual property system of robot and AI society again. 2020; https://www.kottolaw.com/column/200228.html.

[106] Lavagnini S. Artificial Intelligence and Copyright Protection. 2019; https://www.lexology.com/library/detail.aspx?g=4f5fb5fa-b968-4049-8297-0cff617917b5.

[107] Otero RG, Quintais JP. Before the Singularity. Copyright and the Challenges of Artificial Intelligence. 2018; http://copyrightblog.kluweriplaw.com/2018/09/25/singularity-copyright-challenges-artificial-intelligence/.

[108] Vézina B, Peters D. Why We're Advocating for a Cautious Approach to Copyright and Artificial Intelligence. 2020; https://creativecommons.org/2020/02/20/cautious-approach-to-copyright-and-artificial-intelligence/.

[109] Spindler G. Copyright law and artificial intelligence. IIC – International Review of Intellectual Property and Competition Law. 2019, 50, 1049–51.

[110] Toffler A. The Third Wave: The Classic Study of Tomorrow. New York, NY, Bantam, 1980.

[111] Chen KK. Artistic prosumption: cocreative destruction at burning man. American Behavioral Scientist. 2012, 56, 570–95.

[112] Future of Life Institute. AI policy challenges and recommendations. 2020; https://futureo flife.org/ai-policy-challenges-and-recommendations/.

[113] AI control problem; https://en.wikipedia.org/wiki/AI_control_problem.

[114] Bostrom N. Superintelligence: Paths, Dangers, Strategies. 1st edn. Oxford University Press, 2014.

[115] Yampolskiy R. Leakproofing the singularity artificial intelligence confinement problem. Journal of Consciousness Studies. 2012, 19, 194–214.

[116] Singularity; https://en.wikipedia.org/wiki/Ethics_of_artificial_intelligence.

[117] Matsuo Y. Does artificial intelligence surpass humans?. Kadokawa, 2015.

[118] Shanahan M. The Technological Singularity. The MIT Press Essential Knowledge series., 2015.

[119] Helbing D, Frey BS, Gigerenzer G, Hafen E, Hagner M, Hofstetter Y, Van Den Hoven J, Zicari RV, Zwitter A. Will democracy survive big data and artificial intelligence? Sientific American. 2017; https://www.scientificamerican.com/article/will-democracy-survive-big-data-and-artificial-intelligence/.

[120] Ventura T. The Promise & Peril Of The Coming Technological Singularity. 2020; https://me dium.com/predict/the-promise-peril-of-the-coming-technological-singularity-320d3b62cbd.

[121] Colvin J, Mikhael S. The Technological Singularity. 2018; https://www.colvinconsulting.com. au/index.php/insights/the-technological-singularity.

[122] Kurzweil R. The Singularity is Near: When Humans Transcend Biology. Penguin Group., 2005.

[123] Asakawa N. "AI threat theory" 3 Misconceptions in the AI threat theory. 2017; https://www. nikkei.com/article/DGXMZO16386940V10C17A5000000/.

[124] Russell S. Human Compatible: Artificial Intelligence and the Problem of Control. Penguin Books, 2020.

[125] Russell S. Many experts say we shouldn't worry about superintelligent AI. they're wrong. IEEE Spectrum. 2019; https://spectrum.ieee.org/computing/software/many-experts-say-we-shouldnt-worry-about-superintelligent-ai-theyre-wrong.

[126] Shanahan M. Singularity: from Artificial Intelligence to Superintelligence. NTT. 2016; https:// www.nttpub.co.jp/search/books/detail/100002373.html.

[127] Shanahan M. Satori before singularity. Journal of Consciousness Studies. 2012, 19, https:// www.researchgate.net/publication/263156990_Satori_Before_Singularity.

[128] Hori K. Toward an embedded society of artificial intelligence that is not recognized as artificial intelligence. Journal of Information and Communications Policy. 2018, 2, 11–19, https://www.soumu.go.jp/main_content/000592822.pdf.

[129] Brundage M and 58 coauthors. Toward Trustworthy AI Development: Mechanisms for Supporting Verifiable Claims. 2020; https://arxiv.org/pdf/2004.07213.pdf.

Chapter 5
AI in life: from cradle to grave

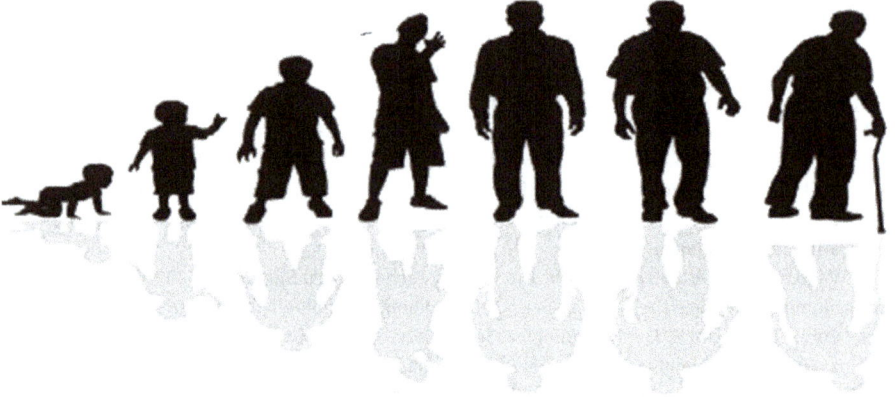

It has ben stated, by the investigation of the longevity person and the research of the molecular biology, that the person can live for up to 120 years with clear understanding of chromosome telomeres and mitochondria, which are deeply involved in longevity, and the precise human body mechanisms protected by functions such as immunity, molecular repair, and regeneration – molecular biology. In the Old Testament (Genesis; chapter 6, section 3), the Lord said, "My spirit will not contend with man forever, for he is mortal; his days will be a hundred and twenty years," although there could be different interpretation for number "120." Human life expectancy has been growing consistently since the nineteenth century. The main reasons are the development of vaccines, improving the safety of childbirth, and improving treatments for diseases leading to death such as cancer and heart disease. The age of the person who has lived the longest so far was 122 years old. According to the latest analysis of world demographics, 122 years old is a limit that can be reached by humans, and it is probably impossible to live any further. Furthermore, the *Tibetan Book of the Dead* (*Bardo Thö-dol*) and even the *Egyptian Book of the Dead*, which is far older than the Tibetan book, clearly mentioned that the longest lifespan for human to achieve is 120 years. Recognizing that human being can live for 120 years as long as possible, the most important and ideal point is that there should not be any different between life expectancy and healthy life expectancy. The former life expectancy is the life expectancy predicted by statistics on how old you will live at the age of 0, whilst the latter healthy life expectancy is the period during which you can live a healthy life without limiting your daily life.

During a life*, even before the birth, AI technology has been assisting extensively. Thinking daily activities from morning till night, even while sleeping, AI technologies

https://doi.org/10.1515/9783110717853-005

have been utilized and ever-increasing applications are proposed and introduced into market. AI has infiltrated our social lives and expanded to AI-based services and tasks, including areas with a large impact on people, such as healthcare.

*see attached figure [http://livingstingy.blogspot.com/2014/02/life-stages.html.]

Modern society is a digital society itself. Our daily lives are surrounded by digital and led to an enriched life. On the other hand, personalization of communication is progressing supported by digital characteristics, and the dilution of family and human relationships is whispered more than before. Even if there is convenient digital knowledge (formal knowledge) that produces results immediately, because its values process and experience knowledge (analog knowledge), life satisfaction is maintained without breaking the old life while enjoying convenience. There are three types people in the digitized society: (i) digital native is typified by a generation that naturally lives stained with digital products and does not feel particularly uncomfortable; particularly children raised in a digital, media-saturated world, require a media-rich learning environment to hold their attention, (ii) digital immigrant who have acquired familiarity with digital systems as an adult and were forced to enter this foreign world from the middle of their lives and (iii) digital refugee who are wondering between digital natives and digital immigrants [1, 2]. In any event, the post-digital era will become soon and at the post-digital time, with remarkable progress in AI technology, AI is deepening while creating image processing, speech recognition, and machine translation against the background of digital technology without much awareness of digital technology, so it is not the end of the digital age, but symbiosis.

5.1 Pregnancy and post-delivery

5.1.1 Infertility treatment

Assisted reproductive technology should include various medical procedures used primarily to address infertility which can include the intracytoplasmic sperm injection (ICSI), in vitro fertilization (IVF), cryopreservation of gametes or embryos, and/or the use of fertility medication. Assisted reproductive technology (ART) mainly belongs to the field of reproductive endocrinology and infertility [3]. About 40 years ago, the world's first IVF embryo graft baby was born (in 1978). Since then, rapid technological innovation has dramatically improved the success rate of IVF. However, in the early 2000s, the success rate of IVF peaked and has not changed much since then. After when micro-insemination (ICSI, developed in 1992) and subsequent improvements in ovulation-inducing agents and embryo freeze–thaw technology were improved, any remarkable and dramatic technological innovations have rarely occurred. The main reason for this is that the quality of eggs and fertilized eggs (embryos) could not be improved or discerned by all means [4]. There could be two possible ways to increase the dramatic success rate of IVF in the future. One is to improve the quality of the eggs and embryos mentioned above, but it is believed that technologies such as regenerative medicine and gene editing are necessary, and there an ethical problem and it seems to be difficult at present. The other technique is to evaluate/examine the quality of eggs and embryos. If we can determine which good embryos can conceive reliably, we could theoretically increase the pregnancy rate per embryo implantation to 100%. In current IVF, the quality of embryos is subjectively evaluated by the human eye [5]. A promising technique is preimpedic genetic testing (PGT), with which some cells of embryos are sampled to analyze genes and inspect them for chromosomal abnormalities. In the future, if genetic analysis technology is promoted, it may be reliable to judge good embryos, although there is still an ethical debate [4]. Based on above background, a method was developed to identify good embryos by AI's image judgment technology, which has rapidly advanced technology in recent years [5]. The growth status of the embryo is constantly monitored, and the subsequent pregnancy rate is analyzed by the big data technology to determine a good embryo. Big data is a field that treats ways to analyze, systematically extract information from, or otherwise deal with data sets that are too large or complex to be dealt with by traditional data-processing application software. Data with many cases (rows) offer greater statistical power, while data with higher complexity (more attributes or columns) may lead to a higher false discovery rate [5]. It is a better technology than PGT because there is no possibility of damaging embryos such as PGT, which takes out some cells. The time-lapse incubator [6, 7] utilizing AI image determination technology will provide in vitro fertilization technique to all IVF patients [4]. Mullin reported that (i) trying to conceive through IVF is a gamble. However, AI may increase the odds of pregnancy, (ii) the pregnancy rate of IVF using new embryos now exhibits a large discrepancy between the rate of around 31% for women under 35 and around 3% for women over 42, (iii) many women who wish pregnant have undergone IVF many times after all. The process is time-consuming, mentally burdening, and expensive, and it was also indicated that an algorithm can select excellent embryos with a high probability of pregnancy [8].

AI-supported fertility system (called Vivola) was introduced [9, 10]. Data on women conceived by IVF are registered for about 1,000 people, and statistical data on IVF such as the average number of eggs and transplants, the duration and cost of treatment can be viewed. A patient can view and search for and reference data leading up to pregnancy through IVF of others whose physical conditions are similar. For extracting co-quality data similar to yours, the value of anti-Müller tube hormone (AMH) is employed which is a hormone secreted from the follicles of the developmental process, with or without diseases related to age and pregnancy. One of the characteristics of infertility treatment is that treatment policies differ depending on the medical institution. For example, there exists a discrepancy among medical clinics in terms of policies and methods for conducting the IVF stimulation the ovaries to raise follicles in order to collect mature eggs. There is a medical institution of the policy of raising a lot of follicles by administering a lot of hormonal drugs, and there is a medical institution of the policy of raising the follicle while approaching the natural cycle without using the hormone drug so much, and it is said that a big difference occurs in the amount of the hormone drug administered by the medical clinics. The tests that can be received also vary depending on the medical institution. In the United States, it is common to perform pre-implantation diagnosis to test for abnormalities in the chromosomes and genes of fertilized eggs [9, 10].

The selection of a viable embryo remains a principal challenge of IVF. This is essential in order to predict outcomes which could lead to shorter time to pregnancy and result in a healthy, live-birth. In 1997, Kaufmann et al. [11] developed an ANN with a predictive power of 59% using only four inputs (age, number of eggs recovered, number of embryos transferred, and whether there was embryo freezing) in an attempt to predict the likelihood of successful IVF. Obstacles to greater sensitivity in this area include the many unknown factors which lead to a successful outcome for IVF which are needed to "teach" for the purpose of ML [12]. It is for this reason that large datasets, such as computer vision, are being used to construct an ANN in order to increase predictive power. Time-lapse imaging of embryos and image data have been used to obtain large datasets. Several studies were presented at the 2018 Annual Congress of the American Society for Reproductive Medicine and the European Society for Human Reproduction and Embryology. More specifically, one of those studies achieved 83% overall accuracy in predicting live birth by looking at 386 time-lapse images of single blastocyst transfers [13]. Another study reviewed 50,392 images from 10,148 embryos and managed to obtain 97.53% accuracy in discriminating between a poor and good blastocyst [14]. Others used pre-treatment characteristics of known cycles to predict first cycle success, which had an accuracy of 81% [15].

The company named "Future Fertility" is the first compant to use AI to grade the viability of woman's harvested eggs for use in IVF and freezing eggs. It was reported that at last count, more than 65,000 so-called social egg freezers in the United States are doing so each year, despite the costs, physical toll, and the uncertain odds. Egg freezing doesn't guarantee a child-bearing future; recent data from one of Europe's largest IVF clinics indicates that while most eggs survive the thawing process, only about a third of them result in successful pregnancies. Earlier as mentioned [8], DL algorithm was used to grade the quality of embryos even better than human fertility technicians. The Toronto-based startup called Future Fertility unveiled the world's first fully automated egg-scoring algorithm [16].

5.1.2 Sperm and embryo culturing

Currently, the use of AI in infertility treatment is remarkable in the selection of sperm used in IVF and micro fertilization. Usually, the sperm used for these is not used as it is after collection, but the one with good form and high motility is sorted, but most of it is done by the hand of the profession called the embryo culturer. In other words, because it depends largely on the human eye, the accuracy varies depending on the technology, experience, knowledge, and so on of the embryo culturer. AI comes into play, but technologies are currently being developed to improve accuracy by accumulating data and expert evaluations of embryo culturers who have long years of experience and have excellent skills when determining sperm, and to help embryo culturers sort sperm [17, 18]. AI assists to determine the motility/impetus and morphology of sperm cells. With the advent of AI, even inexperienced embryo culturers will be able to provide the same skills as skilled embryo culturers and will be able to cover the technical differences between facilities that have been pointed out so far to some extent. In addition, by DL information not only about sperm but also about fertilized eggs, it is also used to determine fertilized eggs suitable for implantation. In combination with time-lapse incubators, we make accurate decisions. AI is feared to take away our human jobs, but in fact, depending on how we use it, it can enrich our lives and be a clue to solving problems that could not be solved before. It is desirable that AI will continue to demonstrate its power not only in the field of fertility treatment, but also in various fields [18].

Visual morphology assessment is routinely used for evaluating of embryo quality and selecting human blastocysts for transfer after IVF. However, the assessment produces different results between embryologists and as a result, the success rate of IVF remains low. To overcome uncertainties in embryo quality, multiple embryos are often implanted resulting in undesired multiple pregnancies and complications. Unlike in other imaging fields, human embryology and IVF have not yet leveraged AI for unbiased, automated embryo assessment. Khosravi et al. [19] postulated that an AI approach trained on thousands of embryos can reliably predict embryo quality without human intervention. An AI approach was implemented based on DNNs to select highest quality embryos using a large collection of human embryo time-lapse images (about 50,000 images) from a high-volume fertility center in the United States. A framework (STORK) was developed based on Google's Inception model. STORK predicts blastocyst quality with an AUC (area under the curve) of >0.98 and generalizes well to images from other clinics outside the United States and outperforms individual embryologists. Using clinical data for 2,182 embryos, a decision tree was created to integrate embryo quality and patient age to identify scenarios associated with pregnancy likelihood. Analysis shows that the chance of pregnancy based on individual embryos varies from 13.8% (age ≥41 and poor quality) to 66.3% (age <37 and good quality) depending on automated blastocyst quality assessment and patient age, concluding that this AI-driven approach provides a reproducible way to assess embryo quality and uncovers new, potentially personalized strategies to select embryos [19]. Until now, the condition of fertilized eggs had to be regularly observed under a microscope by an embryo culturer who is a medical technician. At that time, it was necessary to remove the fertilized egg from the incubator which was a device with almost the same environment as the woman's body, and it was said that touching the outside air at this time adversely affected the fertilized egg. It was announced that it has jointly developed a system that uses AI to determine the developmental status of IVFed eggs in fertility treatment. This system confirms the image of the camera installed in the incubator with the detection software. An AI/DL was used to make decisions. There is no need to remove the fertilized egg from the incubator, and the observation work can be streamlined, reducing the burden on the embryo culture [20].

5.1.3 Pregnant woman support

After a child is born, the examination of the placenta by a pathologist for abnormalities, such as infection or maternal vascular malperfusion, can provide important information about the immediate and long-term health of the infant. Detection of the pathologic placental blood vessel lesion decidual vasculopathy (DV) has been shown to predict adverse pregnancy outcomes, such as preeclampsia, which can lead to mother and neonatal morbidity in subsequent pregnancies. However, because of the high volume of deliveries at large hospitals and limited resources, currently a large proportion of delivered placentas are discarded without inspection. Moreover, the correct diagnosis of DV often requires the expertise of an experienced perinatal pathologist [21]. Accordingly, Clymer et al. [21] had introduced a hierarchical ML approach for the automated detection and classification of DV lesions in digitized placenta slides, along with a method of coupling learned image features with patient metadata to predict the presence of DV. It was mentioned that (i) ultimately, the approach will allow many more placentas to be screened in a more standardized manner, providing feedback about which cases would benefit most from more in-depth pathologic inspection, (ii) such computer-assisted examination of human placentas will enable real-time adjustment to infant and maternal care and possible chemoprevention (e.g., aspirin therapy) to prevent preeclampsia, a disease that affects 2% to 8% of pregnancies worldwide, in women identified to be at risk with future pregnancies, and (iii) the algorithm's ability to discriminate key features and candidate locations within a high-resolution WSI (whole slide image) while keeping false positives minimized; suggesting that the AI approach can allow many more placentas to be screened with fewer pathologists, increasing DV detection for mothers who are at risk for preeclampsia in subsequent pregnancies [21].

There is a collaborative research on development of AI health advise application for obese expectant and nursing mothers to prevent disease during pregnancy and post-delivery [22], with an attempt to prevent the serious illness of pregnant and post-delivery obese women during pregnancy and in the future. Based on reference information on IoT devices (body composition meter, activity sleep meter) and daily life health data (nutrition, mood, etc.) acquired by this application, valuable information can be provided to general expectant and nursing women through automatic advice functions and columns by this application.

5.2 Obstetrics and gynecology (OB/GYN)

Reviewing some important applications of AI in obstetrics and gynecology to discuss whether AI can lead to deeper understanding of several pathophysiological aspects of these specialties and to recommend areas in obstetrics and gynecology where AI may be used to improve healthcare provision, Emini et al. [23] discussed whether AI can lead to a deeper understanding of pathophysiological concepts in obstetrics and gynecology, allowing delineation of some grey zones, leading to improved healthcare provision and concluded that (i) AI can be used as a promising tool in obstetrics and gynecology, as an approach to resolve several longstanding challenges, (ii) AI may also be a means to augment knowledge and assist clinicians in decision-making in a variety of areas in obstetrics and gynecology, (iii) the cardiotocography (CTG) interpretation and fetal physiology can potentially be facilitated by AI, thereby limiting adverse events in obstetrics, and (iv) in the area of gynecology, AI can delineate the complexity of the molecular biology of gynecological cancer and consequently serve the vision of personalized medicine.

The effects of AI have already been felt in the areas of fetal heart monitoring and fertilization [24]. The development of ANN has given an impetus for researchers who aim to prospectively predict outcomes. ANNs are mathematical systems which are reliable, flexible and evaluate multifactorial data. An example of AI utility is its application in assessing cardiotocographs during labor. Intrapartum monitoring is dogged by inconsistencies between different centers and between obstetricians. AI could make assessment more consistent and reduce perinatal and maternal morbidity. Potentially, an intelligence support software could reduce chances of litigation and the economic burden of healthcare particularly in developing countries [25]. A much needed boost for earlier detection of epithelial ovarian cancer (EOC) using noncoding RNAs has been attempted with a neural network model, and initial results suggest that circulating miRNAs have the potential to develop as a noninvasive diagnostic test for ovarian cancer [26]. Attempts have also been made to predict preterm labor by analyzing uterine electrical signals (electrohysterography), a specific type of electromyography [27].

Both AI and augmented reality are being increasingly incorporated into the practice of modern medicine to optimize decision making and ultimately improve patient outcomes. Moawad et al. [28] summarized their reviews that AI has already been incorporated into many areas of medical practice but has been slow to catch on in clinical gynecology; however, several applications of augmented reality are currently in use in gynecologic surgery.

AI/DL is in good at image pattern recognition and therefore of high benefit to doctors who heavily depend on images, like sonologists, radiographers, and pathologists [29]. Although obstetric and gynecologic ultrasounds are two of the commonly performed imaging studies, AI has had little impact on this field so far. Nevertheless, there is huge potential to assist in repetitive ultrasound tasks, such as automatically identifying good acquisitions and immediate quality assure. For this potential to thrive interdisciplinary communication between AI developers and ultrasound professionals is necessary. Drukker et al. [29] explored the fundamentals of medical imaging AI, from theory to applicability, and introduce some key terms to medical professionals in the field of ultrasound. A wider knowledge of AI will help accelerate its integration into healthcare. Han et al. [30] evaluated the feasibility of reidentifying accelerometer-measured physical activity data collect from wearable devices, which have had geographic and protected health information removed, using support vector machines (SVMs) and random forest methods from machine learning. It was mentioned that a further subset is DL in which massive amounts of neuronal networks interpret and use large amounts of data for deeper "cognitive" capabilities. It has also been called a convolutional neural network in part due to its resemblance to the neurons and connections in our cerebral cortexes. DL has led to breakthroughs in healthcare, specifically in radiologic image recognition. As to the future of AI in Ob/Gyn (obstetrics and gynecology) ultrasound, it was mentioned that (i) data is the fundamental requirement for creating a successful DL application, (ii) one of the practical limitations of software developers is ethically and efficiently obtaining de-identified patient data to create such a thing, (iii) one of the biggest players on the AI block is the UK-based company Intelligent Ultrasound, which acquired over 1 million high-quality images from real obstetric scans to develop algorithms for the software ScanNav and the goals of ScanNav are to provide real-time guidance to sonographers by automatically capturing the six correct images as recommended by the UK fetal anomaly screening program and provide an audit showing that all the images were obtained. In a sense, this provides a layer of quality improvement to ensure that optimal patient care is being delivered. It was furthermore indicated that (iv) the software is still in development and some limitations include real-time guidance for probe placement, especially with unique patient considerations such as obesity and (v) a crucial aspect to consider in these situations is patient privacy; while individual data may be de-identified, further advances in ML may be able to identify individuals if appropriate safety measures are not taken with data security [31].

Iftikhar et al. [32] reviewed the pertinent aspects of AI in Ob/Gyn and how these can be applied to improve patient outcomes and reduce the healthcare costs and workload for clinicians. It is well recognized that (1) AI has been used as a tool to interpret fetal heart rate (FHR) and cardiotocography (CTG) to aid in the detection of preterm labor, and pregnancy complications, (2) AI systems can be used as tools to create algorithms identifying asymptomatic women with short cervical length who are at risk of preterm birth and (3) the benefits of using the vast data capacity of AI storage can assist in determining the risk factors for preterm labor using multiomics and extensive genomic data. It was concluded that (i) AI has a promising future in overcoming diagnostic challenges and improving treatment modalities and patient outcomes in Ob/Gyn, (ii) further studies need to be done to decrease bias when creating algorithms and to increase adaptability in the system, enabling the incorporation of new medical knowledge as new technology surfaces, (iii) practitioners must also take safety measures to ensure that the analysis is valid and accurate, (iv) AI is not meant to replace practitioners but rather to serve as an adjunct in decision-making, and (v) ethically, the use of patient records might bridge patient confidentiality since large amounts of data are required to enable AI systems to have access to the large and varied population statistics which are encountered in clinical settings, hence providing realistic and accurate predictions [32]. With the advent of AI, technological innovations in various fields are advancing, and it is expected to be applied to the Ob/Gyn field, which was difficult to mechanize. Hence, the research was proposed to conduct for developing an automatic diagnosis system using AI for fetal heart rate variability, elucidated the bonds between mothers and children by rhythm correlation analysis, identified delivery time, estimated autonomic nervous system and hormone balance by continuous body temperature measurement using a watch thermometer and heart rate meter in gynecological field, and used AI to estimate menstrual dysmenorrhea, fibroids, endometriosis, and so on [33].

5.3 Pediatrics

Breastfeeding is a social issue that more than 50% (in Japan) are those who are possible to breastfeed at 3 months after childbirth despite the wishes of 90% of mothers during pregnancy. In addition, while the UN UNICEF and WHO are actively engaged in activities to recommend breastfeeding as a global issue to be solved, there is a problem that research is not sufficient and information is not sufficient worldwide, and many people are worried. With about breastfeeding, which is also a global issue, we will elucidate the mechanism, nutritional standards, correlation between breast milk and maternal and child health, and provide and establish services to improve social issues and problems. The "Bonyu.lab," which wants breath milk, as the first meal of life to be a richer and happier experience, was developed for analysis of nutritional status of collected breast milk and feedback is provided on "necessary nutrition and diet" supervised by specialists such as midwives and registered dietitians [34].

The *Maternal and Child Health Handbook* (*Boshi-Techou*) [35] was established in Japan. In 1948, it was not long after the Second World War, when children suffered from malnutrition and infectious diseases a lot. At that time, in order to protect and manage the health of pregnant mothers and children born, this handbook was established.

Initially, having a mother and child handbook allowed pregnant and nursing mothers to preferentially distribute milk and sugar. Recently, there is already a movement to digitize maternal and child health handbooks, but there is no meaning simply by digitizing paper notebooks. It is important to digitize and standardize maternal and child health handbooks in a way that can be used with perinatal period (before and after childbirth) electronic medical charts handled by medical institutions. By establishing such a mechanism, personal health records (PHR) can be started from a healthy in early childhood instead of creating an electronic medical record for the first time after becoming ill. In fact, the effects of health management in early childhood and schoolchildren on adult health are significant. For example, it is known that there is a strong correlation between eating habits up to the age of 6 and the risk of suffering from diabetes after adulthood. If the maternal and child health handbook is digitized, the merit is large. For example, standardization the digitization of vaccination information for the prevention of infectious diseases in newborns and infants is one of them. When vaccinated, information such as the manufacturer and production lot is stored as electronic data by reading the barcode and recorded in the maternal and child health handbook. As a result, even if any side effects or reactions are caused after vaccination, the cause can be investigated retroactively to the production lot [35].

AI-based methods have emerged as powerful tools to transform medical care. Although machine learning classifiers (MLCs) have already demonstrated strong performance in image-based diagnoses, analysis of diverse and massive electronic health record (EHR) data remains challenging. Here, Liang et al. [36] have shown that MLCs can query EHRs in a manner being similar to the hypothetico-deductive reasoning used by physicians and unearth associations that previous statistical methods have not found. The model applies an automated NLP system using DL techniques to extract clinically relevant information from EHRs. In total, 101.6 million data points from 1,362,559 pediatric patient visits presenting to a major referral center were analyzed to train and validate the framework. The model demonstrates high diagnostic accuracy across multiple organ systems and is comparable to experienced pediatricians in diagnosing common childhood diseases. It was indicated that (i) the study provides a proof of concept for implementing an AI-based system as a means to aid physicians in tackling large amounts of data, augmenting diagnostic evaluations, and to provide clinical decision support in cases of diagnostic uncertainty or complexity and (ii) although this impact may be most evident in areas where healthcare providers are in relative shortage, the benefits of such an AI system are likely to be universal.

In a relationship between doctor and patient, there could be various combinations showing difficulties of verbal communication; because of (1) patient is to young with a lack of enough vocabulary to convey the physical and emotional problem or desires, (2) elders with developed dementia, or (3) different mother tongues (e.g., in the United States, English is a common language between, for example, Spanish mother tongue doctor and Japanese mother tongue patient) [37]. By collecting and analyzing weeping from more than 20,000 monitor users, an analysis algorithm for cry diagnosis was established [38]. This allows users to know why they are crying by activating the crying diagnostics function from the application and collecting baby's crying with a microphone. In the diagnostic algorithm, the classification and probability analyzed as highly likely are displayed in five categories of "hungry," "sleepy," "uncomfortable," "angry," and "want you to play," and the correct answer rate of more than 80% is recorded according to the feedback results of the monitor user in advance.

Kuniyoshi et al. [39] had developed a developmental simulation of fetuses and newly born infants and the simulation is a three-way interaction simulation of the intrauterine environment including the brain nervous system consisting of motor and sensory field, a musculoskeletal system consisting of about 200 muscles, and amniotic water, based on physiological knowledge, the sensory development of the fetus in the womb. As a result, the possibility of acquisition by learning of body maps and reflex behavior in the fetal period was shown, and new insights were given to existing knowledge. In the simulation of the brain, Izhikevich et al. [40] conducted a large-scale computational experiment of the thalamus cortical system of the mammalian brain by making full use of 1 million spike neurons and 500 million synaptic connections. Normal brain activity has been observed, and it has been reported that it occurred as an interaction between many neurons rather than built-in ones. However, there is no body, only artificial input is given, and it is difficult to interpret the meaning of the generated activity. Naturally, assuming that the development of the body affects the development of the brain [41], it should simulate the process by which input from the body constructs the structure of information processing in the brain. In that sense, the simulation is more essentially reflected [39].

5.4 Daily life

Our everyday lives are influenced by AI technology. If you've used mapping and navigation technology, streaming services, social media or voice-activated digital assistants like Apple Siri or Amazon Alexa, then you know first-hand the impacts of AI on contemporary life. AI is a behind-the-scenes tool that makes technology easy-to-use and highly functional. Without AI, the platforms that dominate our lives would not be as popular. AI's popularity is revealed through survey results [42]. According to the survey [43] it was reported that 78% of those surveyed used social media at home, 77% relied on AI for video and music streaming, while 74% found their way around town using mapping and navigation tools.

Data science and ML have made tremendous progress in anticipating customer behaviors and making personalized recommendations. Amazon credits its recommendation engine with driving 35% of sales. Recommendation engines use AI to collect information from similar customer purchases and then create shopper profiles. The system customizes calls to action and customer loyalty programs. AI applies this collected information to recommendation algorithms. Not only does AI streamline the buying experience, but it offers products to customers before customers even know that they want a product [42, 44]. It was reported that Amazon has been keeping opening high-tech grab-and-go grocery stores to minimize human interaction [45]. The grocery variant uses the same technology Amazon Go does. Sensors track customer movements throughout the store and calculating the total checkout value without a human cashier, which is a feature that seems to be valued during heightened social distancing because of the coronavirus pandemic. AI algorithm can analyze the nutritional balance of items that a customer's cashed grocery items and indicate what are missing for next grocery shop list [45].

Diet affects the human gastrointestinal microbiota. Blood and urine samples have been used to determine nutritional biomarkers. However, there is a dearth of knowledge on the utility of fecal biomarkers, including microbes, as biomarkers of food intake [46-48]. Based on this background, Shinn et al. [49] identified a compact set of fecal microbial biomarkers of food intake with high predictive accuracy. It was found that (i) fecal bacterial biomarkers can be utilized as indicators of consumption of specific foods by healthy adults, being important because they establish proof-of-concept for the use of bacterial biomarkers for predicting food intake, (ii) the constructed panel of microbes, or, more broadly, the analytic approach utilized to establish this panel, lays the foundational work necessary for future studies that aim to predict food intake using fecal microbial biomarkers and (iii) generation of biomarkers from stool samples has the potential to increase the robustness of other biomarkers from urine or blood to bolster objective assessment of dietary intake and compliance of participants in nutrition intervention studies. Based on these findings, it was concluded that (iv) the identification of biomarkers of food intake will complement traditional dietary assessment methods, advance compliance evaluation in nutritional intervention studies, and move us toward diet-microbiota tailored therapies for disease prevention and treatment and (v) ultimately, establishing microbial biomarkers of food intake will facilitate the development of personalized nutrition approaches that modulate the gastrointestinal microbiome for human health benefit.

AI also can assist to manage healthcare, which will be discussed in Chapter 8. As we live longer and technology continues its rapid arc of development, we can imagine a future where machines will augment our human abilities and help us make better life choices, from health to wealth [50].

5.5 NCDs and frailty

In a progressive aging society, daily illnesses typified among elderly should include cold, pressure ulcers, osteoporosis and fractures, sprains and dislocations, cataract, and enlarged prostate. As disorders common in the elder population are low nutrition, aspiration, fall, swelling, incontinence, dehydration, bedridden or life inactivity syndrome. Moreover, the followings are typical diseases among aged population; angina, myocardial infarction, heart failure, hypertension, arrhythmia, chronic renal failure, pneumonia cerebral infarction, dementia, subcortical hemorrhage, Parkinson's disease, intestinal obstruction, osteoarthritis, joint rheumatism, senile skin pruritus, shingles, or diabetes. Basic and common characteristics of the aforementioned diseases and disorders in the elderly can include (1) susceptible to be chronic, (2) accompanied with more than one disease, (3) prone to complications, (4) prone to impaired consciousness, (5) prone to dehydration, (6) no clear symptom appearance, (7) prone to bedridden and dementia, and (8) easy side effect development of the drug.

There is always a need for continuous supervision and quick diagnosis in the case of older patients. The promise of modernizing the health system by delivering an efficient, precise, person-centered, and cost-effective healthcare has been a driver to implement new technologies. However, very little has been delivered to date in terms of direct patients' care and benefits. The ethical issues remain unaddressed and without this regulated the wider AI implementation within the healthcare system cannot proceed. Health professionals should not blindly embrace technological advances but instead take them carefully when discussing algorithm-driven clinical decisions, include ML in multidisciplinary meetings, while additionally learning new skills in statistics and computer science to help develop the clinical algorithms and their evaluation in routine clinical practice. This will require evolving the general healthcare culture, updating current medical curricula and training future doctors with new diagnostic and management concepts. In addition, acceptance of digital clinical decisions should be approved by healthcare regulatory organizations, so that legal and clinical backing is required. Considering the novel and potential impact of AI on future healthcare systems, consideration must be given to legal, ethical, and social implementation with all stakeholders before its implementation including patients, public, and a wide range of healthcare providers to enable its meaningful development [51].

Although most of these diseases and disorders are managed or controlled in medicine to some extent, and some of them are under the AI-powdered medical care (as discussed in Chapter 8), there are still medical concerns such as noncommunicable diseases (NCDs) and frailty.

5.5.1 NCDs

Noncommunicable diseases (NCDs), also known as chronic diseases, tend to be of long duration and are the result of a combination of genetic, physiological, environmental and behavior factors. NCDs are omnipresent, 95% of the world's population are suffering from at least one burden, every third person suffers from more than five physical deficiencies. Chronic diseases cause nearly 40 million deaths per year, which is equivalent to 70% of all deaths globally. The main types of NCDs are cardiovascular diseases (like heart attacks and stroke), cancers, chronic respiratory diseases (such as chronic obstructive pulmonary disease and asthma) and diabetes [52]. Today, NCDs are responsible for 7 out of 10 of deaths worldwide, with 85% of premature NCD deaths occurring in low- and middle-income countries. Along with our global partners, CDC is working to reduce the impact of NCDs like cardiovascular disease, cancer, diabetes, chronic respiratory disease, and injury [53].

The solution for NCDs requires new answers and disruptive solutions, which are collaborative and innovative. AI is a modern tool to enhance understanding of data to find better solutions to a broad range of problems. This is done by extracting patterns and relationships from data, including images, for diagnosis and treatment. The effectiveness of AI requires a strong collaboration between clinicians (the users), engineers (the solution providers), and scientists (the knowledge generators). The motto of this international conference is to bring together these three groups under one roof to discuss exciting possibilities to face the future challenges of humanity [54]. The global challenge against NCDs gained transnational attention over the last years. Statistics illustrate, that approximately 70% of all deaths worldwide can be traced back to chronic diseases; on the other hand, AI is on the rise as the number of implemented software featuring ML skills increased tremendously in the last years. Based on this background, Holaus [55], combining those aspects in order to create synergies of the link between existing approaches with the aim to face NCDs and AI, reviewed firstly specifically only on the most prevalent NCDs and their characteristics before analyzing the rise of ML software from the very beginning and those two fields will be combined with the aim to point out both benefits and risks when implementing AI in the healthcare sector. It was found that (i) the use of such software would definitely cut a three-digit number of billions of Euros of spending on healthcare due to the possibility to early-detect diseases and the hereby arising possibility to treat patients in a more efficient way as until now and (ii) on the other hand, the biggest risk will be, that AI works on huge datasets that need to be set up first – during the setting up process of such data, failures can take place which would have a grave impact on further use.

5.5.2 Frailty

Ageing population worldwide is rapidly accelerating from 461 million people aged over 65 years in 2004 to an estimated 2 billion people by 2050 [56], which has profound implications for the planning and delivery of health and social care. The most problematic expression of population ageing is the clinical condition of frailty (a geriatric syndrome associated with an increased risk of negative outcomes for older people [57]). Frailty develops as a consequence of age-related decline in multiple physiological systems, which collectively results in a vulnerability to sudden health status changes triggered by relatively minor stressor events [58]. It is highly prevalent among residents of residential aged care facilities (also called long term care facilities or nursing homes) [56]. It is estimated that a quarter to a half of people over 85 years are frail and these people have significantly increased risk of falls, disability, long-term care, and death. Importantly, up to three quarters of people over 85 years might not be frail, raising the questions of how frailty develops, how it might be prevented and how it can be detected reliably [58].

The application of technological solutions in health care is a field that is constantly expanding and about which there are great expectations both on the part of users and healthcare professionals and providers, despite the existing reticence in this area known as a technology gap. Worldwide, the number of elderly members of the population and the need to be able to assist them properly is on the increase. This need requires major material and human resources that in turn increase costs. In order to analyze the extent to which technology is present in terms of its relationship with frailty and what technological resources are used to treat it, 80 documents related to research, validation and/or the ascertaining of different types of hardware, software or both were reviewed in prevention, care, diagnosis, and treatment areas [59]. The following scales were used: in the area of diagnosis, Fried's phenotype model of frailty and a model based on trials for the design of devices. The technologies developed that are based on these models accounted for 55% and 45% of cases respectively. In the area of prevention, the results proved similar regarding the use of wireless sensors with cameras (35.71%), and Kinect™ sensors (28.57%) to analyze movements and postures that indicate a risk of falling. In the area of care, results were found referring to the use of different motion, physiological and environmental wireless sensors (46.15%), that is, so-called smart homes. In the area of treatment, the results show with a percentage of 37.5% that the Nintendo® Wii™ console is the most used tool for treating frailty in elderly persons. It is of the utmost importance to continue working to reduce the gap existing between technology, frail elderly persons, and healthcare professionals and providers by bringing together the different views about technology and thus stimulate dialogue, an increased awareness and knowledge about the respective fields in order to engage in collaboration on projects that may reduce costs and improve health and quality of life [59].

The University of Patras have developed a frailty detection model — FrailSafe project — that takes texts of older people into account and considered not only physical, but also cognitive, behavioral, social and psychological domains [60]. A frailty text analysis model is a complex statistical model derived by elaborate AI/ML methods that give computers the ability to learn automatically from text. The model aims to detect a person's frailty status (i.e., non-frail, pre-frail, frail) based on written text and help doctors in their diagnosis and treatment of frailty. It was reported that (i) the frailty text analysis model is capable of predicting the three frailty conditions (non-frail, pre-frail, frail) by an average accuracy of 64%, indicating that an older person's written text for our text analysis model can be used to predict the person's frailty status, and have a 64% chance our prediction of frailty status to be true, which is way better than a random guess with a probability of 33.33% (since three different frailty conditions are predetermined). To further increase the model prediction accuracy, the use of a simplified model was investigated to reduce the possible predictions to only non-frail and frail. The accuracy achieved in this case was at 84%, nearly a 20% increase. It was hence concluded that (i) these results are quite encouraging given that these models are based only on text and (ii) higher levels of prediction accuracy are achieved when combining data from other domains including physical, cognitive, behavioral, social, and psychological [60]. Ambagtsheer et al. [57] determined the effectiveness of AI algorithms in accurately identifying frailty among residents aged 75 years and over in comparison with a calculated electronic Frailty Index (eFI) based on a routinely-collected residential aged care administrative data set drawn from 10 residential care facilities located in Queensland, Australia. A frailty prediction system was designed, based on the eFI identification of frailty, allocating 84.5% and 15.5% of the data to training and test data sets respectively to compare the performance of 18 specific scenarios to predict frailty against eFI based on unique combinations of three ML algorithms (support vector machines [SVM], decision trees [DT] and K-nearest neighbors [KNN]) and six cases (6, 10, 11, 14, 39, and 70 input variables). It was obtained that (i) of 592 eligible resident records, 500 were allocated to the training set and 92 to the test set and (ii) three scenarios (10, 11, and 70 input variables), all based on SVM algorithm, returned overall accuracy above 75%; concluding that there is some potential for AI techniques to contribute toward better frailty identification within residential care; however, potential benefits will need to be weighed against administrative burden, data quality concerns, and presence of potential bias [57].

It is reported that the current pandemic COVID-19 accelerates corona-related frailty [61]. Closing this chapter, it will be worthy to review the sarcopenia, which is strongly related to frailty. Sarcopenia indicates a condition in which muscle mass is decreasing and muscle strength and physical function are decreasing and is considered to be a risk factor for physical dysfunction and falls, especially in the elderly. On the other hand, frail is a so-called frail that indicates a condition in which the body's reserve capacity decreases with aging and it becomes easy to cause health problems. While sarcopenia is targeted at reducing muscle mass, frails are considered to be more conceptual because they include items such as weight loss, malaise, and decreased activity. Both causes include aging, nutritional deficiencies, reduced physical activity, and the merger of various diseases. In addition, the two states are related to each other, such as sarcopenia (muscle weakness) leading to frail (frail state) [62, 63].

References

[1] Prensky M. Digital Natives, Digital Immigrants. On the Horizon. MCB University Press, 2001, Vol. 9, 1–6.

[2] Rikhye R, Cook S, Berge ZL. Digital natives vs. Digital immigrants: myth or reality?. International Journal of Instructional Technology and Distance Learning. 2009, 6, 1–14.

[3] https://en.wikipedia.org/wiki/Assisted_reproductive_technology.

[4] https://www.tenderlovingcare.jp/about/ai_labo.

[5] Breur T. Statistical power analysis and the contemporary "crisis" in social sciences. Journal of Marketing Analytics. 2016, 4, 61–5.

[6] Armstrong S, Bhide P, Jordan V, Pacey A, Marjoribanks J, Farquhar C. Time-lapse systems for embryo incubation and embryo assessment for couples undergoing in vitro fertilisation and intracytoplasmic sperm injection. Cochrane. 2019; https://www.cochrane.org/CD011320/MENSTR_time-lapse-systems-embryo-incubation-and-embryo-assessment-couples-undergoing-vitro-fertilisation.

[7] Probenszky C, Nilselid A-M, Montag M. Time-lapse culture with morphokinetic embryo selection improves pregnancy and live birth chances and reduces early pregnancy loss: a meta-analysis. Reproductive Biomedicine Online. 2017, 35, 511–20.

[8] Mullin E. IVF often doesn't work. could an algorithm Help?. The Wall Street Journal. 2019. https://www.wsj.com/articles/ivf-often-doesnt-work-could-an-algorithm-help-11554386243.

[9] https://project.nikkeibp.co.jp/behealth/atcl/feature/00003/072700140/.

[10] https://prtimes.jp/main/html/rd/p/000000001.000058605.html.

[11] Kaufmann SJ, Eastaugh JL, Snowden S, Smye SW, Sharma V. The application of neural networks in predicting the outcome of *in vitro* fertilization. Human Reproduction. 1997, 12, 1454–7.

[12] Curchoe CL, Bormann CL. Artificial intelligence and machine learning for human reproduction and embryology presented at ASRM and ESHRE 2018. Journal of Assisted Reproduction and Genetics. 2019, 36, 591–600.

[13] Zaninovic N, Rocha CJ, Zhan Q, Toschi M, Malmsten J, Nogueira M, Meseguer M, Rosenwaks Z, Hickman C. Application of artificial intelligence technology to increase the efficacy of embryo selection and prediction of live birth using human blastocysts cultured in a time-lapse incubator. Feril Steril. 2018, 110, e372–3.

[14] Zaninovic N, Khosravi P, Hajirasouliha I, Malmsten JE, Kazemi E, Zhan Q, Toschi M, Elemento O, Rosenwaks Z. Assessing human blastocyst quality using artificial intelligence (AI) convolutional neural network (CNN). Fert Steril. 2018, 110, e89.

[15] Correa N, Brazal S, García D, Brassesco M, Vassena R. Development and validation of an artificial intelligence based algorithm for the selection of an optimal stimulation protocols in IVF patients. Human Reproduction. 2018, 33(Suppl 1). doi: 10.1093/humrep/33.supplement_1.1.

[16] Molteni M. Women May Soon Start Using AI to Tell Good Eggs From Bad. 2019; https://www.wired.com/story/women-may-soon-start-using-ai-to-tell-good-eggs-from-bad/.

[17] https://embryologist.info/ai.html.

[18] https://www.nikkei.com/article/DGXMZO50609900U9A001C1XB0000/.

[19] Khosravi P, Kazemi E, Zhan Q, Malmsten JE, Toschi M, Zisimopoulos P, Sigaras A, Lavery S, Cooper LAD, Hickman C, Meseguer M, Rosenwaks Z, Elemento O, Zaninovic N, Hajirasouliha I. Deep learning enables robust assessment and selection of human blastocysts after in vitro fertilization. npj Digital Medicine. 2019, 2, 21. doi: https://doi.org/10.1038/s41746-019-0096-y.

[20] https://www.sankeibiz.jp/business/news/180526/bsc1805260500005-n1.htm.

[21] Clymer D, Kostadinov S, Catov J, Pantanowitz L, Cagan J, LeDuc P. Decidual vasculopathy identification in whole slide images using multiresolution hierarchical convolutional neural networks. American Journal of Pathology. 2020, 190, 2111–22.

[22] https://prtimes.jp/main/html/rd/p/000000037.000018672.html.

[23] Emini EI, Emin E, Papalois A, Willmott F, Clarkes S, Sideris M. Artificial intelligence in obstetrics and gynaecology: Is this the way forward?. In Vivo. 2019, 33, 1547–51.

[24] Siristatidis C, Pouliakis A, Chrelias C et al. Artificial intelligence in IVF: a need. Systems Biology in Reproductive Medicine. 2011, 57, 179–85.

[25] Desai GS. Artificial Intelligence: The Future of Obstetrics and Gynecology. Journal of Obstetrics and Gynaecology of India. 2018, 68, 326–7.

[26] Elias KM, Fendler W, Stawiski K et al. Diagnostic potential for a serum miRNA neural network for detection of ovarian cancer. Elife 2017, 6, e28932. doi: 10.7554/eLife.28932.

[27] Fergus P, Cheung P, Hussain A et al. Prediction of preterm deliveries from EHG signals using machine learning. PLoS ONE. 2013, 8(10), e77154. doi: 10.1371/journal.pone.0077154.

[28] Moawad G, Tyan P, Louie M. Artificial intelligence and augmented reality in gynecology. Current Opinion in Obstetrics and Gynecology. 2019, 31, 345–8.

[29] Drukker L, Noble JA, Papageorghiou AT. Introduction to artificial intelligence in ultrasound imaging in obstetrics and gynecology. Ultrasound in Obstetrics & Gynecology. 2020. https://doi.org/10.1002/uog.22122.

[30] Han CS, Datkhaeva I. Artificial intelligence in ob/gyn ultrasound. Contemporary Ob/Gyn. 2019, 64, https://www.contemporaryobgyn.net/view/artificial-intelligence-obgyn-ultrasound.

[31] Na L, Yang C, Lo CC, Zhao F, Fukuoka Y, Aswani A. Feasibility of reidentifying individuals in large national physical activity data sets from which protected health information has been removed with use of machine learning. JAMA Netw Open. 2018, 1. doi: 10.1001/jamanetworkopen.2018.6040.

[32] Iftikhar P, Kuijpers MV, Khayyat A, Iftikhar A, De Sa MD. Artificial intelligence: a new paradigm in obstetrics and gynecology research and clinical practice. Cureus. 2020, 12. doi: 10.7759/cureus.7124.

[33] Artificial Intelligence (AI) in Obstetrics and Gynecology Medical Care, 2019; https://www.med.tohoku.ac.jp/about/laboratory/261.html.

[34] https://lovetech-media.com/news/parent/20191209_02/2/.

[35] Health Care Even Before It Was Born – Why Did Intel and Microsoft Standardize Electronic Maternal and Child Handbooks? 2014; https://xtech.nikkei.com/dm/article/FEATURE/20140212/333510/.

[36] Liang H, Tsui BY, Xia H. Evaluation and accurate diagnoses of pediatric diseases using artificial intelligence. Nature Medicine. 2019, 25, 433–8.

[37] Oshida Y. Aesthetics of the Face. Ishiyaku Pb., Tokyo, 2019.

[38] Analysis baby crying with AI. 2018; https://fqmagazine.jp/77260/firstascent/. https://robotstart.info/2018/08/01/angel-ai.html.

[39] Kuniyoshi Y, Sangawa S. Early motor development from partially ordered neural-body dynamics: experiments with a. cortico-spinal-musculo-sleletal model. Biological Cybernetics. 2006, 95, 589–605.

[40] Izhikevich EM, Edelman GM. Large-scale model of mammalian thalamocortical systems. PNAS. 2008, 105, 3593–8.

[41] Pfeifer R, Bongard JC. How the Body Shapes the Way We Think: A New View of Intelligence. MIT press, 2006.

[42] Maksymenko S. How AI technology changes our daily life. Customer Think. 2020; https://customerthink.com/how-ai-technology-changes-our-daily-life/.

[43] Sharifi S. How AI technology affects your lifestyle. 2019; https://www.surveymonkey.com/cu riosity/how-ai-technology-affects-your-lifestyle/.

[44] Horowitz JH. How do we use AI in everyday life? IT Chronicles. 2020; https://itchronicles.com/ artificial-intelligence/how-do-we-use-ai-in-everyday-life/.

[45] Moore C. More cashierless Amazon Go Grocery stores to open. 2020; https://www.foxbusi ness.com/money/more-amazon-go-grocery-stores.

[46] Marchesi JR, Adams DH, Fava F, Hermes GDA, Hirschfield GM, Hold G, Quraishi MN, Kinross J, Smidt H, Tuohy KM, Thomas LV, Zoetendal EG, Hart A. The gut microbiota and host health: a new clinical frontier. Gut. 2016, 65, 330–9.

[47] Knight R, Wegener Parfrey L, Ursell LK, Clemente JC. The impact of the gut microbiota on human health: an integrative review. Cell. 2012, 148, 1258–70.

[48] Wu GD, Chen J, Hoffmann C, Bittinger K, Chen Y-Y, Keilbaugh SA, Bewtra M, Knights D, Walters WA, Knight R, Sinha R, Gilroy E, Gupta K, Baldassano R, Nessel L. Li H, Bushman FD, Lewis JD. Linking long-term dietary patterns with gut microbial enterotypes. Science. 2011, 334, 105–8.

[49] Shinn LM, Li Y, Mansharamani A, Auvil LS, Welge ME, Bushell C, Khan NA, Charron CS, Novotny JA, Baer DJ, Zhu R, Holscher HD. Fecal bacteria as biomarkers for predicting food intake in healthy adults. The Journal of Nutrition. 2020, 285. doi: https://doi.org/10.1093/jn/ nxaa285.

[50] Lau T. When AI becomes a part of our daily lives. Harvard Business Review. 2019; https://hbr. org/2019/05/when-ai-becomes-a-part-of-our-daily-lives.

[51] Mukaetova-Ladinska EB, Harwood T, Maltby J. Artificial intelligence in the healthcare of older people. Arch Psychiatr Ment Health. 2020, 4, 007–013.

[52] WHO. Noncommunicable diseases. World Health Organization. 2018; https://www.who.int/ news-room/fact-sheets/detail/noncommunicable-diseases.

[53] CDC. Global noncommunicable diseases. Centers for Disease Control and Prevention. 2020; https://www.cdc.gov/globalhealth/healthprotection/ncd/index.html.

[54] https://easychair.org/cfp/ICAN-2020.

[55] Holaus J. Artificial intelligence in healthcare. An analysis of the link of AI to health promotion and prevention programs to face and early-detect non-communicable diseases. 2019; https://www.grin.com/document/491644.

[56] Kinsella K, Phillips DR. Global aging: the challenge of success. Population Bulletin. 2005, 60, 5–42.

[57] Ambagtsheer R, Shafiabady N, Dent E, Seiboth C, Beilby JJ. The application of artificial intelligence (AI) techniques to identify frailty within a residential aged care administrative data set. International Journal of Medical Informatics. 2020, 136. doi: 10.1016/ j.ijmedinf.2020.104094.

[58] Clegg A, Young J, Iliffe S, Rikkert MO, Rockwood K. Frailty in elderly people. Lancet (London, England). 2013, 381, 752–62.

[59] Mugueta-Aguinaga I, Garcia-Zapirain B. Is technology present in frailty? technology a back-up tool for dealing with frailty in the elderly: a systematic review. Aging and Disease. 2017, 8, 176–95.

[60] Frailty detection from text analysis. 2018; https://frailsafe-project.eu/news/80-text-analysis.

[61] https://www.nhk.or.jp/politics/articles/feature/57292.html.

[62] Dodds R, Sayer AA. Sarcopenia and frailty: new challenges for clinical practice. Clin Med (Lond). 2016, 16, 455–8.

[63] DerSarkissian C. Sarcopenia with aging. 2020; https://www.webmd.com/healthy-aging /guide/sarcopenia-with-aging.

Chapter 6
AI in QOL (quality of living), QOS (quality of sleeping), and QOD (quality of dying)

In the previous chapter, we discussed how AI can assist or enhance life from the cradle to the grave. This chapter deals with how AI can enrich quality of living, sleep, and dying.

6.1 Quality of living (QOL)

There are numerous definitions on standard of living (SOL) and quality of life (QOL) [1–5]. SOL refers to the level or degree of wealth, comfort, material goods, and necessities available to a certain identifiable class or geographic area such as a nation, city, culture, or socioeconomic group. They may also include measurable aspects of lifestyle, community life, health, freedom, climate, and safety. QOL, on the other hand, is a subjective (while the former is more objective) and intangible term that can measure happiness, health, and comfort of an identifiable group of people. The definitions of these terms can be difficult to tear apart and may overlap in some areas, depending on individuals. But the difference between the two is more than just semantics [2, 4]. Although both are related, in practice they tend to differ widely in their ranking of nations or cities. SOL rankings tend to use a large number of factors to try to estimate how happy people are in a particular group. QOL directly asks people if they are happy [2]. In fact, daily activities should be considered as living, not matter of life. Our society teaches us how to live life by focusing on the outside, making it all about what we can see and what is tangible. The buzzword in this context is lifestyle, which often corresponds to a certain image of life. Livingness also means to allow oneself to strengthen the connection to a place of stillness within and to allow oneself more moments to reflect back on the body and this inner place of stillness, which serves us greatly throughout the hustle and bustle of everyday life. Hence, livingness is an inner quality that needs nurturing and constant advancement and in turn holds and supports us through life by providing an inner strength and certain quality of movement [6]. Definitely, there is a difference between "life" and "living." Life is that throws upon various opportunities on you and gifts several events and moments whereas living is how you grab those chances falling upon and how you find little happiness in those moments and create events to memorize them in your remembrance [7]. Therefore, living means the state of being alive.

From morning till night, our living patterns are somewhat AI-assisted or -controlled to enrich the quality of living. AI technology and ML are increasingly

https://doi.org/10.1515/9783110717853-006

being used to analyze human behavior so that applications can predict what users want and when they want it. These applications make activities like ordering groceries, watching movies, listening to music easier, and others.

People take various actions every day for their lives and/or work. Daily activities can be divided into four categories: necessitated actions, restraint activities, free actions, and others. Sleep, meals (breakfast, lunch, dinner, late at night), personal care (face wash, toilet, bathing, changing clothes, makeup, haircut) or recuperation and hospital visit can be categorized in necessitated actions. Restraint activities can include work-related actions, academic actions (at educational institute), housework (cooking, washing, cleaning, shopping, caring for children, household chores), commuting, social participation (participation in local events and meetings, ceremonial occasions, volunteer activities), while free actions can include conversation and companionship (e-mail communication, SNS, tele-conversation), leisure activities (sports, entertainment, hobbies), mass media contact (TV, radio, newspapers, books, CDs, DVDs), or rest (break, tea/coffee break). Behavior that does not apply to any of the above should go to the others group. A lifelog is a personal record of one's daily life as digital data in a varying amount of detail. The record contains a comprehensive data set of a human's activities. The data could be used to increase knowledge about how people live their lives. The lifelog accumulates long-term continuous recording of human behavior, experiences, cognition, life events, and physical conditions. By continuously recording, accumulating, and analyzing these behaviors, it is available to grasp the behavior patterns, lifestyles, and health conditions characteristic and unique to the person [8]. You may be able to understand your personal interests, preferences, and relationships. In addition, if all daily events are recorded and arranged, it is possible to immediately find out and confirm past actions, remarks, materials, and so on. The lifelog includes not only the information intentionally recorded by users, but also the records automatically left by the Internet or digital devices. In a broad sense, it will also include information recorded by external agencies. At the individual level, it is done for the purpose of improving life, life hacks, recording hobbies, communicating with like-minded people, and there are many smartphone apps and Internet services that support it. At the corporate level, we are in the process of exploring applications in marketing and advertising. Recording an individual's life record for these purposes is called life logging, and the recording itself is called lifelog. It may also refer to services and technologies that record, collect, and reuse lifelog information. Lifelog has been attracting attention due to the advent of easy-to-use recording devices such as smartphones and the development of various sensor devices and storage devices. Recently, some lifelog data has been automatically captured by wearable technology or mobile devices [9].

By allowing computers to perform machine learning, computers that mimic human neural circuits themselves capture the potential characteristics of data and make more accurate and efficient decisions, such as human intelligence. In particular, an image recognition is excellent, but voice assistants that combine speech recognition and natural language processing, and the field of converting from picture to word and from word to picture are also advancing [10]. With advanced sensing technology, essential data accumulation can improve the accuracy of AI (particularly, ML). With advanced sensing technology, numerous AI applications can be developed, including camera installed behind the mirror can monitor and judge your health condition or a pressure meter installed in the bed mattress on the bed can detect the apnea syndrome.

6.1.1 QOL by general healthcare

As AI in healthcare and personal health applications often involves the use of sensitive personal data, a key priority is obviously to use that data responsibly. Data must be kept safe and secure at all the times and processed in compliance with all relevant privacy regulations. It was reported that there are five principles for responsible use of AI in healthcare and healthy living, including (1) well-being, (2) oversight, (3) robustness, (4) fairness, and (5) transparency [11].

The healthcare industry is ripe for some major changes. From chronic diseases and cancer to radiology and risk assessment, there are nearly endless opportunities to leverage technology to deploy more precise, efficient, and impactful interventions at exactly the right moment in a patient's care. AI offers a number of advantages over traditional analytics and clinical decision-making techniques. Learning algorithms can become more precise and accurate as they interact with training data, allowing humans to gain unprecedented insights into diagnostics, care processes, treatment variability, and patient outcomes. At the 2018 World Medical Innovation Forum (WMIF) on AI presented by Partners Healthcare, a leading researchers and clinical faculty members showcased the twelve technologies and areas of the healthcare industry that are most likely to see a major impact from AI within the next decade [12]. They include (1) unifying mind and machine through brain–computer interfaces (BCIs), (2) developing the next generation of radiology tools, (3) expanding access to care in underserved or developing regions, (4) reducing the burdens of electronic health record (EHR) use, (5) containing the risks of antibiotic resistance, (6) creating more precise analytics for pathology images, (7) bringing intelligence to medical devices and machines, (8) advancing the use of immunotherapy for cancer treatment, (9) turning the EHR into a reliable risk predictor, (10) monitoring health through wearables and personal devices, (11) making smartphone selfies into powerful diagnostic tools, and (12) revolutionizing clinical decision-making with artificial intelligence at the bedside. It was concluded that by powering a new generation of tools and systems that make clinicians more aware of nuances, more efficient when delivering care, and more likely to get ahead of developing problems, AI will usher in a new era of clinical quality and exciting breakthroughs in patient care [12].

6.1.2 QOL by checking excretions/secretions

In health management for maintaining and improving QOL, excrement is equally important as food which will be discussed in Chapter 7. There are mainly two types of excrements: that is, liquids (sweat, urine), and solids (feces).

6.1.2.1 Sweat and urine

In addition to water, sweat contains all sorts of substances secreted in the body. These include waste products eaten by bacteria that cause body odor. These substances help to provide an opportunity to elucidate how the body works, regardless of health status [13].

Wearable biosensors have emerged as an alternative evolutionary development in the field of healthcare technology due to their potential to change conventional medical diagnostics and health monitoring. However, a number of critical technological challenges including selectivity, stability of (bio)recognition, efficient sample handling, invasiveness, and mechanical compliance to increase user comfort must still be overcome to successfully bring devices closer to commercial applications. Among many bodily fluids, sweat provides a significant amount of information about a patient's health and is easily available, making it suitable for biosensing noninvasive wearables. Sweat contains important electrolytes, metabolites, amino acids, proteins, and hormones, allowing monitoring of metabolic diseases, physiological conditions, addiction levels, and so on. Based on this background, Parlak et al. [14] introduced the integration of an electrochemical transistor and a tailor-made synthetic and biomimetic polymeric membrane, which acts as a molecular memory layer facilitating the stable and selective molecular recognition of the human stress hormone cortisol. It was reported that (i) the sensor and a laser-patterned microcapillary channel array are integrated in a wearable sweat diagnostics platform, providing accurate sweat acquisition and precise sample delivery to the sensor interface and (ii) the integrated devices were successfully used with both ex situ methods using skin-like microfluidics and on human subjects with on-body real-sample analysis using a wearable sensor assembly. Once humans can track the degree of sweat and the concentration of sweat at the same time, we can measure what the body's moisture state looks like. It was reported that a monitoring sensor was developed to detect the dehydration [15].

The ECHO Smart Patch [16] is designed to catch vital signs in real time 24 h a day, 365 days a year, and can analyze the body's moisture level, electrolyte balance, calorie consumption, lactic acid, and blood sugar levels with a drop of sweat. The sensor body is attached to the body's thorax or abdomen and communicates with a dedicated smartphone app through low-power Bluetooth to alert the body when it is in a critical state. It is a wearable sensor that can be attached and removed, and not only acquires bio data from sweat, but also performs predictive analysis of fitness and health. Comprehensive metabolic panels are the most reliable and common methods for monitoring general physiology in clinical healthcare. Translation of this clinical practice to personal health and wellness tracking requires reliable, non-invasive, miniaturized, ambulatory, and inexpensive systems for continuous measurement of biochemical analytes [16]. A similar wearable device – wearable multiplexed biosensor system [17] – was developed, enabling future evaluation of temporal changes of the sweat biomarkers. It was claimed that, by analyzing sweat, it is possible to notify the user of dehydration during exercise and military training so that the user does not go down, to be able to choose the right exercise for his/her metabolism, and to refrain from salt and sugar in the diet. It can be used not only in sports, but of course in the medical field.

Wearable sweat sensors have the potential to provide continuous measurements of useful biomarkers. However, current sensors cannot accurately detect low analyte concentrations, lack multimodal sensing, or are difficult to fabricate at large scale. Yang et al. [18] developed an entirely laser-engraved sensor for simultaneous sweat sampling, chemical sensing and vital sign monitoring and demonstrated continuous detection of temperature, respiration rate, and low concentrations of uric acid and tyrosine, analytes associated with diseases such as gout and metabolic disorders. The performance of the device was examined in both physically trained and untrained subjects under exercise and after a protein-rich diet. Its utility for gout monitoring in patients and healthy controls through a purine-rich meal challenge was also evaluated. It was mentioned that levels of uric acid in sweat were higher in patients with gout than in healthy individuals, and a similar trend was observed in serum. Nyein et al. [19] introduced microfluidic sensing patches mass fabricated via roll-to-roll (R2R) processes, allowing sweat capture within a spiral microfluidic for real-time measurement of sweat parameters including $[Na^+]$, $[K^+]$, [glucose], and sweat rate in exercise and chemically induced sweat. The patch is demonstrated for investigating regional sweat composition, predicting whole-body fluid/electrolyte loss during exercise, uncovering relationships between sweat metrics, and tracking glucose dynamics to explore sweat-to-blood correlations in healthy and diabetic individuals. It was reported that, by enabling a comprehensive sweat analysis, the presented device is a crucial tool for advancing sweat testing beyond the research stage for point-of-care medical and athletic applications [19].

6.1.2.2 Feces

Imagine a smart toilet that analyzes human excrement in real time every day. You do not have to go to the hospital every six months anymore for checking purpose. Analysis of urine and feces will allow you to immediately check for signs of disease.

With the aging society, the number of certified persons requiring long-term care continues to increase. On the other hand, the chronic labor shortage of caregivers continues, and the efficiency improvement of the site is a major problem. There is a product called "Helppad" [20]. It is a nursing care sensor that can detect the excretion of caregivers just by laying it on a bed, and "smell" is triggered. It is a mechanism that detects urination and defecation with a smell sensor and then informs nursing care staff. In addition, it has the function of managing data in the cloud. Digitally record who did the urine or stool, when and what type of stool, and create excretion patterns by overlapping trends in notifications and records. The technology utilizing smell sensor for diagnosis purposes will be discussed in Chapter 8.

AI of the original algorithm is used for the creation. If the pattern is understood, the timing of changing diapers, which previously relied on the intuition of staff, will be shared, and newcomers will be able to respond according to the data. Since data management is done through web services, there is little burden of learning for staff. The help pad can be said to be a device that improves the efficiency of excretion treatment, which is one of the difficult problems in nursing care, with IoT and contributes to on-site work that is suffering from labor shortages. Even now, stool tests are widely used as screening indicators for colorectal cancer, and urinary infections and urinary tract malignancies are screened by urine qualitative tests. Smart toilets [21] are proposed to analyze human waste in real time every day to catch any sign of disease much faster because of urine analysis and stool analysis.

It is known that the state of defecation is a guide to the daily physical condition of a person. In particular, in elderly facilities, nursing care staff provide various assorted as well as managing and recording defecation for each resident with the aim of preventing dehydration due to intestinal obstruction and diarrhea among the elderly who are residents. However, it is common to visually check and record this defecation management work by hand. A new system was developed [22]. The timing, shape, and size of defecation can be automatically determined and recorded using AI technology, and residents' defecation status can be checked at the station of the facility packed by nursing care staff. Using a camera and LED module built into the back of the toilet seat, the shape of the excreted stool is photographed, and the shape and size of the stool are automatically determined by AI technology. The results of the determination are recorded in conjunction with the date and time of defecation and aggregated into stations. Nursing care staff will be able to provide appropriate nursing care based on this defecation information, for example. By automating defecation records performed visually and by hand, it is also possible to improve the QOL of residents [22].

Technologies for the longitudinal monitoring of a person's health are poorly integrated with clinical workflows and have rarely produced actionable biometric data for healthcare providers. Park et al. [23] described easily deployable hardware and software for the long-term analysis of a user's excreta through data collection and models of human health. The "smart" toilet (a mountable toilet system for personalized health monitoring via the analysis of excreta), which is self-contained and operates autonomously by leveraging pressure and motion sensors, analyses the user's urine using a standard-of-care colorimetric assay that traces red–green–blue values from images of urinalysis strips, calculates the flow rate and volume of urine using computer vision as a uroflowmeter, and classifies stool according to the Bristol stool form scale using deep learning, with performance that is comparable to the performance of trained medical personnel. Each user of the toilet is identified through their fingerprint and the distinctive features of their anoderm, and the data are securely stored and analyzed in an encrypted cloud server. The toilet may find uses in the screening, diagnosis, and longitudinal monitoring of specific patient populations. The system possesses two characteristics [23]. First of all, it has an authentication function of user's anus. When authenticating users in the toilet, facial recognition requires a camera, which causes the user to feel a sense of resistance. The authentication with a camera inside the toilet is a breakthrough in the problem of user authentication in smart toilets. The Stanford's smart toilets use a sensor fusion system that combines authentication with image recognition and a fingerprint sensor inserted into a drainage lever to increase recognition rates. The second point is an application of the deep learning. For creating a high-performance AI, it is extremely important to collect high-quality samples. If you think about it, sanitary ware is a good subject for image recognition. The background is pure white and there is no noise, and the target is yellow or brown. After using it, it will turn white again with toilet paper, so you can easily estimate the usage time only by the image. In addition, by combining the pressure of the sitting sensor and the flowmeter for urine volume, it is likely to be possible to perform more advanced analysis [23].

6.2 Quality of sleep (QOS)

Our bodies are equipped with sleep rhythms of waking up in the morning and sleeping at night, but it is an insomnia that such sleep rhythms are impaired, and we cannot sleep, sleep is shallow and we wake up quickly, or we wake up early in the morning. The first thing to do to cure insomnia is to remove the causes of insomnia and create an environment that makes it easier to fall asleep. If you still cannot sleep, use sleeping pills to help you sleep. There are several characteristic symptoms associated with the insomnia: (1) try to get on the floor and sleep, but cannot sleep, (2) get up many times in the middle of the night, (3) there is no feeling of deep sleep even while sleeping, (4) wake up early in the morning and cannot sleep after that, (5) sleep is shallow and/or (6) short sleep time. The continous follow-up of such symptoms generally lasts for more than 3 weeks. There are three types of insomnia: sleep disorder (the state that cannot sleep easily), "halfway awakening" (state that occurs on the way) and "early morning awakening" (the state that wakes up early in the morning). Causes of insomnia include environmental factors such as noise and vibration, brightness, temperature, and bedding conditions, mental factors, physical factors such as pain, illness, effects of medications and ingredients (caffeine, alcohol, etc.), and age (the older, the less sleepy). Recently, there seem to be a lot of mental factors related to stress. In addition, there are many cases where insomnia is caused by heart disease. Typical mental illnesses with insomnia are accompanied with depression (sleep disorders and early morning wakefulness) and neurosis (sleep disorders and mid-wakefulness are often) [24].

The traditional distractions of jobs, family, and friends have been exacerbated in recent years by irregular work, long commutes, smartphones, and all-night benders, causing unconscious stress, lack of sleep, and modern diseases. It is hence unsurprising that two-thirds of adults in developed nations do not get the nightly 8 h of sleep recommended by the WHO (World Health Organization), which doctors warn is leading us down a cheery path toward chronic diseases, mental health disorders, and dysfunctional relationships [25]. In order to improve the treatment of sleep disorders, the method suggested by the American Society of Sleep Medicine (ASSM) is to analyze by using AI in data on brain waves, blood oxygen levels, heart rate, breathing, and eye and foot movements collected by sleep polygraph tests [25, 26]. In order to improve the sensibility of ML, AI algorithms need to be tested repeatedly. AI algorithms are not calculating like humans in two ways: – algorithms are literal: if you set a goal, the algorithm cannot adjust itself and only understands what it has been told explicitly, and algorithms are black boxes; algorithms can predict extremely precise, but not the cause or the why. Leger et al. [27] mentioned that AI applied to the sleep comprehension and to sleep disorders detection is very promising. However, several major areas of AI in sleep are still quite unexplored. We believe that AI applied to sleep has already shown progress among parallel axes: (i) understanding better how human sleep among the entire night, (ii) understanding how polysomnography may be better analyzed and understood with supervised and non-supervised ML and (iii) improving the quality of sleep with AI/ML.

Sleep polysomnography test records typically, during sleep, brain waves, breathing, leg movements, chin movements, eye movements (REM – rapid eye movement – sleep and non-REM sleep), electrocardiocardiodes, oxygen saturation, chest wall movements, abdominal wall movements, and so on during sleep [28]. Interestingly, we're at a point where technology – something that has long been viewed as a deterrent to healthy sleep patterns – is converging with sleep to allow for healthier rest. Specifically, AI is being leveraged to address serious pain points in the field. Various types of devices have been developed to improve and enhance quality of sleep, based on some testing item(s) of the sleep polysomnography [29].

6.2.1 Checking quality of sleep

SleepScore application is the advanced sleep improvement system featuring noncontact, sonar sleep tracker technology to measure your sleep from your nightstand, plus sleep sounds (or breath) through smartphone's microphone, body movements, and sleep cycle analysis. Details about each stage of your sleep in a score calculated from 0 to 100 are displayed, which you can then track over time and make improvements to always awaken your best [30]. SLEEPON is another advanced sleep tracker with heart rate and blood oxygen (percutaneous arterial oxygen saturation: SpO_2) monitoring functions and provides comprehensive sleep feedback for you to better understand and optimize your sleep [31].

6.2.2 Headband

"Dreem" is a sleep headband device that monitors, analyzes, and claims to enhance quality of sleep. It is a miniaturized and autonomous headband that monitors the quality of sleep and then uses soporific sound through bone conduction to help fall asleep faster, get deeper sleep, and wake up at the optimal time through smart alarm. Dreem is a headband that collects, reports data and also claims to enhance the user's behavior in real time according to the data collected. The sensors include dry EEG (electroencephalogram) electrodes, accelerometer and pulse oximeter. This data is then used to attempt to determine if the wearer is in light sleep, deep sleep, or REM [29, 32].

6.2.3 Armband

Sleep.ai is an armband that detects snoring and then emits a vibration that pushes you onto your side [29].

6.2.4 Bed and mattress

There are several types of bed developed and introduced in market. ReST Bed is a fully customizable mattress that uses sensors to automatically detect and respond to pressure. Essentially, the ReST Bed allows you to set firmness levels for each of the five zones for back and side sleeping. Then, as you switch positions throughout the night, the bed automatically adjusts to the set firmness by detecting body pressure. It seems like an ideal solution for those who toss and turn at night [33]. The Nanit camera is a revolutionary nursery camera that is placed above the crib and streams live HD video of your sleeping baby. But what makes this technology unique is that it tracks sleep insights and analytics, providing parents with sleep behavior monitoring. As a result, both baby and parents get better sleep [34].

The Sleep Number 360 is another cool bed technology that uses AI and ML to provide a better night of sleep. The bed, which works independently of an app, learns your behaviors and responds accordingly. For example, the bed knows when you get in and can turn on a foot warmer at the end of the bed when temperatures are cooler. Because the bed learns your behaviors, like what time you go to bed on Wednesday nights, it can even turn on before you get in bed. The bed can also raise your head slightly when you start snoring, which gives both you and your partner better sleep [35]. HEKA, a product of smart mattress, is the first fully AI-powered mattress. It has the ability to collect data on a person's body pressure distribution, as well as body shape and sleeping positions, to then determine the best pressure distribution for the individual's sleeping positions [36]. Accumulative sleep lack might lead to "sleep debt" and is now a social problem. Under such circumstances, this mattress was developed to improve the quality of sleep and improve health. NEWPEACE AI Motion Mattress achieves ideal sleep through a process of movement, temperature, and AI [37]. The mattress is equipped with an AI sleep program, performing (1) gentle shaking like a hammock or cradle invites drowsy, (2) adjusting the temperature in the mattress and increasing to a body temperature that increases drowsiness, and (3) by obtaining personal sleep data by smart sensing, it leads to sleep that is tailored to each individual. This three-stage process produces the ideal sleep from sleep to awakening [37, 38]. The quality of a good sleep is important for a healthy life. Wu et al. [39] mentioned that recently, several sleep analysis products have emerged on the market, as seen in the above; however, many of them require additional hardware or there is a lack of scientific evidence regarding their clinical efficacy. Hence, a novel method was proposed via clustering of sound events for discovering the sleep pattern. This method extended conventional self-organizing map algorithm by kernelized and sequence-based technologies, and obtained a fine-grained map that depicts the distribution and changes of sleep-related events.

6.3 Quality of dying (QOD)

We found a clear difference in definition between life and living. However, as many manuscripts which are cited in this section are entitled with "quality of death and dying," indicating that (i) there is not clear discrepancy between death and dying and (ii) maybe making a clear difference between two is meaningless [40]. To make it simple, death is the end of life, while dying is the process of approaching death, including the choices and actions involved in the process. Therefore, when the quality of death or the dying process is considered, both should be taken into considerations.

During the past decade, research has examined definitions and conceptualizations of quality of dying and death in different populations. At the same time, there has been a call to clarify the distinctions between quality of dying and death and other end-of-life constructs. Hales et al. [41] reviewed research that examined definitions and conceptualizations of the quality of dying and death, clarified the quality of dying and death construct and its distinction from QOL and quality of care at the end of life, and outlined challenges that remain for health care professionals, researchers, and policy makers. It was reported that (i) the quality of dying and death construct is multidimensional, with seven broad domains (namely, physical experience, psychological experience, social experience, spiritual or existential experience, the nature of health care, life closure and death preparation, and the circumstances of death), (ii) the quality of dying and death is subjectively determined with numerous factors that influence its judgment, including culture, type, and stage of disease, and social and professional role in the dying experience, and (iii) quality of dying and death is broader in scope than either QOL at the end of life or quality of care at the end of life, although there is overlap among these constructs [41].

A high quality of death is regarded as a goal at the end of life and, therefore, an assessment of end-of-life experience is essential. Improving the quality of end-of-life care has become a major agenda for patients, families, and the loved ones of persons near death as well as healthcare professionals, researchers, and policy makers who organize and provide care. Patrick et al. [42] defined the quality of dying and death as the degree to which a person's preferences for dying and the moment of death agree with observations of how the person actually died. Model for evaluating the quality of dying and death was proposed, based on concepts elicited from literature review, qualitative interviews with persons with and without chronic and terminal conditions, and consideration of desirable measurement properties. Expected level of agreement is modified by circumstances surrounding death that may prevent following patient's prior preferences. Qualitative data analysis yielded six conceptual domains (i.e., symptoms and personal care, preparation for death, moment of death, family, treatment preferences, and whole person concerns). These domains encompass 31 aspects that can be rated by patients and others as to their importance prior to death and assessed by significant others or clinicians after death to assess the quality of the dying experience. The proposed model uses personal preferences about the dying experience to inform evaluation of this experience by others after death. This operational definition will guide validation of after-death reports of the quality of dying experience and evaluation of interventions to improve quality of end-of-life care [42].

Palliative care has the potential to play significant role in better quality of dying and death for non-cancer patients. Kim et al. [43] reviewed to determine the definition of quality of dying and death for non-cancer patients. MEDLINE (1990–2015) and Google Scholar (1999–2015) were searched using keyword terms "quality of dying or death," "good or bad death." It was reported that (i) in the 13 definitions of quality of death and dying, the most common terms are related to patient's decision-making, (ii) the most common second terms are related to medical and social support, (iii) final terms are related to psychological support. In order for the social and psychological aspects of death awareness and acceptance to take place, the dying person's suffering should be reduced and they must be relieved of pain and (iv) furthermore, it is rapidly increased percentages of non-cancer patients among those utilizing specialist palliative care services in the countries at the top of the quality of death ranking such as the United Kingdom and the United States. Unlike cancer disease, non-cancer diseases need to longitudinal supportive system. Palliative care will be effective intervention to manage symptoms of non-cancer patients and treat intercurrent medical condition is appropriate. Based on the finding, it was concluded that the quality of death and dying should be combined of psychological, medical, and social support as external role and their patient's decision-making as internal role in palliative care research [43]. The most of articles dealing with the quality of death and dying pay a special to the palliative care. Scientists have found a new medical application for AI: predicting when a seriously ill patient admitted to the hospital will likely die. In hospitals, palliative care teams are charged with improving the QOL of gravely ill patients and making sure their final wishes are carried out. Bennington-Castro [44] mentioned that a research scientist at Stanford University School of Medicine had developed the new AI algorithm, which can predict if a hospital inpatient will die within 3 to 12 months (a window during which palliative care is thought to be most useful) with over 90% accuracy, and also can help by identifying patients who are seriously ill and who may benefit from palliative care. It was further mentioned that importantly, the AI is not like a self-driving car, where human decisions largely take a backseat, rather, it is more like vehicle proximity sensors that flag drivers' attention when necessary. And it is not about replacing a doctor's judgment, the really is about providing extra care [44].

The means to predict mortality using AI could be a transformative factor in the future of palliative health care. AI has the potential to help medical care providers and doctors significantly improve the delivery of patient care in hospice situations. Getting the right kind of treatment at the end-of-life stage is more important than many assume. Not enough treatment – or even inaccurate treatment – can provide a painful experience for patients, and overcare may result in hundreds of thousands of dollars in unnecessary medical bills, even if the patient is covered by insurance. While it is crucial to select the proper medical coverage that includes hospice care regardless of the situation – especially for people over 65 or older, because there are specific plans for specific purposes to help with these medical costs – AI advances may help patients and physicians determine illness sooner to prepare for end-of-life costs and treatments before it is too late [45]. Rajkomar et al. [46] mentioned that technology will soon allow physicians to improve the timing and delivery of patient care. Researchers used AI to scan EHRs and notes doctors left in patient records to detect potential clinical problems and health risks. The AI system predicted patient mortality rate and final diagnoses more accurately and quickly than physicians.

End-of-life medical conversations also often involve language in extremis. As cancer brings a person's life near to its end, they may have lost some of their life-long communicative powers to the disease or its treatments. They may have less ability to speak subtly and indirectly, which is important for politeness. Shallow breathing shortens utterances, and drugs may block word-finding. All of this reinforces an asymmetry in communication that doctors do not always grasp [47]. At the end of a patient's life, there may not be effective medical treatments, just things to discuss and plans to make. This may need a more natural conversation than a medical one, a conversation in which none of the participants may know what the outcome will be. After all, these conversations are not just for doctors; they are for patients, too, and for family members, nursing aides, and housekeeping staff. There are a lot of human beings who have a vested interest in this other human. Although there are critics who do not think AI/ML possesses a role to play in palliative care, AI is helpful, as a discipline that has historically thought of communication as just the art of medicine, to actually think that, no, this is a science, and understanding that science could help us re-engineer the healthcare system to support more meaningful conversations. What it does ultimately is recognize the humanity of things [47].

Weng et al. [48] introduced a new AI model that uses ML to predict the risk of premature death, using banked health data (on age and lifestyle factors) from Brits aged 40 to 69. This study comes months after a joint study which reported results of ML-based data mining of EHRs to assess the likelihood that a patient would die in hospital. One goal of both studies was to assess how this information might help clinicians decide which patients might most benefit from intervention. The FDA [49] is also looking at how AI will be used in health care and posted for a regulatory framework for AI in medical care. As the conversation around AI and medicine progresses, it is clear we must have specific oversight around the role of AI in determining and predicting death. Then there is the issue of unconscious, or implicit, bias in health care, which has been studied extensively, both as it relates to physicians in academic medicine and toward patients [50]. Green et al. [51] mentioned that there are differences; for instance, in how patients of different ethnic groups are treated for pain, though the effect can vary based on the doctor's gender and cognitive load. On the other hand, James [52] reported that these biases may be less likely in black or female physicians. When it comes to death and end-of-life care, these biases may be particularly concerning, as they could perpetuate existing differences. Even though preferences may differ between ethnic groups, bias can still result when a physician may unconsciously not provide all options or make assumptions about what options a given patient may prefer based on their ethnicity [50]. At a growing number of prominent hospitals and clinics around the country, clinicians are turning to AI-powered decision support tools – many of them unproven – to help predict whether hospitalized patients are likely to develop complications or deteriorate, whether they are at risk of readmission, and whether they are likely to die soon. But these patients and their family members are often not informed about or asked to consent to the use of these tools in their care, a STAT examination has found. It was reported that machines that are completely invisible to patients are increasingly guiding decision-making in the clinic [53].

Predicting prognosis is as old as medicine itself. Recent breakthroughs in AI/ML have led to algorithms that promise to answer one of life's ultimate questions about when will die? Analyzing EHR and other data, different algorithms now exist to predict mortality, seemingly with unprecedented accuracy and without the direct input of clinicians or patients [54]. Well intentioned though these efforts are, the rapid dissemination of mortality algorithms by EHR software raises serious ethical concerns, such as the potential use of AI to reduce use of services rather than align care with patient goals or to unintentionally worsen health disparities and raises questions about how these models should be used in practice. In conclusive remarks, it was mentioned that access to large data sets in EHRs and the development of AI algorithms offer the opportunity to ensure that patients with serious illness get the care they need and want. Aligning care with patient priorities and avoiding biases should be the explicit aims of such work. It was recommended to develop AI systems that incorporate patient needs, broadly construed, and not just mortality; working to ensure that data inputs, outputs, and applications are not biased; and involving patients and clinicians in developing the algorithms and building interventions into the system to address identified needs so that the algorithm leads to actions that benefit patients. Research is also needed to elicit patient, family, and health care professional perspectives on the use of mortality algorithms and to identify other outcomes that should be captured. Following these recommendations can help ensure that AI efforts build trust and improve care for the many patients living with serious illness [54].

There is an activity regarding the anti-aging, which prevent, lessen, or slow down the aging process. Most of efforts toward the anti-aging have been concentrated on skin care and nutrition care. It is a bit surprise not to see any activity or thought about the quality of aging (which is somewhat related to QOD).

References

[1]   Fontinelle A. Standard of living vs. Quality of life: What's the difference? Investopedia. 2020; https://www.investopedia.com/articles/financial-theory/08/standard-of-living-quality-of-life.asp#:~:text=Quality%20of%20Life%3A%20An%20Overview,term%20that%20can%20measure%20happiness.

[2]   Spacey J. Standard of living vs Quality of life. Simplicable. 2016; https://simplicable.com/new/standard-of-living-vs-quality-of-life.

[3]   Raza U. Standard of living Vs Quality of life. Nation. 2016; https://nation.com.pk/08-Aug-2016/standard-of-living-vs-quality-of-life.

[4]   Quality of Life vs. Standard of Living; https://gillsocial10.weebly.com/uploads/1/1/2/0/112062845/quality_of_life_vs._standard_of_living.pdf.

[5]   Standard of living Vs. Quality of life. ByMind Fuel Daily; https://www.mindfueldaily.com/live well/standard-of-living-vs-quality-of-life/.

[6]   Andras J. Lifestyle vs. Livingness. Unimed Living; https://www.unimedliving.com/livingness/what-is-the-livingness/lifestyle-vs-livingness.html.

[7]   Gardner K. Six Ways AI Improves Daily Life. 2019; https://www.digitalistmag.com/improving-lives/2019/05/28/6-ways-ai-improves-daily-life-06198539/.

[8]   Lifelog; https://en.wikipedia.org/wiki/Lifelog.

[9]   Koester M. Life Logging or How to Track the Meaning in Life's Miscellaneous. 2016; http://www.markwk.com/2016/10/life-logging-or-tracking-misc.html.

[10] https://www.fmworld.net/connecttips/healthcare/tips000019.html.

[11]   van Houten H. Five guiding principles for responsible use of AI in healthcare and healthy living. 2020; https://www.philips.com/a-w/about/news/archive/blogs/innovation-matters/2020/20200121-five-guiding-principles-for-responsible-use-of-ai-in-healthcare-and-healthy-living.html.

[12]   Bresnick J. Top 12 Ways Artificial Intelligence Will Impact Healthcare. 2018; https://healthitanalytics.com/news/top-12-ways-artificial-intelligence-will-impact-healthcare.

[13]   Best J. Will sweat sensors be the next generation of health tech? –To the future where you can inspect at home. 2019; https://japan.cnet.com/article/35142135/.

[14]   Parlak O, Keene ST, Marais A, Curto VF, Salleo A. Molecularly selective nanoporous membrane-based wearable organic electrochemical device for noninvasive cortisol sensing. Science Advances. 2018, 4, doi: 10.1126/sciadv.aar2904.

[15]   Kloberdanz K. Data against dehydration: This wireless sweat patch powered by jet engine tech could help athletes, air force pilots stay in top shape. Smart Apps. 2018; https://www.ge.com/news/reports/data-dehydration-wireless-sweat-patch-app-help-athletes-air-force-pilot-stay-top-shape.

[16]   https://techable.jp/archives/61305.

[17]   Yokus MA, Songkakul T, Pozdin VA, Bozkurt A, Daniele MA. Wearable multiplexed biosensor system toward continuous monitoring of metabolites. Biosensors & Bioelectronics. 2020, 153, https://www.sciencedirect.com/science/article/abs/pii/S095656632030035X#!.

[18]   Yang Y, Song Y, Bo X, Min J, Pak OS, Zhu L, Wang M, Tu J, Kogan A, Zhang H, Hsiai TK, Li Z, Gao W. A laser-engraved wearable sensor for sensitive detection of uric acid and tyrosine in sweat. Nature Biotechnology. 2020, 38, 217–24.

[19]   Nyein HYY, Bariya M, Kivimäki L, Uusitalo S, Liaw TS, Jansson E, Ahn CH, Hangasky JA, Zhao J, Lin Y, Happonen T, Chao M, Liedert C, Zhao Y, Tai L-C, Hiltunen J, Javey A. Regional and correlative sweat analysis using high-throughput microfluidic sensing patches toward decoding sweat. Science Advances. 2019, 5, doi: 10.1126/sciadv.aaw9906.

[20]   https://project.nikkeibp.co.jp/mirakoto/atcl/wellness/h_vol47/.

[21]   Shankland S. AI toilets will scan your poop to diagnose your ailments. 2018; https://www.cnet.com/news/ai-toilets-scan-your-poop-to-diagnose-medical-ailments/.

[22]   AI toilet automatically determines the shape and size of stools, LYXIL is developing for facilities for the elderly. AI News. 2019; https://monoist.atmarkit.co.jp/mn/articles/1909/26/news045.html.

[23]   Park S-M, Won DD, Lee BJ, Escobedo D, Esteva A, Aalipour A, Ge TJ, Kim JH, Suh S, Choi EH, Lozano AX, Yao C, Bodapati S, Achterberg FB, Kim J, Park H, Choi Y, Kim WJ, Yu JH, Bhatt AM, Lee JK, Spitler R, Wang SX, Gambhir SS. A mountable toilet system for personalized health monitoring via the analysis of excreta. Nature Biomedical Engineering. 2020, 4, 624–35.

[24]   Insomnia. https://www.tsumura.co.jp/kampo/nayami/fumin01.html.

[25]   Macaulay T. Here's how AI can help you sleep. 2020; https://thenextweb.com/neural/2020/03/03/heres-how-ai-can-help-you-sleep/.

[26]   Vleugels A. If your employees don't get enough sleep, that's on you. Future of Work. 2018; https://thenextweb.com/future-of-work/2018/08/27/how-to-sleep-better-technology/.

[27]   Leger D, Christian Guilleminault C. Artificial Intelligence and Sleep. 2020; https://www.sciencedirect.com/journal/sleep-medicine/special-issue/10B5LCK8972.

[28]   Giorgi A. Polysomnography. 2020; https://www.healthline.com/health/polysomnography.

[29]   Alton L. How AI will help you sleep better at night. VB The Machine. 2017; https://venturebeat.com/2017/07/09/how-ai-will-help-you-sleep-better-at-night/.

[30]   What's your sleep score? https://www.resmed.com.au/sleep-score.

[31]   https://www.sleepon.us/.

[32] Next: Wearable neurotechnologies to improve sleep and circadian rhythm regulation. Sharpbrains. 2017; https://sharpbrains.com/blog/2015/09/29/next-wearable-neurotechnologies-to-improve-sleep-and-circadian-rhythm-regulation/.

[33] https://restperformance.com/.

[34] https://www.nanit.com/global/.

[35] https://www.sleepnumber.com/pages/360.

[36] https://www.prnewswire.com/news-releases/heka-launches-the-worlds-first-ai-mattress-which-can-improve-sleep-quality-through-autonomously-adapting-to-individual-body-shapes-and-postures-in-real-time-300606732.html.

[37] https://oceans.tokyo.jp/health/2020-0412-7/.

[38] https://www.excite.co.jp/news/article/Oceans_404087/

[39] Wu H, Kato T, Yamada T, Numao M, Fukui K. Personal sleep pattern visualization via clustering on sound data. AAAI Workshop on Health Intelligence. 2017. https://www.research gate.net/publication/313366888_Personal_Sleep_Pattern_Visualization_via_Clustering_on_ Sound_Data.

[40] Bartlett ET. Differences between death and dying. Journal of Medical Ethics. 1995, 21, 270–6.

[41] Hales S, Zimmermann C, Rodin G. The quality of dying and death. Archives of Internal Medicine. 2008, 168, 912–8.

[42] Patrick DL, Engleberg RA, Curtis JR. Evaluating the quality of dying and death. Journal of Pain and Symptom Management. 2001, 22, 717–26.

[43] Kim M, Cho C, Lee C. A Concept. Analysis of Quality of Dying and Death (QODD) for non-cancer patients. Asian Journal of Human Services. 2015, 9, 96–106.

[44] Bennington-Castro J. AI can predict when we'll die – here's why that's a good thing. MACH. 2018; https://www.nbcnews.com/mach/science/ai-can-predict-when-we-ll-die-here-s-why-ncna844276.

[45] Bay S. How AI could improve the quality of end-of-life care, 2018; https://venturebeat.com/ 2018/06/29/how-ai-could-improve-the-quality-of-end-of-life-care/.

[46] Rajkomar A, Oren E, Chen K, Dai AM, Hajaj N, Hardt M, Liu PJ, Liu X, Marcus L, Sun M, Sundberg P, Yee H, Zhang K, Zhang Y, Flores G, Duggan GE, Irvine J, Le Q, Litsch K, Mossin A, Tansuwan J, Wang D, Wexler J, Wilson J, Ludwig D, Volchenboum SL, Chou K, Pearson M, Madabushi S, Shah NH, Butte AJ, Howell MD, Cui C, Corrado GS, Dean J. Scalable and accurate deep learning with electronic health records. npj Digital Medicine. 2018, 1, doi: 10.1038/s41746-018-0029-1.

[47] Erard M. How a doctor and a linguist are using AI to better talk to dying patients. 2019; https://qz.com/1700778/a-doctor-and-a-linguist-are-using-ai-to-improve-palliative-care/.

[48] Weng SF, Vaz L, Qureshi N, Kai J. Prediction of premature all-cause mortality: A prospective general population cohort study comparing machine-learning and standard epidemiological approaches. PLos One. 2019, 14. doi: 10.1371/journal.pone.0214365. eCollection 2019.

[49] FDA. Make a difference. Submit your comments and let your voice be heard. 2019; https:// beta.regulations.gov/.

[50] Kalaichandran A. AI Could Predict Death. But What If the Algorithm Is Biased? 2019; https:// www.wired.com/story/ai-bias-predict-death/.

[51] Green CR, Anderson KO, Baker TA, Campbell LC, Decker S, Fillingim RB, Kalauokalani DA, Lasch KE, Myers C, Tait RC, Todd KH, Vallerand AH. The unequal burden of pain: confronting racial and ethnic disparities in pain. Pain Medicine. 2003, 4, 277–94.

[52] James SA. The strangest of all encounters: racial and ethnic discrimination in US health care. Cadernos de saude publica, 2017, 33. doi: 10.1590/0102-311X00104416.

[53] Robbins R, Brodwin E. An invisible hand: Patients aren't being told about the AI systems advising their care. 2020; https://www.statnews.com/2020/07/15/artificial-intelligence-patient-consent-hospitals/.

[54] Lindvall C, Cassel CK, Pantilat SZ, DeCamp M. Ethical considerations In the use of AI mortality predictions in the care of people with serious illness. Health Affairs. 2020. https://www.healthaffairs.org/do/10.1377/hblog20200911.401376/full/.

Chapter 7
AI in food industry

The phrase "you are what you eat" is a proverbial saying and the notion that, to be fit and healthy, you need to eat good food [1–3]. However, there should be more profound meaning behind this simple phrase. It might ask your diet habit and philosophy. In this chapter, we will be discussing food industry, biodiversity, protein, agriculture industry, animal husbandry industry, aquaculture industry, food delivery and safety, functional food, and food waste management. All should be related to this phrase to some extent and affected/supported by AI-powered technology.

7.1 Food industry and biodiversity

7.1.1 Food industry

The population of the earth is increasing. Humans keep up with the growing demands of food availability by exploiting plan and animal resources through agriculture, forestry, and fishing. Growing crops, raising, and breeding animals, and harvesting timber and other plants, animals or animal products are necessary for the human race. For answering the global demands on food, some argued that automation, machinery, and mass production are the keys to sustaining a growing population; while others have argued that dealing with the complexity of food-related data requires advanced data science and computer science techniques such as AI and ML [4–7]. For achieving zero hunger and ensuring stable food supply along with food processing automation, segmentation of food items and improving the supply chain [8, 9], there are many companies and organizations that already incorporate AI/ML and AI/DL into food and beverage products and services.

This "multi-tool" will be driven by a combination of advanced software, robotics, and testing capabilities, creating a food safety system that is entirely connected, driven by data, and powerfully accurate [10]. The implementation of AI in the food industry has been growing in the past few years. New technologies involving the use of any branch of AI such as ML, DL, and ANN have changed and improved the different food science areas, such as the production process, quality assessment, and methods for understanding consumers' acceptability. The application of AI in the food industry has led to the development of more reliable, objective, cost-effective, nondestructive, and less time-consuming techniques compared to traditional methods available to the industry [11, 12].

When talking about the food industry, technology is not usually the first thing that comes to mind. But nowadays, technology in the food industry is an essential part of food production and delivery processes. Technology could significantly

https://doi.org/10.1515/9783110717853-007

improve packaging, increasing shelf life and food safety. The quality of food is also improving while production costs are lower. Robotics, machines, drones, and 3D printing are the reality. We already know that among the investments in AI technology, there are significant investments in the food manufacturing sector. For example, AI can more easily predict many issues in agriculture than people can. The benefits of AI in the food industry are: (1) recently, more and more companies are trusting AI improve supply chain management thorough logistics and predictive analytics as well as to add transparency, (2) digitization of the supply chain ultimately drives revenue and provides a better understanding of the situation; AI can analyze enormous amounts of data that are beyond human capability, (3) AI helps businesses to reduce time to market and better deal with uncertainties, and (4) automated sorting will definitely reduce labor costs, increase the speed of the process, and improve the quality of yields [13]. The implementation of AI/ML in food manufacturing and restaurant businesses is already moving the industry to a new level, enabling fewer human errors and less waste of abundant products; lowering costs for storage/delivery and transportation; and creating happier customers, quicker service, voice searching, and more personalized orders. Robotics is still quite a subtle thing to introduce, even for big factories and restaurant businesses, but it will occupy its niche very soon, bringing an obvious benefit in the long run [13]. AI is catching the attention of businesses across many disciplines and sectors with food processing and handling, including following significant applications; sorting products and packages, developing products, food safety compliance, improving the supply chain, ensuring personal hygiene, maintaining cleanliness and cleaning processing equipment, and helping customers with decision-making [14–17].

7.1.2 Biodiversity

Much of the extinction of species is estimated to have been caused by human activity, especially the destruction of animal and plant habitats. It is caused at a high rate of extinction by human consumption of organic resources, especially rainforest destruction [18]. Biodiversity is an essential element of life, the very fabric of the natural capital [19]. The causes of biodiversity loss are population explosions, deforestation, pollution (air pollution, water pollution, soil pollution) caused by human activities, and global warming and climate change. These factors, cumulatively, cause critical damages on the biodiversity. Biodiversity is important as a source of resources that create products with economic value (food, medicine, cosmetics, etc.). This concept of biological resource management is associated with the fear of resource loss due to the decline of biodiversity [20]. In the sense of diverse organisms, biodiversity (biological resources) benefits in food, which include the land animals used for food include vertebrates and insects (insect food); for marine life, a variety of species, including fish, are used for food (e.g., crustaceans, mollusks, algae); other, as a land organism to be eaten, seed plants such as cereals and vegetables, fern plants, and mushrooms (fungi) and the like; further, some of the fungi (yeast, koji mold) and some of the true bacteria (acetic acid bacteria, lactic acid bacteria, natto bacteria) and the like are used in the production of fermented foods [20]. All these diversities help in maintaining the correct balance of nature. But, gradually over the years, there has been a major loss in the biodiversity across the globe. The loss of biodiversity could adversely affect our environment as the balance is lost and the natural food network is disturbed. We must take necessary actions to maintain all the three diversities. Without the proper conservation of this diversity, we could end up in different precarious situations [21, 22].

Every species of plants and animals is important. The existence of both plants and animals has always been vital to humans. Extinction of organisms disrupts the food chain and hence affects the ecosystem. Therefore, humans are very much dependent on plants and animals for survival. Hence precautionary measures should be taken to maintain the stability of the ecosystem. The traditional biodiversity conservation methods have not shown much impact lately. Thus, the use of technology such as ML or AI for biodiversity conservation can help prevent further extinction of plants and animals [23].

7.2 Agriculture industry

There are, in general, two types of agriculture industry: an outdoor field agriculture and an indoor vegetable factory.

7.2.1 Outdoor field agriculture

Agriculture is seeing rapid adoption of AI/ML, both in terms of agricultural products and in-field farming techniques. Agriculture is both a major industry and foundation of the economy. Factors such as climate change, population growth, and food security concerns have propelled the industry into seeking more innovative approaches to protecting and improving crop yield [24–26]. As a result, AI is steadily emerging as part of the industry's technological evolution. Major developments include (1) agricultural robots to handle essential agricultural tasks such as harvesting crops at a higher volume and faster pace than human laborers by weed control (which is a top priority for farmers and an ongoing challenge as herbicide resistance becomes more commonplace) or crop harvesting (for which an automation is emerging in an effort to help address challenges in the labor force), (2) crop/soil monitoring via leveraging computer vision and DL algorithms to process data captured by drones and/or software-based technology to monitor crop and soil health, and (3) predictive analytics (for which ML models are being developed to track and predict various environmental impacts on crop yield such as weather changes) by using satellites for weather prediction, monitoring crop health and its sustainability [24]. Weather tracking and forecasting are important applications of AI in agriculture as it facilitates gathering up-to-date information of prevailing weather conditions such as temperature, rain, wind speed and direction, and solar radiation. According to a research study, 90% of crop losses are due to weather events and 25% of these losses could be prevented by using predictive weather modelling. Weather tracking and forecasting are important applications of AI in agriculture as it facilitates gathering up-to-date information of prevailing weather conditions such as temperature, rain, wind speed and direction, and solar radiation [27]. The amount of data that can potentially be captured by technologies such as drones, and satellites, on a daily basis, will give AI-enhanced agricultural business a new ability to predict changes and identify opportunities. It is fairly predicted that satellite machine vision applications (for weather, crop health, predicting crop yield, etc.) will become more and more commonplace for large industrial farms [27–29]. It was mentioned that cognitive computing, in particular, is all set to become the most disruptive technology in agriculture services as it can understand, learn, and respond to different situations (based on learning) to increase efficiency. The major factors driving the growth of the AI in agriculture market include: (i) the growing demand for agricultural production owing to the increasing population, (ii) rising adoption of information management systems and new advanced technologies for improving crop productivity, (iii) increasing crop productivity by implementing DL techniques, and (iv) growing initiatives by worldwide governments supporting the adoption of modern agricultural techniques [27].

Today, a majority of startups in agriculture are adapting AI-enabled approach to increase the efficiency of agricultural production. The use of AI in agriculture helps the farmers to understand the data insights such as temperature, precipitation, wind speed, and solar radiation. The data analysis of historic values offers a better comparison of the desired outcomes. The best part of implementing AI in agriculture that it should not eliminate the jobs of human farmers rather it will improve their processes. In summary, it can be stated that (1) AI provides more efficient ways to produce, harvest and sell essential crops, (2) AI implementation emphasis on checking defective crops and improving the potential for healthy crop production, (3) the growth in AI technology has strengthened agriculture-based businesses to run more efficiently, (4) AI is being used in applications such as automated machine adjustments for weather forecasting and disease or pest identification, (5) AI can improve crop management practices thus, helping many tech businesses invest in algorithms that are becoming useful in agriculture, and (6) AI solutions have the potential to solve the challenges farmers face such as climate variation, an infestation of pests and weeds that reduces yields [30].

7.2.2 Indoor vegetable factory

According to a forecast released in 2009 by the Food and Agriculture Organization of the United Nations (FAO), the world population will increase to 9.1 billion by 2050 and food production will need to be increased by 1.7 times [31]. It is difficult to simply cultivate 1.7 times more arable land to produce food, and innovation in food production technology is needed to support 9.1 billion people. In the manufacturing industry, productivity can be increased if products can be made in a short period of time using the same equipment. In plant production, since productivity is the speed of growth, there are five essential elements for plant to grow healthy, including electricity, seeds, carbon dioxide, fertilizer, and water [32]. The farmland which can be developed reaches its limit, and fertilizer and agricultural chemicals cannot be used any more. Accordingly, there may be the remaining way to increase the yield by optimizing the cultivation by AI technology. In addition, in terms of not being affected by climate change such as abnormal weather, cultivation at plant factories is likely to open up the possibility of stable supply. It will be built in many parts of the world in place of farmland development. Surrounded by challenges such as population growth and climate change, the image of new agriculture (like a computer-integrated plant factory) that solves them through technology seems to be realistic [33, 34].

There are two types of plant factory. One is an open-type plant factory that controlled the temperature with an air conditioner and it is not possible to control the environment accurately because the temperature difference was 5 degrees depending on the place, and the temperature of tap water varied according to the season. The other is a closed-type plant production machine in which, an engine is installed to control individually twenty parameters (temperature, wind speed, CO_2 concentration, etc.) in three factors (light, air, and feed) that greatly affect plant growth. In the closed-type plant production machine is sealed each layer where the plant grows, the plant environment can be controlled (water temperature ± 0.2 °C, feed rate ± 0.1 L/min, humidity $\pm 2\%$) by the optical supply system, air circulation system and nourishing solution feeding/circulation system [32]. In addition, since the growth weight is also monitored in real time, the progressive growth speed is obviously controlled and detected. AI has no common sense, so it derives environmental conditions that do not exist naturally. It becomes possible to grow vegetables and fruits with excellent flavor with a small amount of water consumption by controlling the plant factory by AI technology developed with the idea of not being bound by existing botany controls the plant factory. In this unique vertical farming method, leafy vegetables and fruits are grown in a literally vertical pillar-like cultivation device. A small amount of water plus minerals and nutrients (without soil) is circulated in the cultivation apparatus to hydroponically grow vegetables. Because water is recycled in factories, water is used only one-twentieth more than growing vegetables on outdoor farms [35–37]. Existing botany is based on the assumption that it is hard to control the environment parameters (such as temperature and water volume); while the indoor plant factory can control all environmental parameters by AI technology. Typical views inside the vertical indoor vegetable factory are shown in Figure 7.1 (a) [35] and (b) [36].

(a)   (b)

Figure 7.1: Typical inside views of vertical vegetable plant factory [35, 36].

Since in the plant factory, factories that use artificial lights are employed, factories are not considered, rather industry. The biggest advantage of the plant factory is the all-year-round production. To achieve this, the biggest mission is to artificially manage natural conditions and harvest the yield of the necessary agricultural products. Plants grow by photosynthesis, which connects the sun's light energy with carbon dioxide. With regard to the most important "light" for its growth, there are two types of light sources in plant factories [38]. The first type is a solar-powered type using natural sunlight. In this solar-powered type, the temperature and humidity of the room fluctuate slightly, so precise control by ICT device is essential. The second one is a completely artificial light type, normally using fluorescent lamps and LED (light-emitting diode) lights. Since the temperature and humidity are almost constant, the inside environment is stable. Parameters which are required to be controlled should include the light intensity, the wavelength, and carbon dioxide gas, indicating much easier environmental management [38, 39].

There are merits and demerits associated with this indoor plant factory. As benefits, (1) vegetables and fruits can be produced in a predetermined plan, (2) since there are few factors to control, it is easy to produce products without experience and knowledge, (3) nutrients and functionality can be easily managed, and (4) processing/running costs are relatively low. On the other hand, there are still many problems that have not been solved by plant factories, including (1) a lump sum of money as initial cost is needed, (2) know-how and equipment technology are not keeping up with anything other than lettuce, and (3) running costs are still high [38, 40]. As to future aspects of the indoor plant factory, several interesting points have been addressed [41, 42]. AI algorithms should become entirely self-learning and be able to adapt to varying climate conditions, greenhouse types, crops and crop varieties. Sensor technology can be advanced for gaining more digital information, especially regarding crop development. Root AI (a mobile robot) operating in indoor farming facilities can pick up products (such as tomatoes) and is able to look at crops and assess their health and conduct simple operations like pruning vines and observing and controlling ripening profiles so that the robot can cultivate crops (initially tomatoes) continuously and more effectively than people. Drones can be fully operated to gather environmental data (instead of sending information detected by sensors) as well as progressive growing conditions [41, 42].

7.3 Animal husbandry

Animal farming is becoming a data-centric business. AI in animal husbandry is used for raising animals for meat, fiber, milk, eggs and other products. With AI, providing day-to-day care and raising livestock has become easier for animal farmers. For example, farmers are making use of wearable AI devices to collect real-time data about them to make necessary decisions. The wearables are helping farmers to get important alerts like when their animals are sick, when they should be vaccinated and when they are ready for insemination. There are numerous ways that AI and ML algorithms are now used to benefit the animal husbandry industry [43]. With the enormous growth in the world population, the farmers are switching to smarter techniques that can aid in regulating the proper use of land, water, and energy to feed the planet and evade the global food crisis. Researchers believe that the answer lies in sensors, robots, and AI. AI is one such technology which needs immediate implementation in the livestock industry. AI will help livestock farms to accumulate and analyze data to accurately predict consumer behavior, like buying patterns and leading trends. With increased investments, farms will be enabled to automate processes, reduce major costs, and improve the quality of livestock products like milk. A technique for monitoring the health of farm animals/dairy cattle with a high degree of accuracy uses a camera and AI to achieve a "smart" cowhouse. Detailed observation by AI-powered image analysis could enable early detection of injuries and illnesses that could impact the quantity and quality of milk production. As AI/ML becomes more common and easily available, it was anticipated that the use of such technology in the dairy industry will automate most of the farm processes while at the same time produce information based on the farm's operational history [44].

The use of AI has been of enormous economic benefit for dairy farmers in many countries through the improvement of their stock. Affordable tools with the ability to continuously monitor the growth rate of livestock animals are highly sought after by the livestock industries. This demand is driven by the potential for these tools to assist in improving animal welfare and production efficiency. Livestock growth is a fundamental measure which can be used for diagnostic purposes in these areas, therefore it is very important to develop a system to automatically determine the growth of individual and groups of livestock animals (e.g., pigs) using welfare friendly and non-invasive methods. A machine vision system was selected to undertake this weight estimation task, whereby pigs' body measurements are extracted from images and used to estimate their weight without physical interference. It was demonstrated that the average weight of groups of pigs can be calculated with sufficient practical accuracy. The precision of the system was also favorable compared to a commercially available system. Therefore, the developed system can be used for practical purposes on commercial farms to determine the average weight and growth of groups of pigs [45]. Ever since man began domesticating animals several thousand years ago, we have always relied on our intuition, collective knowledge, and sensory signals to make effective animal production decisions. So far, this has helped us make significant gains in animal husbandry and farming. Together the growing demand for food and the advancement in sensing technology have the potential to make animal farming more centralized, large scale and efficient. The sensors, big data, and AI/ML are helping animal farmers to lower production costs, increase efficiencies, enhance animal welfare and grow more animals per hectare [46]. Big data plays a key role in applying advanced technologies to animal farming practices and offers a scalable solution to store vast amounts of data on a remote server. Advanced AI and ML algorithms can make use of this extensive data to analyze, predict and notify farmers in case there is something abnormal in a sequential flow from data detected by sensors (to capture) → bid data files (to process) → ML (to analyze) → algorithm (to notify), providing a complete solution [46].

Identifying, predicting, and preventing animal diseases are very crucial in livestock farming. Sensors, big data, and ML algorithms have been employed to successfully diagnose the early onset of several diseases that affect pigs and sheep based on lethargic body movements, slower response times, and decreased activity before the onset of other noticeable disease symptoms [47, 48]. Today, sensors, big data, and ML algorithms have a significant cost advantage over these older detection methods, as illustrated in Figure 7.2 [46].

Identifying a particular animal using facial recognition system among a herd or flock is an important task; particularly in terms of improving animal health outcomes while managing groups of farm animals, which has always been a challenge, especially for large-scale animal farmers. It was also mentioned that applying the concept of the AI-assisted emotional contagion [49] will help animals adopt a quick response to deal efficiently with their surroundings. It helps them collectively move toward something they want or away from danger [50]. Harmonizing the emotional behaviors of individual animals can help other farm-animals develop more empathy

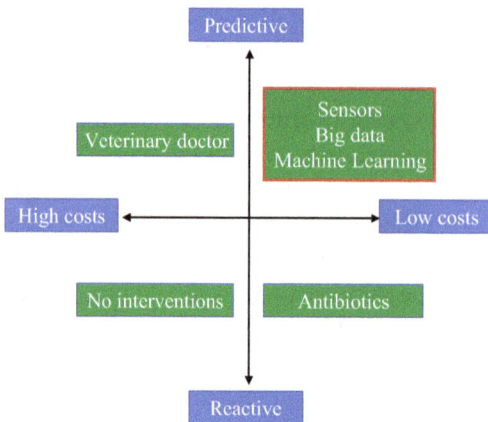

Figure 7.2: Difference between predictive and reactive paradigms of managing diseases among animals [46].

and other desirable traits [46]. This may result in the entire herd developing strong social bonds and improved group coordination [51].

Taking a poultry farming as an example, Connolly [52] pointed out several important points in AI-assisted farming processes.

(1) Three-dimensional printing prosthetics: One of the more inventive ways in which 3D printing can affect the poultry industry is through life-saving techniques. Reproducing feet, legs and even beaks has already been applied for pet birds.

(2) Robots doing the dirty work: One of the most practical applications of digital technology in the poultry industry is that of robots. There are a multitudinous number of repetitive tasks that robots could assist with. Poultry houses require nearly constant attention – cleaning and sanitizing, collecting eggs, and checking birds. This is time-consuming, monotonous work, but it would not bother a robot. Additionally, robots are more precise, thorough, and honest about the work they do compared to their human counterparts.

(3) Drones: The opportunity for drones in chicken houses may seem a little farfetched. There is concern that the drone could make the flock nervous and cause undue stress. Free-range or yard-kept chickens and turkeys that roam fields freely would be a better application for drone technology, which could herd, protect, and monitor them. Adaptation of avian species to drones would probably require training but will most likely succeed outdoors.

(4) Sensors: Sensors probably represent the easiest implement, due to not only lower implementation costs, but also because the benefits are immediately recognized [53]. From a wearable sensor perspective, researchers – and even farmers – could gain a lot of insight into the health and well-being of broilers, layers, turkeys and ducks.

(5) AI: AI technologies have become the backbone of many other technologies. Robots, for example, use AI in the processing plant to improve efficiencies. Automating a procedure such as chicken deboning requires recognition of the shape and size of each chicken and individual adaptation. AI is the perfect technology for this application. Consider that a computer can analyze the difference in density and structure of meat versus bone, thereby making the most precise cut possible. This is a great example of combined technologies: robots perform the work that AI instructs them to do based on the data that sensors collect. This technology has allowed a significant challenge within the layer industry to be overcome. Layers, of course, are designed to produce eggs for consumption. To replace laying hens, farmers have to incubate some eggs, but they cannot tell until they are hatched which are male and which are female. The ability to sex the eggs is crucial and using terahertz spectroscopy, the system can identify male eggs immediately after laying and sell them as unfertilized eggs for the farmer, allowing for significant cost savings within the layer industry.

(6) Augmented reality (AR): augmented (or enhanced) reality is the ability to see things that the human eye cannot, using the non-visible spectrums of light, or to overlay information, including data interpretation, alongside what the person sees. The possible uses of the technology are wide-ranging, but so far, there are a few examples of real commercial applications. The application of AR allows trimmers in factories to see how to cut the chicken carcass and accurately remove defective parts of the meat.

(7) Virtual reality (VR): The most obvious application for VR in the poultry industry is training, particularly processing. It could teach line workers in the processing plant the ideal way to trim meat from birds. Applied to free-range layer houses, it could teach employees how to walk through the house without frightening the birds, find errant eggs, and check on hens.

(8) Blockchain (a method of synchronizing and recording transaction information and other data by combining cryptographic techniques such as public key cryptography on many computers that make up a decentralized network): Blockchain's opportunity in the poultry industry is its ability to resolve food safety and transparency issues. Blockchain can secure digital records and monitor supply chain management, ensuring traceability of the poultry products sold in stores. Blockchain can be used to monitor all aspects of the food supply chain, from farmers and producers to processors and distributors.

(9) IoT: IoT connects many of the sensors in a hen house to a smartphone, and it offers technologies for improved efficiencies within the poultry sector through the use of combined sensor applications using cloud-based technologies and smart farming applications.

(10) The big data advantage: As we can collect more information on animals, including the bacteria in their digestive tract and how they respond to nutrition at the gene level, it becomes clear that farmers are learning how to manage vast amounts

of data as much as they previously understood how to manage their animals. It is estimated that the world poultry production will increase 120% from 2010 to 2050. In order to meet this demand, feed conversion ratios and other production efficiencies must continue to improve. The incorporation of digital technologies, such as those listed above, will greatly aid in these efficiencies and help poultry producers to rise to the demands and meet the increasing needs of a global population [52].

7.3.1 Livestock industry

In the recent times, with requirements of the better yield of farm animals, AI has emerged as a tool that empowers farmers in monitoring, forecasting, as well as optimizing the farm animal growth. Tackling parasites, biosecurity, and diseases, monitoring farm animal along with farm management are some of the thrust areas in livestock industry where the use of AI technology can pay rich dividends. At present AI is being used in: (1) health monitoring, (2) preventing disease outbreaks and optimize nutrition, (3) robotic system to deliver vaccines, and (4) detection of estrus [43, 54–57],].

7.3.2 Dairy industry

Healthy and productive cows are something that is obviously very crucial for the success of a dairy farm. From collar that is applied the cow, the behaviors of the cow can be detected. As a result, it can be known whether the cow is standing or laying, eating, rumination, drinking, walking, all of the standard behaviors of the cow. The collar can provide insights on the health, heat stress, feeding, efficiency, and estrus of the cow. The data is then wirelessly transmitted back to a cloud using these antennas where it can be accessed on a computer and even on a cellphone. In this situation, it is grabbing all these types of data from cows 24 h a day and it is able to synthesize that data into actual insights for farmers [58]. The feed is the most important factor that affects the production level of a cow. AI systems can provide accurate monitoring of the amount of feed that is provided to the cow and help to increase the production level. For example, there is an application that uses a motion-sensing device to transmit the movement of the cow to an AI-driven system. The sensor data, when aligned with real-world behavior, can help the AI system detect when the cow is walking, drinking, or eating. Small dairy barns can be easily taken care of, but when it comes to huge barns, it becomes impossible to keep up with every cow on an individual basis. With facial recognition, AI can help identify each cow uniquely. Unique identification of cows helps farmers provide better treatment to the cows [43].

7.3.3 Poultry industry

AI can help address many challenges in poultry industry. Human interference in farm can be a source of infection plus in many processes machines and robotics can be much more efficient as compared to Humans. AI can very well reduce the error, improve efficiency of farming, and maximize profit of a farm. We will cover one by one how AI can create an impact in the poultry industry. Promising advancements in poultry industry by AI technology can include as follows [59].

(1) Farm management: For example, robots can be programmed to collect information like temperature, humidity, ammonia levels inside the farm, this information can be further processed and results can be produced. These results will now allow the machines to take decision related to ventilation by itself. This process which is repetitive yet very important for the efficiency and profitability of the farm can be managed efficiently by data analysis and ML. Like humans, even animals suffer from nutritional deficiencies. AI machines can help identify the decreased growth of a chicken. The machines can be trained to differentiate between healthy and infected chickens. AI-enabled robots can help poultry farmers in many ways. Robots can do repetitive work like feeding birds, collecting eggs and removing manure. Tasks like collecting, counting, and packing eggs are becoming completely automated, reducing the need for close supervision by humans. Another task that a robot can perform is shifting a hatched chick from a broiler shed to the layer shed. Robots can also keep the birds moving for an added health benefit. Thus, a robot can perform various duties for poultry farming and prove to be a cost-saving attribute to the farmers. AI systems can monitor the environment of a shed and adjust conditions accordingly. AI systems can determine the accuracy of fertility in the early stage of incubation. AI can first learn which eggs are fertile and which are not by scanning the eggs, and then algorithms can be created that can determine the accuracy of fertility [43].

(2) Disease management: Disease management is most essential part of farming practices and managing diseases (such as bird flu or avian diseases [60]) through machines can be a very complex process knowing the variation in symptoms and possibility of unknown number of diseases. AI, however, can simplify the process in near future by assisting in diagnosis [61]. The infected birds may have symptoms like nasal and saliva secretions apart from high temperature (influenza). Pictures or visuals sent from more precise drone in free range and 3D models in control houses are fed to specially built symptoms detector analyzer to recognize the disease at an early stage as the infection is contagious and leads to reduced production. It also helps farmers in saving the life of chicks as clogging blocks the nose of the bird leading to death. By annotating the secretions on the beak of chicken, devices are trained to identify the difference between a healthy bird and influenced bird. Tools help to differentiate precisely the target shape and texture of the secretion [62, 63].

(3) Trials and evaluation of nutrition and medicinal products/genetics studies: AI can be very efficient in collecting data and processing it while conducting trials or comparative studies of various products. Through machines and sensors large number of data points can be monitored at single point of time which will take a lot of time for humans to evaluate. AI will not only help to collect data but also to evaluate results through analytics and provide with comparative results with statistical analysis. Similarly impact of different kind of feed formulations can be evaluated at a pace which is impossible for humans to perform. Programming and robotics can also help in improving genetics of the breed. It can help in identifying breed characteristics and ease off the decision-making in selecting right breed characteristics while studying genetics. These processes can reduce the trial cost significantly and provide with accurate trial results. Research and development is one of the costliest affairs in any industry and AI can help to manage that cost efficiently and also allow company to do multiple studies in single trial that would otherwise take multiple efforts when performed by humans.

(4) Detect behavioral diseases like cannibalism (or aggressive pecking): Pecking behavior often leads to the plumage of laying hens and this can lead to health and welfare conditions of the flock. Death occurs within 10 min of pecking, so early detection becomes the need of the hour. Building machine which gives alerts to the farmers regarding cannibalization within the timeframe of destruction will enhance the productivity of the poultry. By using box annotating tool, machines can be trained to detect the behavior of feather pecking in birds and its severity. It also enables to identify the injured birds [60, 62, 63].

(5) Post farm activities: This is one area where AI is already been used in many developed countries and have created huge impact in efficiency of poultry processing industry. Technology like artificial vision / machine vision are evaluated by meat processing companies for sorting of broiler parts, for identifying diseased carcass. Machines can easily differentiate between density of bones and muscle mass; thus, processes like deboning are perfect for AI to operate. Layer farmers are using AI for grading of egg and to identify quality eggs at much higher speed. Machine vision when combined with smart automation has allowed processing industry to improve the quality of carcass and packaging of broiler meat. Many successful studies have been done by using spectral-line imaging system for presorting application of broiler chicken with high accuracy. AI can positively impact whole industry to improve efficiency of process and make chicken and egg affordable for large proportion of population.

(6) Miscellaneous uses: AI is a vast technology and has multiple uses which can directly or indirectly impact poultry industry. Technologies like smart automation and machine vision can be used by nutrition companies, feed manufacturers, and vaccine and medicine manufacturers to improve accuracy, efficiency, and further enhance the quality control in their manufacturing facilities. It can help overcome challenges of data collection and analytics faced by large integrators

due to scattered distribution of farms across geographies. AI will enable them to fetch analyzed data from multiple farms located at different location and also allow them to monitor them from distance. This will help them to better control the complete production system be it breeder farms, layer farms, or broiler farms. AI will also help to propagate any new technology at a much faster pace as demonstration of technology will become much easier, and convincing new stakeholders to adapt technology will take less time [59].

Drones can be used in poultry industry. Although there is concern that these drones could make flock nervous and impart stress, yet it can be used for free-range or yard farms where poultry birds can roam freely [60]. These drones work as nannies for these non-descript birds (large scale). Regular monitoring of aviculture with the help of drones is helpful for security of flock. Drones take pictures of the flock throughout the day at regular interval and send it immediately to the analyzing system that builds the database for monitoring the flock [62]. In the modern egg industry, robots are commonplace. Among other tasks, robots feed the birds, transport, handle and pack eggs, and manage shed ventilation. In the chicken meat processing sector, robots perform tasks such as automatic transfer of carcasses and detection of defective carcasses. The utilization of robots is expected to increase in the future as other capabilities are developed that improve the ability to remotely monitor birds. The most advanced robotic systems may incorporate some form of AI, aided by computer vision capability. Rapid advances are occurring in the application of robotics to the poultry industry. Robotics has contributed to improved production efficiency, particularly from the perspective of reducing labor costs per bird. Due to the intensive nature of most poultry enterprises, the industry is ideally suited to continue to benefit from the incorporation of robotics. As I see it, a major challenge for the future is how to apply robotics to monitor bird health and welfare [64].

7.3.4 Gibier

Gibier is a French word for the meat of natural wild birds and beasts (such as deer and boar) obtained through hunting, and it is a food culture that has developed since ancient times as a traditional dish of aristocrats in Europe. Recently, the damage to crops caused by wild birds and beasts has become serious. The number of hunters responsible for capture is gradually decreasing due to the aging of the population. Under such circumstances, innovative traps for hunting equipped with IoT sensors are attracting attention. With the latest technology, the day may come when wild boars and deer will be able to make a single effort to the extent that they do not destroy ecosystems. Wild boars and deer commonly used in gibier are also harmful beasts that damage crops, and the promotion of the use of gibier leads to both local resource utilization and problem solving [65, 66]. Gibier-AI is a project that, in cooperation with gibier-related businesses in Tottori Prefecture in Japan, aiming to realize safe and secure distribution of gibier by applying the latest technologies such as AI and IoT, and to expand the use of gibier [67]. New developments in the gibier business include a technology to detect capture with sensors and notify administrators and to develop an AI-based meat quality determination support system with collecting a large amount of data on meat images along with evaluation results.

7.4 Marine products industry

Fish are being depleted faster than they can be generated. Hence to meet desires for fish consumption, there are mainly two types of a fish industry: a natural ocean fish catching and aquaculture (or cultivation/hatchery factory).

7.4.1 Natural ocean fish catching

It is said that fixed net fishing, as one of the sustainable fishing methods, is environmentally friendly from the viewpoint of fishery resource conservation, while it is not possible to increase the catch by the efforts of fishermen. However, there is also a problem that fishermen cannot control the type of fish that can be caught. Global rules stipulate the protection of bluefin tuna resources. The catch of bluefin tuna smaller than 30 kg is restricted as a protection, but it may enter the fixed net in large quantities. It is impossible to choose fishing types on fixed nets, so we are conducting research on how to deal with them [68]. By utilizing AI and IoT, tomorrow's fishing ground and catch can be predicted [69–71]. The new type of the marine products industry should be shifted from the tacit (or experience) knowledge-based to the explicit knowledge-based one. In other words, it should be an evolution from things that rely on an individual's long-standing experience and intuition to things that can be data-converted and shared so that they can be used by computers [72].

There are several AI-assisted technologies to catch fish more effectively and efficiently.

1) Improved fish finder system. Currently, it is being attempted to predict landing data using AI with regard to fixed nets (fixed network AI). Past acoustic data (by fish finder) and landing data (fish name, quantity, unit price, amount, etc., for each net) are learned by AI, and when the current acoustic data is entered, the landing data is predicted, and not only fishery personnel but also distribution companies cooperate and receive the evaluation [72].

2) Regional Meteorological Observation System. This system collects at least temperature, precipitation, wind direction wind speed, and sunshine time. These

data can be interfaced to AI data analyzer to predict possible region of a particular type of fish shoal [73].

3) Earth observation satellite. Fish group visit prediction can be realized by use of the earth observation satellite data [73].

4) Visualization of sea temperature and currents by AI technology: Seamless marine environment information from the coast to the open ocean using a numerical ocean model, such as sea temperature, salinity, and current speed can be predicted at a resolution of 1.2 km mesh up to two weeks in the future. As a result, it is possible to predict sudden tides, red tides, and sudden sea temperature changes that cause significant damage to aquaculture facilities and fixed nets and can be used to prevent damage. The other is "Fishing Field Navi," which provides quasi-real-time data on sea temperature and currents, which are important information when selecting fishing sites [74].

5) Underwater drones: Underwater drone can detect fish group in limited area of ocean [75], as shown in Figure 7.3 [76].

Figure 7.3: Typical underwater drone [76].

7.4.2 Aquaculture

Fisheries provide a significant source of protein for over half of the world's human population, yet the impacts of historical overfishing and climate change challenge the future productivity of the world's oceans. Traditional fisheries management rests on the assumption that the future will look like the past, however, with advances in AI and burgeoning data resources, scientists have new tools for exploring a greater range of future scenarios, including climate change [77, 78]. The use of AI in aquaculture provides actionable insights to optimize the expenses on fish farms. Fish farms provide half of all the fish for human consumption. Free-floating aquapods are used for farming fish. The aquapods can accommodate thousands of fish. However, what happens when the aquapods need repair? To repair the aquapod manually is a time-consuming task. However, robots can complete the task of repairing aquapods in a safer and more cost-effective way. Underwater robots can easily examine and repair the nets of aquapods. Drones can provide applications for aquaculture both above and beneath the water. Monitoring offshore fish farms and inspecting underwater nets for damage and holes can be easily done by drones. Drones can also provide fish stock information and track environmental changes. Sensors can be used in aquaculture to collect data such as oxygen levels, pH, salinity, and pollution level of water. Detection of the hunger level of the fish by sensors can help farmers or even robots to feed them accordingly. Automated recirculation systems can circulate the water according to the information collected by sensors [43]. Marine aquaculture presents an opportunity for increasing seafood production in the face of growing demand for marine protein and limited scope for expanding wild fishery harvests. However, the global capacity for increased aquaculture production from the ocean and the relative productivity potential across countries are unknown. Gentry et al. [79] had mapped the biological production potential for marine aquaculture across the globe using an innovative approach that draws from physiology, allometry and growth theory. Even after applying substantial constraints based on existing ocean uses and limitations, it was found that vast areas in nearly every coastal country are suitable for aquaculture. The development potential far exceeds the space required to meet foreseeable seafood demand; indeed, the current total landings of all wild-capture fisheries could be produced using less than 0.015% of the global ocean area. It was further demonstrated that suitable space is unlikely to limit marine aquaculture development and highlights the role that other factors, such as economics and governance, play in shaping growth trajectories; suggesting that the vast amount of space suitable for marine aquaculture presents an opportunity for countries to develop aquaculture in a way that aligns with their economic, environmental and social objectives [79].

Due to the increase in the world population and the diet diversification, fish production and consumption are increasing year by year. Because the world's fishery resources are already depleted, it is not possible to increase the catch of natural resources, and the aquaculture industry is attracting attention. Under such circumstances, attempts are underway to analyze the conditions in the sea that humans cannot grasp with AI. Understanding changes in the environment and the state of fish is the key to stabilizing the management of the aquaculture industry. However, about 90% of the world's marine fishery resources are caught near or beyond the limit of biologically sustainable levels, and the catch of natural resources cannot be increased any further. Therefore, large-scale aquaculture in the offshore area is attracting attention. Fishing ground for aquaculture is generally located on a coast where the sea is calm, eye-reaching, and easy to manage, but by moving the fishing ground offshore, it is possible to make effective use of the waters that could not be cultivated in the future, and to install larger-scale ginger [80–83]. Many environmental factors (including such as temperature, salinity, weather conditions, tidal wind direction and speed, carbon dioxide, and moon age) are thought to affect the appetite of fish. For efficient aquaculture, these elements were analyzed with AI to accurately determine the feeding that was in the condition of the fish of the day. It is AI's special field to find correlations among vast amounts of data. Feeding, which has been operated based on the "fisher's intuition" until now, is systematized and optimized by AI [83].

Fish weight can be managed in real time from underwater camera footage [84]. In the current aquaculture, the body length and weight measurement of the fish in order to manage the growth state of the fish during cultivation and estimate the appropriate feeding amount and catch amount, or measured by directly ground with a net, a method of measuring by manually plotting the measurement point of the fish to be measured by sending a frame in the pass is the main, it is said that the efficiency of the measurement work by IT is required. The new solution automatically calculates body length and weight using AI/ML from images of farmed fish. The increased demand for fish has put a strain on resources and sustainable practices among fisheries, requiring the innovative use of existing and new technologies. Fortunately, there is great potential to produce this protein source sustainably, particularly through the advent of technology. As mentioned in the livestock industry [52, 53], similar points in AI-assisted aquaculture are described [85–90].

Føre et al. [91] developed a precision fish farming (PFF) to improve accuracy, precision, and repeatability in farming operations, to facilitate more autonomous and continuous biomass/animal monitoring, to provide more reliable decision support, and to reduce dependencies on manual labor and subjective assessments, and thus improve staff safety. To help in defining PFF, it is useful to envision fish farming as several cyclical operational processes realized in four phases where bioresponses in the cage are observed (Observe phase) and interpreted (Interpret phase), resulting in a foundation for making decisions (Decide phase) on which actions to enforce (Act phase) that in turn elicit a bio-response in the fish, as illustrated in Figure 7.4, where the inner cycle represents the present state-of-the-art in industry, with manual actions and monitoring, and experience-based interpretation and decision-making; while the outer cycle illustrates how the introduction of PFF may influence the different phases of the cycle.

According to the figure, first, the farmer observes the fish via direct visual observation or with data acquisition tools such as cameras, the outcome of which is qualitative or quantitative information on the bio-responses of the fish. The farmer then uses primarily subjective experience to interpret this information, yielding a perception of the current state and condition of the fish. These interpretations are then used as a foundation for making decisions concerning farming operations and management, which are then put into action by manually induced actions on the cage. Such decisions may be made based on the estimated present states or expected future states of the system, representing manual versions of the feedback

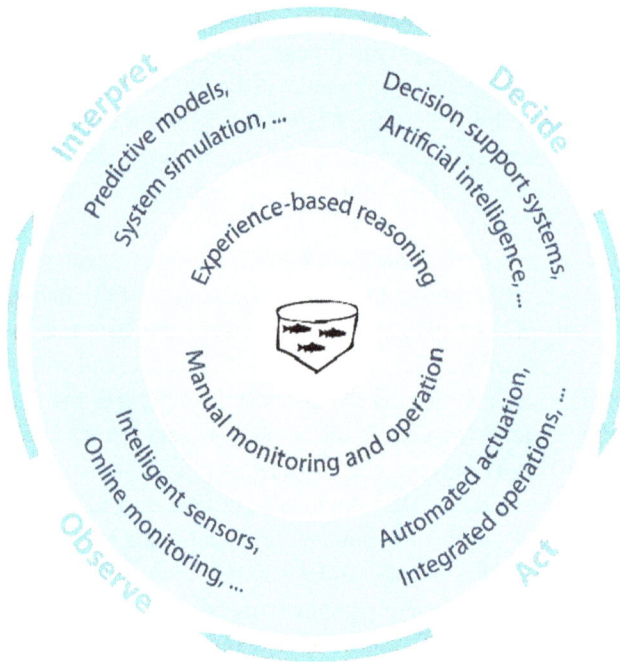

Figure 7.4: A cyclical representation of the precise fish farming (PFF) [91].

and feed-forward principles in control engineering respectively. Methods and tools for fish farming that apply technological solutions and/or automation principles to one, several, or all the different phases of farming operations may be considered PFF approaches. The ultimate result of applying PFF to a particular operation will therefore be that the elements in that operation belonging to the different phases of fish farming operations are shifted from an experience-based to a knowledge-based regime (i.e., by moving from the center toward the outer edge in the figure).

7.5 Supply mechanism and drone delivery

AI application and implementation are quickly realized as in the supply chain management (SCM), including supply chain planning, warehouse management, autonomous vehicles for logistics and shipping, supplier selection and supplier relationship management with ML and predictive analytics, increasing inventory and pricing accuracy for retail, improving demand forecasting for manufacturing, and optimizing distribution network for logistics [92–97].

AI-powered drones have been developed for automated delivery purposes, where users can easily order items for delivery by drone, from an application. The services of the package delivery business expand in remote islands, depopulated areas, mountainous areas, and so on, and deliver daily goods, pharmaceuticals, and so on to areas where land transportation is difficult, and package delivery in regions including cities [98, 99]. Due to an immense and rapid increase in processing power, costs of storage and availability of digital data in the recent years, the utilization of complex AI algorithms have become feasible for drones, and first and solid solutions are already on the market. If AI develops as rapidly as we have seen in the recent years, we will soon find highly automated and comprehensive solutions that will further increase the added value of using drones. But companies must consider that drones and AI only make sense if it saves the user money/time – in some cases, traditional computer vision (in combination with ML and DL) still might be an easier and better solution [100].

In the wake of the pandemic spread of COVID-19 virus, the supply chain (a series of flows from procurement of raw materials to production, delivery, and sales) has been shifted to a supply net (in which effective diversification of manufacturers and suppliers can be formed in a shape of net-formation). We learn from the corona incidence that supply chains that rely on a particular country or region can cause problems not only for companies, but also for consumers [101]. Loten [102] pointed out that global logistics companies are expected to increase their use of digital technology, as manufacturers, retailers, and other businesses reset global supply chains battered by the coronavirus outbreak.

7.6 Functional foods

The definitions of functional foods are numerous, often rather broad and not particularly accurate: (i) the food has been enriched with substances that improve health and prevent a number of diseases, (ii) functional foods are those with an added health benefit beyond traditional nutritional value, (iii) functional foods are defined as any modified food or food ingredient that may provide a health benefit beyond the traditional nutrients it contains, (iv) foods that, by virtue of physiologically active components, provide benefits beyond basic nutrition and may prevent disease or promote health, (v) it is defined as accepted and tolerated experimental foods prepared by fortifying commercially available fruit or vegetable products with GRAS (generally recognized as safe) phytochemicals that are unique in structure, class pattern, or metabolism, (vi) a modified food that claims to improve health or well-being by providing benefit beyond that of the traditional nutrients it contains. Functional foods may include such items as cereals, breads, beverages that are fortified with vitamins, some herbs, and nutraceuticals, (vii) functional foods are foods that have a potentially positive effect on health beyond basic nutrition. Proponents of functional foods say they promote optimal health and help reduce the risk of disease [103–105].

Nutraceuticals are considered to be fortified or enriched foods providing all the essential nutrients required for maintaining the optimal health. Numerous studies have suggested the potential association of nutraceuticals with diet and health, as demonstrated in Figure 7.5 [106]. Gupta et al. [103] reviewed on the classification of nutraceuticals on the basis of the availability of the nutraceutical from the natural sources which is further being processed and proving its medical health benefits in the market and mentioned that the functional nutraceuticals acting as dietary supplements suggested to have the capacity to significantly contribute to the modulation of the complex mechanisms of disease pathology with a view being that they may be essential in optimizing *in vivo* defenses and help in maintaining wellness.

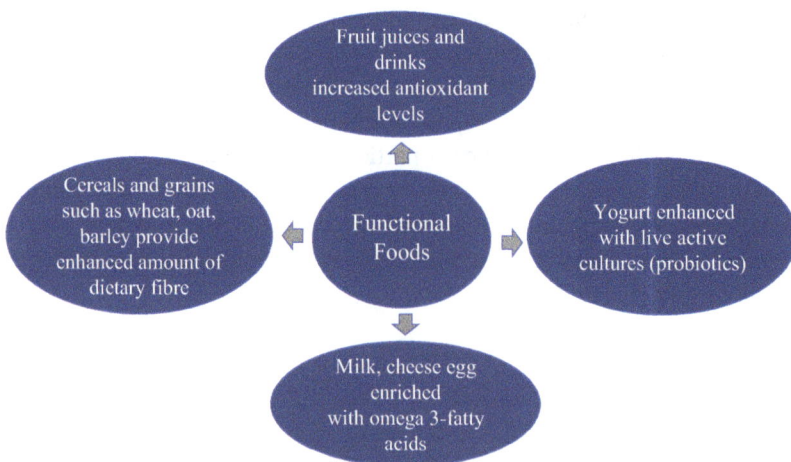

Figure 7.5: Some examples of functional food products exerting the biological activities [106].

There are several works reporting on the AI's contribution on functional food science. Increasing interest in constituents and dietary supplements has created the need for more efficient use of this information in nutrition-related fields. Yordi et al. [107] obtained optimal models to predict the total antioxidant properties of food matrices, using available information on the amount and class of flavonoids present in vegetables. A new dataset using databases that collect the flavonoid content of selected foods has been created. Structural information was obtained using a structural-topological approach called TOPological Sub-Structural Molecular (TOPSMODE). Different AI algorithms were applied, including ML methods. It was demonstrated to show the effectiveness of the models using structural-topological characteristics of dietary flavonoids, based on the proposed models that can be considered, without overfitting, effective in predicting new values of Oxygen Radical Absorption capacity (ORAC), except in the multilayer perceptron (MLP) algorithm. It was concluded that (i) the best optimal model was obtained by the Random Forest (RF) algorithm and (ii) the developed in silico methodology allowed to confirm the effectiveness of the obtained models, by introducing the new structural-topological attributes, as well as selecting those that most influence the class variable [107].

AI technology characterizes in vitro and human efficacy data for a safe immuno-modulatory functional ingredient. The severe impact that chronic and low grade inflammation exerts on human health has been decidedly revealed, emerging as the principal causative factor underlying most chronic diseases; while acute inflammatory responses are vital for the resolution of infection or physical trauma, failure to appropriately resolve inflammation ultimately results in tissue damage, leading to conditions such as psoriasis, arthritis and colitis, increased risk of developing metabolic diseases, and, indeed, cancer [108–111]. In order to functionally assess the effects of interventions designed to counteract the development of frailty and its associated state of inflammation, standardized, relevant testing methods are required. Kennedy et al. [111] utilized AI to identify and validate a natural peptide network (NPN), derived from rice, demonstrating anti-inflammatory activity in vitro. Using an ML approach, rice NPN by identifying a number of constituent bioactive peptides was further characterized. It was reported that (i) rice NPN significantly attenuated TNF-α (tumor necrosis factor) production within 4 weeks of administration, (ii) significant improvements were demonstrated on oral glucose tolerance testing and a decrease in serum low-density lipoprotein with a concomitant increase in high-density lipoprotein, (iii) significant physical gains observed, for example, with the chair stand and short physical performance battery tests and (iv) an anti-inflammatory rice NPN may have significant benefits in age-related inflammation.

The food business recently is doing a deep dive with AI/ML. A fleet of start-ups is hoping to exponentially improve the process of ingredient discovery and development, trying to optimize seed cultivation and livestock output, and improving customization of commercial diets. Established giants are trying to keep pace by using AI/ML in product R&D as well as across the spectrum of operations and relationships [112]. Flavor, the conjunction of taste and smell, is not a sensation that yields easily to analysis. Unlike sights and sounds, which can be captured by cameras and microphones, there is no widespread way to measure flavor. What people experience when they eat has heretofore been largely ineffable and uncomputable [113]. Analytical flavor systems have developed a methodology that allows an AI platform to infer the distribution of consumer preference from the reviews of professional panelists on the gastrograph system to reliably measure flavor [114]. The gastrograph is a smartphone app whose central feature is a wheel with 24 spokes (see Figure 7.6), where each sliver represents a discrete category of sensory experience, such as "meaty," "bitter," or "mouthfeel." Research participants indicate the intensity of their perceptions of a product along these specifications. It also gathers demographic and lifestyle data.

Although there are several application software to create customized eating plans with AI technology, as Buss [112] introduced Cathy Kapica's statement that especially thinking about food, there is an irrational aspect as well as a huge emotional

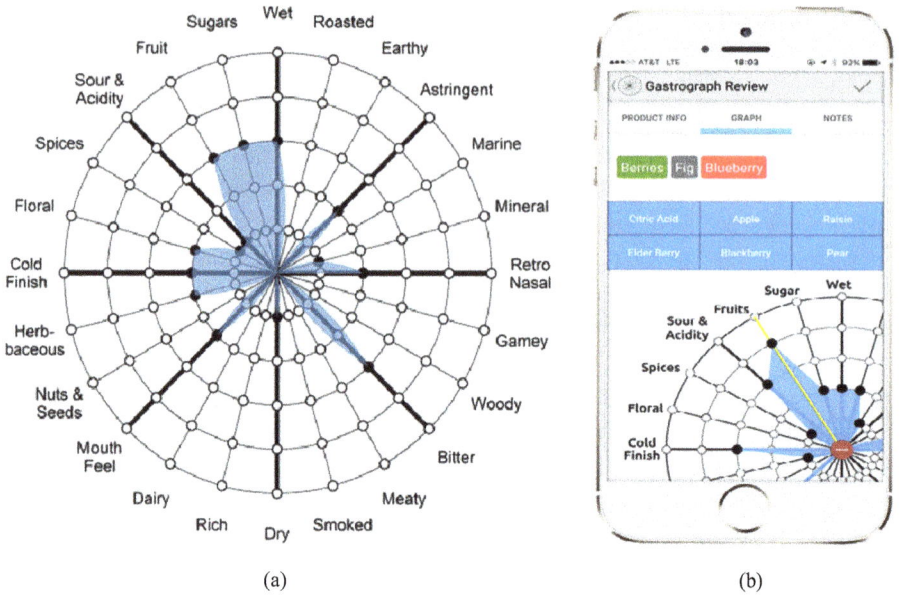

Figure 7.6: Gastrograph with 24 distinct spokes (a) and iPhone interface (b) [112].

component that no algorithm is going to be able to capture, so those things that cannot be got from AI until a fully operative quantum computer available.

7.7 Food safety and waste

Although AI-powered food industry has been well developed, there are still needs for improvements in the traceability of food products (e.g., track-and-trace systems using RFID: radio-frequency identification) tags, QR (quick response) codes, cloud data storage, blockchain) or automate risk factor measurement (e.g., temperature logs), and in testing technology (e.g., loop-mediated isothermal amplification). These improvements are extremely important for modernizing the food safety system. Indeed, as we articulate below, these developments – which require basic investments in data infrastructure, sensing technology, and testing – may be more consequential for the future of food safety than AI technology that generates splashy news coverage [115].

Kovalenko [13] pointed out that one of the most valid points of using AI and robots in food production is that the robots are sterile. This significant benefit is a huge factor in lowering the number of foodborne diseases. Producing foods effectively and efficiently under controlled AI-powered technologies is one great promising answer to ever-growing human needs. But at the same time, we should not overlook the food waste issue [116]. In terms of maintaining food freshness and food waste control, the expiration date of food should be managed. Food expiration dates are the main cause of retail waste and the cost is tremendous, both to business and to the environment. Food waste represents today a global issue of extreme urgency. FAO has reported that globally, about one-third of food produced for human consumption is lost or wasted each year from the farm to the refrigerator, representing about 1.3 billion tons. The economic price tag is estimated at nearly $1 trillion annually. This is damaging to everyone, not only from an economic and social point of view, but from also an environmental one. Producing and disposing of food that is not consumed leads to heavy waste of resources, as well as unnecessary polluting emissions. Data science and AI can help to change this detrimental situation, ensuring a major cut to waste throughout the supply chain, from the manufacturer to the individual consumer [117, 118].

Chuprina [119] proposed some ways to reduce food waste with AI:
1) While some solutions analyze the ripeness of the fruits, another figure out what microbes could increase crop growth without the involvement of synthetic fertilizers.
2) Farmers could get rid of field trials, benefiting from advantages of the AI, which will significantly save money.
3) If farm-based food supply chains will use visual imagery technology, the food inspection process will be much easier.
4) AI food tracking will enable us to sell food before it becomes a waste, connecting farmers with restaurants or people buying food more efficient.

Similarly, Feindt [120] advised, using AI and ML, the following key points should be taken carefully:
1) Improve demand forecasting. The basis of minimizing food waste is to forecast as accurately as possible, and with AI grocery retailers can now significantly improve their accuracy by taking into account hundreds of forecast factors such as weather, day of week, or time of year. The clearer and more accurate the signal, the more successful grocery retailers can be in reducing the environmental impact of waste.
2) Improve availability while minimizing wastage. By understanding the risk of lost sales and wastage, retailers can now ensure the right balance when managing inventory – individually for every item in every store. Not only does this cut food waster, it also makes the stock ordering process a less labor-intensive task and can improve shelf presentation.
3) Setting the right price. AI can put grocery retailers in a better position to set the right price than ever before. Retailers need to find the right balance between selling as a profit and clearing stock that is going out of date, and AI can make a real difference in this process by automatically setting prices based on information such as time of day amount of goods left and expiry date. By gradually reducing items rather than following rigid markdown products sell.

4) Sensing and avoiding transportation disruption. By analyzing a range of factors including weather, time of day, and expiration date of products, ML can advise how to react to potential disruptions or help to avoid them together. For example, if rough seas could potentially lead to a port being closed. ML can help grocery retailers navigate the problem by advising another route to take; a delay to a boatful of bananas could lead to them having a shorter shelf life, or even perishing before they reach the store. Avoiding these kinds of logistical problems could dramatically cut waste.

7.8 AI-supported food technology

Today, there are more than 7 billion humans on this earth. In order for all of us to live a healthy life, it is necessary to secure a huge amount of meat and grains. However, the amount produced currently cannot be covered by the food production system today, and the problem such as starvation always occurs somewhere in the world. In addition, the world's population is ever-growing, and it is difficult to continue to supply food only with existing equipment and food supply system and industry. As indicated in Figure 7.7, by ever-increasing world population, it is anticipated that protein will be shortage of supply by 2030 [121].

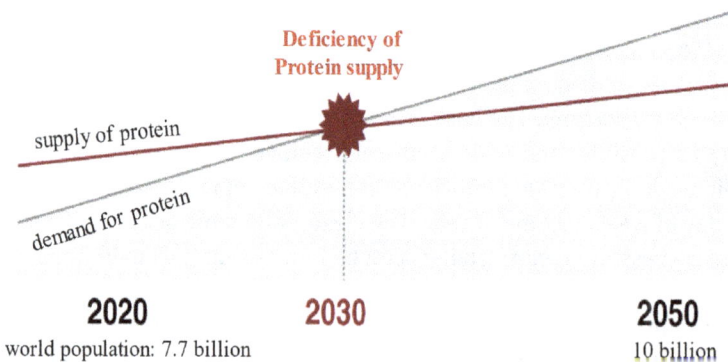

Figure 7.7: Balance between protein supply and its demand [122].

Therefore, food technology is trying to solve various problems related to food while producing food using technologies such as robots and AI. In addition to growing existing livestock and agriculture, the food technology is also developing the next generation of food industries such as insect food and artificial meat [122]. Food technology is the food science and possesses other functions such as the selection, preservation, processing, packaging, distribution, and use of safe food. Related fields include analytical chemistry, biotechnology, biology, biochemistry, engineering, nutrition, quality control, and food safety management.

Growing beef and pork sold in supermarkets requires a great deal of time and effort. Therefore, it takes about 6 months for pigs and about 30 months for cows before it is shipped as meat from the ranch. However, if it is an insect such as crickets, the growth is completed in a short period of time from 1 week to a month. In addition, the small amount of feed and water required for growth is also an advantage of insect food. For example, to produce 1 kg of protein, beef requires 10 kg of feed and more than 20,000 L of water. However, in the case of crickets, the feed required to obtain the same amount of protein is 1.7 kg, and the water is only 4 L. In addition, cows and pigs have 40% food, while crickets have almost 100%. From this, we can see the high production efficiency of crickets. Besides crickets, various insects such as mealworms and silkworms have been studied as ingredients. An insect diet that is easy to produce and rich in nutrients, even in small quantities, is also expected as the space food of the future [121].

In our society, there are vegetarians and vegans who are resisting eating animals, those who follow religion and do not eat beef or pork, and others who are allergic to eating fish and animals. Instead of animal proteins, they have kept them healthy by consuming vegetable proteins such as soybeans. Hence, for those who are vegetarians, are sensitive to allergic reaction, or relied on halal foods, they can take protein from artificial meat made of plants. No matter how much food producers create, they will have to dispose of it if consumers do not buy it. However, by utilizing AI/DL technologies, it is possible to calculate food demand and logistics costs in advance to achieve optimal production targets. In addition, if the robot cooks the food, it will be possible to provide it quickly, safely, and inexpensively.

One of the most representative areas of food tech is molecular gastronomy. Molecular gastronomy is made possible by subdividing ingredients into molecular units, crushing them, and then making them into bubbles, so that they can be cooked without compromising taste or aroma [123, 124]. For example, the grinding of ingredients using liquid nitrogen and the reproduction of meaty foods using plant-based ingredients such as soybeans (aka meat/soybean) are performed.

7.9 You are what you eat

All concerns in this section are strongly related to personalized foods and nutrition.

AI could enable us to utilize vast datasets of detailed agricultural information for the improvement of our food crops, faster than ever before. Better standards and data collection are essential to enabling innovations in food. A lot of people are applying ML around flavor and nutrition in recipe development. This could take off even more with better data sharing and standards to map flavors and aromas, opening up new possibilities for "personalized" foods and recipes that offer better flavor, nutrition, and stainability. Personalized food on the individual level is a lofty goal. To be done well, it would require large amounts of personal data, much of which might be complicated by questions of privacy and security. It is more likely, in the foreseeable future, that AI would be used to predict broad trends in consumer trends and tastes [125].

The idea that there is one universal diet for everyone is biologically impossible. A good diet needs to be individualized. And that is where AI comes in. Unfortunately, most studies are observational and rely on food logs or simply the patient's memory which tend to be very inaccurate [126]. In order to understand health dynamics while considering individual variability and implementing personalized nutrition, efforts should focus on devising predictive methods that monitor the individual's health responses to food. Initially, studies were faulty because of the hassle to obtain such accurate and large amounts of data which prevented researchers from using ML. An ML is a method of data analysis that automates analytical model building. In the past few years a couple studies were able to do this using primarily glycemic responses and other clinical data making results more accurate. With ML, we could all have our own personalized virtual health coaches in a couple years. AI-personalized diets are the beginning of an exciting new era [126]. Chu [127] introduced the European Association for the Study of Obesity (EASO) has become the latest member of an EU-funded project that looks to harness AI to develop personalized nutrition concepts. The H2020 PROTEIN project was launched to gather European public and private sector organizations pooling resources and knowledge to aid consumers in making healthier choices based on needs, behaviors, and preferences. ML is disrupting the food industry. From how farmers determine which seeds to plant, to what toppings your local mom and pop pizza-joint serves, there is an AI-powered solution at work that is revolutionizing what and how we eat. After about 200,000 years of managing our own basic nutritional needs – poorly, in most cases – the grown-ups are finally putting machines in charge. IBM recently unveiled its "5 in 5" for 2019 – five innovations the company believes will change our lives in the next five years. This year's theme seems to be "feed the people," because of its heavy focus on the food industry. And, since it is IBM, all but one relies on ML as the core technology [128].

Before closing this chapter, let us take a look at the PLU (price look up) code. The PLU code is managed by an organization called IFPC (International Federation for Produce Coding) and is used in the United States, Canada, New Zealand, South America, Norway, and so on to identify bulk produce sold in grocery stores and supermarkets. Produce labeled with PLU codes on fruits or vegetables eliminates the need for grocery store checkers and customers to visually identify different varieties, which can make checkout and inventory control easier, faster, and more accurate – something that is important when varieties of produce look similar but have different prices, such as organic and conventional (non-organic) varieties [129]. Basically, there are four PLU code groups: (1) numbers starting with 3 or 4 in four digits given to conventionally grown/cultivated using chemical fertilizers and pesticides, (2) numbers starting with 9 in five digits, assigned for organically grown without chemical fertilizers or pesticides, and (3) numbers starting with 8 in five digits, designated for genetically modified crops [130]. Figure 7.7 shows typical examples of fruits labeled with PLU codes [131].

Figure 7.8: Examples of PLU codes [131].

Eating habit is quite personal matter. If it is bad, it would cause the lifestyle-related diseases including obesity, diabetes, hypertension, hyperlipidemia, hyperuricemia, colon cancer, pulmonary squamous cell carcinoma, chronic bronchitis, emphysema, alcoholic liver damage, and periodontal disease. All diseases are related to eating (bad) habit to some extent. As seen before, some portions of the habit can be improved and bettered by nutrition controlled by AI technology. I start this chapter with the phrase "you are what you eat" and will close this chapter with this phrase. If allowed, I would like to introduce my own healthy eating rule, with which I have been maintaining same weight for last more than a half century with excellent other physical parameters. Years ago, I have ordered myself a simple rule that everyone can follow easily if you have a willing to do so. The rule is "30 ingredients per week." That's it! The ingredients can include staple food (rice, bread, pasta, etc.), side dish (meat, fish, etc.), vegetables, and condiment or seasoning (sugar, salt, vinegar, pepper, sauce, miso, other flavors). For one meal, you go ahead to count numbers of these ingredients that you are taking. However, you cannot count the same stuff twice or more in the same week. For example, tomato in your today's sandwich for lunch cannot be counted if you ate pasta with tomato sauce last night. Tomato is the tomato on matter how you look it. But, it sounds still an easy challenge, does not it? You can count differently for having pork, beef, or chicken as a meat. Similarly, tuna, mackerel, codfish, or others can be counted separately.

Just imagine when you step in a grocery store and start from the veggie section, as shown in Figure 7.8, it should be easy to choose instantly 10 or so different kinds of veggies.

Figure 7.9: Typical veggie section at grocery store.

Besides, you need to use all kinds of seasoning. But, remember, you can count, for example, salt once for a week no matter how much you use for almost very meals. If you start to take a record of numbers of ingredients, say from Monday to keep it for a week. Even so, it is easy to come up 20 counts by Thursday or so. However, it is getting hard to add another 10 counts of unused ingredients for the rest of that week. If you continue doing this, what really might happen in and to your body will be, under an action of metabolic memory, as follows: (1) intend to avoid the repeating same type of meals; (2) soon or later, your body keeps telling you what your body needs for food; and (3) hope that smart AI can learn what I have been exercising for more than 50 years, and it should be benefit to many people who are suffering from lifestyle-related discomfort.

References

[1]   https://www.theidioms.com/you-are-what-you-eat/.
[2]   https://www.phrases.org.uk/meanings/you-are-what-you-eat.html.
[3]   https://gillianmckeith.com/books/eat/.
[4]   Kurilyak S. Coronavirus Update: Food and Artificial Intelligence. 2020; https://blog.produvia. com/artificial-intelligence-ai-in-food-industry-ec8e925fa35e.
[5]   Chuprina R. Machine Learning and AI in Food Industry: Solutions and Potential. 2019; https://www.datasciencecentral.com/profiles/blogs/machine-learning-and-ai-in-food-industry-solutions-and-potential.
[6]   Machine Learning and Artificial Intelligence in the Food Industry. SPD. 2019; https://medium. com/@spd.group/machine-learning-and-artificial-intelligence-in-the-food-industry -598f78471106.
[7]   Koksal I. How AI Determines The Diet Plans. 2020; https://www.forbes.com/sites/ilker koksal/2020/03/07/how-ai-determines-the-diet-plans/#cc936f13ed7f.

[8]   Applications of Artificial Intelligence (AI) in the Food Industry. 2012; https://morioh.com/p/
      6f346ea78169.
[9]   Garver K. 6 Examples of Artificial Intelligence in the Food Industry. 2018; https://foodindus
      tryexecutive.com/2018/04/6-examples-of-artificial-intelligence-in-the-food-industry/.
[10] Fontanazza M. 2020 Expectations: More Artificial Intelligence and Machine Learning,
      Technology Advances in Food Safety Testing. 2020; https://foodsafetytech.com/news_arti
      cle/2020-expectations-more-artificial-intelligence-and-machine-learning-technology-
      advances-in-food-safety-testing/.
[11] Fuentes S. 2020; https://www.mdpi.com/journal/foods/special_issues/Implementation_Arti
      ficial_Intelligence_Food_Science_Food_Quality_Consumer_Preference_Assessment.
[12] Kakani V, Nguyen VH, Kumar BP, Kim H, Pasupuleti VR. A critical review on computer vision
      and artificial intelligence in food industry. Journal of Agriculture and Food Research. 2020, 2.
      doi: https://doi.org/10.1016/j.jafr.2020.100033.
[13] Kovalenko O. Machine learning and AI in food industry: solutions and potential. SPD. 2020;
      https://spd.group/machine-learning/machine-learning-and-ai-in-food-industry/#:~:text=
      Technology%20in%20the%20Food%20Industry,-
      When%20talking%20about&text=Artificial%20Intelligence%20and%20Machine%20Learn
      ing,well%20as%20in%20food%20manufacturing.
[14] Sharma S. How Artificial Intelligence is Revolutionizing Food Processing Business? 2019;
      https://towardsdatascience.com/how-artificial-intelligence-is-revolutionizing-food-
      processing-business-d2a6440c0360.
[15] Utermohlen K. 4 Applications of Artificial Intelligence in the Food Industry. 2019; https://
      heartbeat.fritz.ai/4-applications-of-artificial-intelligence-ai-in-the-food-industry-
      e742d7c02948.
[16] Walker J. AI in Food Processing – Use Cases and Applications That Matter. 2019; https://
      emerj.com/ai-sector-overviews/ai-in-food-processing/.
[17] Sebastian J. Artificial Intelligence: A Real Opportunity in Food Industry. 2018; https://www.
      foodqualityandsafety.com/article/artificial-intelligence-a-real-opportunity-in-food-industry/.
[18] Ehrlich P. Extinction: The Causes and Consequences of the Disappearance of Species.
      Ballantine Books, 1985.
[19] Biodiversity. 2020; https://www.worldbank.org/en/topic/biodiversity.
[20] Biodiversity; https://ja.wikipedia.org/wiki/%E7%94%9F%E7%89%A9%E5%A4%9A%E6%
      A7%98%E6%80%A7.
[21] Importance of Biodiversity. Exploring the Environment. https://www.toppr.com/guides/biol
      ogy/biodiversity-and-conservation/types-of-biodiversity/.
[22] Shah A. Why Is Biodiversity Important? Who Cares? 2014; https://www.globalissues.org/arti
      cle/170/why-is-biodiversity-important-who-cares.
[23] Joshi N. Conserving Biodiversity with Artificial Intelligence. 2019; https://www.technologyfor
      you.org/conserving-biodiversity-with-artificial-intelligence/.
[24] Faggella D. AI in Agriculture – Present Applications and Impact. 2020; https://emerj.com/
      ai-sector-overviews/ai-agriculture-present-applications-impact/.
[25] Artificial Intelligence holds the promise of driving an agricultural revolution at a time when
      the world must produce more food using fewer resources. https://www.intel.com/content/
      www/us/en/big-data/article/agriculture-harvests-big-data.html.
[26] Potter R. How can I start with AI in agriculture? 2020; https://becominghuman.ai/how-can-i-
      start-with-ai-in-agriculture-29f7969ae1d9.
[27] Rapid adoption of artificial intelligence in agriculture. 2019; https://www.futurefarming.com/
      Smart-farmers/Articles/2019/8/Rapid-adoption-of-artificial-intelligence-in-agriculture
      -461266E/.

[28] Tolamise JK. The Role Of Artificial Intelligence In Agricultural Sectors. 2020; https://datafloq.com/read/the-role-of-artificial-intelligence-in-agricultural-sectors/8700.

[29] The Future of Artificial Intelligence and Agriculture. 2018; https://medium.com/@ODSC/the-future-of-artificial-intelligence-and-agriculture-540c39208df6.

[30] Gupta J. The Role of Artificial intelligence in Agriculture Sector. 2019; https://customerthink.com/the-role-of-artificial-intelligence-in-agriculture-sector/.

[31] https://www.afpbb.com/articles/-/2645112?pid=4659075.

[32] https://ledge.ai/aisum-plantx/.

[33] Hatou K, Kamio Y, Hashimoto Y. Computer integrated plant factory based on artificial intelligence. IFAC Proceedings Volumes. 1992, 25, 175–80.

[34] http://www.optronics-media.com/news/20181025/53676/.

[35] https://xtech.nikkei.com/atcl/nxt/column/18/00908/081900004/.

[36] PYMNTS. Farming's AI, Data Driven (And Vertical) Tech Revolution, 2020; https://www.pymnts.com/news/merchant-innovation/2020/crop-one-ai-data-driven-vertical-farming/.

[37] https://www.hydroponicanswers.com/VerticalFarming-Page9.html.

[38] https://smartagri-jp.com/smartagri/157.

[39] https://ampmedia.jp/2018/10/10/agritech_food-crisis/.

[40] Meyers J. Advances in autonomous control for plant management. Manufacturing Automation. 2020; https://www.automationmag.com/advances-in-autonomous-control-for-plant-management/.

[41] Sikkema A. Competition for growing vegetables using AI has started. 2020; https://www.miragenews.com/competition-for-growing-vegetables-using-ai-has-started/.

[42] Shieber J. Your vegetables are going to be picked by robots sooner than you think. 2018; https://www.pinterest.com/pin/403916660325696273/.

[43] Joshi FN. Moo-ving The Animal Husbandry Industry Forward With AI. Cognitive World. 2019; https://www.forbes.com/sites/cognitiveworld/2019/12/18/moo-ving-the-animal-husbandry-industry-forward-with-ai/#fcf18527f41d.

[44] Singh R. Application Of Artificial Intelligence (AI) For Livestock & Poultry Farm Monitoring. 2019; https://www.pashudhanpraharee.com/application-of-artificial-intelligence-ai-for-livestock-poultry-farm-monitoring/.

[45] AI Business. Identifying Animal Growth Using Artificial Intelligence. 2016; https://ai.business/tag/livestock/.

[46] Neethirajan S. The role of sensors, big data and machine learning in modern animal farming. Sensing and Bio-Sensing Research. 2020, 29. doi: https://doi.org/10.1016/j.sbsr.2020.100367.

[47] VanderWaal K, Morrison RB, Neuhauser C, Vilalta C, Perez AM. Translating big data into smart data for veterinary epidemiology. Frontiers in Veterinary Science. 2017. doi: https://doi.org/10.3389/fvets.2017.00110.

[48] Fernández-Carrión E, Martínez-Avilés M, Ivorra B, Martínez-López B, Ramos ÁM, Sánchez-Vizcaíno JM. Motion-based video monitoring for early detection of livestock diseases: the case of African swine fever. PLoS. 2017. doi: https://doi.org/10.1371/journal.pone.0183793.

[49] Tsai J, Bowring E, Marsella S, Wood W, Tambe M. A study of emotional contagion with virtual characters. International Conference on Intelligent Virtual Agents. 2012, 81–8.

[50] Paul ES, Harding EJ, Mendl M. Measuring emotional processes in animals: the utility of a cognitive approach. Neuroscience and Biobehavioral Reviews. 2005, 29, 469–91.

[51] Špinka M. Social dimension of emotions and its implication for animal welfare. Applied Animal Behaviour Science. 2012, 138, 170–81.

[52] Connolly A. Flocking to digital: The future of poultry technology. 2017; https://www.alltech.com/blog/flocking-digital-future-poultry-technology.

[53] Sensors and AI are finding their way into the barnyard. Technology Quarterly. 2019; https://www.economist.com/technology-quarterly/2019/09/12/sensors-and-ai-are-finding-their-way-into-the-barnyard.

[54] https://animalagtech.com/start-ups-transforming-the-livestock-industry/.

[55] 13 ag-tech innovators transforming livestock production. The Poultry Sie. 2020; https://thepoultrysite.com/news/2020/01/13-ag-tech-innovators-transforming-livestock-production.

[56] Ku L. 9 Companies Revolutionizing Livestock Management. https://www.plugandplaytechcenter.com/resources/9-companies-revolutionizing-livestock-management/.

[57] Bedord L. 13 Breakthrough Technologies Support Sustainable, Efficient Livestock Industry. 2020; https://www.agriculture.com/news/livestock/13-breakthrough-technologies-support-sustainable-efficient-livestock-industry.

[58] Holcomb J. New Artificial Intelligence Technology Changing Dairy Industry. https://www.farm-monitor.com/new-artificial-intelligence-technology-changing-dairy-industry/.

[59] Verma K, Arya A. How Artificial Intelligence will reshape future of Poultry Industry. Benison Media. 2020; http://benisonmedia.com/how-artificial-intelligence-will-reshape-future-of-poultry-industry/.

[60] Usha. Can Artificial Intelligence build Future of Poultry Industry? 2018; https://medium.com/oclavi/can-artificial-intelligence-build-future-of-poultry-industry-7b3e7a091222.

[61] Laursen L. Machine vision sees into chickens' futures. IEEE Spectrum. 2012; https://spectrum.ieee.org/biomedical/diagnostics/machine-vision-sees-into-chickens-futures.

[62] Jawad H. Artificial intelligence in poultry industry. Technology Times. 2019; https://www.technologytimes.pk/2019/09/06/artificial-intelligence-poultry-industry/.

[63] McDougal T. Smart farms that can feed chickens and detect bird flu. Health. 2018; https://www.poultryworld.net/Health/Articles/2018/5/Smart-farms-that-can-feed-chickens-and-detect-bird-flu-288723E/.

[64] Cronin G. Robotics in the poultry industry. Poultry Hub. http://www.poultryhub.org/production/husbandry-management/housing-environment/robotics-in-the-poultry-industry/.

[65] https://www.soumu.go.jp/main_content/000692534.pdf.

[66] https://emira-t.jp/topics/8199/.

[67] https://r.m-miura.jp/gibierai/.

[68] https://ascii.jp/elem/000/001/574/1574787/.

[69] https://b2b-ch.infomart.co.jp/news/detail.page?IMNEWS4=634172.

[70] https://www.nikkei.com/article/DGXLZO17571060S7A610C1L41000/.

[71] https://japan.cnet.com/release/30198823/.

[72] https://diamond-rm.net/sales-promotion/41467/.

[73] https://www8.cao.go.jp/space/goodpractice/r02/r02_jirei02_2.pdf.

[74] https://media.fringe81.com/n/n2f15d907f36f.

[75] https://www.itmedia.co.jp/news/articles/1705/16/news114.html.

[76] http://karapaia.com/archives/52239730.html.

[77] Hill H. Artificial intelligence helps manage global fisheries. Climate Change. 2019; http://darwinproject.mit.edu/artificial-intelligence-helps-manage-global-fisheries/.

[78] Memarzadeh M, Britten GL, Worm B, Boettiger C. Rebuilding global fisheries under uncertainty. Proc. National Academy of Science USA. 2019. doi: https://doi.org/10.1073/pnas.1902657116.

[79] Gentry RR, Froehlich HE, Grimm D, Kareiva P, Parke M, Rust M, Gaines SD, Halpern BS. Mapping the global potential for marine aquaculture. Nature Ecology & Evolution. 2017, 1, 1317–24.

[80] https://www.toyota-tsusho.com/press/detail/200521_004622.html.

[81] https://wired.jp/2020/03/03/umitron-and-future-of-aquaculture/.

[82] https://gyoppy.yahoo.co.jp/originals/56.html.

[83] https://ledge.ai/offshore-fishfarming/.

[84] https://www.itmedia.co.jp/enterprise/articles/1804/25/news082.html.

[85] Rishi HR. AI in aquaculture: shaping the future with Observe Technologies. 2019; https://www.governmenteuropa.eu/ai-in-aquaculture/94557/.

[86] 8 digital technologies disrupting aquaculture. 2017; https://www.alltech.com/blog/8-digital-technologies-disrupting-aquaculture.

[87] Van Beijnen J, Yan G. A practical guide to using AI in aquaculture. 2020; https://thefishsite.com/articles/a-practical-guide-to-using-ai-in-aquaculture.

[88] Artificial intelligence makes fishing more sustainable by tracking illegal activity. 2019; https://theconversation.com/artificial-intelligence-makes-fishing-more-sustainable-by-tracking-illegal-activity-115883.

[89] Girard P, Du Payrat T. An inventory of new technologies in fisheries. Greening the Ocean Economy. 2017; https://www.oecd.org/greengrowth/GGSD_2017_Issue%20Paper_New%20technologies%20in%20Fisheries_WEB.pdf.

[90] Bradley D, Merrifield M, Miller KM, Lomonico S, Wilson JR, Gleason MG. Opportunities to improve fisheries management through innovative technology and advanced data systems, Fish and Fisheries, 2019; https://onlinelibrary.wiley.com/doi/full/10.1111/faf.12361.

[91] Føre M, Frank K, Norton T, Svendsen E, Alfredsen JA, Dempster T, Eguiraun H, Watson W, Stahl A, Sunde LM, Schellewald C, Skøien KR, Alver MO, Berckmans D. Precision fish farming: A new framework to improve production in aquaculture. Biosystems Engineering. 2018, 173, 176–93.

[92] Souza K. The Supply Side: Artificial intelligence is slowly shaping the future of retail. 2020; https://talkbusiness.net/2020/02/the-supply-side-artificial-intelligence-is-slowly-shaping-the-future-of-retail/.

[93] Utilization of AI in the supply chain field. 2019; https://translate.google.com/translate?hl=en&sl=ja&u=https://www.fpt-software.jp/the-supply-chain-sector-how-ai-has-penetrated-into-this-emerging-market/&prev=search&pto=aue.

[94] Korolov M. AI in the supply chain: Logistics gets smart. 2018; https://www.cio.com/article/3269513/ai-in-the-supply-chain-logistics-get-smart.html.

[95] 6 Applications of Artificial Intelligence for your Supply Chain. https://medium.com/@Kodiak Rating/6-applications-of-artificial-intelligence-for-your-supply-chain-b82e1e7400c8.

[96] Machine Learning, AI Are Most. Impactful supply chain technologies. MH&L. 2019. https://www.mhlnews.com/technology-automation/article/22055616/machine-learning-ai-are-most-impactful-supply-chain-technologies.

[97] Artificial intelligence works its way into supply chains. https://www.zdnet.com/article/artificial-intelligence-works-its-way-into-supply-chains/.

[98] Bisen VS. How AI Based Drone Works: Artificial Intelligence Drone Use Cases. https://medium.com/vsinghbisen/how-ai-based-drone-works-artificial-intelligence-drone-use-cases-7f3d44b8abe3.

[99] Dedezade E. Jobs of the future: how self-piloted AI drones are creating exciting new opportunities. 2019; https://news.microsoft.com/europe/features/jobs-of-the-future-how-self-piloted-ai-drones-are-creating-exciting-new-opportunities/.

[100] Schroth L. Drones and Artificial Intelligence. 2018; https://droneii.com/drones-and-artificial-intelligence.

[101] From supply chain to supply net; https://www.nhk.or.jp/kaisetsu-blog/300/434400.html.

[102] Loten A. Logistics firms fast-track cloud, AI projects after Covid-19 lays bare supply-chain gaps. CIO Journal. 2020. https://www.wsj.com/articles/logistics-firms-fast-track-cloud-ai-projects-after-covid-19-lays-bare-supply-chain-gaps-11596757388.

[103] Link R. What Are Functional Foods? All You Need to Know. 2020; https://www.healthline. com/nutrition/functional-foods#benefits.

[104] Tur JA, Bibiloni MM. Functional foods. Encyclopedia of Food and Health. 2016, 157–61. doi: https://doi.org/10.1016/B978-0-12-384947-2.00340-8.

[105] Hasler CM. Functional foods: Benefits, concerns and challenges – a position paper from the american council on science and health. The Journal of Nutrition. 2002, 132, 3772–81.

[106] Gupta S, Parvez N, Sharma P. Nutraceuticals as functional foods. Journal of Nutritional Therapeutics. 2015, 4, 64–72.

[107] Yordi EG, Koelig R, Matos MJ, Martinez AP, Caballero Y, Santana L, Quintana MP, Molina E, Uriarte E. Artificial intelligence applied to flavonoid data in food matrices. Foods. 2019, 8(11), 573. doi: https://doi.org/10.3390/foods8110573.

[108] Hunter P. The inflammation theory of disease: The growing realization that chronic inflammation is crucial in many diseases opens new avenues for treatment. EMBO Reports. 2012, 13, 968–70.

[109] El-Gabalawy H, Guenther LYNC, Bernstein CN. Epidemiology of immune-mediated inflammatory diseases: incidence, prevalence, natural history, and comorbidities. The Journal of Rheumatology. 2010, 85, 2–10.

[110] Baker RG, Hayden MS, Ghosh S. NF-κB, inflammation, and metabolic disease. Cell Metabolism. 2011, 13, 11–22.

[111] Kennedy K, Keogh B, Lopez C, Adelfio A, Molloy B, Kerr A, Wall AM, Jalowicki G, Holton TA, Khaldi N. An artificial intelligence characterised functional ingredient, derived from rice, inhibits TNF-α and significantly improves physical strength in an inflammaging population. Foods. 2020, 9, 1147. doi: https://doi.org/10.3390/foods9091147.

[112] Buss D. Food companies get smart about artificial intelligence. Food Technology Magazine. 2018. https://www.ift.org/news-and-publications/food-technology-magazine/issues/2018/ july/features/food-industry-using-artificial-intelligence-and-machine-learning.

[113] Zeeberg A. The AI that knows exactly what you want to eat. The Atlantic. 2018. https://www. theatlantic.com/health/archive/2018/12/gastrograph-flavor-goes-digital/577270/.

[114] Shah D, Ahn R, Cohen J. Predicting Consumer Preference From Reviews of Professional Tasting Panels on the Gastrograph Sensory System. 2017. https://www.semanticscholar.org/ paper/Predicting-Consumer-Preference-From-Reviews-of-on-6-Shah-Ahn /9282006efe8457cb532bbe677c1873519237cb1a.

[115] Altenburger KM, Ho DE. Artificial Intelligence and Food Safety: Hype vs. Reality. Food Safety Magazine. 2020; https://www.foodsafetymagazine.com/magazine-archive1/december-2019january-2020/arfivicial-intelligence-and-food-safety-hype-vs-reality/.

[116] Moher D. Food expiration dates are about to undergo a revolution. https://www.israel21c. org/food-expiration-dates-are-about-to-undergo-a-revolution/.

[117] AI and Data Science working to reduce food waste; https://proaxxes.com/data-science-and-artificial-intelligence-can-make-a-real-contribution-to-cutting-food-waste-2/.

[118] Rejcek P. Food Waste Is a Serious Problem. AI Is Trying to Solve It. https://singularityhub. com/2019/11/03/food-waste-is-a-serious-problem-ai-is-trying-to-solve-it/.

[119] Chuprina R. Machine Learning and AI in Food Industry: Solutions and Potential. 2019; https://www.datasciencecentral.com/profiles/blogs/machine-learning-and-ai-in-food-industry-solutions-and-potential.

[120] Feindt M. Artificial Intelligence & Machine Learning Could Prevent £ 144 Million of Food Waster Annually; https://itsupplychain.com/artificial-intelligence-machine-learning-could-help-uks-top-eight-grocery-retailers-prevent-144-million-of-food-waste-annually/.

[121] Crickets save the planet?; https://www.muji.com/jp/ja/feature/food/460936.

[122] FoodTech – changing the food industry of the future. 2021; https://studyu.jp/feature/theme/foodtec/.

[123] Burke R, Kelly A, Lavelle C, Vo Kientza HT. Handbook of Molecular Gastronomy. 2021, CRC Press.

[124] Barham P, Skibsted LH, Bredie WLP, Frøst MB, Møller P, Risbo J, Snitkjær P, Mortensen LM. Molecular gastronomy: a new emerging scientific discipline. Chemical Reviews. 2010, 110, 2313–65.

[125] Koksal I. Kitchen Disruption: Better Food Through Artificial Intelligence. 2019; https://www.forbes.com/sites/ilkerkoksal/2019/08/03/kitchen-disruption-better-food-through-artificial-intelligence/#1e6ea1667707.

[126] Bone K. Nutrition Meets Artificial Intelligence. 2019; https://medium.com/datadriveninvestor/nutrition-meets-artificial-intelligence-b15847d6cc19.

[127] Chu W. EASO on board in AI-powered personalised nutrition project. 2020; https://www.nutraingredients.com/Article/2019/03/12/EASO-on-board-in-AI-powered-personalised-nutrition-project.

[128] Greene T. AI will soon decide what we eat. Artificial Intelligence. 2019; https://thenextweb.com/artificial-intelligence/2019/02/13/ai-will-soon-decide-what-we-eat/.

[129] https://www.ifpsglobal.com/PLU-Codes.

[130] https://www.hortidaily.com/article/6031762/plu-codes-this-is-what-the-stickers-on-fruit-and-vegetables-mean/.

[131] What are PLU codes found on fruits and vegetables? https://www.livechennai.com/detailnews.asp?newsid=21309.

Coffee break: information diabetes

Human spends most of life in the exchange of intellectual information. Human mechanism exhibits not only the substance-metabolism system and the energy-metabolism system, but also an information-metabolism system. Just as the human body is formed and maintained by metabolizes, each person forms his/her intellectual world by information metabolism, maintains it, and develops it. Information through the body of an organism is first interpreted in various ways in the unconscious. Then, the unformed signal becomes a condensed set, it is returned as *significant* (signifier) and *signifié* (signified), and it becomes a symbol [1]. To some extent, the correlation between symbols is remembered as a connection between words. The relation becomes the language of the consciousness world between these symbols, and we deepen the understanding of the world so to speak through the network of the relation of the language – in a word, the language circuit. And the spirit is formed by the language. Psychiatry and AI are deeply involved in each other. Psychiatry has explored the structure of the human spirit from a medical point of interest. The hypothesis is made about the movement of man's spirit and the deep one, the treatment policy is decided, and it heads to the patient to the disease of the spirit. This knowledge is also important in the construction of AI. New knowledge is understood, judged, and hopefully digested in existing knowledge systems. If it is accepted as true, it will eventually become common sense. However, when new knowledge comes in again, the conflict with common sense begins. If what has been judged to be true is not true, a new system of knowledge is required. This repetition is repeated with the times. That is, when a new fact is discovered by the chance of something, it will encourage the discovery of another new fact. In daily life, when the existing knowledge system collapses, it is required to change to a new knowledge system. The same is the case in the world of science, where change is always required. Doubting common sense is a fundamental act for science that seeks to reveal the truth.

There are three types of information: information to be sent, information to be received, and information to be co-created. Each of them is normally accompanied with the quality and the quantity of information. In medical and dental field, the EBL (evidence-based-learning) has been practiced for years, and the reliability of adopted evidential information becomes crucial. It is generally believed that there are six types of evidence sources, and they are listed in order from the highest reliable evidence to the least reliable evidence as follows: (1) clinical test by the double-blind method, using placebo, (2) clinical studies, based on statistical experimental protocols, without placebo, (3) clinical studies on chronological approach (vertical study form), (4) clinical studies on comparison of multi-group of examinees for a short period of time (horizontal study form), (5) case study on novel technique or idea, and (6) retrospect reports on clinical results. Unfortunately, number of published articles on these categories are declining order from the above; namely the most popular case is the retrospect reports [2]. Nevertheless, here is data on journal information. Entirely

https://doi.org/10.1515/9783110717853-008

in the world, there are about 2,00,0000 journal issues (weekly, monthly, or quarterly), in which about 20,000 journals have articles on biomedicine, in which about 500 journals contain extensively dental articles. Suppose 10% of 500 journal articles are relevant to your professional expertise, you still have 50 articles to cover in every period combined of weekly, monthly, and quarterly. Recently, evidence-based medicine (EBM) and evidence-based dentistry (EBD) are receiving more special attention, so that reliability of each source becomes more important.

As transmitted information, there are analogue (interpretive) information and digital (direct) information. Furthermore, the transmitted information has two distinct effective natures: one is a message with intrinsic information (the function of information acts on things and systems outside the human body, and the function and efficiency are improved) and a massage with extrinsic nature (the function of information acts on human sensory organs, delights, and entertains). As seen in Figure 1 [3, 4], AI field along with AI-related technologies are involved in the Quadrant I. Even in the transmission of information, when the content becomes artistic expression or design, in the old days, it was too complicated, but the information esthetics was established based on the concept close to cybernetics [5], which was a groundbreaking new discipline for a rational and objective understanding of artistic expression, bringing innovative ideas to the art of the computer age.

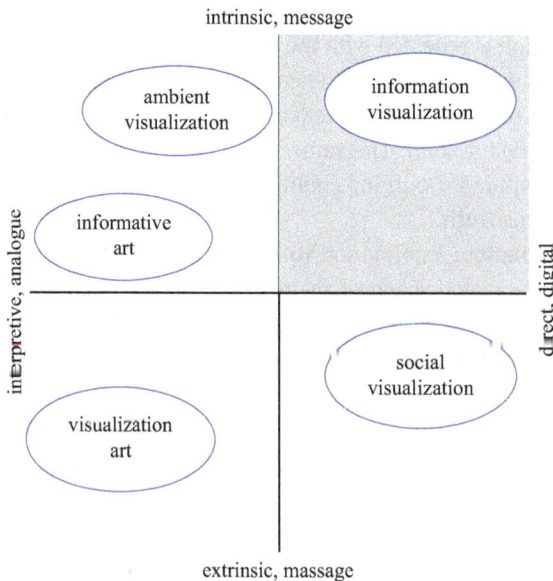

Figure 1: Information mapping [3, 4].

We are facing various issues. They should include informaion society with asymmetry, a sending /receiving mechansim of information that is perfect for people with strong self-manifestations scuh as SNS tools thta can comminicate with others (and an unspecified number) with a concealed face as well as anonymous. Furthermre, there probelms that come from delayes in the response of the brian, including interpretation, understanding, and digesting compared to the speed of the information flowed. The entropy of physics, also translated as "messiness," represents the degree of disorder or randomness of matter and energy, and in nature entropy does not decrease as a whole and continues to increase over time (this is called the "second law of thermodynamics" of physics). There is a concept called the information entropy [6]. In the field of information science, this entropy is taken as "uncertainty of the event," and it is thought that the decrease in uncertainty by certain information is "amount of information" of the information. In other words, it is considered a measure of how much information the source is available, meaning the relative value of uncertainty before and after receiving the information, specifically refers to concepts such as "clutter," "irregularity," "ambiguity," etc.

There are various types of information around us: fake information that is shed regardless of whether you believe it or not, information that flies around in the echo chamber that feels pleasant only in the environment, or information that only pleasantly feels while received, information that comes in unilaterally without time to judge success or failure, information transmitted without sufficient chewing of the contents of the sale, or secondary information, etc. Transmitted irresponsibly is a modern society with a lot of information that crosses each other. The turbulence state is seen in the content of the information and the flow way now. The information itself may have become paralyzed. *Infodemic* (which is a portmanteau of "information" and "epidemic") is the spread of false information (fake news) on the web (especially social media such as Twitter, Facebook, Blog, LINE, Skype, mixi, Google+, Instagram, Tumblr, or LinkedIn), which many people truly receive and panic about, causing social upset. Infodemic was coined in 2003 and has seen renewed usage in the time of COVID-19.

Under this chaotic situation, we are luckily equipped with a mental filter (adjustment of the amount of emotion) function that can ignore information that is not directly related to the goal at once or our own survival. Such function is called the cognitive inhibition, which refers to the mind's ability to tune out stimuli that are irrelevant to the task/process at hand or to the mind's current state. Cognitive inhibition can be done either in whole or in part, intentionally or otherwise [7, 8]. However, if this function is fragile, it will not be filtered by the mental filter, and information will drip and climb into consciousness. Then, a huge amount of signals will enter the brain which is the receiving device of information through the sensory organ (i.e., five senses). If you need to pay attention to all this, you will not be able to understand and organize incoming information. Then, the above-mentioned mental filter device is necessary for the brain, and the information processing of the majority in the brain is not conscious because of that. The fragile condition of this filter is known as the cognitive disinhibition

[9]. It is believed that, over the age of 50, the filtering system (ability not to put non-important information into the brain and to feel physical distance and psychological distance) declines sharply.

The human body has a mechanism to make good use of less nutrition (blood sugar), but unfortunately it does not have a mechanism to throw away too much nutrition. The dietary habit has been shifted to more preference foods such as salt, sweet, and oily foods that has not been used in meals, leading to an irony. Now, people are entering an era of satiation, and the amount of nutrition they consume has exploded. Diabetes mellitus (pancreatic insulin hypofunction) is a disease born from mismatch between the mechanism of the human body and the amount of nutrition ingested. In this respect, man has not learned anything, and has repeatedly destroyed the body with his "favorite thing." Let's replace this nutrition (blood sugar) with information. This analogy is not so unreasonable because information certainly hits the nourishment taken from the outside for the spirit. Looking back on the entire history of humanity in the past, it is self-clear how skillfully people have used less information. Furthermore, even in an era when the amount of information was overwhelmingly small compared to the present age, it can be said that the human brain has evolved so that it can make the most of a small amount of information when knowing the intelligent conclusions taken by its predecessors. The brain's large storage capacity but low output suggests that it is not possible to process large amounts of information at one time, but it is good for storing and utilizing less information. And, if a not so big change of the gene of the brain does not happen, the physiological mechanism of the brain will not have changed basically even now. However, the modern person is troubled by all complications (mental disorder, the thought ability decrease, and the language and movement which missed common sense in the medical treatment room, etc.) by the flood of information overflowing and others by the decrease of the filter function, the decrease in the judgment power, and the flood of information, etc. This phenomenon is similar to diabetes mellitus or diabetes complications, and is called as information diabetes [10].

AI is fed with a huge amount of information. Perhaps AI will not become informational diabetes due to this, but there is no doubt that problems in the digestive organs such as indigestion will be affected by the quality of the information received and fed. Again, the phrase "you are what you eat" should be recalled.

References

[1]   Bradley S. An Introduction To Semiotics – Signifier And Signified. 2016, https://vanseode
       sign.com/web-design/semiotics-signifier-signified/.
[2]   Oshida Y. Bioscience and Bioengineering if Titanium Materials. Elsevier, Oxford, UK, 2007.

[3]   Lau A, Moere AV. Towards a Model of Information Aesthetics in Information Visualization. IV '07: Proceedings of the 11th International Conference Information Visualization. 2007, 87–92; https://doi.org/10.1109/IV.2007.114.

[4]   Takada M. Environment and Tourism. Zinbun-Shokan. 2005.

[5]   Bense M. Information Aesthetics. 1997.

[6]   Shannon E, Weaver W. The Mathematical Theory of Communication. The University of Illinois Press, 1967.

[7]   Gorfein DS, MacLeod CM. Inhibition in Cognition. American Psychological Association. 2007.

[8]   Noreen S, MacLeod MD. What Do We Really Know about Cognitive Inhibition? Task Demands and Inhibitory Effects across a Range of Memory and Behavioural Tasks. PLoS One. 2015, 10; doi: 10.1371/journal.pone.0134951.

[9]   Nash S, Henry JD, McDonald S, Martin I, Brodaty H, Peek-O'Leary M-A. Cognitive disinhibition and socioemotional functioning in Alzheimer's disease. Journal of the International Neuropsychological Society. 2007, 13, 1060–4.

[10] Yoshinaga Y. Where does inspiration come from? Sousisha. 2004.

Chapter 8
AI in practice of medicine

8.1 Introduction

In recent years, applications of AI technology have progressed rapidly in various fields; while medical care is a special area that deals directly with human life, the effects of AI are slightly different from other fields [1]. Until now, AI has been a very limited "intelligence" that can only be answered under certain rules, but the advanced development of DL technology since around 2010 has allowed programs to build judgment criteria by autonomously learning from given data. This phenomenon implies that AI can build and classify its own criteria, even if humanity has not been able to provide clear standards. When this mechanism is brought to medical care, it is called "medical AI" [2]. In medical care, even if we focus solely on diseases, there should be at least four judgments, such as risk assessment, disease diagnosis, treatment selection, and prognosis evaluation. Moreover, these judgments are complicated by differences in circumstances for each individual, and the judgment is usually very difficult. If we build criteria based on accumulated big patient data and, for example, present optimal treatments for each patient, it can be imagined that the benefits of medical AI are very significant for both patients and medical professionals [2–4]. There are four major fields in medicine, namely (1) prevention, (2) diagnosis, (3) treatment, and (4) prognosis. In this chapter our discussion will move in this order, accompanied by AI-powered or AI-assisted examples on individual clinical cases. In the last section, we will deal with precision medicine as well as edge-AI application. Disease occurs always accompanied by either clearly indicated or hidden symptom(s), which is medical appearance of elemental reason(s) causing the disease. If such elemental reason(s) can be translated and expressed by either physical or chemical parameters which should be digitized, the thus digitized data can be sent to AI technology, which can eat only 1s and 0s. In the following, all interesting AI activities have been developed on this basis.

Sreedharan et al. [5] conducted a comprehensive overview of the top 100 most-cited articles relating to the use of AI in medicine over the past 70 years and reported that, with respect to keywords used in the top 100 articles, the term "artificial intelligence" is marked the highest with 46 articles, followed by natural language processing (20), ML (18), data mining (14), ANN (13), and so on. Frost et al. [6] mentioned that AI market for healthcare applications in the world for year 2014 was 663.8 million dollars, and it jumped 10-fold to 6,662 million dollars for year 2021 due to several factors: (i) excellent treatment and patient outcomes in various studies using AI as a decision support tool coupled with drastic reduction in diagnostic and treatment costs have been the principal drivers for this market, (ii) relatively new market with huge growth potential due to easier integration within healthcare space, and (iii) future shift to fee-for-value model for healthcare positively impacting

https://doi.org/10.1515/9783110717853-009

this market. Daley [7] mentioned that AI simplifies the lives of patients, doctors, and hospital administrators by performing tasks that are typically done by humans, but in less time and at a fraction of the cost. AI has countless applications in healthcare. Whether it is being used to discover links between genetic codes, to power surgical robots, or even to maximize hospital efficiency, AI has been an indispensable boon to the healthcare industry. One of the world's highest growth industries, the AI sector was valued at about 600 million dollars in 2014 and is projected to reach 150 billion dollars by 2026 [7].

AI has the potential to transform how healthcare is delivered. It can increase productivity and the efficiency of care delivery and allow healthcare systems to provide more and better care to more people. AI can help improve the experience of healthcare practitioners, enabling them to spend more time in direct patient care and reducing burnout. Spatharou et al. [8] indicated that there are six core areas where AI possesses a direct impact on the patient, including (1) chronic care management, (2) self-care/prevention/wellness, (3) triage and diagnosis, (4) diagnostics, (5) clinical decision support, and (6) care delivery. A study on how automation and AI are likely to affect the future of work concluded that automation will affect most jobs across sectors, but the degree varies significantly, and healthcare is one of the sectors with the lowest overall potential for automation – only 35% of time spent is potentially automatable, and this varies by type of occupation. Figure 8.1 shows the share of hours currently worked that could be freed up by automation by 2030 for a wide range of healthcare occupations in selected European countries (including France, Germany, Hungary, Italy, Portugal, Sweden, and the UK). This does not reflect the potential for further disruption through other factors, such as personalization, that may revolutionize healthcare by focusing on a "segment of one" [8].

In terms of predictive analytics and image recognition, AI may soon become more effective than physicians, who cannot handle millions of images in any reasonable time frame [9]. This has led to some concern that AI-based systems will replace physicians and especially radiologists. With regard to current physicians' positioning, Krittanawong [10] described that physicians in everyday clinical practice are under pressure to innovate faster than ever because of the rapid, exponential growth in healthcare data. Physicians can analyze big data, but at present it requires a large amount of time and sophisticated analytic tools such as supercomputers; meanwhile, AI handling big data could assist physicians in shortening processing times and improving the quality of patient care in clinical practice. Physicians diagnose diseases based on personal medical histories, individual biomarkers, simple scores (e.g., CURB-65: confusion, urea, respiratory rate, and blood pressure criteria in community-acquired pneumonia; or MELD: model for end-stage liver disease), and their physical examinations of individual patients. In contrast, AI can diagnose diseases based on a complex algorithm using hundreds of biomarkers, imaging results from millions of patients, aggregated published clinical research from PubMed, and thousands of physicians' notes from electronic health records (EHRs). There could be a certain impact of AI medicine on physicians. In general, it seems a

Occupation | Share of hours
percent

Occupation	Percent
Medical equipment preparers	48
Medical assistants	32
Occupational health and safety technicians	30
Pharmacy technicians	29
Medical and clinical laboratory technicians	29
Dental assistants	26
Pharmacists	23
Medical records and health information technicians	23
Radiation therapists	21
Medical and clinical laboratory technologists	21
Dietitians and nutritionists	19
Speech-language pathologists	18
Audiologists	17
Nurse anaesthetists	16
Ophthalmic medical technicians	16
Occupational therapy assistants	15
Optometrists	15
Emergency medical technicians and paramedics	14
Magnetic resonance imaging technologists	13
Physical therapists	12
Family and general practitioners	12
Physicians and surgeons, all other	12
Obstetricians and gynaecologists	11
Nursing assistants	10
Anaesthesiologists	10
Oral and maxillofacial surgeons	10
Therapists, all other	10
Internists, general	9
Exercise physiologists	8
Nurse practitioners	8
Recreational therapists	8
Health diagnosing and treating practitioners, all other	8
Occupational therapists	8
Licensed practical and licensed vocational nurses	7
Podiatrists	7
Surgeons	7
Healthcare practitioners and technical workers, all other	7
Genetic counselors	7
Clinical, counseling, and school psychologists	6
Paediatricians, general	6
Opticians	6
Home health aides	5
Nurse midwives	5
Psychiatrists	4
Dental hygienists	3
Orthotists and prosthetists	3
Chiropractors	2

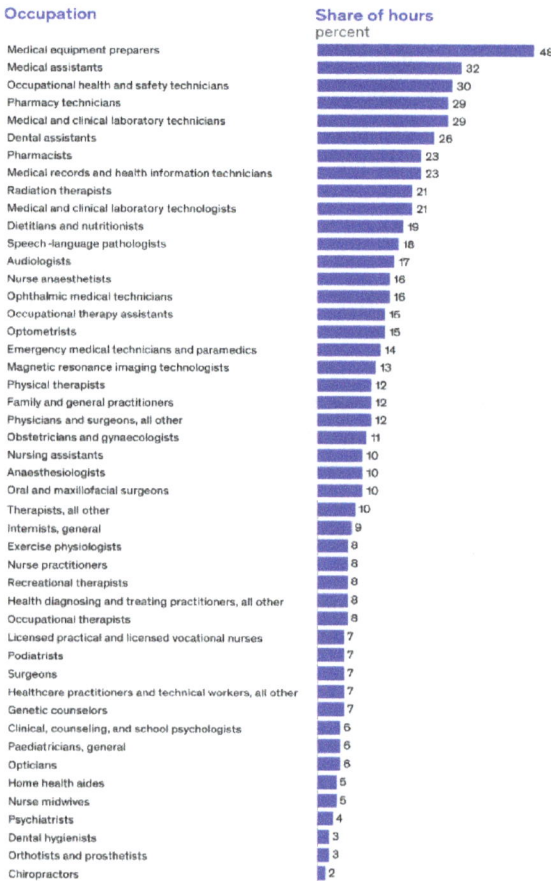

Figure 8.1: Areas of impact for AI in healthcare [4].

reasonable prediction that AI is likely to augment physicians rather than replace them [1]. There are several limitations of AI, such as that AI cannot engage in high-level conversation or interaction with patients to gain their trust, reassure them, or express empathy, all important parts of the doctor–patient relationship [9], until the time comes when this may change at some point in the future with AI being able to make a medical conversation. It is still true that physicians are still needed for traditional physical exams, especially in areas such as neurology, which require high-level patient–physician interaction and critical thinking [1, 10]. AI will interpret even the most complex clinical images as accurately as today's most experienced radiologists [11] and become a routine part of radiologists' daily lives, making their work more efficient, accurate, and valuable [12], resulting in eventually replacing radiologists [11, 12]. Major categories of AI applications in healthcare area can involve diagnosis and treatment recommendations, patient engagement and adherence, and administrative activities [13]. Most AI technologies possess immediate relevance to the healthcare field, including ML, DL, NLP, ES, and RPA.

The patient can start his/her day by stepping on an internet-connected scale that would monitor changes in weight for possible signs of fluid retention – a hallmark of heart failure – then strap on a smart watch or other sensor to track steps and activity level, as well as use a phone app to log specific symptoms, such as shortness of breath. All of the biological data would be transferred directly to the EHRs. The doctor then could alert the patient that he/she is approaching danger and take steps to avert it, allowing medical care to be given proactively rather than reactively. This kind of AI-assisted interaction in medicine has been years in the making [14]. Although mining and managing valuable medical data are crucial for developing new drugs, preventative medicine, and proper diagnosis, unfortunately, such highly valuable information can sometimes get lost among the forest of trillions of data points. In order to stop data hemorrhaging, there are AI companies helping the healthcare industry stay afloat in an ocean of data [14]. These include (i) Tempus (a massive data library for personalized healthcare treatments; it uses AI to sift through the world's largest collection of clinical and molecular data in order to personalize healthcare treatments. Tempus is currently using its AI-driven data to tackle cancer research and treatment), (ii) KenSci (combining big data and AI for hospitals to predict clinical, financial, and operational risks; combines big data and AI to predict clinical, financial, and operational risks by taking data from existing sources to foretell everything from who might get sick to what is driving up a hospital's healthcare costs), (iii) Proscia (a digital pathology platform using AI to detect patterns in cancer cells and help pathology labs eliminate bottlenecks in data management; it uses AI-powered image analysis to connect data points that support cancer discovery and treatment), (iv) H2O.ai (it uses AI throughout the health system to mine, automate, and predict processes, to predict ICU transfers, to improve clinical workflows, and even to pinpoint a patient's risk of hospital-acquired infections), (v) IBM (it helps healthcare professionals harness their data to optimize hospital efficiency, better engage with patients, and improve treatment. Watson is currently applying its skills to everything, from developing personalized health plans to interpreting genetic testing results and catching early signs of disease), (vi) DeepMind (it is an AI software being used by hospitals all over the world to help move patients from testing to treatment more efficiently. This program notifies doctors when a patient's health deteriorates and can even help in the diagnosis of ailments by combing its massive dataset for comparable symptoms. By collecting symptoms of a patient and inputting them into the DeepMind platform, doctors can diagnose quickly and more effectively), and (vii) iCarbonX (it uses AI and big data to look more closely at human life characteristics in a way they describe as digital life. By analyzing the health and actions of human beings in a "carbon cloud," the company hopes its big data will become so powerful that it can manage all aspects of health. The iCarbonX believes its technology can gather enough data to better classify symptoms, develop treatment options, and get people healthier). [14].

Finally, even though AI may reach the point where it can conduct real-time CT scans or other physical scans, physicians will still be needed for interpretation in ambiguous and challenging cases. AI-based systems are based on precedence in the case of ML and DL, but such algorithms can underperform in novel or unusual cases of drug side effects or treatment resistance where there is no prior example to build on. For these reasons it can be concluded that AI-based systems will support the skills of physicians and are unlikely to replace the traditional physician–patient relationship [1].

8.2 AI medicine versus ES medicine

Expert systems (ES) are one of the prominent areas of research in the field of AI and an early product of the whole AI endeavor. In fact, ES represent the most successful demonstration of the capabilities of AI, and they are the first truly commercial application of the work done in the field of AI. ES are computer programs that simulate the thought process of a human expert to solve complex decision problems in a specific domain. It began as a special branch of AI in the late 1960s to early 1970s but has grown dramatically in the past few years [15, 16]. Table 8.1 differentiates and compares several elements between AI and ES [15, 17–20].

Healthcare systems produce tremendous amounts of information (patient, demographic, clinical, and billing data), which are susceptible to analysis by intelligent software and need new techniques to extract new knowledge. A variety of medical ES tools are available and can function as intelligent assistants to clinicians, helping in diagnostic processes, laboratory analysis, treatment protocol, and teaching of medical students and residents [21, 22]. So far the following three ES systems have been successfully triumphed in the world of medicine: MYCIN (was designed at Stanford University in the early 1970s by E. Shortliffe and was an early backward-chaining ES that used AI to identify bacteria causing severe infectious diseases in the blood, such as bacteremia and meningitis, and to recommend antibiotics, with the dosage adjusted for patient's body weight – the name derived from the anti biotics themselves, as many antibiotics have the suffix "-mycin" [24]), PUFF (was designed in the late 1970s with the collaboration of R. Fallat (specialist in pulmonary diseases), and interprets measurements from respiratory tests administered to patients in the pulmonary (lung), enabling to diagnose lung diseases [24]), and CADUCEUS (was originally from the University of Pittsburgh by H. Pople and is used for the diagnosis of internal medicine. CADUCEUS has been described as the "most knowledge-intensive expert system" in existence [25]). These ES are very useful and can facilitate practice for professionals. Despite this, many experts in the world of medicine have expressed their concern about the idea that in the future the entire decision-making process could be in the hands of machines, causing the role of

Table 8.1: Differences between AI and ES [15, 17–20].

	Artificial intelligence: AI	Expert system: ES
Definition	AI is the simulation of human intelligence in machines that are programmed to imitate the human capabilities of thinking, sensing, and learning.	ES refers to computer programs that simulate the thought process of a human expert to solve complex decision problems in a specific domain.
In general	AI manages more comprehensive issues of automating a system. This computerization should be possible by utilizing any field such as image processing, cognitive science, neural systems, and machine learning.	An ES is an AI software that uses knowledge stored in a knowledge base to solve problems that would usually require a human expert, thus preserving a human expert's knowledge in its knowledge base.
Goal	AI is the study of systems that act in a way that it would appear to be intelligent to any observer.	Experts systems represent the most successful demonstration of the capabilities of AI.
Components	Components of AI include natural language processing (NLP), knowledge representation, reasoning, problem solving, and machine learning.	ES are typically composed of the inference engine, knowledge base, user interface, and knowledge acquisition module.
Conditions	Requires a learning machine with large amounts of sample data and configuration.	The burden on designers (humans) is high. Advanced rules can be difficult to define.
Applications	AI systems are used in a wide range of industries, from healthcare to finance, automotive, data security, social media, travel and transport, etc.	ES provide expert advice and guidance in a wide variety of activities, from computer diagnosis to delicate medical surgery.

doctors to become irrelevant. However, it must be clarified that ES in medicine are not designed to replace doctors [25].

In ES, it is common to perform "inference" by means of local computations on such large but sparse networks. In general, non-probabilistic methods are used to handle uncertainty when propagating the effects of evidence, and it has appeared that exact probabilistic methods are not computationally feasible. Based on this background, Lauritzen et al. [26], being motivated by an application in electromyography, exploited a range of local representations for the joint probability distribution, combined with topological changes to the original network termed "marrying" and "filling-in." It was reported that the resultant structure allows efficient algorithms for transfer between representations, providing rapid absorption and propagation of evidence. These models typically possess a bipartite causal graph structure modeling dependencies between diseases and findings [27].

8.3 Preventive medicine

8.3.1 Prevention

Healthcare is health management along with maintenance and promotion of health. In the Orient, the healthy condition is planned to be maintained or promoted, based on the concept of curing or "non-illness," indicating the physical state before getting sick by enhancing a self-healing power. There is a saying that "a good doctor is one who detects and deals with abnormalities at the pre-illness stage, not after the onset of illness." On the other hand, in Western medicine, pre-illness tends to be overlooked, and treatment is started only after the onset of illness. In the West, it is easy to think that the role of medical care is to treat after injury or illness. But, in recent years, attention has also been paid to the maintenance and promotion of health and preventive medicine in the West, too.

When it comes to healthcare, the system itself is often sicker than the patients. Even the terminology is misleading. Today, going to the doctor is about sick care more than healthcare; hence, it is reactive, not proactive [28]; however, as seen later, the situation is gradually and certainly changing toward the more healthcare-oriented atmosphere. AI has yet to transform healthcare, cutting costs by making providers more efficient and improving the health of patients. The promise of AI to do just that – by augmenting human activities, not replacing them – is real [29]. Just as humans are better equipped to understand the world when they take in high-quality facts, so too are algorithms. This is a special problem in healthcare, where data are often fragmented, stored, and held in a form designed for humans, not computers, to understand. Today, the healthcare industry is turning to AI for back-office work, automating tasks to make them less tedious and more efficient. There are no magical algorithms than can read a patient's chart and tell doctors with certainty what is wrong and what the treatment should be. Although the data also have to be unbiased, there are problems of ethics and liability [29]. Preventive medicine can include (i) the prevention of potential diseases, (ii) maintaining the current healthy condition, and (iii) the prediction of diseases. AI is used to provide optimal information to individuals through data collection and analysis, and to improve the convenience of the tool itself and promote its continuous use. AI-equipped apps automatically diagnose daily diet, sleep, and exercise and provide health advice suitable for each individual, as discussed in Chapter 6. One of the factors that contributed to the development of healthcare AI in the field of prevention is the spread of smart devices. The diversification of the types of acquired data by wearable devices, as mentioned later, the increase in the amount of data, and the efficiency of "transmitting", managing, and sharing data by smart devices of existing vital measuring instruments – all these factors have supported the development of data that can be utilized in the AI field.

In order to collect such valuable intrabody information, there should be well-structured communication systems to collect, transfer to each other, and communicate in different scale of areas to be covered. Normally, as seen in Figure 8.2, there are three typical networks in the public domain: WAN (wide area network; in comparison to a MAN, WAN is not restricted to a geographical location, although it might be confined within the bounds of a state or country; the Internet is an example of a worldwide public WAN), MAN (metropolitan area network; it is a larger network that usually spans several buildings in the same city or town), and LAN (local area network; it is a group of computers and network devices connected together, usually within the same building, and the connections must be high speed and relatively inexpensive). Between WAN and MAN, there is RAN (radio access network), which has been in use since the beginning of cellular technology. This network uses a base station and antennas to cover a region, and it includes 3G, 4G, and 5G network connections. Between MAN and LAN, there is CAN (campus area network), which is a network that covers multiple LANs in a smaller geographical area like a university. The range of a CAN is about 1–5 kilometers. There are still smaller devices than LAN – namely, PAN (personal area network), which is a network that connects devices within a person's workspace. This includes devices like laptops, smartphones, and tablets. We can use a PAN for connectivity between these devices or to connect to another network type (LAN or WAN) where one device becomes the gateway for all other devices. PAN networks can be wired (for example, USB) or wireless. A wireless personal area network (WPAN) is a PAN that uses short-distance low-power wireless technology (e. g., Bluetooth or Zigbee). A WPAN covers anywhere between a few centimeters to a few meters. BAN (body area network) is a wireless network for devices near or inside the body. BAN devices include not only wearable technology, such as smart watches or earbuds, but also those inside the body, such as implants. This network type is also known as a wireless BAN (WBAN), body sensor network (BSN), or medical BAN (MBAN). The latest standard for WBANs is IEEE 802.15.6. Finally, a nanonetwork is a set of small devices (a few hundred nanometers or a few micrometers) that can perform only simple tasks. These devices could be sensors, actuators, or data storage or computing devices. Nanonetworks allow new applications in fields such as environmental research, military technology, and the biomedical field. IEEE created the P1906.1 workgroup to develop a common framework for nano-scale and molecular communication [30, 31].

Wearable technologies can be innovative solutions for healthcare problems. Wu et al. [32] mentioned that some wearable technology applications are designed for prevention of diseases and maintenance of health, such as weight control and physical activity monitoring. Wearable devices are also used for patient management and disease management. The wearable applications can directly impact clinical decision-making. Some believe that wearable technologies could improve the quality of patient care while reducing the cost of care, such as patient rehabilitation outside of hospitals. The big data generated by wearable devices is both a challenge and an opportunity for researchers who can apply more AI techniques on these data in the future. Most wearable technologies are still in their prototype stages. Issues such as

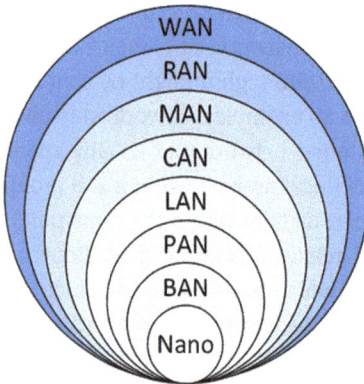

Figure 8.2: Interrelationship among various scales of networks [30].

user acceptance, security, ethics, and big data concerns in wearable technology still need to be addressed to enhance the usability and functions of these devices for practical use.

When healthcare practitioners must make life-or-death decisions, the quality of information at their disposal is critical. Having more specific data – and being able to access it in real time – leads to more informed decisions. The internet of medical things (IoMT) makes this possible through an infrastructure of connected medical devices, software applications, and health systems powered by 5G wireless technology and edge computing, which enables connected devices to process data closer to where it is created [33–36]. The rise in the number of connected medical devices comes at a critical time as barriers to access health services continue to increase [35, 36]. Dimitrov [37] mentioned that since many patients have increased their accesses to high-speed Internet and smartphones, they have started to use mobile apps to manage various health needs, and these devices and mobile apps are now increasingly used and integrated with telemedicine and telehealth via the medical IoT (aka, IoMT). After reviewing this medical IoT, it was concluded that (i) a new category of the personalized preventative health coaches (digital health advisors: DHA) will emerge; (ii) these workers will possess the skills and the ability to interpret and understand health and well-being data; (iii) they will help their clients avoid chronic and diet-related illness, improve cognitive function, achieve improved mental health, and achieve improved lifestyles overall; and (iv) as the global population ages, such roles will become increasingly important. Figure 8.3 illustrates the interconnections between various medical sectors in the IoMT atmosphere [37].

Intrabody communication (IBC) is a promising wireless communication technology that provides data from the human body to the medical and sports devices. In fact, in the last two decades, lots of wearable electronics devices have been developed: various body implants, medical devices to control health parameters, and technologies for sport activities which monitor athletes' performances are examples of this technology revolution [38–40]. There are two primary methods for signals to be

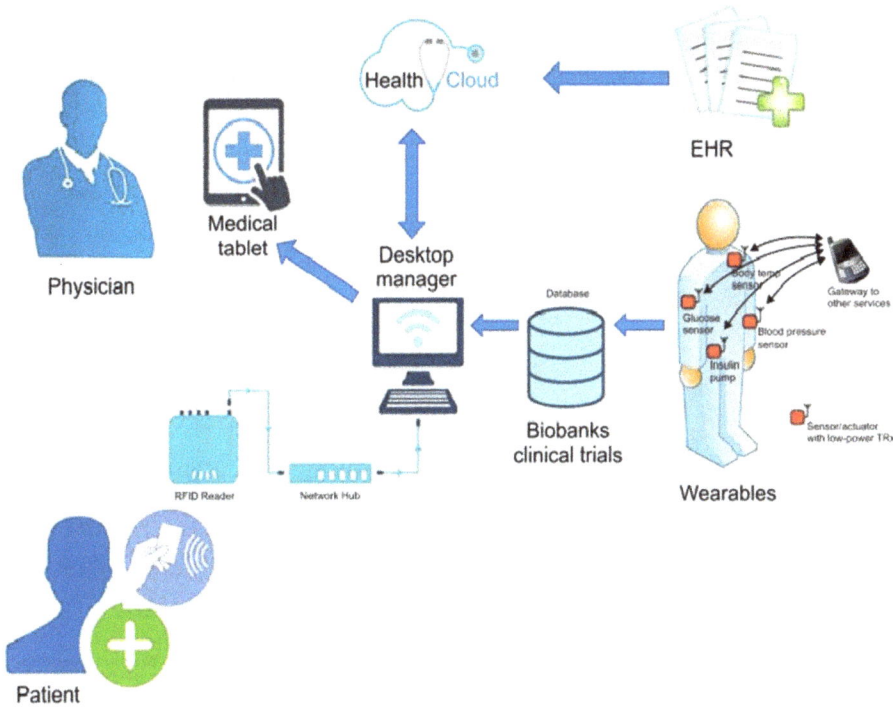

Figure 8.3: Interconnections between various medical sectors in the IoMT atmosphere [37].

coupled into the body for IBC (capacitive coupling and galvanic coupling) and two less popular approaches (inductive coupling and the classical RF coupling) [38]. In capacitive coupling, a signal is applied to a transceiver electrode, and an electric field is built up. This approach mainly aims to maximize the coupling between the transceiver and the human body, thus reducing the interference because of ambient noise. In galvanic coupling, a signal is controlled by the applied current flow, while the human body could be considered as the waveguide. In comparison to capacitive coupling, the propagation of the current in galvanic coupling through the body does not involve any return path. Therefore, this method is less susceptible to noise. In inductive coupling, the electromagnetic coupling is used to provide a communication link to implanted devices, by placing the external coil close to the patient that couples to a coil implanted below the skin surface. The implant is powered by the coupled magnetic field. The best power transfer is achieved in inductive coupling when it is used in large transmitting and receiving coils. Therefore, this method could be used effectively if space is not a constraint. RF-based IBC is one of the alternative techniques that can increase the bandwidth and enable two-way data communication [38]. Figure 8.4 shows typical wearable intrabody sensors: (a) a simplified map of the biosignals to be measured from the human body through wearables systems and the corresponding sensing point [41] and (b) examples of wearables and the associated medical capabilities [33]. Besides aforementioned intrabody information, there are still more devices sensing blood glucose, tissue/organ measurement, nutrition, intraoral, sweat, urine, brain, sleep, emotion recognition, or stress.

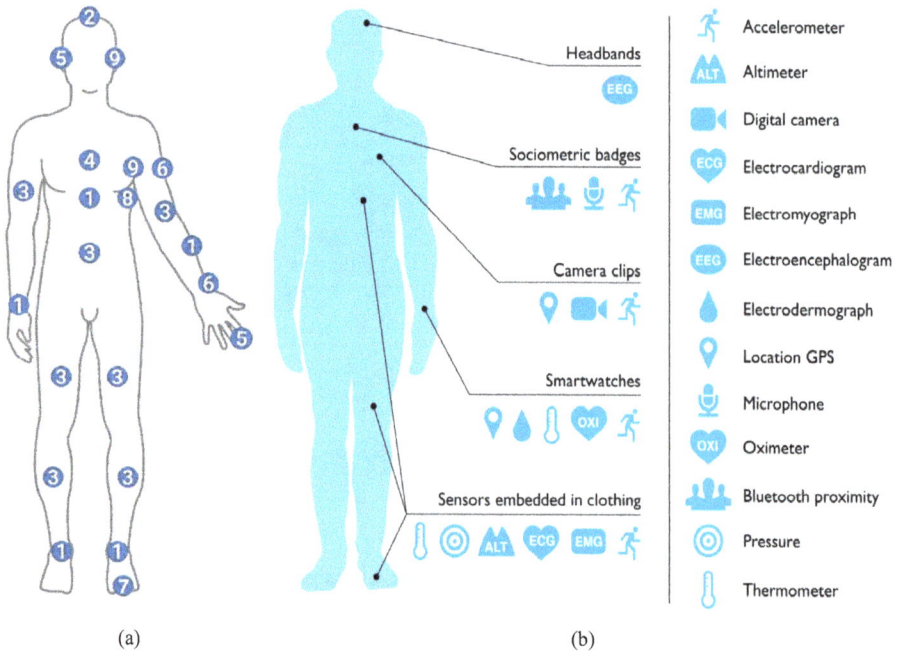

Figure 8.4: (a) Simplified map of the biosignals through wearables systems, where 1, ECG (electrocardiogram); 2, EEG (electroencephalogram); 3, EMG (electromyogram); 4, respiration; 5, blood gas; 6, blood pressure; 7, interface pressure; 8, resistance; and 9, temperature [41]. (b) Examples of wearables and the associated medical capabilities [33].

8.3.2 Prediction

AI technology can not only prevent diseases but also predict potential risks of diseases and injuries. To briefly explain the advantages of AI, it will be the fastest "predictive" machine that will learn by bringing in information processing power and new information one after another. In healthcare, such AI-based prediction systems are emerging. AI predicts the risk of three lifestyle-related diseases: type II diabetes, hypertension, and dyslipidemia. Based on health check-up data of more than 1 million people, healthcare service and big data analyzing capability have been jointly developing a useful health guidance for the prevention of serious diseases. The developing system can be modified for being applicable to dementia and cancer [42, 43]. The AI-assisted gait-analysis system, designed particularly for elders, was developed to analyze a 5-m walking video and to visualize the risk of falling in walking and how to help [44].

8.4 Diagnosis

8.4.1 Introduction

One of the areas where AI in healthcare has shown the most promise is diagnostics. Early diagnosis is one of the most important factors in the ultimate outcome of a patient's care. AI/Dl algorithms are being used to shave down the time it takes to diagnose serious illnesses. The way AI rapidly processes large amounts of information and arrives at likely causes for symptoms can drastically reduce the diagnosis–treatment–recovery cycle for many patients. The effects of this are already being felt in several areas [45]. The amount of the healthcare data industry is increasing exponentially. It was reported that the volume of global healthcare data was around 153 exabytes in 2013 [46] (an exabyte equals 1 billion gigabytes; 5 exabytes is equal to all the words ever spoken by humans [47]), and a 48% annual growth rate was projected, meaning that the figure would reach 2,314 exabytes by 2020 [46].

The medical diagnostic process, defined as a mapping from a patient's data (normal and abnormal history, physical examination, and laboratory data) to a nosology of disease states, starts with a differential diagnosis (DDx), consisting of a ranked list of possible diagnoses, that is used to guide further assessments and possible treatments [48]. The traditional approach of diagnosis by individual physicians has a high rate of misdiagnosis. Hence, Barnett et al. [49] assessed how the diagnostic accuracy of groups of physicians and trainees compares with the diagnostic accuracy of individual physicians. They demonstrated that diagnosis accuracy is around just 60% when only one medical provider is involved in the decision-making process, concluding that a collective intelligence approach was associated with higher diagnostic accuracy compared to individuals, including individual specialists whose expertise matched the case diagnosis, across a range of medical cases [49]. Ideally, medical diagnosis can be viewed as a complete information game in which the medical provider has all the information needed for diagnosis, and the entirety of this information is used to draw conclusions. However, there are many factors that prevent this from happening in the real world, including (i) patient holding back pertinent information that they perceive as unimportant or irrelevant, (ii) cost of acquiring information, whether that be the cost of a test or the time it takes to further talk to the patient, given that a typical medical provider's visit lasts only 15 min, (iii) reliance on recency and availability bias due to sheer number of diseases that are possible, and the need to quickly arrive at a diagnosis, (iv) lack of appropriate diagnostic tools to gather complete information about the patient for certain diseases such as mental health, chronic fatigue, and Lyme disease, and (v) patient accessing different disconnected areas of healthcare, resulting in an incomplete view of their medical record [27, 50].

Recently, many researchers have conducted data mining over medical data to uncover hidden patterns and use them to learn prediction models for clinical decision-making and personalized medicine. While such healthcare learning models can achieve encouraging results, they seldom incorporate existing expert knowledge into their frameworks, and hence prediction accuracy for individual patients can still be improved. However, expert knowledge spans across various websites and multiple databases with heterogeneous representations and hence is difficult to harness for improving learning models. In addition, patients' queries at medical consult websites are often ambiguous in their specified terms and hence the returned responses may not contain the information they seek. To these ends, Xue et al. [51] designed a knowledge extraction framework that can generate an aggregated dataset to characterize diseases by integrating heterogeneous medical data sources and proposed an end-to-end DL-based medical diagnosis system (DL-MDS) to provide disease diagnosis for authorized users, based on the integrated dataset. In real life, symptoms are often tricky to spot even by the best experts; however, diagnostic mistakes are acknowledged as the most frequent and harmful medical errors [52]. Between 12 and 18 million Americans every year will experience some sort of misdiagnosis; even symptoms never lie [53].

AI/ML possesses a potential to analyze large datasets and extract meaningful insights, and this proves to be helpful in both radiology and pathology. Images from MRI machines, CT scanners, and X-rays can contain large amounts of complex data that can be difficult and time-consuming for human providers to evaluate [54]. AI/DL technology and its implementation into day-to-day clinical imaging are poised to transform the practice of radiology [55]. AI can provide clinical decision support to radiologists and improve the delivery of care to patients. With regard to image processing, DL algorithms can help select and extract features from medical images as well as help create new features. With respect to image interpretation, DL algorithms can help identify and classify disease patterns from images and help the radiologist suggest suitable care pathways for a patient in consultation with other physicians involved in the care of the patient [1, 55].

There are several driving forces that push forward the use of DL in radiology and other diagnostic practices, including (i) the continued growth of computing power and storage technologies, (ii) declining cost of hardware, (iii) rising cost of healthcare, (iv) the shortage of healthcare workers, and (v) an abundance of medical data to train models [52]. A lot of AI algorithms have been proposed and developed by computer science scientists for the detection and diagnosis of cancer, diseases of lung, rheumatoid arthritis, diabetic retinopathy, diseases of heart, Alzheimer's disease, hepatitis, dengue, liver disease, and Parkinson's disease [56]. AI-based methods have emerged as powerful tools to transform medical care. Although the ML classifiers (MLCs) have already demonstrated strong performance in image-based diagnoses, analysis of diverse and massive EHRs data remains challenging [13, 57]. Medical conditions have grown more complex, and with a vast history of electronic medical records (EMR) building, the likelihood of case duplication is high. Although someone today with a rare illness is less likely to be the only person to have suffered from any given disease, the inability to access cases from similarly symptomatic origins is a major roadblock for physicians. It was further mentioned that the implementation of AI to not only help find similar cases and treatments but also factor in chief symptoms and help the physicians ask the most appropriate questions helps the patient receive the most accurate diagnosis and treatment possible [57].

8.4.2 Diagnosis

Diagnostics is the most invested area in the AI-assisted healthcare field. In the field of diagnostic support using AI, in addition to data from blood tests, bio-data such as genetic data and medical images such as X-rays are used. Among them, medical imaging is the most actively used. Figure 8.5 demonstrates contribution of DL in different diagnostics and indicates that a clear majority of DL is used in imaging analysis, which makes sense given that images are naturally complex and high volume [58].

Figure 8.5: The distribution of data sources for deep learning [58].

Until now, reading and diagnosing medical images obtained using MRI, CT, endoscopy, echo, and so on have been performed by radiologists and endoscopists. Pastorello et al. [59] mentioned that DL algorithms (in particular, CNNs) are used in vision problems for (i) the classification of images/frames according to some pre-set labels. In this case, a large number of images (usually in the range of millions) are used, together with previously known "labels" to train the model. Such a model does learn how to match images with labels and can then apply this learning to new images, (ii) finding the position of objects in images (segmentation). The NN in this case learns to find the pixels associated with an object in an image, (iii) generating new images starting from a pool of original pictures. In this case the use of generative models allows the production of new (often realistic) images. In healthcare and medical imaging processing, classification and segmentation are the main applications of CNNs [59].

The possibility of AI-based diagnostic imaging is attracting attention for the following three characteristics [60]. (1) *Leveraging time and resources*: Recently, the number of images per test has increased due to improved equipment quality and inspection throughput, as well as where the number of medical images that need to be read by one specialist is concerned. Therefore, AI that can process large amounts of images efficiently assists doctors in making decisions by screening and marking in advance [61]. (2) *Improved accuracy*: Attention is focused on the possibility of improving the accuracy of a doctor's diagnosis by assisting in the diagnosis of the doctor. In a screening test for lung cancer using AI developed by Google to support doctors' diagnosis, it is said that it was able to reduce false negatives by 5% and false positives by 11% compared to tests performed by six radiologists [62]. (3) *Early diagnosis*: As can be seen from the results of lowering false negatives discussed previously, it is also expected that we may be able to detect super-early signs of the disease. In addition, DL models developed by MIT's Computer Science and AI Laboratories and Massachusetts General Hospital have been designed to predict breast cancer in patients 5 years before the onset of the disease, aiming to improve prognosis through early diagnosis and early treatment of diseases [63].

Despite the background of actual clinical use and activation of regulatory approvals, there are two major challenges in the development of diagnostic imaging AI [60].

(1) *Lack of data*: Medical images are difficult to obtain from the viewpoint of personal information protection, and so on, and may hinder technological development. In the United States, consortiums such as the Lune Image Database Consortium and the Image Database Resource Institute are leading the way in constructing case databases and providing them to technologically advanced private companies and research institutions. In addition, there are cases where startups and hospitals work directly together to conduct research and development [64]. In recent years, pseudo-teacher data creation algorithms such as GAN have been used for mask images that take time when created by doctors and for generating rare data that are difficult to acquire in the first place.

(2) *Annotation costs*: The annotation of medical images (tagging teacher data) is a highly specialized task, so it is necessary to employ healthcare professionals, and it is expected that the financial burden will be considerably greater for startups. The NIH and its supporting research institute, DeepMind under Google, are releasing annotated data to the public as open data to boost R&D [60].

Complex medical devices such as CT, MRI, and ultrasound machines are controlled by instructions sent from a host PC. bnormal or anomalous instructions introduce many potentially harmful threats to patients, such as radiation overexposure, manipulation of device components, or functional manipulation of medical images. Threats can occur due to cyberattacks, human errors such as a technician's configuration mistake, or host PC software bugs. Mahler et al. [65] and his researchers have developed a new AI technique that will protect medical devices from malicious operating instructions in a cyberattack as well as other human and system errors. The new architecture in the CT domain was evaluated using 8,277 recorded CT instructions, the CF (context free) layer was evaluated using 14 different unsupervised anomaly detection algorithms, and then the CS (context sensitive) layer for four different types of clinical objective contexts was further evaluated using five supervised classification algorithms for each context. It was reported that (i) adding the second CS layer to the architecture improved the overall anomaly detection performance from an F1 score of 71.6%, using only the CF layer, to between 82% and 99%, depending on the clinical objective or the body part, and (ii) the CS layer enables the detection of CS anomalies using the semantics of the device's procedure, an anomaly type that cannot be detected using only the CF layer [65]. The potential risks of cyberattacks on medical devices have been pointed out. In addition to the possibility of preventing such threats, the study can also address human errors such as mis-setting engineers and software bugs in host PCs [65].

It was reported that, for decades, perfumers and scientists have struggled to predict the relationship between a molecule's structure and its scent [66]. While scientists can look at a wavelength of light and identify what color it is, when it comes to scents, scientists cannot simply look at a molecule and identify its odor. Researchers from the Google Brain Team are hoping AI might change that. The researchers created a data set of nearly 5,000 molecules identified by perfumers, who labeled the molecules with descriptions ranging from "buttery" to "tropical" and "weedy." The team used about two-thirds of the data set to train its AI (a graph NN or GNN) to associate molecules with the descriptors they often receive. The researchers then used the remaining scents to test the AI, and it passed. The algorithms were able to predict molecules' smells based on their structures [66]. Imam et al. [67] reported about a neural algorithm for the rapid online learning and identification of odorant samples under noise, based on the architecture of the mammalian olfactory bulb and implemented on the Intel Loihi neuromorphic system. As with biological olfaction, the spike timing-based algorithm utilizes distributed, event-driven computations and rapid (one shot) online learning. Spike timing-dependent plasticity rules operate iteratively over sequential gamma-frequency packets to construct odor representations from the activity of chemosensor arrays mounted in a wind tunnel. Learned odorants then are reliably identified despite strong destructive interference. Noise resistance is further enhanced by neuromodulation and contextual priming. Lifelong learning capabilities are enabled by adult neurogenesis. The algorithm is applicable to any signal identification problem in which high-dimensional signals are embedded in unknown backgrounds.

There are many examining methods for detecting diseases such as blood, urine, blood pressure, chest X-rays, and electrocardiogram (ECG). Under these developments, research to make use of odors as a new method of diagnosing diseases is becoming active [68]. It measures biological gases from the human body, such as skin gas and air, and identifies diseases at an early time. Biogas measurement is also attracting attention as a "noninvasive" method. Examples that have already been introduced include "Exhalation NO (nitric monoxide) test" in asthma and "urea aeration test method" of Pylori bacteria (which is strongly related to ammonia gas) testing that causes badness in the stomach and duodenum, and their usefulness is also medically recognized. The Bio-Sniffer has been developed to detect a special gas (trimethylamine) which is originated from metabolism disorders. Using MSS (membrane-type surface sensor), medical applications include routine health check-up and detection of odors specific to cancer patients. The sensors that make new disease diagnosis are not only artificial devices, as mentioned earlier. For example, as an early cancer detection method, "sniffing using living organisms" is attracting attention. Nematodes with an excellent sense of smell possess advantages of the characteristics (chemical chemotaxix) that approached the urine of cancer patients and separated from the urine of healthy people. It is a very simple method that can be tested with a drop of urine, and previous demonstrations have shown that nematodes respond to about 10 types of cancers, including stomach cancer, colorectal cancer (CRC), pancreatic cancer, eating tract cancer, gallbladder cancer, bile duct cancer, prostate cancer, breast cancer, lung cancer, and cecum cancer. It reacts with high sensitivity even at early stages such as stages 0 and 1, and the judgment sensitivity is as high as 93.8%. The detecting method is called as "N-NOSE." Besides nematodes, similarly, cancer detection dogs are well known as organisms that sniff out cancer odors [68].

8.4.3 Diagnosis cases

8.4.3.1 Cancer

WHO [69] reported that between 30% and 50% of all cancer cases are preventable, and the prevention offers the most cost-effective long-term strategy for the control of the cancer. Basically, there are quite a variety of cancers from head to toe, including head area (brain tumor), head and neck area (head and neck cancer, salivary gland cancer, and thyroid cancer), chest area (lung cancer, small cell lung cancer, breast cancer, and mesothelioma), liver, gallbladder, pancreas area (pancreatic cancer, liver cancer, and biliary cancer), digestive area (esophageal cancer, stomach cancer, gastrointestinal (GI) stromal tumor – GIST, small bowel cancer, and CRC), urinary area (kidney cancer, renal pelvis and urethritis cancer, bladder cancer, and prostate cancer), female-specific (cervical cancer, ovarian cancer, and uterine sarcoma), blood and lymph area (malignant lymphoma, leukemia, chronic lymphocytic leukemia – CLL, and multiple myeloma), and systemic (skin cancer, melanoma, sarcoma, and carcinoma of unknown primary). According to WHO data for year 2018 [70], cancer is a leading cause of death worldwide, accounting for an estimated 9.6 million deaths in 2018. The most common causes of cancer death are lung (1.76 million deaths), colorectal (862,000 deaths), stomach (783,000 deaths), liver (782,000 deaths), and breast (627,000 deaths) cancers.

Lung cancer is the world's deadliest oncology disease; it leads the list of cancer-related mortality and is second only to skin cancer in the prevalence rate. As with other malignancies, early detection may be lifesaving [32]. Unfortunately, lung cancer symptoms are very similar to those of pneumonia or bronchitis. This is the reason why it is spotted only in advanced stages in about 70% of cases [71]. Johnson and co-researchers of Northwestern Medicine developed a DL model, with which 42,000 chest CT scans were trained to exhibit better result at diagnosing lung cancer than radiologists with 8 years of expertise. The algorithm was able to find malignant lung modes 5–9.5% more often than human specialists [72]. The odds are that before long AI systems will assist radiologists in analyzing large numbers of CT images, thus contributing to successful treatment and increasing survival rate [32]. Lung cancer screening with chest CT reduces lung cancer death. The Centers for Medicare and Medicaid Services (CMS) eligibility criteria for lung cancer screening with CT require detailed smoking information and miss many incident lung cancers. An automated DL approach based on chest radiograph images may identify more smokers at high risk for lung cancer who could benefit from screening with CT. Lu et al. [73] developed and validated a convolutional neural network (CXR-LC) that predicts long-term incident lung cancer using data commonly available in the EMR (chest radiograph, age, sex, and whether currently smoking). Measurements were continued up to 12-year lung cancer incidence predicted by CXR-LC. It was found that the CXR-LC model had better discrimination (area under the receiver-operating characteristic curve: AUC) for incident lung cancer than CMS eligibility, concluding that the CXR-LC model identified smokers at high risk for incident lung cancer, beyond CMS eligibility and using information commonly available in the EMR.

CRC is one of the most common cancer diagnoses and is a leading cause of cancer-related deaths worldwide. AI is emerging as a tool with which gastroenterologists can improve colon polyp detection rates, characterization, and management. Colonoscopy remains the standard strategy for screening for CRC around the world due to its efficacy in both detecting adenomatous or pre-cancerous lesions and the capacity to remove them intra-procedurally. Computer-aided detection and diagnosis (CAD) – AI/DL technology – leads to a promising solution to human biases in performance by guarantying decision support during colonoscopy. The application of CAD on real-time colonoscopy helps in increasing the adenoma detection rate and therefore contributes to reduce the incidence of interval cancers, thus improving the effectiveness of colonoscopy screening on critical outcome such as CRC-related mortality [74]. The AI assistant system is expected to improve the quality of automatic polyp detection and classification. It could also help prevent endoscopists from missing polyps and make an accurate optical diagnosis. These functions provided by AI could result in a higher adenoma detection rate and decrease the cost of polypectomy for hyperplastic polyps. In addition, AI has good performance in the staging, diagnosis, and segmentation of CRC [75]. The effect of colonoscopy on CRC mortality is limited by several factors, among them a certain miss rate, leading to limited adenoma detection rates (ADRs). The effect of an automatic polyp detection system based on DL on polyp detection rate and ADR was investigated [76, 77]. In an open, non-blinded trial, consecutive patients were prospectively randomized to undergo diagnostic colonoscopy with or without the assistance of a real-time automatic polyp detection system providing a simultaneous visual notice and sound alarm on polyp detection. The primary outcome was ADR. Of 1,058 patients included, 536 were randomized to standard colonoscopy and 522 were randomized to colonoscopy with computer-aided diagnosis. It was obtained that (i) AI system significantly increased ADR (29.1% vs. 20.3%) and the mean number of adenomas per patient (0.53 vs. 0.31), due to a higher number of diminutive adenomas found (185 vs. 102), while there was no statistical difference in larger adenomas (77 vs. 58), and (ii) the number of hyperplastic polyps was also significantly increased (114 vs. 52), concluding that (iii) in a low prevalent ADR population, an automatic polyp detection system during colonoscopy resulted in a significant increase in the number of diminutive adenomas detected, as well as an increase in the rate of hyperplastic polyps [76, 77].

Despite population screening and availability of several stool-based, non-invasive screening methods, over 20% of CRC in the United States are metastatic at the time of diagnosis. Blood-based methods using cell-free DNA (cfDNA) are under development as an alternative to stool-based tests. However, early-stage detection of cancer using tumor-derived mutations in cfDNA (circulating tumor DNA, or ctDNA) has proven challenging because of the small proportion of cfDNA derived from tumor tissue (tumor fraction, ctDNA/cfDNA ratio) in early-stage disease. Niehous et al. [78] introduced using an AI-driven approach based on ML to discover signatures in cfDNA potentially reflective of both tumor and immune contributions; this may represent a promising direction for the early detection of cancer. It was found that (i) in a cohort heavily weighted toward early-stage cancer (82% stage I/II), this method achieves a sensitivity in cross-validation of 81% (Clopper-Pearson 95% confidence interval, 77–84%) at 85% specificity, (ii) sensitivity generally increased by tumor stage. Stratification by sequencing batch was required for reliable generalization; (iii) further analyses revealed susceptibility to additional confounders, including variation in preanalytical and analytical processes such as institution-specific blood collection protocols, and (iv) down-sampling the dataset to balance with respect to such confounders can reduce sensitivity at 85% specificity by 20–30%. Based on these findings, it was concluded that (v) an ML approach using a single analyte was able to achieve high sensitivity and specificity in a predominantly early-stage CRC cohort, and (vi) the observation of systematic technical and site-specific biases warrants similar confounder analyses in other retrospective studies [78]. Because of limited biomarkers identified as prognostic predictors for stage III colon cancer, Jiang et al. [79] developed a computer-aided approach which combined CNN with machine classifier to predict the prognosis of stage III colon cancer from routinely hematoxylin and eosin (H&E) stained tissue slides, using 101 cancers. Gradient boosting colon is an independent machine prognostic predictor which allows stratification of stage III colon cancer into high- and low-risk recurrence groups, and poor and good prognosis groups directly from the H&E tissue slides. It was indicated that findings could provide crucial information to aid treatment planning during stage III colon cancer [79]. A new modeling tool may be able to help doctors assess which treatments are best for individual patients with CRC. The AI program analyzes a patient's disease details (such as the stage of cancer and other chronic conditions) and compares those details to other CRC cases to predict the patient's chance of surviving past 10 years. Predicting a cancer patient's survival time lends valuable insight into the best course of treatment for both the long and the short term, helping to determine what is likely most suitable [80]. Bibault et al. [81] introduced an accurate and explainable prediction of the risk to die within 10 years after CRC diagnosis, by incorporating the tumor features and the patient medical and demographic information. It was reported that (i) during the follow-up, 2,359 patients were diagnosed with CRC. Median follow-up was 16.8 years (14.4–18.9) for mortality. In total, 686 patients (29%) died from CRC during the follow-up, (ii) the dataset was randomly split into a training (n = 1,887) and a testing (n = 472) dataset, and (iii) the AUC operating characteristic was 0.84 (±0.04) and accuracy was 0.83 (±0.04), with a 0.5 classification threshold. The model is available online for research use, concluding that (iv) the model has high predictive performances at the individual scale [81]. AI tools will be used regularly in the clinical setting to guide endoscopy (including colonoscopy), esophagogastroduodenoscopy, capsule enteroscopy, and endoscopic ultrasound and endoscopic retrograde cholangiopancreatography [81].

Gastric cancer (GC) is the fifth most common malignancy and the third leading cause of cancer-related mortality worldwide [82]. Although advanced GC has a poor 5-year survival rate of <25% [83], early detection can substantially improve both morbidity and survival; for example, nationwide mass screening programs for gastric neoplasia in Japan have reduced the related mortality [84]. There are few reports regarding the accuracy of endoscopy in detecting cancer. Hosokawa et al. [85] investigated the difference in the false-negative rates for cancer detection between gastroscopy and colonoscopy using the records of a population-based cancer registry and concluded that, for routine examinations, surveillance after negative endoscopy should be discussed based on the difference in false-negative rates between gastroscopy and colonoscopy. Detecting early GC is difficult, and it may even be overlooked by experienced endoscopists. Recently, AI-based on DL through CNNs has enabled significant advancements in the field of gastroenterology. However, it remains unclear whether a CNN can outperform endoscopists. Ikenoyama et al. [86] evaluated whether the performance of a CNN in detecting early GC is better than that of endoscopists. The CNN was constructed using 13,584 endoscopic images from 2,639 lesions of GC. Subsequently, its diagnostic ability was compared to that of 67 endoscopists using an independent test dataset (2,940 images from 140 cases).

It was found that (i) the average diagnostic time for analyzing 2,940 test endoscopic images by the CNN and endoscopists were 45.5 ± 1.8 s and 173.0 ± 66.0 min, respectively, (ii) the sensitivity, specificity, and positive and negative predictive values for the CNN were 58.4%, 87.3%, 26.0%, and 96.5%, respectively, (iii) these values for the 67 endoscopists were 31.9%, 97.2%, 46.2%, and 94.9%, respectively, and (iv) the CNN had a significantly higher sensitivity than the endoscopists (by 26.5%; 95% confidence interval, 14.9–32.5%). Based on these findings, it was concluded that (v) the CNN detected more early GC cases in a shorter time than the endoscopists. Hirasawa et al. [87] developed a CNN to automatically detect GC in endoscopic images. It was reported that (i) the CNN correctly diagnosed 71 of 77 GC lesions with an overall sensitivity of 92.2%, and 161 non-cancerous lesions were detected as GC, resulting in a positive predictive value of 30.6%, (ii) 70 of the 71 lesions (98.6%) with a diameter of 6 mm or more as well as all invasive cancers were correctly detected, (iii) all missed lesions were superficially depressed and differentiated-type intramucosal cancers that were difficult to distinguish from gastritis even for experienced endoscopists, and (iv) nearly half of the false-positive lesions were gastritis with changes in color tone or an irregular mucosal surface, indicating that (v) the constructed CNN system for detecting GC could process numerous stored endoscopic images in a very short time with a clinically relevant diagnostic ability [87]. AI developed by Riken and National Cancer Center is able to correctly detect cancer in 80% of stomach cancer images while displaying 95% accuracy for healthy tissue images. In order to train AI for accurate identification of stomach cancer, a team of researchers employed 100 endoscopic images of early-stage stomach cancer and 100 images of normal stomach tissue to test the AI. As a result, the AI took only 0.004 s to judge whether an endoscopic image showed stomach cancer in the early stages or in normal stomach tissue [88]. Stratifying GC risk and endoscopy findings in high-risk individuals may provide effective surveillance for GC. Nakahira et al. [89] developed a computerized image-analysis system for endoscopic images to stratify the risk of GC. The performance of the AI system was evaluated by the prevalence of GC in each group using an independent validation dataset of patients who underwent endoscopic examination and H. pylori serum antibody testing. In total, 12,824 images from 454 patients were included in the analysis. It was obtained that (i) the time required for diagnosing all the images was 345 seconds; (II) the AI system diagnosed 46, 250, and 158 patients as low risk, moderate risk, and high risk, respectively; (iii) the prevalence of GC in the low-, moderate-, and high-risk groups was 2.2%, 8.8%, and 16.4%, respectively; and (iii) three experienced endoscopists also successfully stratified the risk. However, interobserver agreement was not satisfactory (kappa value of 0.27, indicating fair agreement), concluding that the AI system detected significant differences in the prevalence of GC among the low-, moderate-, and high-risk groups, suggesting its potential for stratifying GC risk [89]. AI for GI endoscopy is an important and rapidly growing area of research [90, 91]. Much initial work in AI for endoscopy has focused on detection and optical diagnosis of colon polyps. However, AI has the potential to aid clinical decision-making in many other aspects of gastroenterology. A great deal of perspiration and collaboration will be required to accumulate the annotated big data that will be a key factor for the giant leaps ahead in the field of AI-assisted endoscopy [91].

Liver cancer is the second leading cause of cancer-related deaths worldwide, and the incidence rate has been growing on a global scale [92], while most cancers tend to exhibit the stable incidence or declining trends [93]. The liver is a cone-shaped organ located in the upper right area of the abdominal cavity, below our diaphragm, and on top of our stomach, intestines, and right kidney. It weighs about three pounds and is dark reddish-brown. It receives oxygenated blood from the hepatic artery and nutrient-rich blood from the hepatic portal vein. At any given time, our liver has about 13% of blood flowing through its two main lobes, each of which contains eight segments with 1,000 small lobules. The liver's lobules are connected to small ducts which flow into larger ducts and eventually form the common hepatic duct, which moves bile made by the liver to our gallbladder and duodenum, which is the beginning of the small intestine. The liver's main function is to excrete bile and regulate the levels of most substances that circulate in the blood. Cancer can form in the liver itself or the intrahepatic bile duct [94]. AI/DL can mine clinically useful information from histology. In GI and liver cancer, such algorithms can predict survival and molecular alterations. Once pathology workflows are widely digitized, these methods could be used as inexpensive biomarkers, although clinical translation requires training interdisciplinary researchers in both programming and clinical applications [95].

Early-stage diagnosis and treatment can improve survival rates of liver cancer patients. Dynamic contrast-enhanced MRI provides the most comprehensive information for differential diagnosis of liver tumors, but MRI diagnosis is affected by subjective experience, so DL may supply a new diagnostic strategy. Zhen et al. [96] employed CNNs to develop a DL system to classify liver tumors based on enhanced MR images, unenhanced MR images, and clinical data, including text and laboratory test results. It was found that (i) DL achieves a performance on a par with three experienced radiologists on classifying liver tumors in seven categories, (ii) using only unenhanced images, CNN performs well in distinguishing malignant from benign liver tumors (AUC, 0.946), (iii) new CNN combining unenhanced images with clinical data greatly improved the performance of classifying malignancies as hepatocellular carcinoma (HCC; AUC, 0.985), metastatic tumors (0.998), and other primary malignancies (0.963), and the agreement with pathology was 91.9%, (iv) these models mined diagnostic information in unenhanced images and clinical data by deep NN, which were different to previous methods that utilized enhanced images, and (v) the sensitivity and specificity of almost every category in these models reached the same high level compared to three experienced radiologists. It was, therefore, concluded that (vi) trained with data in various acquisition conditions, the DL system that integrated these models could be used as an accurate and time-saving assisted-diagnostic strategy for liver tumors in clinical settings, even in the absence of contrast agents, and (vii) the DL system therefore has the potential to avoid contrast-related side effects and reduce economic costs associated with current standard MRI inspection practices for liver tumor patients [96]. Regarding the diagnosis of liver tumors using ultrasound images, 4- or 5-class classifications, including the discrimination of HCC, metastatic tumors, hemangiomas, liver cysts, and focal nodular hyperplasia, have been reported using AI. A combination of the radiomic approach with AI is also becoming a powerful tool for predicting the outcome in patients with HCC after treatment, indicating the potential of AI for applying personalized medical care. However, ultrasound images show high heterogeneity because of differences in conditions during the examination, and a variety of imaging parameters may affect the quality of images; such conditions may hamper the development of ultrasound-based AI [97]. AI's application in the area of hepatology is especially relevant for the study of HCC, as this is a very common tumor, with particular radiological characteristics that allow its diagnosis without the need for a histological study [98]. However, the interpretation and analysis of the resulting images is not always easy, in addition to which the images vary during the course of the disease, and prognosis and treatment response can be conditioned by multiple factors. The vast amount of data available lend themselves to study and analysis by AI in its various branches, such as DL and ML, which play a fundamental role in decision-making as well as overcoming the constraints involved in human evaluation. Radiomics uses computers to extract a large amount of information from different types of images, form various quantifiable features, and select relevant features using AI algorithms to build models, in order to predict the outcomes of clinical problems (such as diagnosis, treatment, and prognosis). The study of liver diseases by radiomics will contribute to early diagnosis and treatment of liver diseases and improve survival and cure rates of liver diseases. This field is currently in the ascendant and may have great development in the future [99]. Genomic medicine company Lucence Diagnostics [100] developed AI algorithms for improving diagnosis and treatment of liver cancer to combine the imaging and molecular data from liver cancer patients into smarter software tools that help physicians make better treatment decisions. Liver cancer is the second leading cause of avoidable cancer deaths globally, and hepatitis viruses contribute to the bulk of this disease. The incidence rate of liver cancer is also rising faster than any other cancer in both men and women in the United States [101]. The best chance of cure is surgery, and good characterization of the extent and type of the disease is critical for surgical planning. Imaging tests such as ultrasound, CT, and MRI play a crucial role in the visualization of liver tumors. Fusing imaging data with sequencing data that include both cancer mutations and viral DNA will create a unique opportunity for AI-based approaches to advance liver cancer care. The aforementioned AI algorithms will evaluate a dataset of more than 5,000 patients to identify image changes and patterns that are linked to diagnostic and treatment outcomes in liver cancer [100].

According to the WHO [69], breast cancer is the most common oncology disease among women that leads to around 627,000 deaths annually. To save lives, many countries have introduced screening programs aiming to detect the cancer at an early stage. The diagnostic procedures vary from country to country. For instance, US women go for a mammogram (X-ray of the breast) once every 2 years, and each image is analyzed by a single radiologist. British women are screened once every 3 years, but with two experts providing results. Though neither approach is perfect, double reading shows better accuracy [70]. Despite the existence of screening programs worldwide, the interpretation of mammograms is affected by high rates of false positives and false negatives. McKinney et al. [102] presented an AI system that is capable of surpassing human experts in breast cancer prediction. To assess its performance in the clinical setting, we curated a large representative dataset from the United Kingdom and a largely enriched dataset from the United States. We show an absolute reduction of 5.7% and 1.2% (the United States and the United Kingdom) in false positives and 9.4% and 2.7% in false negatives. It was further reported that, in an independent study of six radiologists, the AI system outperformed all of the human readers: the AUC-ROC for the AI system was greater than the AUC-ROC for the average radiologist by an absolute margin of 11.5%. Koh [103] reported that the AI alone showed 88.8% sensitivity in breast cancer detection, whereas radiologists alone showed 75.3%. When radiologists were aided by AI, the accuracy increased by 9.5–84.8%. The comparison between AI and human expert, when dealing with fatty breasts, was 90% and 78%. The studies are still in their early stages with more clinical trials needed. For now, models can serve as an additional reader to automatically produce the second opinion [70]. Potentially, they will plug a growing shortage of trained radiologists [104]. Endometrial cancer starts when cells in the endometrium (the inner lining of the uterus) start to grow out of control. Cells in nearly any part of the body can become cancer, and they can spread to other parts of the body. Basen-Engquist et al. [105] determines the prevalence of physical activity and obesity and their relationship to QOL, PF (physical functioning), fatigue, and pain in endometrial cancer survivors. Surveys were mailed to 200 survivors of endometrial cancer diagnosed within the last 5 years; 61% were returned. Surveys assessed physical activity, height and weight, comorbid health problems, PF, fatigue, and pain. It was obtained that (i) in all, 22% exercised in the past month at the level of current public health recommendations, 41% reported no physical activity, and 38% reported some activity, (ii) a total of 16% were overweight and 50% were obese, (iii) both lower body mass index (BMI) and higher physical activity were related to better PF, and (iv) higher physical activity was related to less fatigue, primarily for patients of normal BMI, suggesting that (v) the endometrial cancer survivors' obesity and inactivity contributes to poorer quality of life. Rossi et al. [106] evaluated the acceptability and validity of the Fitbit Alta™ physical activity monitor for socioculturally diverse endometrial cancer survivors, and it was reported that (i) the Fitbits were well accepted by 25 participants and the physical activity data indicated an insufficiently active population and (ii) physical inactivity and sedentary behavior are common among breast cancer survivors. Nguyen et al. [107] evaluated wearable activity trackers as behavioral interventions to increase physical activity and reduce sedentary behavior within this population and reported that wearable technique programs have the potential to provide effective, intensive, home-based rehabilitation. To catch cancer earlier, it is needed to predict who is going to get it in the future. The complex nature of forecasting risk has been bolstered by AI tools, but the adoption of AI in medicine has been limited by poor performance on new patient populations and neglect of racial minorities [108].

A team of scientists from MIT's Computer Science and Artificial Intelligence Laboratory and Jameel Clinic demonstrated a DL system to predict cancer risk using just a patient's mammogram. The model showed significant promise and even improved inclusivity: it was equally accurate for both white and black women, which is especially important given that black women are 43% more likely to die from breast cancer. To further test these updates across diverse clinical settings, the scientists evaluated AI algorithm "Mirai" on new test sets from Karolinska in Sweden and Chang Gung Memorial Hospital in Taiwan and found it obtained consistent performance. The team also analyzed the model's performance across races, ages, and breast density categories in the Massachusetts General Hospital test set, and across cancer subtypes on the Karolinska dataset, and found it performed similarly across all subgroups. "Mirai" not only performs risk assessments from mammographic images but can also provide higher prediction accuracy by taking into account clinical risk factors such as age and family history when they are available as input information. In addition, it is characterized by algorithmic design that is not affected by realistic fluctuations in the clinical environment, such as differences in mammography equipment, and is making progress toward realizing the world standard breast cancer prediction tool [108].

AI technology is involved in diagnosing, treating, and prognosing of prostate cancer with various extents. Prostate-specific antigen (PSA) testing is a simple "biological" examination, limiting the options for AI deployment. Determining the PSA level in a patient's blood is done through a laboratory test. As this is a chemical process, the options of deploying AI in the very process of PSA level determination are limited. The way in which AI can help is by using the PSA value as an input to an algorithm and determining the risk of having a positive prostate biopsy. Such algorithms often include other input data besides PSA level, for example, patient age, white blood count in urinalysis, (estimated) prostate volume, and the status of the digital rectal examination (DRE). Results show that just by including this extra patient information and training an NN to become an expert in classifying prostate cases, increased sensitivity versus a regular PSA test can be achieved [109]. As the prostate MRI reading process is relatively complicated and involves many steps, it offers a wide range of opportunities for the application of AI. Prostate carcinoma is one of the most prevalent cancers worldwide. Multiparametric magnetic resonance imaging (mpMRI) is a non-invasive tool that can improve prostate lesion detection, classification, and volume quantification. AI/ML can rapidly and accurately analyze mpMRI images [110, 111]. ML could provide better standardization and consistency in identifying prostate lesions and enhance prostate carcinoma management [110]. Besides AI contribution to the (mp)MRI, there are still other tasks which are supported by AI technology, including transrectal ultrasound (TRUS) detection and classification, positron emission tomography (PET) for cancer staging, tracking the effects of radiotherapy, and the detection of metastases, pathology, treatment (radiotherapy, brachytherapy), and surgery [109]. After the treatment has been ended, there are many ways to keep track of the status of a patient. Most of those coincide with methods for the diagnosis trajectory. Several studies [112–115] support that the (integrated) radiomic-clinicopathologic nomogram (RadClip) is a better prognosticator of biochemical recurrence-free survival and adverse pathology than other standard tools.

Skin diseases are the fourth most frequent cause of disability worldwide, while skin cancer is the world's most common malignancy, hitting 20% of people by age 70. Luckily, 99% of the cases are curable if they are spotted and treated on time. And that is where AI can play a meaningful role. Similar to radiologists, dermatologists largely rely on visual pattern recognition [32, 52]. Skin cancer, the most common human malignancy, is primarily diagnosed visually, beginning with an initial clinical screening and followed potentially by dermoscopic analysis, a biopsy, and histopathological examination [116]. Esteva et al. [116] demonstrated classification of skin lesions using a single CNN, trained end to end from images directly, using only pixels and disease labels as inputs, and CNN was trained on 129,450 clinical images, consisting of 2,032 different skin diseases, to detect cancer. It was found that the algorithm reached the accuracy demonstrated by dermatologists. DL/CNN may facilitate melanoma detection, but data comparing a CNN's diagnostic performance to larger groups of dermatologists are lacking. Hence, Hasenssle et al. [117] trained CNN architecture and validated using dermoscopic images and corresponding diagnoses. In a comparative cross-sectional reader study a 100-image test set was used (level-I: dermoscopy only; level-II: dermoscopy plus clinical information and images). Main outcome measures were sensitivity, specificity, and AUC of ROC for diagnostic classification (dichotomous) of lesions by the CNN versus an international group of 58 dermatologists during level-I or -II of the reader study. It was concluded that, for the first time of comparison of a CNN's diagnostic performance to a large international group of 58 dermatologists including 30 experts, (i) most dermatologists were outperformed by the CNN, (ii) the CNN correctly detected melanomas in 95% of cases, while the accuracy of dermatologists was 86.6%, and (iii) irrespective of any physicians' experience, they may benefit from assistance by a CNN's image classification [117]. One of the most successful DL architectures (i.e., CNN) has consistently demonstrated its outstanding performance in medical research. Han et al. [118] trained an algorithm with 220,680 images of 174 disorders and validated it using Edinburgh (1,300 images; 10 disorders) and Seoul National University datasets (2,201 images; 134 disorders). The algorithm could accurately predict malignancy, suggest primary treatment options, render multiclass classification among 134 disorders, and improve the performance of medical professionals. The results suggest that (i) the algorithm may serve as augmented intelligence that can empower medical professionals in diagnostic dermatology and (ii) AI proved its capability to distinguish between melanoma and birthmarks at human expert level. Besides enhancing the speed and accuracy of diagnosis, there are plans to run CNN algorithms on smartphones for non-professional skin exams [118]. This can encourage people to visit dermatologists for lesions that might be ignored otherwise [32].

8.4.3.2 Diabetes

Diabetes is a disease that occurs when the blood glucose (or blood sugar) is too high. Blood glucose is a source of energy and comes from the food. Insulin, a hormone made by the pancreas, helps glucose from food get into cells to be used for energy. Sometimes, a body does not make enough (type 1 diabetes: T1D) or any insulin or does not use insulin well (type 2 diabetes: T2D); glucose then stays in the blood and does not reach cells. Over time, having too much glucose in your blood can cause complications [119], including (1) eye disease (due to changes in fluid levels, swelling in the tissues, and damage to the blood vessels in the eyes), (2) foot problems (caused by damage to the nerves and reduced blood flow to your feet), (3) gum disease and other dental problems (because a high amount of blood sugar in your saliva helps harmful bacteria grow in your mouth. The bacteria combine with food to form a soft, sticky film called plaque. Plaque also comes from eating foods that contain sugars or starches. Some types of plaque cause gum disease and bad breath. Other types cause tooth decay and cavities), (4) heart disease and stroke (caused by damage to your blood vessels and the nerves that control your heart and blood vessels), (5) kidney disease (due to damage to the blood vessels in your kidneys. Many people with diabetes develop high blood pressure, and can also damage your kidneys), (6) nerve problems or diabetic neuropathy (caused by damage to the nerves and the small blood vessels that nourish your nerves with oxygen and nutrients), (7) sexual and bladder problems (caused by damage to the nerves and reduced blood flow to the genitals and bladder), and (8) skin conditions, some of which are caused by changes in the small blood vessels and reduced circulation. People with diabetes are also more likely to have infections, including skin infections [119].

Recent consumer technologies are helping the diabetic community to take great strides toward truly personalized, real-time, data-driven management of this chronic disease [32, 120]. The consumer technologies include smartphone apps, wearable devices, and sensors. Dudde et al. [121] developed a good example as the wearable artificial endocrine pancreas for diabetes management, which is a closed-loop system formed by a wearable glucose monitor and an implanted insulin pump. Closed-loop control (CLC) for the management of type 1 diabetes (T1D) is a novel method for optimizing glucose control, and strategies for individualized implementation are being developed. Brown et al. [122] analyzed glycemic control in an overnight CLC system designed to "reset" the patient to near-normal glycemic targets every morning. It was concluded that (i) overnight-only CLC increased the time in the target range over 24 h and decreased the time in hypoglycemic range over 24 h in a supervised outpatient setting and (ii) a pilot extension study at home showed a similar nonsignificant trend. Researchers [123] also explored the possibilities of using Google Glass to simplify the daily life of people with diabetes mellitus. With the increasing cost of healthcare, wearable devices and systems could have the potential to facilitate self-care through monitoring and prevention. Lee et al. [124] developed a wearable bioelectronic technology to provide non-invasive monitoring of sweat-based glucose level.

The management of diabetes and its complication is very tough job as there are several factors that keep blood sugar level in control. The application of AI in diagnosis or monitoring of diabetes and its complication may improve the patient's quality of life. Computer-assisted diagnosis, decision support systems, ES, and implementation of software may assist physicians to minimize the intra- and inter-observer variability. The application of AI facilitates interpretation of results with high accuracy and maximum speed [125]. Peripheral diabetic neuropathy is the foremost cause of disability in diabetic patients. The early diagnosis can be done by morphometric parameters of corneal nerves. Foot amputation is one of the advanced stages of diabetic neuropathy. Microalbuminuria (MA) is an independent predictor of cardiovascular and renal disease, development of overt nephropathy, and cardiovascular mortality in patients with T2D. Marateb et al. [126] introduced a new method for MA diagnosis using clinical parameters usually monitored in type 2 diabetic patients without the need of an additional measurement of urinary albumin with designing an expert-based fuzzy MA classifier in which rule induction was performed by particle swarm optimization. The rule induction was done by particle swarm optimization. Statistic featuring was done with multiple logistic regressions. Age, gender, BMI, DD, systolic BP, FBS, HbA1c, Bs2hpp, CHOL, LDL, HDL, and TG were used as input parameters. The performance of MA classifier was evaluated using 10-fold cross-validation. The minimum sensitivity, specificity, precision, and accuracy of the proposed fuzzy classifier system with feature extraction were 95%, 85%, 84%, and 92%, respectively [126]. Decline in kidney function is variable in diabetic patients; hence, it is difficult to predict determinants of diabetic nephropathy. Cho et al. [127] applied various ML techniques such as feature selection methods and support vector machine (SVM) classification and developed a new visualization system which uses a nomogram approach. The proposed method can predict the onset of diabetic nephropathy about 2–3 months before the actual diagnosis. Additionally, the visualization system provides physicians with intuitive information for risk factor analysis [127]. The performance of a comprehensive computer-aided diagnosis (CAD) system was evaluated for diabetic retinopathy (DR) screening, using a publicly available database of retinal images, and to compare its performance with that of human experts [127]. They applied previously developed CAD system to 1,200 digital color fundus photographs. It was found that (i) the system achieved an area under the ROC curve of 0.876 for successfully distinguishing normal images from those with DR with a sensitivity of 92.2% at a specificity of 50%, and (ii) these compare favorably with the two experts, who achieved sensitivities of 94.5% and 91.2% at a specificity of 50%, suggesting that (iii) the performance of a comprehensive DR screening system on an independent, publicly available dataset. The performance of the system on this dataset is comparable with that of human experts [128].

Gestational diabetes mellitus (GDM) can cause adverse consequences to both mothers and their newborns. The GDM is the metabolic abnormality of sugar (glucose) in the bloodstream that is higher than normal and found for the first time during pregnancy, and it is one of the most common complications affecting up to 15% of pregnant women [129, 130]. Early detection and intervention are desirable because gestational diabetes is associated with the risk of stillbirth and premature birth, but it is said that there are many cases where it is not discovered until late pregnancy [130]. Shen et al. [131] investigated the implementation of a well-performing AI algorithm in GDM diagnosis in a setting which requires fewer medical equipment and staff to establish an app based on the AI algorithm. An AI model including nine algorithms was trained on 12,304 pregnant outpatients. It was obtained that (i) AUC-ROC of external validation dataset for SVM, random forest, AdaBoost, k-nearest neighbors (kNN), naive Bayes (NB), decision tree, logistic regression (LR), eXtreme gradient boosting (XGBoost), and gradient boosting decision tree (GBDT) were 0.780, 0.657, 0.736, 0.669, 0.774, 0.614, 0.769, 0.742, and 0.757, respectively, (ii) SVM also retained high performance in other criteria, and (iii) the specificity for SVM retained 100% in the external validation set with an accuracy of 88.7%. Based on these results, it was concluded that (iv) the first study supported the GDM diagnosis for pregnant women in resource-limited areas, using only fasting blood glucose value, patients' age, and a smartphone connected to the Internet, (v) the study proved that SVM can achieve accurate diagnosis with less operation cost and higher efficacy, and (vi) the study (referred to as GDM-AI study, i.e., the study of AI-based diagnosis of GDM) also shows that the app has a promising future in improving the quality of maternal health for pregnant women, precision medicine, and long-distance medical care, recommending that future work should expand the dataset scope and replicate the process to validate the performance of the AI algorithms [132]. Liu et al. [132] investigated GDM diagnosed by early and standard OGTTs and determined adverse maternal and neonatal outcomes associated with early GDM diagnosis. A total of 522 participants completed both the early and standard OGTTs. It was found that pregnant women in the GDM group had higher glycated hemoglobin, C-peptide, and homeostasis model assessment of insulin resistance in the late gestational period, and it was concluded that early-onset GDM diagnosed by OGTT (oral glucose tolerance test) at 18 to 20 gws is associated with maternal and neonatal metabolic disorders and adverse pregnancy outcomes. Previously, when we prevented gestational diabetes, we were only focusing on pregnant women with obesity or pregnant women with a history of gestational diabetes or those with elevated blood glucose during early pregnancy, and now we have expanded the scope of target population through the prediction model, which will enable more high-risk pregnant women to gain potential benefits [133]. Wu et al. [130] established effective models to predict early GDM during the first trimester of pregnancy. Pregnancy data for 73 variables during the first trimester were extracted from the EMR system. Based on an ML-driven feature selection method, 17 variables were selected for early GDM prediction. To facilitate clinical application, 7 variables were selected from the 17-variable panel. Advanced ML approaches were then employed using the 7-variable data set and the 73-variable data set to build models predicting early GDM for different situations, respectively. It was found that (i) a total of 16,819 and 14,992 cases were included in the training and testing sets, respectively. Using 73 variables, the deep NN model achieved high discriminative power, with AUC values of 0.80, (ii) the 7-variable logistic regression (LR) model also achieved effective discriminate power (AUC = 0.77), (iii) low BMI ($\leq$17) was related to an increased risk of GDM, compared to a BMI in the range of 17–18 (minimum risk interval) (11.8% vs. 8.7%), and (iv) lipoprotein(a) demonstrated a promising predictive value (AUC = 0.66), indicating that ML models achieved high accuracy in predicting GDM in early pregnancy [130]. GDM is typically diagnosed at 24–28 weeks of gestation, but earlier detection is desirable as this may prevent or considerably reduce the risk of adverse pregnancy outcomes. Hence, Atzi et al. [134] utilized an ML approach to predict GDM on retrospective data of 588,622 pregnancies in Israel for which comprehensive EHRs were available. It was mentioned that the models predict GDM with high accuracy even at pregnancy initiation (AUC-ROC: 0.85), substantially outperforming a baseline risk score (AUC-ROC: 0.68), and concluded that the models may allow early-stage intervention in high-risk women, as well as a cost-effective screening approach that could avoid the need for glucose tolerance tests by identifying low-risk women.

8.4.3.3 Cardiovascular disease

Heart disease is one of the lifestyle-related diseases that increase the risk as we get older, and we know that the risk of onset increases due to aging of the circulatory system. Heart disease and the aging of the circulatory system are major causes of lowering QOL in middle age and beyond, since once developed, it can also cause sequelae. Heart disease is said to have a greater influence on lifestyle than genetic characteristics. For this reason, lifestyle can be expected to have a great effect on the prevention of heart disease and cardiovascular diseases (CVDs) by improving based on medical point of seeing. CVD is a major threat to human health and the leading cause of death worldwide. The incidence of CVD caused 17.6 million deaths in 2016, an increase of 14.5% from 2006 to 2016. Unfortunately, the mortality and morbidity rates of CVD are increasing year by year, especially in developing regions.

Studies have shown that approximately 80% of CVD-related deaths occur in low- and middle-income countries. Besides, these deaths occur at a younger age than in high-income countries. Of all heart diseases, coronary heart disease (CHD, aka heart attack) is by far the most common and the most fatal. In the United States, for example, it is estimated that someone has a heart attack every 40 s and about 805,000 Americans have a heart attack every year [135–137]. Vascular diseases affect the arteries, veins, or capillaries throughout the body and around the heart, and they include [138] peripheral artery disease (which causes arteries to become narrow and reduces blood flow to the limbs), aneurysm (a bulge or enlargement in an artery that can rupture and bleed), atherosclerosis (in which plaque forms along the walls of blood vessels, narrowing them and restricting the flow of oxygen rich blood), renal artery disease (which affects the flow of blood to and from the kidneys and can lead to high blood pressure), Raynaud's disease (which causes arteries to spasm and temporarily restrict blood flow), peripheral venous disease, or general damage in the veins that transport blood from the feet and arms back to the heart, which causes leg swelling and varicose veins, ischemic stroke (in which a blood clot moves to the brain and causes damage), venous blood clots (which can break loose and become dangerous if they travel to the pulmonary artery), blood clotting disorders (in which blood clots form too quickly or not quickly enough and lead to excessive bleeding or clotting), and Buerger's disease (which leads to blood clots and inflammation, often in the legs, and which may result in gangrene) [138].

Diseases and conditions that affect the heart include [138] angina (a type of chest pain that occurs due to decreased blood flow into the heart), arrhythmia (or an irregular heartbeat or heart rhythm), congenital heart disease (in which a problem with heart function or structure is present from birth), coronary artery disease (which affects the arteries that feed the heart muscle), heart attack (or a sudden blockage to the heart's blood flow and oxygen supply), heart failure (wherein the heart cannot contract or relax normally), dilated cardiomyopathy (a type of heart failure, in which the heart gets larger and cannot pump blood efficiently), hypertrophic cardiomyopathy (in which the heart muscle walls thicken and problems with relaxation of the muscle, blood flow, and electrical instability develop), mitral regurgitation (in which blood leaks back through the mitral valve of the heart during contractions), mitral valve prolapse (in which part of the mitral valve bulges into the left atrium of the heart while it contracts, causing mitral regurgitation), pulmonary stenosis (in which a narrowing of the pulmonary artery reduces blood flow from the right ventricle (pumping chamber to the lungs) to the pulmonary artery (blood vessel that carries deoxygenated blood to the lungs)), aortic stenosis, a narrowing of the heart valve that can cause blockage to blood flow leaving the heart, atrial fibrillation (an irregular rhythm that can increase the risk of stroke), rheumatic heart disease, a complication of strep throat that causes inflammation in the heart and which can affect the function of heart valves, and radiation heart disease (wherein radiation to the chest can lead to damage to the heart valves and blood vessels) [138]. Wearable devices have been developed to do cardiovascular monitoring and enable mHealth applications in cardiac patients [32]. Winokur et al. [139] developed a low-power wearable ECG monitoring system entirely from discrete electronic components and a custom-printed circuit board. Yang et al. [140] developed a wearable patch-style heart activity monitoring system (HAMS), which was used for recording ECG signal. The device was characterized by minimizing the electromagnetic interference (EMI) by removing the cables. In the same subjects who were exposed under stress and non-stress, the questionnaire was given out, the amount of the stress hormone was measured by blood test and the ECG signal can be also recorded. Wearable technology can assess patient heart activity outside of a laboratory or clinical environment. It is possible to perform heart assessments during a wide range of everyday conditions without interfering with a patient's activity tasks. For example, researchers designed a textile-based wearable device for the unobtrusive recording of ECG, respiration, and accelerometric data, and to assess the 3D sternal seismocardiogram (SCG) in daily life [32]. He et al. [141] developed a portable and continuous BCG monitor that is wearable at the ear, comprises the form factor of a hearing aid, and is wirelessly connected to a PC for data recording and analysis. With the ear as an anchoring point, the device uses a MEMS tri-axial accelerometer to measure BCG at the head. Morphological differences exist between head BCG and traditional BCG, but the principal peaks (J waves) and their vectors are preserved. The frequency of J waves corresponds to heart rate, and when used in conjunction with an ECG ventricular depolarization R wave, the timing of J waves yields the RJ interval. It was mentioned that there was a linear correlation between the RJ interval and the heart's pre-ejection period during hemodynamic maneuvers, thus revealing important information about cardiac contractility and its regulation. The implantable cardioverter defibrillator (ICD) is able to reduce sudden arrhythmic death in patients who are considered to be at high risk. However, the arrhythmic risk may be increased only temporarily as long as the proarrhythmic conditions persist, left ventricular ejection fraction remains low, or heart failure prevails [142]. The wearable cardioverter defibrillator (WCD) represents an alternative approach to prevent sudden arrhythmic death until either ICD implantation is clearly indicated or the arrhythmic risk is considered significantly lower or even absent. The WCD is also indicated for interrupted protection by an already implanted ICD, temporary inability to implant an ICD, and lastly refusal of an indicated ICD by the patient. The WCD is not an alternative to the ICD but a device that may contribute to better selection of patients for ICD therapy. The WCD has the characteristics of an ICD but does not need to be implanted, and it has similarities with an external defibrillator but does not require a bystander to apply lifesaving shocks when necessary. The WCD was introduced into clinical practice about 8 years ago, and indications for its use are currently expanding [142].

Current approaches to predict cardiovascular risk fail to identify many people who would benefit from preventive treatment, while others receive unnecessary intervention. ML offers opportunity to improve accuracy by exploiting complex interactions between risk factors. Hence, Weng et al. [143] assessed whether ML improves cardio-vascular risk prediction. Prospective cohort study used routine clinical data of 378,256 patients from UK family practices, free from CVD at the outset. Four ML algorithms (random forest, logistic regression, gradient boosting machines, NNs) were compared to an established algorithm (American College of Cardiology guidelines) to predict first cardiovascular event over 10 years. It was obtained that (i) 24,970 incident cardiovascular events (6.6%) occurred, (ii) compared to the established risk prediction algorithm (AUC 0.728, 95% CI 0.723–0.735), ML algorithms improved prediction: random forest + 1.7% (AUC 0.745, 95% CI 0.739–0.750), logistic regression + 3.2% (AUC 0.760, 95% CI 0.755–0.766), gradient boosting + 3.3% (AUC 0.761, 95% CI 0.755–0.766), NNs + 3.6% (AUC 0.764, 95% CI 0.759–0.769), and (iii) the highest achieving (NNs) algorithm predicted 4,998/7,404 cases (sensitivity 67.5%, PPV 18.4%) and 53,458/ 75,585 non-cases (specificity 70.7%, NPV 95.7%), correctly predicting 355 (+7.6%) more patients who developed CVD compared to the established algorithm, indicating that ML significantly improves accuracy of cardiovascular risk prediction, increasing the number of patients identified who could benefit from preventive treatment, while avoiding unnecessary treatment of others [143]. Dawes et al. [144] determined if patient survival and mechanisms of right ventricular failure in pulmonary hypertension could be predicted by using supervised ML of 3D patterns of systolic cardiac motion. Two hundred and fifty-six patients (143 women; 63 years ± 17) with newly diagnosed pulmonary hypertension underwent cardiac MR imaging, right-sided heart catheterization, and 6-min walk testing with a median follow-up of 4.0 years. Semiautomated segmentation of short-axis cine images was used to create a three-dimensional model of right ventricular motion. It was found that (i) at the end of follow-up, 36% of patients (93 of 256) died, and one underwent lung transplantation, (ii) poor outcome was predicted by a loss of effective contraction in the septum and free wall, coupled with reduced basal longitudinal motion. When added to conventional imaging and hemodynamic, functional, and clinical markers, three-dimensional cardiac motion improved survival prediction (AUC-ROC, 0.73 vs. 0.60, respectively) and provided greater differentiation according to difference in median survival time between high- and low-risk groups (13.8 vs. 10.7 years, respectively). It was, accordingly, concluded that an ML survival model that uses three-dimensional cardiac motion predicts outcome independent of conventional risk factors in patients with newly diagnosed pulmonary hypertension. The silver lining is that heart attacks are highly preventable and simple lifestyle modifications (such as reducing alcohol and tobacco use; eating healthily and exercising) coupled with early treatment greatly improves its prognosis. It is, however, difficult to identify high-risk patients because of the multifactorial nature of several contributory risk factors such as diabetes, high blood pressure, and high cholesterol. This is where ML and data mining come to the rescue [136]. Mordecai [136] reviewed the development of a screening tool for predicting whether a patient has 10-year risk of developing CHD using different ML techniques. It was concluded that the model can be used as a simple screening tool by inputting data such as age, BMI, systolic and diastolic blood pressures, heart rate and blood glucose levels after which the model can run and it outputs a prediction. Poplin et al. [145] demonstrated that DL can extract new knowledge from retinal fundus images. Using DL models trained on data from 284,335 patients and validated on two independent datasets of 12,026 and 999 patients, they predicted cardiovascular risk factors not previously thought to be present or quantifiable in retinal images, such as age (mean absolute error within 3.26 years), gender (AUC = 0.97), smoking status (AUC = 0.71), systolic blood pressure (mean absolute error within 11.23 mmHg) and major adverse cardiac events (AUC = 0.70). It was also shown that the trained DL models used anatomical features, such as the optic disc or blood vessels, to generate each prediction. T2D is one common chronic disease caused by insulin secretion disorder that often leads to severe outcomes and even death due to complications, among which CHD represents the most common and severe one [145]. In order to identify the ones with high risks of CHD complication, Fan et al. [146] first curated a dataset of 1,273 T2D patients including 304 and 969 ones with or without CHD, respectively, and then trained an AI model using randomly selected 4/5 of the dataset and use the rest data to validate the performance of the model. It was shown that the model achieved an AUC of 0.77 (5-fold cross-validation) on the training dataset and 0.80 on the testing dataset, indicating that an AI model can determine the risk of T2D patients to develop to CHD, which has potential value in providing early warning personalized guidance of CHD risk for both T2D patients and clinicians.

Vital signs are usually recorded at 4–8 h intervals in hospital patients, and deterioration between measurements can have serious consequences. Hence, Hernandez-Silveira et al. [147] assessed agreement between a new ultra-low power, wireless and wearable surveillance system for continuous ambulatory monitoring of vital signs and a widely used clinical vital signs monitor and examined the system's ability to automatically identify and reject invalid physiological data. It was found that the tested wearable device provided reliable heart rate value for about 80% of the patients and the overall agreement between the new device and clinical monitor was satisfactory because the comparison was statistically significant, concluding that (i) overall agreement between digital patch and clinical monitor was satisfactory, as was the efficacy of the system for automatic rejection of invalid data and (ii) wireless monitoring technologies, such as the one tested, may offer clinical value when implemented as part of wider hospital systems that integrate and support existing clinical protocols and workflows. Capabilities such as heart rate monitoring may be useful in hospitalized patients as a means of enhancing routine monitoring or as part of an early warning system to detect clinical deterioration. Kroll et al. [148] evaluated the accuracy of heart rate monitoring by a personal fitness tracker (PFT) among hospital inpatients. A prospective observational study was conducted on 50 stable patients in the intensive care unit who each completed 24 h of heart rate monitoring using a wrist-worn PFT. Accuracy of heart rate recordings was compared with gold standard measurements derived from continuous electrocardiographic (cECG) monitoring. The accuracy of heart rates measured by pulse oximetry (Spo2.R) was also measured as a positive control. It was obtained that a wrist-worn PFT device can be used to monitor the heart rate of patients even though the collected heart rates were slightly lower than the standard of cECG monitoring. It was, therefore, concluded that (i) PFT-derived heart rates were slightly lower than those derived from cECG monitoring in real-world testing and not as accurate as Spo2.R-derived heart rates, (ii) performance was worse among patients who were not in sinus rhythm, and (iii) further clinical evaluation is indicated to see if PFTs can augment early warning systems in hospitals. Heat stroke, as well, can be potentially damaging for people while exercising in hot environments. To prevent this dangerous situation, Chen et al. [149] developed a wearable heat-stroke-detection device (WHDD) with early notification ability. The device utilized several physical sensors (such as galvanic skin response – GSR, heart pulse, and body temperature) to acquire medical data from exercising people. The device included risk evaluation functional components that were based on fuzzy theory to detect the features of heat stroke for users. If a dangerous situation is detected, then the device will activate the alert function to remind the user to respond adequately to avoid heat stroke. Timely risk assessment based on ECGs – the quickest and simplest test of heart activity – may significantly decrease mortality and prevent heart attacks [52]. With more than 300 million ECGs preformed globally each year, algorithms obtain a huge data pool for learning. Multiple studies show that AI already not only spots current abnormalities from ECGs but predicts future risks as well. The variability present between these consecutive heartbeats is indicative of one's potential risk for an adverse cardiovascular event. The ML algorithms behind RiskCardio were trained using ECG data from documented patients. Through this process, the AI was trained to identify pairs of adjacent heartbeats as being normal or indicative of heart failure risk. It was reported that, within the first 15 min of a patient having ACS (acute coronary syndrome), there was sufficient data to assess the likelihood of cardiovascular death within 30, 60, 90, or 365 days for ACS patients using RiskCardio technology [150]. AI is expected to save human experts considerable time and cut the number of misdiagnoses. Paired with low-cost hardware, deep learning algorithms may potentially enable the use of ECG as a diagnostic tool in places where cardiologists are rare or absent [52].

8.4.3.4 Lung

An early diagnosis of tuberculosis (TB) is crucial but challenging for resource-poor countries. Kuo et al. [151] developed the TBShoNet system by transferring DL for TB detection on chest X-ray images captured by phone camera. The NN was pretrained on a database containing 250,044 chest X-rays with 14 pulmonary labels, which did not include TB. The model was then recalibrated for chest X-ray photographs by using simulation methods to augment the dataset. TBShoNet provides a method to develop an algorithm that can be deployed on phones to assist healthcare providers in areas where radiologists and high-resolution digital images are unavailable. The

TBShoNet model was built by connecting the pretrained model to an additional 2-layer NN trained on augmented chest X-ray images. Then 662 chest X-ray photographs taken by 5 different phones (TB: 336; normal: 326) were used to test the model performance. It was reported that the sensitivity and specificity for TB classification were 81% and 84%, respectively. Chronic obstructive pulmonary disease is a type of lung disease caused by chronically poor airflow that makes breathing difficult. As a chronic illness, chronic obstructive pulmonary disease typically worsens over time, so extensive, long-term pulmonary rehabilitation exercises and patient management are required. Tey et al. [152] developed a remote rehabilitation system for a multimodal sensor-based application for patients who have chronic breathing difficulties. The process involves the fusion of sensory data-captured motion data by stereo-camera and photo-plethysmogram signal by a wearable photoplethysmography (PPG) sensor that are the input variables of a detection and evaluation framework. The system included a set of rehabilitation exercises specific for pulmonary patients, and provided exercise tracking progress, patient performance, exercise assignments, and exercise guidance. Patients in the study could receive accurate pulmonary exercises guidance from the sensory data.

Acute respiratory distress syndrome (ARDS) is a serious respiratory condition with high mortality and associated morbidity. Le et al. [153] developed and evaluated a novel application of gradient boosted tree models trained on patient health record data for the early prediction of ARDS. Total 9,919 patient encounters were retrospectively analyzed from the Medical Information Mart for Intensive Care III (MIMIC-III) data base. XGBoost gradient boosted tree models for early ARDS prediction were created using routinely collected clinical variables and numerical representations of radiology reports as inputs. XGBoost models were iteratively trained and validated using 10-fold cross-validation. It was found that on a hold-out test set, algorithm classifiers attained AUC-ROC values of 0.905 when tested for the detection of ARDS at onset and 0.827, 0.810, and 0.790 for the prediction of ARDS at 12-, 24-, and 48-h windows prior to onset, respectively, suggesting that supervised ML predictions may help predict patients with ARDS up to 48 h prior to onset [153]. Pulmonary embolism (PE), a deadly disease in which blood clots are clogged in blood vessels in the lungs, is the cause of up to 200,000 deaths annually in the United States. Many studies have known that early detection of PE improves lifesaving, and various technological developments have been promoted to make AI-led triage beneficial for this workflow. Imbio and Aidoc have joined forces to develop a complete PE solution to radiologists and interventionalists, both in terms of the suite of relevant algorithms and a seamless workflow native to each type of physician [154]. Aidoc's AI solution for providing real-time notifications for patients with suspected PE is already FDA-cleared, while Imbio is developing imaging analysis intended for the automatic calculation of RV/LV ratio. Automating and presenting RV/LV ratio measurements to PE intervention teams has the ability to aid in patient care, and combining this with Aidoc's always-on AI platform is intended to trigger targeted alerts to the relevant physicians, enhancing the synchronization of intervention teams and the quality of treatment provided to the patients. According to CDC report [155], about 6.2 million adults in the United States have heart failure, and in 2018, heart failure was mentioned on 379,800 death certificates (13.4%). The PE, a sign of decreased cardiac function in which excess fluid is stored in the lungs, is used as a judgment material for treatment policies. In daily medical care, the level of pulmonary edema is often judged from X-ray images, and certain standards are required for how to catch the features. A patient's exact level of excess fluid often dictates the doctor's course of action, but making such determinations is difficult and requires clinicians to rely on subtle features in X-rays that sometimes lead to inconsistent diagnoses and treatment plans. To better handle that kind of nuance, a group led by researchers at MIT's Computer Science and Artificial Intelligence Lab [156] has developed a ML model that can look at an X-ray to quantify how severe the edema is, on a four-level scale ranging from 0 (healthy) to 3 (very, very bad). The system determined the right level more than half of the time, and correctly diagnosed level 3 cases 90% of the time. The PE diagnosis would help and augment doctors manage not only acute heart issues but also other conditions like sepsis and kidney failure that are strongly associated with edema.

8.4.3.5 Brain

Patients with brain and spinal cord injuries need exercises to improve motor recovery. Often, these patients are not qualified to monitor or assess their own conditions, and they need healthcare provider guidance. Therefore, there is a need to transmit physiological data to clinicians from patients in their home environment [32]. It was known that patients who practice functional movements at home in conjunction with outpatient therapy show higher improvement in motor recovery. However, patients are not qualified to monitor or assess their own condition that must be reported back to the clinician. Burns et al. [157] reviewed the wearable AI technology for in-home health monitoring, assessment, and rehabilitation of patients with brain and spinal cord injuries to transmit physiological data to clinicians from patients in their home environment.

   Stroke, predominantly a condition of advanced age, is a major cause of acquired disability in the global population. Conventional treatment paradigms in intensive therapy are expensive and sometimes not feasible because of social and environmental factors [158]. Stroke, predominantly a condition of elderly, is a major cause of acquired disability in the global population and puts an increasing burden on healthcare resources. Clear evidence for the importance of intensity of therapy in optimizing functional outcomes is found in animal models, supported by neuroimaging and behavioral research and strengthened by recent meta-analyses from multiple clinical trials. However, providing intensive therapy using conventional treatment paradigms is expensive and sometimes not feasible because of social and environmental factors. A need for cost-effectiveness increased intensity of practice and suggests potential benefits of tele-health (TH) as an innovative model of care in physical therapy has been addressed [158]. Based on this background, Burridge et al. [158] overviewed the TH and presented evidence that a web-supported program, used in conjunction with constraint-induced therapy (CIT), can increase intensity and adherence to a rehabilitation regimen. It was also described how wearable sensors can monitor activity and provide feedback to patients and therapists, indicating that wearable sensor technologies and devices with embedded inertial and mechanomyographic sensors and TH programs have the potential to provide most effective, intensive, home-based stroke rehabilitation. Stroke or the sudden death of brain cells due to lack of oxygen is the second major cause of death and the third leading cause of long-term disability globally. This dangerous condition requires immediate diagnosis and treatment. Statistics [159] show that patients who receive professional help within three hours after the first symptoms typically make a better and faster recovery. But, unfortunately, emergency medical service (EMS) workers overlook roughly 15% of strokes which leads to delays in critical care and increases risks of fatal outcomes [52, 159].

Intracranial hemorrhage (ICH) requires prompt diagnosis to optimize patient outcomes. Arbabshirani et al. [160] hypothesized that ML algorithms could automatically analyze CT of the head, prioritize radiology worklists and reduce time to diagnosis of ICH. Total 46,583 brain CT scans were collected to create a model capable of flagging the signs of ICH – the deadliest type of stroke with 40% mortality within 30 days and hard disabilities in most survivors. They implemented the algorithm into routine care and tested it for three months. In some cases, this translated to a decrease of diagnostic time by 96%. It was concluded that the AI algorithm can prioritize radiology worklists to reduce time to diagnosis of new outpatient ICH by 96% and may also identify subtle ICH overlooked by radiologists, indicating the positive impact of advanced ML in radiology workflow optimization.

Acute stroke caused by large vessel occlusions (LVOs) requires emergent detection and treatment by endovascular thrombectomy. However, radiologic LVO detection and treatment is subject to variable delays and human expertise, resulting in morbidity. Imaging software using AI/ML, a branch of AI, may improve rapid frontline detection of LVO strokes. Murray et al. [161] conducted a systematic review of AI in acute LVO stroke identification and triage and characterized LVO detection software. It was concluded that AI may improve LVO stroke detection and rapid triage necessary for expedited treatment. Stroke is a very common and serious medical problem. Ischemic strokes happen when a blood clot blocks blood flow to the brain, causing brain cells to die. A hemorrhagic stroke happens when a brain artery breaks and blood flows freely into the brain. Telling the difference between the two types of stroke is very important as their treatments are quite different, and the longer a stroke goes without treatment, the more brain cells die. AI has allowed great progress in identifying situations or items in the technological world, but it has rarely been used in medicine. Recently, however, there have been a few reports. For example, AI has been used to detect bone fractures from X-ray images. Kummer [162] tested AI on CT images of ischemic and hemorrhage strokes to see whether AI could tell the difference between the two types. The online AI program Google Teachable Machine was used for these experiments. First, multiple images of the two types of strokes were selected from freely available online sources. Then the images were cleaned to focus on the brain only. The images were organized into categories: ischemic versus hemorrhagic, large versus small, and training versus validation sets. The training set was used to teach the AI program, while the validation set was used to test it and reported that trained algorithms correctly identify the type of stroke in 77.4% of cases. Potentially, AI trained by neuroradiologists may deliver a reliable "second opinion" to non-expert medical service providers so that they can make fast decisions and minimize damages [162].

Stroke is a challenging disease to diagnose in an emergency room (ER) setting. While an MRI scan is very useful in detecting ischemic stroke, it is usually not available due to space constraint and high cost in the ER. Clinical tests like the Cincinnati Pre-hospital Stroke Scale (CPSS) and the Face Arm Speech Test (FAST) are helpful tools used by neurologists, but there may not be neurologists immediately available to conduct the tests [163]. Based on this background, Yu et al. [163] developed a novel multimodal DL framework to achieve computer-aided stroke presence assessment over facial motion weaknesses and speech inability for patients with suspicion of stroke showing facial paralysis and speech disorders in an acute setting. It was reported that (i) experiments on video dataset collected on actual ER patients performing specific speech tests show that the proposed approach achieves diagnostic performance comparable to that of ER doctors, attaining a 93.12% sensitivity rate while maintaining 79.27% accuracy, and (ii) each assessment can be completed in less than 4 min, suggesting that (iii) this demonstrates the high clinical value of the framework, and (iv) the work, when deployed on a smartphone, will enable self-assessment by at-risk patients at the time when stroke-like symptoms emerge. CINA Head are integrated with Canon's Automation Platform for seamless detection and prioritization of ICH and LVOs from CT scans [164]. Avicenna provides Canon with its FDA-approved CINA Head triage AI solution for neurovascular emergencies, which detects two of the leading causes of stroke – ICH and LVO – from CT-scan imaging Using a combination of DL and ML technologies, CINA Head automatically detects and prioritizes acute ICH and LVO cases within 20 s, seamlessly alerting the radiologist within their existing systems and workflow. It was reported that (i) CINA's ICH detection capability was validated using data from 814 cases conducted at more than 250 imaging centers across the United States, with 96% accuracy, 91.4% sensitivity and 97.5% specificity, and (ii) the product's LVO detection capability was validated based on 476 cases, with 97.7% accuracy 97.9% sensitivity and 97.6% specificity [164].

The diagnosis of posttraumatic stress disorder (PTSD) is usually based on clinical interviews or self-report measures. Both approaches are subject to under- and over-reporting of symptoms. Marmar et al. [165] developed a classifier of PTSD based on objective speech marker features that discriminate PTSD cases from controls. Speech samples were obtained from warzone-exposed veterans, 52 cases with PTSD and 77 controls, assessed with the Clinician-Administered PTSD Scale. Individuals with major depressive disorder (MDD) were excluded. Audio recordings of clinical interviews were used to obtain 40,526 speech features which were input to a random forest (RF) algorithm. It was found that (i) the selected RF used 18 speech features and the receiver operating characteristic curve had an AUC of 0.954, (ii) at a probability of PTSD cut point of 0.423, Youden's index was 0.787, and overall correct classification rate was 89.1%, (iii) the probability of PTSD was higher for markers that indicated slower, more monotonous speech, less change in tonality, and less activation, and (iv) depression symptoms, alcohol use disorder, and traumatic brain injury (TBI) did not meet statistical tests to be considered confounders. It was, then, concluded that (v) the study demonstrates that a speech-based algorithm can objectively differentiate PTSD cases from controls, (vi) the RF classifier had a high AUC, and (vii) further validation in an independent sample and appraisal of the classifier to identify those with MDD only compared with those with PTSD comorbid with MDD is required [165].

The onset of COVID-19 has highlighted both the immediate value and long-term need for a scalable solution to identify behavioral health symptoms. Representative panel surveys conducted by the Centers for Disease Control and Prevention (CDC) to assess mental health during the pandemic among adults found that 40.9% of respondents reported at least one adverse mental or behavioral health condition, including symptoms of anxiety disorder or depressive disorder (30.9%), a considerable increase compared with the same period in 2019 [166]. Ellipsis Health [167], pioneer of the first commercial-grade voice-based vital signs to quantify and manage depression and anxiety symptoms at scale, has announced the general availability of the Rising Higher mobile app. Rising Higher is a clinical decision support tool that enables healthcare providers and payers to remotely monitor symptoms in high-risk patient populations. Rising Higher creates behavioral health vital signs by detecting anxiety or depression symptoms via dual acoustic and semantic-based assessments of patient speech – providing critical behavioral health screening as well as monitoring between clinical encounters to improve health outcomes. Healthcare providers alone do not have the bandwidth to solve the problem at scale – but by leveraging the ubiquity of smartphones, behavioral health measurement is now available anytime, anywhere through our Rising Higher app. This groundbreaking development in the quantification of depression and anxiety symptoms is revolutionizing patient care and improves QOL for those seeking help for mental illness [167].

8.4.3.6 Neurodegenerative disease
Neurological diseases like Parkinson's and Alzheimer's disease affect over one million people. As they are progressive it is important to diagnose them as soon as possible.

The retina at the back of the eye is basically an outpost of the brain and the only part of the central nervous system we can see directly from the outside and it is known that, in Alzheimer's disease and Parkinson's disease, the retina is affected. AI for diagnosing neurological disease such as Parkinson's and Alzheimer's from eye scans will be tested and scaled in the NHS (National Health Service in England). AI tools have already been developed to interpret the OCT (optical coherence tomography) images and detect common eye diseases. AI tools are going to be further developed to capture signs of neurological disease, using the vast quantities of OCT scans required. The aim of the project is to use NHS data to teach computers how to detect early signs of neurological disease via retinal imaging. Ultimately, the project will help to catch those at risk earlier, before other symptoms develop [168].

The complexity of the molecular mechanisms underlying neuronal degeneration and the heterogeneity of the patient population present massive challenges to the development of early diagnostic tools and effective treatments for these diseases. AI/ML is enabling scientists, clinicians and patients to address some of these challenges. Myazczynska et al. [169] discussed how ML can aid early diagnosis and interpretation of medical images as well as the discovery and development of new therapies. A unifying theme of the different applications of ML is the integration of multiple high-dimensional sources of data, which all provide a different view on disease, and the automated derivation of actionable insights. It was pointed out that (i) ML and NLP are forms of AI that enable robust interrogation of multiple datasets to identify previously undiscovered patterns and relationships in the data, (ii) ML approaches have been applied to the study of neurodegenerative diseases and show promise in the areas of early diagnosis, prognosis and development of new therapies, (iii) a substantial number of ML algorithms exist, and choosing the correct algorithm to apply to different types of data is crucial to obtain reliable results, (iv) neuroimaging was the first area of neurology to benefit from the application of ML approaches to improve diagnosis; more recently, application of ML methods to motor function and language feature analysis has shown promise in decreasing the time taken to perform clinical assessments, (v) the application of ML to longitudinal patient data collection and EHRs has the potential to inform prognosis prediction and patient stratification, and (vi) large collections of curated datasets and robust assessment of ML methods will be needed to achieve full integration of ML into diagnostic and prognostic neurology practice and the design of future therapeutics [169]. Georgia State University researchers [170] are working to use AI/DL to learn more about how mental illness and other disorders affect the brain, by combining different types of brain imaging data to capture patterns that are indicative of brain disorders. It was mentioned that the project works to combine different kinds of imaging data such as how brain structures look, how the brain is functioning over time or how the brain is connected or wired. Advances in brain imaging mean researchers have access to significantly more data than in the past, but the relationships among modalities, or types of data captured, are complex and poorly understood, so that it is needed to characterize these relationships by merging and analyzing data from multiple sources. Using DL, the research team will develop and train algorithms on thousands of existing datasets. By examining the data along two spectra, mood and psychosis, they can determine which modalities or brain regions are most relevant to specific disorders. The hope is to develop multimodal biomarkers that healthcare experts could use for diagnosis of mental health disorders such as schizophrenia, depression or bipolar disorder [170].

8.4.3.7 Parkinson's disease

Parkinson's disease (PD) is a progressive neurodegenerative disorder that affects predominately dopamine-producing ("dopaminergic") neurons in a specific area of the brain called substantia nigra. Symptoms start gradually, sometimes starting with a barely noticeable tremor in just one hand. Tremors are common, but the disorder also commonly causes stiffness or slowing of movement. Data on PD's statistics [171] indicate that (i) nearly one million will be living with PD in the United States by 2020, which is more than the combined number of people diagnosed with multiple sclerosis, muscular dystrophy and Lou Gehrig's disease (or amyotrophic lateral sclerosis), (ii) approximately 60,000 Americans are diagnosed with PD each year, (iii) more than 10 million people worldwide are living with PD, (iv) incidence of PD increases with age, but an estimated 4% of people with PD are diagnosed before age 50, and (v) men are 1.5 times more likely to have PD than women. A group of neurological disorders that display very similar symptoms to PD are included under the general term of parkinsonism. There are two types of PD: primary parkinsonism and secondary parkinsonism [172].

(1) Primary parkinsonism: Most patients (about 80–85%) diagnosed with Parkinson's disease have what is called primary parkinsonism or idiopathic Parkinson's disease (meaning that the disease has unknown cause). This type tends to respond well to drugs that work by increasing or substituting dopamine molecules in the brain. (2) Secondary parkinsonism (parkinsonian syndrome or atypical parkinsonism): The remaining types of Parkinson's are termed secondary or atypical parkinsonism or Parkinson's Plus. In these cases, the cause of the disease is generally known and although it is very difficult to differentiate idiopathic Parkinson's disease and secondary parkinsonism, a key difference is that patients with secondary parkinsonism do not respond well to dopaminergic medications such as levodopa [173].

AI/ML has found numerous applications in computer-aided diagnostics, monitoring, and management of neurodegenerative movement disorders of parkinsonian type. These tasks are not trivial due to high inter-subject variability and similarity of clinical presentations of different neurodegenerative disorders in the early stages. Belić et al. [172] conducted a comprehensive, high-level overview of applications of AI through ML algorithms in kinematic analysis of movement disorders, specifically

PD, based on articles published between January 2007 and January 2019, within on-line databases, including PubMed and Science Direct, with a focus on the most re-cently published studies. It was pointed out that (i) different ML algorithms showed promising results, particularly for early PD diagnostics, and (ii) although publica-tions demonstrated the potentials of collecting data from affordable and globally available devices, to fully exploit AI technologies in the future, more widespread collaboration is advised among medical institutions, clinicians, and researchers, to facilitate aligning of data collection protocols, sharing and merging of data sets. Kim et al. [174] identified the extent to which AI could be used in the diagnosis of PD from ioflupane-123 (^{123}I) single-photon emission computed tomography (SPECT) dopamine transporter scans using transfer learning. A data set of 54 normal and 54 abnormal ^{123}I SPECT scans was amplified 44-fold using a process of image augmen-tation, resulting in a training set of 2,376 normal and 2,376 abnormal images. This was used to retrain the top layer of the Inception v3 network. The resulting NN func-tioned as a classifier for new ^{123}I SPECT scans as either normal or abnormal. A completely separated set of 45 ^{123}I SPECT scans was used for final testing of the net-work. It was found that the AUC-ROC in final testing was 0.87, which corresponded to a test sensitivity of 96.3%, a specificity of 66.7%, a positive predictive value of 81.3% and a negative predictive value of 92.3%, using an optimum diagnostic thresh-old. It was, therefore, concluded that (i) the study has provided proof of concept for the use of transfer learning, from convolutional NNs pretrained on nonmedical im-ages, for the interpretation of ^{123}I SPECT scans, and (ii) the technique is likely to be applicable to many areas of diagnostic imaging [174].

To manage PD, wearable devices offer huge potential to collect rich sources of data that provide insights into the diagnosis and the effects of treatment interven-tions. Ten-second whole-hand-grasp action is widely used to assess bradykinesia se-verity, since bradykinesia is one of the primary symptoms of Parkinson's disease [32]. Bradykinesia is one of the primary characteristic symptoms of PD. Lin et al. [175] re-ported, using an attitude-estimation algorithm to extract the parkinsonian bradykinesia parameters, that clinical experiment with 15 PD patients and 5 age-matched healthy controls demonstrated that the predicted bradykinesia scores by proposed model corre-lated well with the judgments of neurologists, indicating that the proposed quantifica-tion model demonstrated the greater goodness-of-fit compared with the related works. A variety of clinical scales are available to assess dyskinesia severity in PD patients; however, such assessments are subjective, they do not provide long-term monitoring, and their use is subject to inter- and intra-rater variability. An objective dyskinesia score was developed using a motion capture system to collect patient kinematic data [176]. It was remarked that the portable wearable technology can be used remotely to monitor the full-body severity of dyskinesia, necessary for therapeutic optimization, es-pecially in the patients' home environment [176].

Current speech therapy for people with PD focuses on increasing loudness, slowing rate, and improving speech clarity. Often, it is difficult for people with PD to remember to use speech therapy techniques during everyday communication environments [177]. A wearable SpeechVive was developed by J. Huber (Purdue University speech scientist) to watch people with PD struggle to speak, which often led to social isolation and depression. SpeechVive's mission is to improve the quality of communication and quality of life for people with PD by empowering care partners and clinicians with technologies and support to restore the ability to communicate [178]. "Eyes are as eloquent as the tongue." Disease progression is characterized by nerve cell decay that thins the walls of the retina, the layer of tissue that lines the back of the eyeball. The disease also affects the microscopic blood vessels, or microvasculature, of the retina. These characteristics present an opportunity to leverage the power of AI to examine images of the eyes for signs of Parkinson's disease [179]. Diaz et al. [180] reported that a type of AI called SVM learning was developed, using pictures of the back of the eye from both patients with PD and control participants. The results indicated that the ML networks can classify PD based on retina vasculature, with the key features being smaller blood vessels. The proposed methods further support the idea that changes in brain physiology can be observed in the eye. Those traditional imaging approaches with MRI, CT, and nuclear medicine techniques can be very costly; in contrast, the new approach uses basic photography with equipment commonly available in eye clinics to get an image. The images can even be captured by a smartphone with a special lens. Furthermore, the approach may also have applications in identifying other diseases that affect the structure of the brain, such as Alzheimer's disease and multiple sclerosis [180].

8.4.3.8 Alzheimer's disease

Alzheimer's disease (AD) is a disease in which the brain is going to atrophy. Major neuro-cognitive disorder due to AD is the symptom and is a type of dementia whose main symptoms are cognitive decline and personality changes, accounting for 60–70% of dementia. Symptoms are progressive cognitive deficits (which could include memory disorders, disorientation disorders, learning disabilities, attention disorders, visuospatial cognitive disorders, problem-solving abilities, etc.) that interfere with life. As the severity increases, it becomes impossible to eat, change clothes, and communicate, and eventually leads to being bedridden. Unlike cerebrovascular dementia, which progresses in a step-like manner (i.e., the symptoms clearly worsen at a certain point in time), it is characteristic that it progresses gradually. Persecutory delusions and hallucinations may appear during the course of symptoms. Problematic behaviors such as ranting, violence, wandering, and filth (so-called BPSD: behavior and psychological symptoms of dementia) may be seen, which poses great difficulties in long-term care and is the greatest opportunity to see a medical institution [181–183]. According to CDC data [183], in 2014, as many as 5 million Americans were living with Alzheimer's disease, and this number is projected to nearly triple to 14 million people by 2060. WHO [184] reported that AD affects 44 million people worldwide; this is projected to triple by year 2050.

Modern AI research is producing tools that will help to solve this challenge. The AI/ML, in particular, is developing algorithms that can accurately learn models of real-world phenomena, or categorical differences among those phenomena, from representative data. Progress in this area has recently accelerated dramatically, due to the discovery of new algorithms in ANNs, the availability of powerful hardware capable of vast parallel processing, and the wide proliferation of data across a variety of domains, including healthcare. In particular, recent advances are showing that AI can be used to assess and monitor AD and other types of dementia, using no more than simple measures of a speaker's voice [185]. In terms of dementia diagnosis, there have been increasing applications of various ML approaches, most commonly with imaging data for diagnosis and disease progression [186, 187]. Reviewing 37 published articles on evidence on the state of the art of studies investigating and the prognosis of dementia using ML and microsimulation techniques, Dallora et al. [187] mentioned that (i) the current research is focused on the investigation of the patients with mild cognitive impairment (MCI) that will evolve to Alzheimer's disease, using ML techniques, (ii) microsimulation studies were concerned with cost estimation and had a populational focus, and (iii) neuroimaging was the most commonly used variable. It was, then, concluded that (iv) prediction of conversion from MCI to AD is the dominant theme in the selected studies and (v) most studies used ML techniques on neuroimaging data. The use of big data in dementia research has also recently emerged [188]. Including a total of 5,041 case patients in Taiwan and 13,902 control patients who were matched with age, gender, and index date, Chen et al. [189] employed logistic regression to identify the comorbid associations between dementia and various kinds of illnesses. It was mentioned that (i) advanced stratification analyses revealed that some comorbid illnesses associated with dementia only exists in subgroups of patients with specific age, gender, or prescription for dementia drugs and (ii) the mined characteristics would be helpful in managing patients with dementia. Alipour et al. [190] diagnosed AD using blood-based biomarkers and neuropsychological tests with DM techniques. An alternative strategy for the traditional drug development (which is costly, time-consuming, and burdened by a very low success rate) is drug repositioning, redirecting existing drugs for another disease [190]. The information related to AD pathogenesis was obtained from the OMIM and PubMed databases. Drug-target data was extracted from the DrugBank and Therapeutic Target Database. Zhang et al. [191] generated a list of 524 AD-related proteins, 18 of which are targets for 75 existing drugs – novel candidates for repurposing as anti-AD treatments and developed a ranking algorithm to prioritize the anti-AD targets, which revealed CD33 and MIF as the strongest candidates with seven existing drugs. It was suggested that, by the systematic 'omics' mining, (i) drugs with novel anti-AD indications, including drugs modulating the immune system or reducing neuroinflammation that are particularly promising for AD intervention, and (ii) furthermore, the list of 524 AD-related proteins could be useful not only as potential anti-AD targets but also considered for AD biomarker development.

Identifying individuals destined to develop Alzheimer's dementia within time frames acceptable for clinical trials constitutes an important challenge to design studies to test emerging disease-modifying therapies. Although amyloid-β protein is the core pathologic feature of Alzheimer's disease, biomarkers of neuronal degeneration are the only ones believed to provide satisfactory predictions of clinical progression within short time frames. Based on this background, Mathotaarachchi et al. [192] proposed an ML-based probabilistic method designed to assess the progression to dementia within 24 months, based on the regional information from a single amyloid PET scan. It was mentioned that (i) the proposed method was designed to overcome the inherent adverse imbalance proportions between stable and progressive MCI individuals within a short observation period, and (ii) the novel algorithm obtained an accuracy of 84% and an under-receiver operating characteristic curve of 0.91, outperforming the existing algorithms using the same biomarker measures and previous studies using multiple biomarker modalities, suggesting that (iii) with its high accuracy, this algorithm has immediate applications for population enrichment in clinical trials designed to test disease-modifying therapies aiming to mitigate the progression to Alzheimer's disease dementia [192]. ML methods have successfully been used to predict the conversion of MCI to AD, by classifying MCI converters (MCI-C) from MCI nonconverters (MCI-NC), although most existing methods construct classifiers using data from one particular target domain (e.g., MCI), and ignore data in other related domains (e.g., AD and normal control – NC) that may provide valuable information to improve MCI conversion prediction performance. Accordingly, Cheng et al. [193] developed a novel domain transfer learning method for MCI conversion prediction, which can use data from both the target domain (i.e., MCI) and auxiliary domains (i.e., AD and NC). Specifically, the proposed method consists of three key components: (1) a domain transfer feature selection component that selects the most informative feature subset from both target domain and auxiliary domains from different imaging modalities; (2) a domain transfer sample selection component that selects the most informative sample subset from the same target and auxiliary domains from different data modalities; and (3) a domain transfer SVM classification component that fuses the selected features and samples to separate MCI-C and MCI-NC patients. It was reported that the proposed method can classify MCI-C patients from MCI-NC patients with an accuracy of 79.4%, with the aid of additional domain knowledge learned from AD and NC [193]. Recorded changes in the language and speech of aging individuals offer a new means of quantifying neurodegeneration. Korcovelos et al. [194] mentioned that, by analyzing linguistic features such as parts of speech, word length, word frequency, and acoustic variables, automated techniques in computational linguistics make it possible to classify groups of differing linguistic ability. It was concluded that (i) binary classification between groups was between 13% and 21% more accurate than baseline values, and four-way classification was 14.9% better and (ii) linguistic features yielded better predictions than did the addition of acoustic features. Dodge et al. [195] determined whether unobtrusive long-term in-home assessment of walking speed and its variability can distinguish those with MCI from those with intact cognition and mentioned that (i) walking speed and its daily variability may be an early marker of the development of MCI and (ii) these and other real-time measures of function may offer novel ways of detecting transition phases leading to dementia.

8.4.3.9 Ophthalmology

Many ophthalmology practices are already using ML and DL to revolutionize vision care. The immediate impact has been observed in the field of retinal diseases. Schlegl et al. [196] developed and validated a fully automated method to detect and quantify macular fluid in conventional OCT (optical coherence tomography) images. The clinical dataset for fluid detection consisted of 1,200 OCT volumes of patients with neovascular age-related macular degeneration (AMD, n = 400), diabetic macular edema (DME, n = 400), or retinal vein occlusion (RVO, n = 400) acquired with Zeiss Cirrus (Carl Zeiss Meditec, Dublin, CA) (n = 600) or Heidelberg Spectralis (Heidelberg Engineering, Heidelberg, Germany) (n = 600) OCT devices. A method based on DL to automatically detect and quantify intraretinal cystoid fluid (IRC) and subretinal fluid (SRF) was developed. The performance of the algorithm in accurately identifying fluid localization and extent was evaluated against a manual consensus reading of 2 masked reading center graders. It was reported that (i) the newly designed, fully automated diagnostic method based on DL achieved optimal accuracy for the detection and quantification of IRC for all three macular pathologies with a mean accuracy (AUC) of 0.94 (range, 0.91–0.97), a mean precision of 0.91, and a mean recall of 0.84, (ii) the detection and measurement of SRF were also highly accurate with an AUC of 0.92 (range, 0.86–0.98), a mean precision of 0.61, and a mean recall of 0.81, with superior performance in neovascular AMD and RVO compared with DME, which was represented rarely in the population studied, and (iii) high linear correlation was confirmed between automated and manual fluid localization and quantification, yielding an average Pearson's correlation coefficient of 0.90 for IRC and of 0.96 for SRF. Based on these findings, it was concluded that (iv) DL in retinal image analysis achieves excellent accuracy for the differential detection of retinal fluid types across the most prevalent exudative macular diseases and OCT devices, (v) furthermore, quantification of fluid achieves a high level of concordance with manual expert assessment, and (vi) fully automated analysis of retinal OCT images from clinical routine provides a promising horizon in improving accuracy and reliability of retinal diagnosis for research and clinical practice in ophthalmology [196].

AI systems used in the field of ophthalmology is versatile. Liu et al. [197] presented a computer vision-based framework for the automatic localization and diagnosis of slit-lamp images by identifying the lens region of interest and employing a DL/CNN. It was mentioned that the qualitative and quantitative experimental results demonstrate that the proposed method offers exceptional mean accuracy, sensitivity and specificity: classification (97.07%, 97.28%, and 96.83%) and a three-degree grading area (89.02%, 86.63%, and 90.75%), density (92.68%, 91.05%, and 93.94%) and location (89.28%, 82.70%, and 93.08%). Kim et al. [198] developed ML models that have strong prediction power and interpretability for diagnosis of glaucoma based on retinal nerve fiber layer thickness and visual field (VF). It was mentioned that (i) the developed prediction models show high accuracy, sensitivity, specificity, and AUC in classifying among glaucoma and healthy eyes, and (ii) it will be used for predicting glaucoma against unknown examination records. Liu et al. [199] established a DL system for detection of GON (glaucomatous optic neuropathy) using retinal fundus images and glaucoma diagnosis with CNN (GD-CNN) that enables to be generalized across populations. It was concluded that application of GD-CNN to fundus images from different settings and varying image quality demonstrated a high sensitivity, specificity, and generalizability for detecting GON, suggesting that automated DL system could enhance current screening programs in a cost-effective and time-efficient manner. Ruiz et al. [200] validated a recently developed program for automatic and objective detection of keratoconus based on Scheimpflug tonometry (Keratoconus Assistant: KA) by applying it to a new population and comparing it with other methods described in the literature. It was indicated that KA is effective at detecting early keratoconus and agrees with trained clinical judgment. Brown et al. [201] implemented and validated an algorithm based on DL to automatically diagnose plus disease from retinal photographs. It was concluded that (i) the fully automated algorithm diagnosed plus disease in ROP (retinopathy of prematurity) with comparable or better accuracy than human experts, and (ii) the has potential applications in disease detection, monitoring, and prognosis in infants at risk of ROP.

8.4.3.10 Otorhinology

Bur et al. [202] reviewed the AI with ML and NLP, as it applies to otolaryngology and reported that (i) in the near future, otolaryngologists are key stakeholders in the development and clinical integration of meaningful AI technologies that will improve patient care, and (ii) high-quality data collection is essential for the development of AI technologies, and otolaryngologists should seek opportunities to collaborate with data scientists to guide them toward the most impactful clinical questions. AI is quickly making its way into our specialty. Both otolaryngologists and audiologists will soon be incorporating this technology into their clinical practices. Hughes et al. [203] mentioned that (i) ML algorithms have also been central to the development of multiple assistive technologies that can help patients to overcome or alleviate disabilities. For example, in the context of hearing loss, significant advances in automated transcription apps, driven by ML algorithms, have proven particularly useful in recent months for patients who find themselves unable to lipread due to the use of face coverings to prevent the spread of COVID-19, and (ii) in addition to their role in general image classification,

CNNs are likely to play a significant role in the introduction of ML in healthcare, especially in image-heavy specialties such as otolaryngology.

Senaras et al. [204] proposed an automated otoscopy image analysis system called "Autoscope," which was the first system designed to detect a wide range of eardrum abnormalities by using high-resolution otoscope images and report the condition of the eardrum as "normal" or "abnormal." In order to achieve this goal, a preprocessing step was developed to reduce camera-specific problems, detect the region of interest in the image, and prepare the image for further analysis. Subsequently, a new set of clinically motivated eardrum features (CMEF) was designed. Furthermore, the potential of the visual MPEG-7 (moving and picture expert groups) descriptors was evaluated for the task of tympanic membrane image classification, and fused the information extracted from the CMEF and state-of-the-art computer vision features, which included MPEG-7 descriptors and two additional features together, using a state-of-the-art classifier. It was reported that 247 tympanic membrane images with 14 different types of abnormality were used, and Autoscope was able to classify the given tympanic membrane images as normal or abnormal with 84.6% accuracy [204].

8.4.3.11 Blood disorder

Wearable trackers have drawn interest from health professionals studying blood disorders. Early detection of essential hypertension can support the prevention of CVD, a leading cause of death. The traditional method of identification of hypertension involves periodic blood pressure measurement using brachial cuff-based measurement devices. While these devices are noninvasive, they require manual setup for each measurement, and they are not suitable for continuous monitoring. Wearable devices capable of measuring physiological signals such as heart rate, galvanic skin response, and skin temperature have recently become popular, although these signals are not accurate and are prone to noise due to different artifacts. Based on this background, Ghosh et al. [205] introduced wearable devices to enable to detect hypertension with physiological signals and mentioned that physiological signals extracted from wearable devices can distinguish between these two groups with high accuracy. Some of the most widely used wearable devices are applications for evaluating and monitoring blood pressure, including cuffless blood pressure sensors, wireless smartphone-enabled upper arm blood pressure monitors, mobile applications, and remote monitoring technologies. They have the potential to improve hypertension control and medication adherence through easier logging of repeated blood pressure measurements, better connectivity with healthcare providers, and medication reminder alerts [206]. Patients with orthostatic hypotension have pathologic hemodynamics related to changes in body posture. Fujikawa et al. [207] developed a novel cephalic laser blood flow meter that can be worn on the tragus to investigate the hemodynamics upon rising from a sitting or squatting posture. The relationship between cephalic hemodynamics and cerebral ischemic symptoms in 63 subjects in sitting, squatting, and standing positions using the new device was evaluated. It was reported that (i) transient decrease in blood pressure within 15 s after rising to an erect position possibly causes dizziness, syncope, and fall, and (ii) subjects exhibiting dizziness upon standing showed a significant decrease in the cephalic blood flow (CBF) and indirect beat-to-beat systolic blood pressure, as monitored by the Finometer (Finapres Medical Systems, Amsterdam, The Netherlands), and a significant correlation was observed between the drop ratio (drop value on rising/mean value in the squatting position) of CBF and that of systolic blood pressure, concluding that (iii) the novel wearable CBF meter is potentially useful for estimating cephalic hemodynamics and objectively diagnosing cerebral ischemic symptoms of subjects in a standing posture. Blood flow contains abundant physiological information, but it is hard to measure blood flow during exercise using conventional blood flowmeters because of their size, weight, and use of optic fibers. To resolve these disadvantages, Iwasaki et al. [208] developed a micro integrated laser Doppler blood wearable flowmeter using microelectromechanical systems technology, which is also capable of stable measurement signals even during movement. Sight Diagnostics developed a compact, desktop machine (called OLO) system which analyzes single-use cartridges manually loaded with drops of the patient's blood and can deliver "lab-grade" complete blood count tests from only a finger prick of blood, using computer vision and ML technology, to diagnose a broad range of common medical conditions [209].

8.4.3.12 ASD and ADHD

More than 3.5 million people in the United States are on the autism spectrum [210]. Most of these children struggle to make eye contact, recognize facial expressions, and engage in social interactions. However, it is important for autistic children to recognize and classify their emotions, such as anger, disgust, fear, happiness, sadness, and surprise [71]. Recent advances in computer vision and wearable technology have created an opportunity to introduce mobile therapy systems for autism spectrum disorders (ASD) that can respond to the increasing demand for therapeutic interventions; however, feasibility questions must be answered first. Daniels et al. [211] studied the feasibility of a prototype therapeutic tool for children with ASD using Google Glass, examining whether children with ASD would wear such a device, if providing the emotion classification will improve emotion recognition, and how emotion recognition differs between ASD participants and neurotypical controls (NC). It was found that (i) all 43 children were comfortable wearing the Glass, (ii) ASD and NC participants who completed the computer task with Glass providing audible emotion labeling (n = 33) showed increased accuracies in emotion labeling, and the logistic regression classifier achieved an accuracy of 72.7%, and (iii) further analysis suggests that the ability to recognize surprise, fear, and neutrality may distinguish ASD cases from NC, concluding that (iv) the feasibility study supports the utility of a wearable device for social affective learning in ASD children and demonstrates subtle differences in how ASD and NC children perform on an emotion recognition task. The incidence of ASD has been rising; however, ASD-risk biomarkers remain lacking. Ramirez-Celis et al. [212] specifically focused on maternal autoantibody-related ASD (MAR ASD) to create and validate a serological assay to identify ASD-specific maternal autoantibody patterns of reactivity against eight previously identified proteins (CRMP1, CRMP2, GDA, NSE, LDHA, LDHB, STIP1, and YBOX) that are highly expressed in developing brain and determine the relationship of these reactivity patterns with ASD outcome severity. It was reported that uses ML subgroup discovery to identify with 100% accuracy MAR ASD-specific patterns as potential biomarkers of risk for a subset of up to 18% of ASD cases in this study population.

Wearable technology can also assist with the screening, diagnosis, and monitoring of psychiatric disorders such as depression [32]. The analysis of cognitive and autonomic responses to emotionally relevant stimuli could provide a viable solution for the automatic recognition of different mood states, in both normal and pathological conditions. Valenza et al. [213] introduced a methodological application describing a novel system based on wearable textile technology and instantaneous nonlinear heart rate variability (HRV) assessment, able to characterize the autonomic status of bipolar patients by considering only ECG recordings. It was shown that (i) the system achieves much higher accuracy than the traditional techniques and (ii) the inclusion of instantaneous higher-order spectra features significantly improves the accuracy in successfully recognizing depression from euthymia. A wearable depression monitoring system is proposed with an application-specific system-on-chip solution. Roh et al. [214] designed the system-on-chip to accelerate the filtering and feature extraction of HRV from the ECG. It was further mentioned that (i) for user's convenience, the system interfaces with smartphones through Bluetooth communication and (ii) with the features of the HRV and Beck depression inventory (BDI), the smartphone application trains and classifies the user's depression scale with 71% accuracy. Luo et al. [215] reported that collaborative research teams (Northwestern University and Harvard University) have revealed that AI-enhanced precision medicine approaches have identified new subtypes of ASD. Unlike the classical approach of classifying from manifested symptoms, this subtype has a clear cause and mechanism at the molecular biology level and is attracting attention as a multidimensional evidence-based subtype. It was reported that parental dyslipidemia was strongly associated with some ASD in offspring, and changes in the blood lipid profile of infants who were actually diagnosed was later confirmed, indicating the potential for effective early screening and early intervention for some ASD [215].

Scientists have known that heart rate is linked to depression, but until recently they have been unable to understand exactly how one is related to the other [216]. In part, this is because while heart rates can fluctuate quickly, depression both arrives and leaves over a longer period, with most treatments taking months to take effect. This makes it difficult to see whether or not changes in one's depressive state might be related to heart rate. There are two innovative elements, including the continuous registration of heart rate for several days and nights, and the use of the new antidepressant ketamine, which can lift depression more or less instantly; these allows us to see that average resting heart rate may change quite suddenly to reflect the change in mood. It was found that (i) those with depression had both a higher baseline heart rate and a lower heart rate variation, (ii) on average, depressed patients had a heart rate which was roughly 10–15 beats per minute higher than in controls, and (iii) after treatment with ketamine, re-measuring the heart rates indicated that both the rate and the heartrate fluctuation of the previously depressed patients had changed to be closer to those found in the controls.

Attention-deficit hyperactivity disorder (ADHD) is one of the most common mental disorders affecting children. ADHD also affects many adults. Symptoms of ADHD include inattention (not being able to keep focus), hyperactivity (excess movement that is not fitting to the setting), and impulsivity (hasty acts that occur in the moment without thought). An estimated 8.4% of children and 2.5% of adults have ADHD. ADHD is often first identified in school-aged children when it leads to disruption in the classroom or problems with schoolwork. It can also affect adults. It is more common among boys than girls [217]. Scientists using AI/ML with data from hundreds of children who struggle at school identified clusters of learning difficulties which did not match the previous diagnosis the children had been given. The researchers from the Medical Research Council Cognition and Brain Sciences Unit at the University of Cambridge say this reinforces the need for children to receive detailed assessments of their cognitive skills to identify the best type of support [218]. Chen et al. [219] developed a multichannel DNN (mcDNN) classification model based on multiscale brain functional connectome data and demonstrate the value of this model by using ADHD detection as an example. In the retrospective case-control study, existing data from the Neuro Bureau ADHD-200 dataset consisting of 973 participants were subjected to study. It was obtained that (i) in the cross-validation, the mcDNN model using combined features (fusion of the multiscale brain connectome data and PCD) achieved the best performance in ADHD detection with an AUC of 0.82 (95% confidence interval – CI: 0.80, 0.83) compared with scDNN models using the features of the brain connectome at each individual scale and PCD, independently, and (ii) in the hold-out validation, the mcDNN model achieved an AUC of 0.74 (95% CI: 0.73, 0.76). Based on these findings, it was concluded that (iii) an mcDNN model was developed for multiscale brain functional connectome data, and its utility for ADHD detection was demonstrated, and (iv) by fusing the multiscale brain connectome data, the mcDNN model improved ADHD detection performance considerably over the use of a single scale [219].

8.4.3.13 Schizophrenia

Schizophrenia is a chronic brain disorder that affects less than 1% of the US population. When schizophrenia is active, symptoms can include delusions, hallucinations, disorganized speech, trouble with thinking, and lack of motivation. However, with treatment, most symptoms of schizophrenia will greatly improve and the likelihood of a recurrence can be diminished [220]. Genetic predisposition and environmental, social, and psychological factors all affect neurodevelopmental abnormalities and target features, causing brain dysfunction and improper balance of chemicals and finally leading to developing schizophrenia. In the literature, there are substantial ML attempts to classify schizophrenia based on alterations in resting-state (RS) brain patterns using functional MRI (fMRI). Most studies with classification accuracies > 80% are based on small sample datasets, which may be insufficient to capture the heterogeneity of schizophrenia, limiting generalization to unseen cases. Hence, Kalmady et al. [221] utilized RS fMRI data collected from a cohort of antipsychotic drug treatment-naive patients meeting DSM IV criteria for schizophrenia (n = 81) as well as age- and sex-matched healthy controls (n = 93). An ensemble model – EMPaSchiz (read as "Emphasis"; standing for "Ensemble algorithm with Multiple Parcellations for Schizophrenia prediction") – was developed to predict from several "single-source" models, each based on features of regional activity and functional connectivity, over a range of different a priori parcellation schemes. It was reported that (i) EMPaSchiz yielded a classification accuracy of 87% (versus chance accuracy of 53%), which outperforms earlier ML models built for diagnosing schizophrenia using RS fMRI measures modeled on large samples (n > 100), (ii) the EMPaSchiz was first to be reported that has been trained and validated exclusively on data from drug-naive patients diagnosed with schizophrenia, indicating that the method relies on a single modality of MRI acquisition and can be readily scaled-up without needing to rebuild parcellation maps from incoming training images [221].

Subtle features in people's everyday language may harbor the signs of future mental illness. ML offers an approach for the rapid and accurate extraction of these signs. Schizophrenia is often associated with disrupted brain connectivity. However, identifying specific neuroimaging-based patterns pathognomonic for schizophrenia and related symptom severity remains a challenging open problem requiring large-scale data-driven analyses emphasizing not only statistical significance but also stability across multiple datasets, contexts and cohorts. Accurate prediction on previously unseen subjects, or generalization, is also essential for any useful biomarker of schizophrenia [222]. In order to build a predictive model based on functional network feature patterns, Gheiratmand et al. [222] studied whole-brain fMRI functional networks, both at the voxel level and lower resolution supervoxel level. Targeting Auditory Oddball task data on the FBIRN fMRI dataset (n = 95) with considerations on node-degree and link-weight network features, evaluated stability and generalization accuracy of statistically significant feature sets in discriminating patients versus controls. It was found that (i) whole-brain link-weight features achieved 74% accuracy in identifying patients and were more stable than voxel-wise node degrees, (ii) link-weight features predicted severity of several negative and positive symptom scales, including inattentiveness and bizarre behavior, and (iii) the most-significant, stable and discriminative functional connectivity changes involved increased correlations between thalamus and primary motor/primary sensory cortex, and between precuneus (BA7) and thalamus, putamen, and Brodmann areas BA9 and BA44. Precuneus, along with BA6 and primary sensory cortex, was also involved in predicting severity of several symptoms, indicating that the proposed multi-step methodology may help identify more reliable multivariate patterns allowing for accurate prediction of schizophrenia and its symptoms severity [223].

8.4.4 AI diagnosis versus doctor diagnosis

8.4.4.1 Comparison
Recent advances in ML and DL offer considerable promise for medical diagnostics. There are studies indicating that AI-assisted diagnosis is evaluated equivalent to results done by pathologists, whilst other studies describing better performance by AI technology than human pathologists.

Liu et al. [224] evaluated the diagnostic accuracy of DL algorithms versus healthcare professionals in classifying diseases using medical imaging, based on the systematic review and meta-analysis on Ovid-MEDLINE, Embase, Science Citation Index, and Conference Proceedings Citation Index, published from Jan 1, 2012, to June 6, 2019. Studies compared the diagnostic performance of DL models and health-care professionals based on medical imaging, for any diseases. An out-of-sample external validation was done in 25 studies, of which 14 made the comparison between DL models and healthcare professionals in the same sample. It was reported that (i) AI has become more accurate of identifying disease diagnosis in images and has become a more viable source of diagnostic information and (ii) comparing DL models and healthcare professionals within the same sample indicates that the diagnostic performances were found to be equivalent. With advances in AI, DL may become even more efficient in identifying diagnosis in the next few years [224, 225]. It was reported that Caption Guidance AI technology enables nurses without prior ultrasound experience to obtain diagnostic-quality images in patients with implanted electrophysiological devices [226]. Eight nurses without prior ultrasound experience each acquired 10 standard transthoracic echocardiography (TTE) views in 30 patients (for a total of 240 total patients) guided by the novel Caption Guidance software. A pacemaker or ICD was present in 27 of these patients. On the same day, trained cardiac sonographers obtained the same 10 TTE views without using the DL algorithm. A panel of five Level 3 echocardiographers independently assessed the diagnostic quality of the exams acquired by the nurses and sonographers. It was mentioned that the descriptive analysis evaluated performance of the software for patients with implanted EP devices, focusing on right ventricular (RV) size and function. Nurses using Caption Guidance software acquired TTEs of sufficient quality to make qualitative assessments of RV size and function in greater than 80% of cases for patients with implanted EP devices, and greater than 90% in patients without them. Sonographers performed similarly: descriptively, there was no significant difference between nurse- and sonographer-acquired scans for all comparisons (i.e., patients with EP (electrophysiology) devices, patients without EP devices, and overall) [226], indicating that the software performs well in patients with implanted EP devices, supporting the overall conclusions of the pivotal study demonstrating that Caption Guidance can guide novice ultrasound users to obtain diagnostic-quality TTEs generally.

DL image reconstruction has the potential to disrupt the current state of MRI by significantly decreasing the time required for MRI examinations. Recht et al. [227], using DL technology, accelerated MRI to allow a 5-min comprehensive examination of the knee without compromising image quality or diagnostic accuracy. The DL model was trained using dedicated multisequence training, in which a single reconstruction model was trained with data from multiple sequences with different contrast and orientations. After training, data from 108 patients were retrospectively under-sampled in a manner that would correspond with a net 3.49-fold acceleration of fully sampled data acquisition and a 1.88-fold acceleration compared with our standard 2-fold accelerated parallel acquisition. An interchangeability study was performed, in which the ability of six readers to detect internal derangement of the knee was compared for clinical and DL-accelerated images. It was found that (i) a high degree of interchangeability was found between standard and DL-accelerated images, (ii) in particular, results showed that interchanging the sequences would produce discordant clinical opinions no more than 4% of the time for any feature evaluated, and (iii) the accelerated sequence was judged by all six readers to have better quality than the clinical sequence. Based on these findings, it was concluded that (iv) an optimized DL model allowed acceleration of knee images that performed interchangeably with standard images for detection of internal derangement of the knee [227].

Recent advances in ML yielded new techniques to train deep NNs, which resulted in highly successful applications in many pattern recognition tasks such as object detection and speech recognition. Kooi et al. [228] conducted a head-to-head comparison study between a state-of-the art in mammography CAD system, relying on a manually designed feature set and a CNN, aiming for a system that can ultimately read mammograms independently. Both systems are trained on a large data set of around 45,000 images and results show the CNN outperforms the traditional CAD system at low sensitivity and performs comparable at high sensitivity. It was mentioned that the system based on DL performed at the level of a radiologist.

Mishra et al. [125] pointed out that early diagnosis of any chronic disease is very helpful in minimizing compilations of disease. It helps in deciding treatment protocols and there are various diagnosis and treatment protocols which prove that AI is a boon in healthcare. The computer assisted diagnosis, decision support systems, ES and implementation of software may assist physicians to minimize the intra and inter-observer variability. There has been an exponential growth in the application of AI in health and in pathology, due to the innovation of DL technologies that are specifically aimed at cellular imaging and practical applications that could transform diagnostic pathology. Serag et al. [229] reviewed the different approaches to DL in pathology, the public grand challenges that have driven this innovation and a range of emerging applications in pathology. It was noted that (i) the translation of AI into clinical practice will require applications to be embedded seamlessly within digital pathology workflows, driving an integrated approach to diagnostics and providing pathologists with new tools that accelerate workflow and improve diagnostic consistency and reduce errors and (ii) AI and computational pathology will continue to mature as researchers, clinicians, industry, regulatory organizations and patient advocacy groups work together to innovate and deliver new technologies to healthcare providers: technologies which are better, faster, cheaper, more precise, and safe [40, 229].

Chronic kidney damage is routinely assessed with a semi-quantitative manner by scoring the amount of fibrosis and tubular atrophy in a renal biopsy sample. Although image digitization and morphometric techniques can better quantify the extent of histologic damage, more widely applicable ways are needed to stratify kidney disease severity. Kolachalama et al. [230] trained six CNN models to use these images as inputs and the training classes as outputs, respectively. For comparison, separate classifiers were trained using the pathologist-estimated fibrosis score as input and the training classes as outputs, respectively. It was concluded that the study demonstrates a proof of principle that deep learning can be applied to routine renal biopsy images. Clearly, AI-based systems have the potential to augment clinical decision-making for nephrologists and ML along with imaging analytics from renal biopsies can help to predict how long a kidney will function adequately in patients with chronic kidney damage [230, 231].

Causal knowledge is vital for effective reasoning in science, as causal relations, unlike correlations, allow one to reason about the outcomes of interventions. Algorithms that can discover causal relations from observational data are based on the assumption that all variables have been jointly measured in a single dataset. In many cases this assumption fails. Previous approaches to overcoming this shortcoming devised algorithms that returned all joint causal structures consistent with the conditional independence information contained in each individual dataset. Based on this background, Dhir et al. [232] developed and extended the so-called bivariate causal discovery algorithms to the problem of learning consistent causal structures from multiple datasets with overlapping variables belonging to the same generating process, providing a sound and complete algorithm that outperforms previous approaches on synthetic and real data. By fusing old, overlapping and incomplete datasets this new method, inspired by quantum cryptography, paves the way for researchers to glean the results of medical trials that would otherwise be too expensive, difficult or unethical to run [233]. Breast cancer (BC) is a heterogeneous disease where genomic alterations, protein expression deregulation, signaling pathway alterations, hormone disruption, ethnicity and environmental determinants are involved. Due to the complexity of BC, the prediction of proteins involved in this disease is a trending topic in drug design. Accordingly, López-Cortés et al. [234] proposed accurate prediction classifier for BC proteins using six sets of protein sequence descriptors and 13 ML methods. After using a univariate feature selection for the mix of five descriptor families, the best classifier was obtained using multilayer perceptron method (ANN) and 300 features. It was mentioned that the performance of the model is demonstrated by AUC-ROC of 0.980 ± 0.0037, and accuracy of 0.936 ± 0.0056 (3-fold cross-validation), indicating that the powerful model predicts several BC-related proteins that should be deeply studied to find new biomarkers and better therapeutic targets.

A DL system that is better than most pathologists at determining the aggressiveness of prostate cancer was developed to examine biopsies the same way a pathologist does [235]. By means of deep learning, the system examined thousands of images of biopsies to learn what a healthy prostate is, and what more or less aggressive prostate cancer tissue looks like. The AI system has now been trained with 5,759 biopsies from more than 1,200 patients. It was reported that (i) when the performance of the algorithm is compared with that of 15 pathologists from various countries and with differing levels of experience, this AI system performed better than ten of them and was comparable to highly experienced pathologists, and (ii) an additional advantage of such a computer system is that it is consistent and can be used anywhere; the treatment of a patient no longer depends on the pathologist looking at the tissue [235]. Adenocarcinoma of the prostate is the second most common cancer diagnosed in men, with more than one million new cases diagnosed annually, researchers noted. Currently, the main way to diagnose prostate cancer is assessment of biopsy tissue, which includes core needle biopsy (CNB). AI algorithm was able to identify prostate cancer from tissue slides significantly more accurately than expert pathologists and previous algorithms, suggesting that the AI-based algorithm could be used as a tool to automate screening of prostate CNBs for primary diagnosis, assess signed-out cases for quality control purposes, and standardize reporting to improve patient management [235].

In 2015, misdiagnosing illness and medical error accounted for 10% of all US deaths [236], due to incomplete medical histories and loads of large cases. Bejnordi et al. [237] assessed the performance of automated DL algorithms at detecting metastases in hematoxylin and eosin–stained tissue sections of lymph nodes of women with breast cancer and compared it with pathologists' diagnoses in a diagnostic setting. A training data set of whole-slide images from two centers in the Netherlands with (n = 110) and without (n = 160) nodal metastases verified by immunohistochemical staining were provided to challenge participants to build algorithms. Algorithm performance was evaluated in an independent test set of 129 whole-slide images (49 with and 80 without metastases). The same test set of corresponding glass slides was also evaluated by a panel of 11 pathologists with time constraint (WTC) from the Netherlands to ascertain likelihood of nodal metastases for each slide in a flexible 2-hour session, simulating routine pathology workflow, and by 1 pathologist without time constraint (WOTC). It was obtained that (i) the AUC for the algorithms ranged from 0.556 to 0.994, (ii) the top-performing algorithm achieved a lesion-level, true-positive fraction comparable with that of the pathologist WOTC (72.4%) at a mean of 0.0125 false-positives per normal whole-slide image, (iii) for the whole-slide image classification task, the best algorithm (AUC, 0.994) performed significantly better than the pathologists WTC in a diagnostic simulation (mean AUC, 0.810) and (iv) the top five algorithms had a mean AUC that was comparable with the pathologist interpreting the slides in the absence of time constraints (mean AUC, 0.960 for the top five algorithms vs. 0.966 for the pathologist WOTC). Based on these findings, it was concluded that (v) in the setting of a challenge competition, some DL algorithms achieved better diagnostic performance than a panel of 11 pathologists participating in a simulation exercise designed to mimic routine pathology workflow; algorithm performance was comparable with an expert pathologist interpreting whole-slide images without time constraints, and (vi) an AI model using algorithms and DL diagnosed breast cancer at a higher rate than 11 pathologists [237].

8.4.4.2 Second opinion

In today's world, people routinely rely on the advice of algorithms for all aspects of their lives, from mundane tasks like choosing the most efficient navigation route home, to significant financial decisions regarding how to invest their retirement savings. Because of the ubiquity of algorithms, people have become increasingly comfortable relying on those algorithms – a tendency known as automation bias [238, 239]. AI era may change the traditional style and principle of obtaining a second opinion. A second opinion is to ask a doctor at a different medical institution for a second (or alternative) opinion on the progress of treatment, the next stage of treatment selection, and so on, separately from the doctor currently receiving the treatment so that the patient can choose a convincing treatment. Some people think that a second opinion is about changing doctors, transferring clinics, and receiving treatment, but that is not the case. First, it is a second opinion to ask other doctors for their opinions. You may not be satisfied with the diagnosis and treatment policy explained by your doctor. You may think whether there could be another cure. By receiving a second opinion, you can consider the opinion of the doctor in charge from a different angle, and even if the same diagnosis or treatment policy is explained, the understanding of the disease may be deepened. In addition, if another treatment is proposed, the range of choices will be expanded, so that you can be more satisfied with the treatment. Depending on your condition and progress, you may not have time to spare and you may want to start treatment as early as possible, so preparing for a second opinion starts with checking with your current doctor about your current condition and treatment needs. Well, this story may be changed since AI (non-human)-doctor is involved in this process. AI-doctor knows detailed information about the disease based fed big data, whilst human-doctor's knowledge about the disease is to some extent personalized. Accordingly, second opinions do not necessarily need to be human-formulated opinions. In the era of big data and AI, different algorithms that are based on dissimilar data and assumptions can offer second opinions and introduce more options to users. In fact, algorithmic second opinions may be more objective than human-formulated second opinions because a human reviewing the first algorithmic opinion is herself affected by automation bias. Given our significant automation bias, great care must be taken to ensure objective human second opinions. As we have been discussing in the above that the increasing focus of AI in radiology has led to replacing radiologists, these suggestions raise the question of whether AI-based systems will eventually replace physicians in some specializations or will augment the role of physicians without actually replacing them [240–242].

Concilio (the European leader in personalized healthcare support) believes that there is no reason why a patient – already suffering from anxiety – should be confronted with what amounts to an obstacle course in order to obtain a second medical opinion. Concilio's second medical opinion, based on our AI technology and tele-consultations, combines a few simple clicks for administrative purposes with care provided by a team of specialist doctors recommended by their peers. In this way, patients do not take any chances when faced with illness. The Concilio's second medical opinion service is aimed to provide support for patients, to avoid any source of conflict with their initial doctor, and to provide quick, easy and trouble-free access to a complementary team comprising leading medical experts [243]. Recommended by health authorities the world over, second medical opinions help bring about a sharp reduction in diagnostic errors, reduce the number of unnecessary procedures and disseminate best medical practices.

Human brains have limitations in terms of the volume of data they can store and process. AI may address this, accelerating time to diagnosis and treatment. With smart algorithms, physicians get a second pair of eyes to detect a problem that can be overlooked due to weariness, distractions, lack of expertise or other human factors. In addition to this current situation, although the entire society is aging demanding more care, the sector fails to grow equally. In the coming years, we'll see more diagnostic solutions utilizing DL algorithms to bring enormous improvements to patient care. But who will make a final decision and bear responsibility? [52]. Apparently, who is the boss, AI-doctor or human-doctor?

8.4.5 Iatrogenesis

The term iatrogenesis means "brought forth by a healer," from the Greek ἰατρός (ia-tros, "healer") and γένεσις (genesis, "origin"); as such, in its earlier forms, it could refer to good or bad effects [240]. Iatrogenesis is the causation of a disease, a harmful complication, or other ill effect by any medical activity, including diagnosis, intervention, error, or negligence. Sociologist I. Ilyich considers iatrogenic disease step by step into clinical iatrogenesis, social iatrogenesis, cultural iatrogenesis and electrical iatrogenesis, and has a considerable influence on the development of medical sociology [241]. Some iatrogenic events are obvious, like amputation of the wrong limb, whereas others, like drug interactions, can evade recognition. In fact, about 200,000 to 420,000 people a year died due to side effects of the drug (adverse drug reactions) and medical error in the United States [242]. When this is compared to traffic fatalities of 40,000 to 45,000 in the United States, it should be 4 to 10 times as many people died [243]. Moreover, in a 2013 estimate, about 20 million negative effects from treatment had occurred globally [244]. In 2013, an estimated 142,000 persons died from adverse effects of medical treatment, up from an estimated 94,000 in 1990 [245]. There is a clear discrepancy among side effect, side reaction and adverse reaction of dosed drugs, which will be discussed in Chapter 10.

8.5 Triage

8.5.1 Triage in general

The triage should be positioned in between of diagnosis and treatment from point view of its nature and context. Triage is the medical process of determining the priority of patients' treatments by the severity of their condition or likelihood of recovery with and without treatment. This determines patient treatment efficiently when resources are insufficient for all to be treated immediately; influencing the order and priority of emergency treatment, emergency transport, or transport destination for the patient. One of the hallmarks of an emergency department (ED), compared with an urgent care, or a physician's office, is that patients are seen by the clinicians in order of acuity, not arrival. When there are plenty of caretakers and few patients, determining the order is less important; everyone gets seen quickly. But when there are more patients than there are staff, some form of prioritization is needed to prevent a critically ill patient deteriorating in the waiting room. Thus, triage is a critical need when there are not enough clinicians to go around. And as our departments become more and more crowded, our waits longer, there appears to be a greater and greater need for triage [246].

At the emergency site, it is difficult to accurately grasp and convey what happened to the patient and the situation at that time. Therefore, a smartphone app and AI are connected to quickly and accurately share the condition of patients undergoing emergency transportation and to shorten the time to the start of treatment to improve the lifesaving rate and reduce sequelae. According to the system developed at Jikei Medical University, Tokyo, AI analyzes the information obtained from interviews and vital sign measurement and carries out triage. Based on the results, the plan is to make it possible to quickly select the delivery destination, taking into consideration the acceptance system of medical institutions. The "Join" is an app for communicating between multiple healthcare professionals. It has a chat function that allows you to have real-time conversations on your smartphone, and you can also share medical images such as X-ray CT and MRI, ECG, and images in the operating room [247]. In cerebrovascular disease, the prognosis of a patient depends on the rapid treatment at the onset. For example, in the case of cerebral infarction, it is highly possible that the sequelae can be alleviated by administering tPA (tissue plasminogen activator) within 4.5 h from the onset or performing endovascular treatment with a thrombectomy device within 8 h. Join supports team medical care when such quick decisions are required. The newly developed mechanism "Cloud ER" is an emergency medical support system for patients with suspected brain and CVDs and aims to increase the lifesaving rate and reduce sequelae by performing correct triage. The "Cloud ER" uses AI to analyze two types of data. The first is vital signs such as blood pressure, pulse, and ECG measured with a wristband-type terminal worn on the patient. The second is the interview information collected by the smartphone app. Wristband-type terminals will be designed so that blood pressure can be measured with a cuffless, and accuracy can be maintained even during body movements, keeping in mind that it will be used during emergency transportation. Analyze this information using AI. While quickly selecting the optimal destination, medical personnel at the receiving facility will share patient information via the "Join." As a result, the system can be established to start optimal treatment immediately after arrival and realize a mechanism to instruct emergency personnel on the measures to be taken during transportation [247].

Miles et al. [248] assessed the accuracy of ML methods in their application of triaging the acuity of patients (who contacted the ambulance service or turned up to at the emergency department) presenting in the emergency care system (ECS). The data sources included MEDLINE, CINAHL, PubMed and the grey literature were searched in December 2019. It was found that (i) there was a total of 92 models (from 25 studies) included in the review, (ii) there were two main triage outcomes: hospitalization (56 models), and critical care need (25 models); for hospitalization, NNs and tree-based methods both had a median C-statistic of 0.81, for critical care need, NNs had a median C-statistic of 0.89, concluding that (iii) ML methods appear accurate in triaging undifferentiated patients entering the ECS [248]. Kang et al. [249] developed and validated an AI algorithm based on DL to predict the need for critical care during EMSs. The algorithm was established using development data from the Korean national emergency department information system, which were collected during visits in real time from 151 emergency departments (EDs). The study subjects comprised adult patients who visited EDs. It was found that (i) the number of patients in the development data was 8,981,181, and the validation data comprised 2604 EMS run sheets from two hospitals, (ii) the AUC-ROC of the algorithm to predict the critical care was 0.867 and (iii) this result outperformed the Emergency Severity Index (0.839), Korean Triage and Acuity System (0.824), National Early Warning Score (0.741), and Modified Early Warning Score (0.696), concluding that (iv) the AI algorithm accurately predicted the need for the critical care of patients using information during EMS and outperformed the conventional triage tools and early warning scores [249]. Farahmand et al. [250] evaluated the application of AI in patients presenting with acute abdominal pain to estimate emergency severity index version 4 (ESI-4) score without the estimate of the required resources. It was found that (i) totally, 215 patients who were triaged by the emergency medicine specialist were enrolled in the study, (ii) triage Levels 1 and 5 were omitted due to low number of cases and in triage Level 2, all systems showed fair level of prediction with NN being the highest; whilst in Level 3, all systems again showed fair level of prediction, and (iii) however, in triage Level 4, decision tree was the only system with fair prediction. Based on these finding, it was concluded that (iv) the application of AI in triage of patients with acute abdominal pain resulted in a model with acceptable level of accuracy and (v) the model works with optimized number of input variables for quick assessment.

Katz [251] mentioned that triage, today, is the ultimate in front line medical care, providing both the medical practitioner and patient with the coordinates they need to ensure that care is provided to the right person, on time; however, while we are several hundred years and at least three centuries ahead of the origination of the word "triage," the fact that doctors and medical centers remain under pressure and triage, especially in the recent pandemic, has never been more critical. Against this backdrop AI has become an increasingly powerful tool for ER triage. The algorithms and intelligence of AI have evolved significantly over the past few years to become more reliable and are designed to support physicians as they juggle increasingly challenging workloads. The science that sits behind the algorithm is deep and highly complex. This is not just technology and data, it is DL, NN and ML. It is using data sets and insights to create algorithms that are capable of identifying the different layers of patient triage so that physicians can ensure that patients are accurately categorized and cared for. This requires significant volumes of clean data that can be used to ensure that AI is not just capable but also of value in an ER setting. This means that AI in triage has to follow rigorous process, testing, and modeling to get the best results. AI in the form of a MLed algorithm correctly triaged the vast majority of postoperative patients to the ICU in its first proof-of-concept application in a university hospital setting. The accuracy of this computer-generated algorithm is leading surgeons to envision active use of AI in the real-time acquisition of clinical information from a patient's EMR to more reliably determine whether a patient needs intensive or routine postoperative care. The algorithm evaluated included 87 clinical variables and 15 specific criteria related to the appropriateness of admission to the ICU within 48 h of surgery. It was reported that the accuracy rate was around 82%, which indicated, according to the study, that AI could be a solid partner in supporting physicians during the triage process [252].

To be of value to radiologists, AI must be in the clinical workflow. AI may help with patient care in many ways. It may reduce the turn-around-time (TAT) of reports by reducing the time radiologists spend analyzing cases. Or, it may reduce the TAT of the most clinically important reports by prioritizing studies so that the studies deemed positive by AI triage rise to the top of the worklist for review [253]. In order to shorten the time from the onset of the disease to the start of treatment, there is a movement to leave the "analysis" of medical images to AI. It can be used for "triage" to prioritize treatment for stroke and diabetic retinopathy, and automatically record the progression of diseased organs and therapeutic effects. There are several FDA-cleared and/or FDA-approved AI-triage soft systems available.

The FDA approved an application that uses AI to alert physicians of a potential stroke, signaling a notable shift in the way the agency reviews clinical decision support software used for triage [254]. The application, called Viz.AI Contact, uses an AI algorithm to analyze CT scans and identify signs of a stroke in patients. The application notifies a neurovascular specialist via smartphone or tablet when it has identified a potential blockage in the brain, reducing the time it takes for a specialty provider to review the scans. Medical startup Nines [254], which has developed an AI-based triage tool that has received clearance from the FDA, is making that tool available for free to all until June 30 to help address the growing burden on radiology diagnostics departments as COVID-19 continues to reshape the healthcare landscape in the United States. NinesAI is designed to identify possible emergent cases of ICH and mass effect conditions in patients, helping radiologists prioritize cases to review for further study. NinesAI is a supplemental tool, providing an early signal that some CT scans merit further investigation by trained radiologists, but even that can help tremendously in decreasing workload and eliminating manual early steps that are time-consuming [254, 255]. Infermedica (the Poland-founded health tech startup) introduced an AI-driven platform for preliminary diagnosis and triage and closed a $10 million funding round to support R&D and boost symptom-checking features and clinical decision support analysis [256]. Rochester, Minn.-based Mayo Clinic announced a new collaboration with Diagnostic Robotics to implement its AI platform that predicts patients' hospitalization risk and triages them to appropriate care. The platform uses AI tech, EHR data, and a questionnaire to perform clinical intake of patients visiting emergency departments and urgent care clinics as well as patients at home. Once the patient completes the questionnaire, Mayo Clinic staff can review the individual's self-reported condition, potential diagnoses, and their hospitalization risk score to support the physician's decision-making process [257]. The new system, developed by a company called Epic, analyzes a patient's EMRs and their current vital signs to predict the level and immediacy of care needed, which can inform decisions of whether to treat a patient on site or refer to a higher-acuity setting. The next step in its evaluation will be a pilot program at Stanford University. The questions of cost and potential return-on-investment lie down the road. However, it is likely the efficiencies would be greater in hospital-based urgent care, where the cost would be absorbed by fees associated with transferring patients to other settings in-house [258].

AI virtual assistants (symptom checkers) are a convenient and valuable resource for users to better understand the underlying cause(s) of their symptoms and to receive advice on the most appropriate point of care [259–261]. Typically, symptom checkers cater to three healthcare needs of a patient: (1) first is the provision of information, wherein a patient may seek to know more about the symptoms or conditions that they know or think they have, (2) secondly, a patient may want to know whether their symptoms require treatment or further investigation; this is medical triage and involves directing patients to the most suitable location within an appropriate time frame, and (3) finally, patients may want to understand the conditions that might be responsible for their symptoms. This corresponds to diagnosis or "differential diagnosis" and is typically performed by an experienced medical practitioner [259]. Based on this background, Baker et al. [259] developed an AI virtual assistant which provides patients with triage and diagnostic information and found that the AI system is able to provide patients with triage and diagnostic information with a level of clinical accuracy and safety comparable to that of human doctors. Robotics, chatbots, AI and ML algorithms are the new norm in healthcare. These innovations and other factors are the basis of a so-called "new medicine," that is predictive, proactive, and data-based practice. Technological advancement, and the related shift from products to services/solutions, democratizing access, explosion of medical data – this all puts pressure on medical businesses to transform [262]. Triage's major, unpredictable challenge is the human factor, which is prone to doubts and errors. Over-triaging, when a doctor doubts their own evaluation and recommends over-treatment, is not unusual. This results in people being sent to unnecessary, expensive, and time-consuming intensive care treatment. Under- and over-triaging can lead to irreversible, negative consequences. Medical facilities can avoid both by using AI to triage patients. AI models are shown to be effective in predicting risk of complications, possibility of cardiac arrest, and likelihood of CT scans identifying medical problems. Triage ML has proven to be an effective tool. Levin et al. [266] evaluated an electronic triage system (e-triage) based on ML that predicts likelihood of acute outcomes enabling improved patient differentiation. It was obtained that (i) the E-triage predictions had an area under the curve ranging from 0.73 to 0.92 and demonstrated equivalent or improved identification of clinical patient outcomes compared with ESI at both EDs, (ii) E-triage provided rationale for risk-based differentiation of the more than 65% of ED visits triaged to ESI level 3, (iii) matching the ESI patient distribution for comparisons, e-triage identified more than 10% (14,326 patients) of ESI level 3 patients requiring up triage who had substantially increased risk of critical care or emergency procedure (1.7%

ESI level 3 vs. 6.2% up triaged) and hospitalization (18.9% vs. 45.4%) across EDs, concluding that (iv) E-triage more accurately classifies ESI level 3 patients and highlights opportunities to use predictive analytics to support triage decision-making [263].

8.5.2 Triage in COVID-19 era

The pandemic COVID-19 continues to reshape the healthcare landscape and to force to alter the operational style thereof in the world.

At the current stage, fast, accurate and early clinical assessment of the disease severity is vital. To support decision-making and logistical planning in healthcare systems, Yan et al. [264] employed a database of blood samples from 404 infected patients to identify crucial predictive biomarkers of disease severity. For this end, ML tools selected three biomarkers that predict the survival of individual patients with more than 90% accuracy: lactic dehydrogenase (LDH), lymphocyte and high-sensitivity C-reactive protein (hs-CRP). It was found that (i) in particular, relatively high levels of LDH alone seem to play a crucial role in distinguishing the vast majority of cases that require immediate medical attention and (ii) this finding is consistent with current medical knowledge that high LDH levels are associated with tissue breakdown occurring in various diseases, including pulmonary disorders such as pneumonia, suggesting that (iii) a simple and operable formula to quickly predict patients at the highest risk, allowing them to be prioritized and potentially reducing the mortality rate. Prompt identification of patients suspected to have COVID-19 is crucial for disease control. Wang et al. [265] developed a DL algorithm, based on the chest CT, for rapid triaging in fever clinics. The U-Net-based model was trained on unenhanced chest CT scans obtained from 2,447 patients between Feb 1, 2020, and March 3, 2020 (1,647 patients with RT-PCR-confirmed COVID-19 and 800 patients without COVID-19), to segment lung opacities and alert cases with COVID-19 imaging manifestations. The ability of AI to triage patients suspected to have COVID-19 was assessed in a large external validation set. It was found that (i) in the external validation set, using radiological reports as the reference standard, AI-aided triage achieved an area under the curve of 0.953, with a sensitivity of 0.923, specificity of 0.851, a positive predictive value of 0.790, and a negative predictive value of 0.948, (ii) with regard to the identification of increases in lesion burden, AI achieved a sensitivity of 0.962 and a specificity of 0.875, and (iii) the agreement between AI and the radiologist panel was high. Based on these findings, it was concluded that (iv) a DL algorithm for triaging patients with suspected COVID-19 at fever clinics was developed and externally validated, (v) given its high accuracy across populations with varied COVID-19 prevalence, integration of this system into the standard clinical workflow could expedite identification of chest CT scans with imaging indications of COVID-19 [265].

The COVID-19 pandemic has created unique challenges for the US healthcare system due to the staggering mismatch between healthcare system capacity and patient demand. In the setting of the COVID-19 pandemic, AI can be used to carry out specific tasks such as pre-hospital triage and enable clinicians to deliver care at scale [266]. Recognizing that the majority of COVID-19 cases are mild and do not require hospitalization, Partners HealthCare (now Mass General Brigham) implemented a digitally automated pre-hospital triage solution to direct patients to the appropriate care setting before they showed up at the emergency department and clinics, which would otherwise consume resources, expose other patients and staff to potential viral transmission, and further exacerbate supply and demand mismatching. Although the use of AI has been well-established in other industries to optimize supply and demand matching, the introduction of AI to perform tasks remotely that were traditionally performed in-person by clinical staff represents a significant milestone in healthcare operations strategy. Lai et al. [266] focused on the use of simple AI in the form of RPA as a digital pre-hospital triage tool for population health management. RPA technology is considered a form of expert system and one of the foundational tools in AI by leading experts. Simple AI can have powerful effects in managing big problems, including the COVID-19 pandemic. Our COVID-19 RPA tool emulates the decision-making ability of a human expert designed to navigate complex triage problems on massive scale using if-then algorithmic branching logic rules. The expertise in our COVID-19 screener is found in the depth and complexity of work that went into the design of the clinical pathways (reassurance vs. quarantine vs. refer patient to the nurse hotline or a telemedicine consult vs. refer to respiratory clinics or the ED) to siphon through the algorithm to reach a triage end-decision. It was mentioned that although the use of AI has been well-established in other industries to optimize supply and demand matching, the introduction of AI to perform tasks remotely that were traditionally performed in-person by clinical staff represents a significant milestone in healthcare operations strategy [266].

8.6 Treatment, precision medicine, pain management

In this section, we will discuss mainly AI-assisted treatments, precision medicine, and pain management.

8.6.1 Treatment cases

Before getting into details of treatments, it would be worthy to consider how a doctor can predict the treatment outcome of an individual patient. Traditionally, the effectiveness of medical treatments is studied by randomized trials where patients are randomly divided into two groups: one of the groups is given treatment, and the other a placebo. However, a reliability on treatment effectiveness is in doubt with this traditional method. Ryynänen et al. [267] developed a new method. It was claimed that, using modeling, the method makes it possible to compare different treatment alternatives and to identify patients who will benefit from treatment. Relying on AI, the method is based on causal Bayesian networks and it is also possible to replace some randomized trials with modeling. It was further reported that (i) the method was used to evaluate treatment effectiveness in obstructive sleep apnea, (ii) the study showed that in patients with sleep apnea, the continuous positive airway pressure (CPAP) treatment reduced mortality and the occurrence of myocardial infarctions and cerebrovascular insults by 5% in the long term, and (iii) for patients with heart conditions, CPAP was less beneficial.

8.6.1.1 Cancer

Cancer treatment is generally divided into two major methods: (1) local therapy and (2) systemic therapy. (1) Local therapy is focused on the cancer itself and is used when the lesion is limited, including (i) surgery to remove the lesion, (ii) radiation therapy that attacks cancer by irradiating it, or (iii) photodynamic therapy, which is a type of laser treatment. (2) Systemic therapy is applied when multiple lesions can be confirmed or when cancer cells are eroding the whole body, including (i) chemotherapy using anticancer drugs or (ii) immune cell therapy that activates and uses self-immune cells. In addition to the above major treatments, immunotherapy is expected to be the "fourth treatment" in recent years. Although immunotherapy is under research, the immunotherapy that has been confirmed to be effective is limited to some immune checkpoint inhibitors. Some immunotherapies that are used in free practice have not been proven to be effective, so it is necessary to check carefully [268].

Currently, radiologists employ a manual process, called image segmentation, to take CT and MRI scans and use them to create a map of the patient's anatomy with clear guidelines of where to direct the radiation. This procedure takes normally about 4 h. Google DeepMind with University College London Hospital explored using AI to treat patients with head and neck cancers, to develop tools to automatically identify cancerous cells for radiology machines. It can perform the same procedure in 1 h [269]. Oncology is a rapidly evolving discipline with many complex treatment and data challenges. IBM offers medical imaging solutions that can help clinicians deliver more consistent care and tools for researchers looking to conduct efficient clinical trials: Watson for Oncology – WFO [270]. IBM Watson for Clinical Trial Matching (CTM) is a cognitive computing solution that uses NLP to help increase the efficiency and accuracy of the clinical trial matching process [271]. This solution helps providers locate suitable protocols for their patients by reading the trial criteria and matching it to the structured and unstructured patient characteristics when integrated with the EMR. It is also designed to determine which sites have the most viable patient population and identify inclusion and exclusion criteria that limit enrollment. Beck et al.

[272] organized a project for collaboration among Highlands Oncology Group (HOG), Novartis and IBM Watson Health to explore the use of CTM in a community oncology practice. HOG is in Northeast Arkansas and has 15 physicians and 310 staff members working across 3 sites. During the 16-week pilot period, data from 2,620 visits by lung and breast cancer patients were processed by the CTM system. Using NLP capabilities, CTM read the clinical trial protocols provided by Novartis, and evaluated the patient data against the protocols' inclusion and exclusion criteria. Watson excluded ineligible patients, determined those that needed further screening, and assisted in that process. It was obtained that (i) in an initial pre-screening test, the HOG clinical trial coordinator (CTC) took 1 h and 50 min to process 90 patients against 3 breast cancer protocols, (ii) conversely, when the CTM screening solution was used, it took 24 min, suggesting a significant reduction in time of 86 min or 78%, and (iii) Watson excluded 94% of the patients automatically, providing criteria level evidence regarding the reason for exclusion, thus reducing the screening workload dramatically. Hence, it was concluded that (iv) IBM Watson CTM can help expedite the screening of patient charts for clinical trial eligibility and therefore may also help determine the feasibility of protocols to optimize site selection and enable higher and more efficient trial accruals [271, 272].

Oral carcinoma accounts for roughly 3% of all cancers diagnosed annually in the United States, with nearly 400,000 new cases being diagnosed annually worldwide. Oral carcinoma most often occurs in people over age 40 and affects more than twice as many men as women. Most oral cancers are related to tobacco use, alcohol use, or both. Infection by the human papillomavirus, which is very common in oropharyngeal carcinomas, is a less common cause of oral carcinomas. Researchers at Case Western Reserve University and partners in the United States and India [273] use advanced computer vision and ML techniques to identify cancer and immune cells on digitized images of oral squamous cell carcinoma tissue slides and then recognize spatial patterns among those cells. This technology allows computerized vision to recognize patterns and quantify features that simply are beyond the human visual system but are powerful indicators of tumor biology. These algorithms help oncologists and pathologists to determine which cancers are more versus less aggressive.

Whole-genome sequencing (WGS) brings comprehensive insights to cancer genome interpretation. WGS is the mapping out of a person's unique DNA. A doctor or genetic counselor could use WGS to see if a patient has a genetic disorder or is at risk for a disease. WGS results are placed into three categories: (1) single-gene disorders (sometimes also called Mendelian disorders) are diseases that are caused by a mutation in the DNA for one gene, and an example of these diseases is Sickle Cell Anemia(link is external), (2) multifactorial disorders are diseases associated with DNA changes in more than one gene, and this often includes diseases like obesity and diabetes and often is highly influenced by your environment, and (3) pharmacogenomic profiles use an individual's genetic code to determine how they will respond to a drug so that a doctor can prescribe the correct amount, and this is an example of personalized medicine [274, 275]. WGS using NGS technology is a comprehensive method for analyzing entire genomes. Genomic information has been instrumental in identifying inherited disorders, characterizing the mutations that drive cancer progression, and tracking disease outbreaks. Advantages of WGS include (i) it provides a high-resolution, base-by-base view of the genome, (ii) it captures both large and small variants that might be missed with targeted approaches, (iii) it identifies potential causative variants for further follow-up studies of gene expression and regulation mechanisms, and (iv) it delivers large volumes of data in a short amount of time to support assembly of novel genomes. While this method is commonly associated with sequencing human genomes, the scalable, flexible nature of NGS technology makes it equally useful for sequencing any species, such as agriculturally important livestock, plants, or disease-related microbes [276]. To explore the clinical value of WGS, Staaf et al. [277] sequenced 254 triple-negative breast cancers (TNBCs) for which associated treatment and outcome data were collected between 2010 and 2015 via the population-based Sweden Cancerome Analysis Network-Breast (SCAN-B) project. It was reported that applying the HRDetect mutational-signature-based algorithm to classify tumors, 59% were predicted to have homologous-recombination-repair deficiency (HRDetect-high): 67% explained by germline/somatic mutations of BRCA1/BRCA2, BRCA1 promoter hypermethylation, RAD51C hypermethylation or biallelic loss of PALB2, indicating that the population-based study advocates for WGS of TNBC to better inform trial stratification and improve clinical decision-making. In the field of medicine, the use of AI is also highly expected with the spread of 5G. As with 5G, it is expected to increase the processing effect of AI data. In order to cure cancer from inside the body, immune cells must play an active role. In order to achieve this and make it work effectively, it is necessary to handle a huge amount of data. It is no exaggeration to say that even if there is only one patient, it will not end forever if the doctor treats it one by one. If AI makes it possible to utilize big data and process it efficiently, it will lead to great improvements. Furthermore, if the results can be communicated to doctors via 5G, the speed of cancer treatment is likely to be overwhelmingly faster than it is today. With the advent of 5G and AI, advanced cancer treatment is about to reach a major turning point [278].

8.6.1.2 Cardiovascular disease

AF (atrial fibrillation) can be considered an electrical storm in the atria in which all synchronized activities have disappeared, and the necessary coordination of the contraction stops. This can lead to a number of symptoms including palpitations, shortness of breath, and in some patients, heart failure. AF is also a major risk factor for stroke, as stagnant blood in the atrium during AF may lead to the formation of a clot that can migrate to the brain. AF increases the likelihood of stroke by four to five times and AF-induced strokes that are generally associated with more severe damage," according to the company's press release [279]. Volta Medical announced that it has obtained FDA clearance for its revolutionary VX1 AI software [280]. This is the first FDA clearance for an AI based tool in interventional cardiac electrophysiology. VX1 is a ML- and DL-based algorithm designed to assist operators in the real-time manual annotation of 3D anatomical and electrical maps of the human atria during AF or atrial tachycardia. These annotations help physicians locate heart regions harboring a specific electrogram abnormality, known as spatiotemporal dispersion. This innovative detection and patient-specific localization of abnormal regions may eventually help physicians better decide where they need to intervene to either burn (radiofrequency) or freeze (cryotherapy) faulty electrical pathways. There are several smart watch app to detect cardiovascular problems. The Apple Watch Series 4 is the very first direct-to-consumer product that enables users to get an ECG directly from their wrist. The app that permits the readings provides vital data to physicians that may otherwise be missed. Rapid and skipped heartbeats are clocked and users will now receive a notification if an irregular heart rhythm (or AF) is detected. The number of accessories and add ones that technology companies are releasing for the Apple Watch is also beginning to crossover into the health industry. Companies, such as AliveCor have released custom straps that allow a clinical grade wearable ECG that replaces the original Apple Watch band. Although the strap may be rendered useless with the Series 4, for any of the earlier watches, the strap may prove a useful attachment to identify AF. In addition, earlier this year, Omron Healthcare made the news when they deployed a new smart watch, called Omron HeartGuide. The watch can take a user's blood pressure on the go while interpreting blood pressure data to provide actionable insights to users on a daily basis [281].

8.6.1.3 Brain

Alex Fornito, a cognitive neuroscientist at Monash University, in Australia, scans brains using diffusion MRI, which tracks water molecules to trace anatomical connectivity and uses algorithms to segment the brain into different regions, forming a network of nodes and the links between them. Using network analysis, the particular dysfunctions of ADHD, schizophrenia, and Alzheimer's disease can be identified [282]. Brain implants that deliver electrical pulses tuned to a person's feelings and behavior are being tested in people for the first time. Two teams funded by the US military's research arm, the), have begun preliminary trials of "closed-loop" brain implants that use algorithms to detect patterns associated with mood disorders [283]. These devices can shock the brain back to a healthy state without input from a physician. The general approach, using a brain implant to deliver electric pulses that alter neural activity, is known as deep-brain stimulation (DBS). It is used to treat movement disorders such as Parkinson's disease, but has been less successful when tested against mood disorders. Early evidence suggested that constant stimulation of certain brain regions could ease chronic depression, but a major study involving 90 people with depression found no improvement after a year of treatment. The groups are developing their technologies in experiments with people with epilepsy who already have electrodes implanted in their brains to track their seizures. The researchers can use these electrodes to record what happens as they stimulate the brain intermittently – rather than constantly, as with older implants. The comments from Wayne Goodman (a psychiatrist at Baylor College of Medicine in Houston, Texas) was introduced, hoping that closed-loop stimulation will prove a better long-term treatment for mood disorders than previous attempts at DBS – partly because the latest generation of algorithms is more personalized and based on physiological signals, rather than a doctor's judgment [283].

DBS is already used to subdue the shakes and tremors of people with Parkinson's disease. Electrodes are implanted into a specific part of the brain, connected via wires under the skin to a pacemaker-like stimulator in the chest. That pacemaker sends out electrical signals that stifle the parts of the brain that are causing tremors. Researchers are beginning to test whether similar devices, or new types of implants, could help people with other complex neurological conditions. At the same time, a handful of projects devoted to creating the next generation of brain implants. RAM (restoring active memory) aims to use implants to improve soldier's memories after TBI. Another, called the Systems Based Neurotechnology for Emerging Therapies (SUBNETS) program, is developing devices to treat PTSD, chronic pain, anxiety, epilepsy, depression, and Alzheimer's disease [284].

8.6.1.4 Alzheimer's disease

Caregiving is another key target of dementia technology. While the potential of robots to complement or even replace human caregivers has attracted much recent attention, research has been underway for many years. Pineau et al. [285] described a mobile robotic assisting system for elderly individuals with mild cognitive and physical impairments, as well as support nurses in their daily activities. It was concluded that, in an assisted living facility, the robot successfully demonstrated that it could autonomously provide reminders and guidance for elderly residents. Recent projects have coupled social presence robots with remote monitoring using sensors. Coradeschi et al. [286] presented a system called GiraffPlus system consisted of a network of home sensors that can be automatically configured to collect data for a range of monitoring services; a semi-autonomous telepresence robot; a sophisticated context recognition system that can give high-level and long-term interpretations of the collected data and respond to certain events; and personalized services delivered through adaptive user interfaces for primary users. It was claimed that the system performs a range of services including data collection and analysis of long-term trends in behaviors and physiological parameters (e.g., relating to sleep or daily activity); warnings, alarms and reminders; and social interaction through the telepresence robot; the latter is based on the Giraff telepresence robot, which is already in place in a number of homes. Moreover, some devices for participation in recreational activities in care homes [287].

In terms of care management, the IoT facilitates the connection between items in the real world and computer systems, providing data more easily, efficiently and economically [186, 288]. It is crucial for decision makers, such as health care service providers, governments, clinics, etc., to enable to assess different types of IoT applications in relation to the specific nature of dementia. Dimirioglou et al. [288] used the AHP (analytic hierarchy process) and attempts to develop a multicriteria model in order to evaluate the potential of various IoT technologies applications in dementia care. Six IoT-based healthcare services were selected and compared against two conventional services (i.e., family-based healthcare and assisted living facility), in terms of their effectiveness, safety and patient perspectives. An AHP questionnaire was structured and data was collected and analyzed from a group of 12 experts in dementia. It was obtained that (i) the results indicate the potential of IoT technologies; however, the importance of conventional dementia care services is still highly appreciated, and (ii) the design and development of IoT-based services for dementia patients should take into consideration the fact that cognitive dysfunction is an obstacle for using new technologies, thus further development is necessary and new functionality need to be implemented for the IoT to be competitive. Future directions in dementia therapy are predicted to include nanotechnology either to repair brain damage or for drug delivery. New means of data collection, ubiquitous monitoring technologies, population surveillance, data mining and modeling may all play a key role in accelerating the development pipeline of preventive therapeutics for dementia [186].

8.6.1.5 Dementia

Dementia is a name for a group of symptoms that commonly include problems with memory, thinking, problem-solving, language and visual perception. For people with dementia, these symptoms have progressed enough to affect daily life and are not a natural part of aging, as they're caused by different diseases. All these diseases cause the loss of nerve cells, and this gets gradually worse over time, as these nerve cells cannot be replaced. As more and more cells die, the brain shrinks (atrophies) and symptoms sharpen. Which symptoms set in first depends on which part of the brain atrophies – so people are impacted differently [289, 290].

In the age of big data and AI technologies, the goal of more precise, early diagnosis and prediction of the progression of dementia may not be very far away. However, solving a huge biomedical problem via AI could have a profound impact on human privacy, rights and dignity [291]. Contrary to cancer, heart disease and HIV (human immunodeficiency virus), there is still no reliable method for early diagnosis, or treatment to prevent or reverse dementia. Recent health innovations and dementia research are gaining momentum using AI algorithms. Referring to systems that respond autonomously to their environment, AI/ML could identify "biomarkers," common patterns, measurable biological molecules or functions, which are indicative of pre-onset dementia. Early diagnosis may set the first stage in combating dementia, but AI could present additional possibilities. As AI has been leveraged as a diagnostic tool, it can be also utilized as a treatment tool as well. As AI technologies progress, interdisciplinary efforts by neuroscientists and AI experts create maps of nerve cell connections in the brain such as the EU's Human Brain Project (HBP), the United States' Brain Research Initiative, Japan's Brain Project, Neuralink (founded by Elon Musk) or Kernel (founded by Bryan Johnson) which are working toward an understanding of how, on the structural and functional level, millions of neurons come together to make the brain work, generate and store information. More complex scenarios emerge when considering whether AI-powered brain-interface technologies could be utilized as a dementia treatment to restore brain information. The possibilities for misuse of AI technologies therefore pose fundamental challenges to basic human rights, dignity and even personhood [291].

8.6.1.6 Otorhinology

Wathour et al. [292] assessed whether CI (cochlear implant) programming by means of a software application using AI may improve CI performance. Two adult CI recipients who had mixed auditory results with their manual fitting were selected for an AI-assisted fitting. Even after 17 months CI experience and 19 manual fitting sessions, the first subject had not developed open set word recognition. The second subject, after 9 months of manual fitting, had developed good open set word recognition, but his scores remained poor at soft and loud presentation levels. It was reported that (i) for first patient, a first approach trying to optimize the home maps by means of AI-proposed adaptations was not successful whereas a second approach based on the use of Automaps (an AI approach based on universal, i.e. population based group statistics) during 3 months allowed the development of open set word recognition, (ii) for second patient, the word recognition scores improved at soft and loud intensities with the AI suggestions, and (iii) the AI-suggested modifications seem to be atypical, concluding that the two case studies illustrate that adults implanted with manual CI fitting may experience an improvement in their auditory results with AI-assisted fitting.

8.6.1.7 Schizophrenia

Avatar therapy (AT) is a therapist-assisted, computer-based intervention in which therapists facilitate a conversation between patients who are experiencing persistent persecutory auditory hallucinations or auditory verbal hallucinations – AVH (hearing voices) and a visual representation of one of the persecutory voices (the avatar). In the approach, people who hear voices have a dialogue with a digital representation (avatar) of their presumed persecutor, voiced by the therapist so that the avatar responds by becoming less hostile and concedes power over the course of therapy [293]. The therapy is divided over a 24-week span. There are 3 main sessions, that incrementally help patients control their symptoms and hallucinations: (1) changes in beliefs about voices (specifically related to omnipotence and malevolence) and appraisal of the voice relationship (specifically relative power and assertiveness) are likely mechanisms of action, (2) the second phase specifically targets improvements in self-concept and development of a more positive identity, work that is consistent with recent approaches emphasizing the importance of self-esteem and self-compassion in working with distressing voices, and (3) finally, given that anxiety processes are seen as central in the maintenance of distressing voices and that AVATAR therapy involves exposure to a distressing stimulus (voice content and image), reductions in anxiety may also be an important mechanism of action. Basically, in these studies, the treatment is administered through very experienced/ specialized doctors (who played two roles – the mediator/supporter and the avatar) – something that cannot be scaled because of the difficulty to provide specialized care for a high number of patients. Additionally, these weekly treatments are generalized simulation rather than treatment optimized based on the patient's history, which is not ideal either. Most of the main scalability and effectiveness issues can be resolved using NLP, the category of AI that deals with understanding and processing the human language. As a bonus, deep video synthesis (or deep fakes) can be used to augment the avatar process itself, and further the idea of bringing these persecutors to life. The idea is to use a chatbot like structure that uses NLP to initiate dialogue and generate responses for the avatar and the mediator. Using capabilities such as sentiment analysis, and harnessing data from previous sessions, the entire process will be much more efficient and deployable by any non-specialized professional. Theoretically, if this were to be deployed, the treatment time and directly correlated facility space used by schizophrenia patients throughout the world will be significantly reduced (by almost half) [293].

Fernández-Caballero et al. [294] discussed the future of alternative treatments that take advantage of a social and cognitive approach with regards to pharmacological therapy of auditory verbal hallucinations (AVH) in patients with schizophrenia. AVH are the perception of voices in the absence of auditory stimulation and represents a severe mental health symptom. VR/AR and BCIs are technologies that are growing more and more in different medical and psychological applications. A quarter of people with psychotic conditions experience persistent AVH, despite treatment. Craig et al. [295] investigated the effect of the avatar therapy on AVH, compared with a supportive counseling control condition. Through the single-blind, randomized controlled trial at a single clinical location, there were participants were aged 18 to 65 years, who had a clinical diagnosis of a schizophrenia spectrum (ICD10 F20–29) or affective disorder (F30–39 with psychotic symptoms), and had enduring AVH during the previous 12 months, despite continued treatment. Participants were randomly assigned (1:1) to receive the avatar therapy or supportive counseling with randomized permuted blocks (block size randomly varying between two and six). Assessments were done at baseline, 12 weeks, and 24 weeks, by research assessors who were masked to therapy allocation. The primary outcome was reduction in auditory verbal hallucinations at 12 weeks, measured by total score on the Psychotic Symptoms Rating Scales Auditory Hallucinations (PSYRATS–AH). Between November 1, 2013, and January 28, 2016, 394 people were referred to the study, of whom 369 were assessed for eligibility. It was found that (i) of these people, 150 were eligible and were randomly assigned to receive either avatar therapy (n = 75) or supportive counseling (n = 75), (ii) 124 (83%) met the primary outcome, (iii) the reduction in PSYRATS–AH total score at 12 weeks was significantly greater for avatar therapy than for supportive counseling (mean difference – 3.82 [SE 1.47], 95% CI –6.70 to –0.94), and (iv) there was no evidence of any adverse events attributable to either therapy. Based on these findings, it was indicated that the targeted therapy was more effective after 12 weeks of treatment than was supportive counseling in reducing the severity of persistent auditory verbal hallucinations, with a large effect size [295]. Effective treatment strategies for schizophrenia remain very challenging and many treatment-resistant patients will suffer from persistent auditory verbal hallucinations (AVH). While clozapine is the gold-standard medication for this complex population, many will not respond to this molecule. For these ultra-resistant patients, limited options are available. Cognitive-behavioral therapy (CBT) is the most widely used psychological intervention, though it offers modest effects. With the interpersonal dimension of AVH being recognized, AT, a novel experiential treatment enabling patients to create an avatar of their persecutor and allowing them to gain control over their symptoms, was developed and tested [296]. It was noted that AVH symptomatology was significantly improved. Dellazizzo et al. [296] described a case report showcasing the beneficial results of AT for even the most severe and symptomatic cases of schizophrenia by introducing a single case of the patient, who has been afflicted with the persistency of all his voices for almost 20 years.

8.6.1.8 Organ transplantation

Organ transplantation is the surgical implantation of an organ or section of an organ into a person with an incapacitated organ. Summarizing data on organ transplantation [297–300], 113,668 candidates were waiting for organ and tissue donation in 2018, 113,000 in 2019, and 110,000 in 2020; while 39,357 transplant operations were performed in 2018, 39,718 in 2019, and 40,000 in 2020. About 20 waiting-listed candidates die every day for these years. Matching between donor and recipient is a crucial issue. Besides, success of organ transplantation depends on a number of factors: how old and how healthy the donor is, how old and how healthy the recipient is, how good a biological match can be found, how ready the patient is to receive it [301]. Heinrich [301] mentioned that AI/ML can answer the following questions: (i) what if a surgeon could use better mathematics to predict how long a donated organ would last before transplanting it into her/his patient? (ii) what if the patient could know exactly how much better an organ from a better donor would be, if s/he waited? and (iii) what if that patient could also know exactly how long, if she/he passes her/his turn this time, s/he'd have to wait for an optimal donor? Health Canada searched for technological approaches in AI/DL to predict the success of possible donor–recipient matches and transplant outcomes to support evidence-based decision-making about organ donation and transplantation. AI experts are needed to use DL to continually enhance the predictive ability of these tools toward improved circulatory determined death donation rates and to identify the best potential donor-transplant recipient matches [302]. AI and biotechnology are both on an exponential growth trajectory, with the potential to improve how we experience our lives and even to extend life itself. The matching power of AI means eight lives could be saved by just one deceased organ donor; innovations in biotechnology could ensure that organs are never wasted [303]. But with enough patients and willing donors, big data and AI make it possible to facilitate far more matches than this one-to-one system allows, through a system of paired kidney donation. Patients can now procure a donor who is not a biological fit and still receive a kidney, because AI can match donors to recipients across a massive array of patient–donor relationships [303].

Briceño [304] reviewed the current status on organ transplantation, based on the fact that many decisions in organ transplantation can now be addressed in a more concisely manner with the support of the classifiers. It was reported as follows. Any aspect of organ transplantation (image processing, prediction of results, diagnostic proposals, therapeutic algorithms or precision treatments) consists of a set of input variables and a set of output variables. AI classifiers differ in the way they establish relationships between the input variables, how they select the data groups to train patterns and how they are able to predict the possible options of the output variables. There are hundreds of classifiers to achieve this goal. The most appropriate classifiers to address the different aspects of organ transplantation are ANNs, decision tree classifiers, random forest, and NB classification models. There are hundreds of examples of the usefulness of AI in organ transplantation, especially in image processing, organ allocation, D-R (donor–recipient) matching, precision pathology, real-time immunosuppression, transplant oncology, and predictive analysis. Accordingly, it is anticipated that, in the coming years, clinical transplant experts will increasingly use DL-based models to support their decisions, especially in those cases where subjectivity is common [304]. A key issue in the field of kidney transplants is the analysis of transplant recipients' survival. By means of the information obtained from transplant patients, it is possible to analyze in which cases a transplant has a higher likelihood of success and the factors on which it will depend. In general, these analyses have been conducted by applying traditional statistical techniques, as the amount and variety of data available about kidney transplant processes were limited. However, two main changes have taken place in this field in the last decade. First is the digitalization of medical information through the use of

EHRs, which store patients' medical histories electronically. This facilitates automatic information processing through specialized software. Second, medical Big Data has provided access to vast amounts of data on medical processes. The information currently available on kidney transplants is huge and varied by comparison to that initially available for this kind of study. This new context has led to the use of other non-traditional techniques more suitable to conduct survival analyses in these new conditions. Specifically, this paper provides a review of the main ML methods and tools that are being used to conduct kidney transplant patient and graft survival analyses [305].

There are facts on kidney transplantations based on available sources [306–308], indicating that (i) there are currently 121,678 people waiting for lifesaving organ transplants in the United States. Of these, 100,791 await kidney transplants (as of January 2016), (ii) the median wait time for an individual's first kidney transplant is 3.6 years and can vary depending on health, compatibility and availability of organs and (iii) in 2014, 17,107 kidney transplants took place in the United States. Of these, 11,570 came from deceased donors and 5,537 came from living donors. It was further reported that, on average, (iv) over 3,000 new patients are added to the kidney waiting list each month, (v) 13 people die each day while waiting for a life-saving kidney transplant, (vi) every 14 min someone is added to the kidney transplant list, and (vii) in 2014, 4,761 patients died while waiting for a kidney transplant and another, 3,668 people became too sick to receive a kidney transplant. Common causes of end-stage kidney disease include diabetes, chronic and uncontrolled high blood pressure, chronic glomerulonephritis – an inflammation and eventual scarring of the tiny filters within your kidneys (glomeruli), and polycystic kidney disease [309]. It is also known that kidney transplant surgery carries a risk of significant complications, including: blood clots and bleeding, leaking from or blockage of the tube (ureter) that links the kidney to the bladder, infection, failure or rejection of the donated kidney, an infection or cancer that can be transmitted with the donated kidney, and/or death, heart attack and stroke [309].

In order to better predict outcomes of graft survival in kidney donation and ultimately save more lives for those waiting on the list, Pahl [310] compared ML methods to kidney donor risk index (KDRI aka KDPI; kidney donor risk index also known as kidney donor profile index) for the ability to predict graft failure by 12, 24, and 36 months after deceased donor kidney transplantation (DDKT). The ML model, an ensemble of thousands of randomly generated decision trees, was trained with the same data initially used to develop KDRI. It was found that (i) ML methods trained with the readily available recipient and donor variables performs significantly better than KDRI/KDPI when predicting graft failure by 12, 24, and 36 months after DDKT, (ii) when comparing equal prediction failure rates of 10%, ML methods successfully predicted 126% more successful DDKTs (an additional 2,148) than KDRI/KDPI from 1995 to 2005, and (iii) over the entire ROC curve, the ML methods performed statistically significantly better c-statistic than KDRI/KDPI in all predictions. Based on these results, it was concluded that (iv) using ML methods, many high-KDRI kidney offers resulted in thousands of successful patient outcomes without increasing risk of predicted graft failure, (v) the ML methods provided a significant improvement over KDRI for the assessment of kidney offers and give clinical professionals an improved basis for making the critical decisions, and (vi) the present work lays the foundation for future ML methods in organ transplantation and describes the steps to measure, analyze, and validate future models. Burlacu et al. [311] conducted a review to depict current research and impact of AI/ML algorithms on dialysis and kidney transplantation. It was concluded that (i) although guidelines are reluctant to recommend AI implementation in daily practice, there is plenty of evidence that AI/ML algorithms can predict better than nephrologists: volumes, Kt/V (which is a number used to quantify hemodialysis and peritoneal dialysis treatment adequacy), and hypotension or cardiovascular events during dialysis, (ii) altogether, these trials report a robust impact of AI/ML on quality of life and survival in the kidney disease stage G5 patients, and (iii) in the coming years, one would probably witness the emergence of AI/ML devices that facilitate the management of dialysis patients, thus increasing the quality of life and survival.

8.6.1.9 AI for patients with chronic disease(s)

One Drop has expanded its consumer offering beyond diabetes management with data-driven tools and personalized support for people living with diabetes, prediabetes, high blood pressure, high cholesterol, or any combination of these conditions. One Drop (a leader in digital solutions for people living with diabetes and other chronic conditions) announced the launch of its Digital Membership, a new direct-to-consumer subscription service providing a whole-person approach to chronic condition self-care [312]. It is further reported that with the new Digital Membership, One Drop will expand its proactive, preventative AI-powered solution to support people living with diabetes, prediabetes, high blood pressure, high cholesterol, or any combination of these conditions. It is characterized that, unlike common mobile health applications, such as telemedicine and remote monitoring and management (RMM) which provides assistance after a problem has occurred, One Drop helps users prevent problems before they happen and build better habits for life.

8.6.1.10 AI for people with disabilities

As Nuem et al. [313] pointed out that design innovation is needed in health and medicine, issues such as diverse patient needs, an aging population and the impact of globalization on disease should be emphasized. The concept "social design" [314, 315] which is the application of design methodologies in order to tackle complex human issues, placing the social issues as the priority, is becoming to receive a special attention in medicine, particularly healthcare of the people with disabilities. Design has a potential to envision alternative futures for healthcare through new forms of innovation. Valentine et al. [315] proposed a strategic framework for fostering a culture of design thinking for social innovation in healthcare. Drawing upon the theory of design (and its thinking), in conjunction with global and national healthcare strategies and policies, we critically reflect on pedagogical approaches for enhancing the curriculum in design as a means of discussing the need for new thinking in health. Based on the above, it was suggested (i) new mechanisms of knowledge acquisition, application, and exploration are needed to address the complex challenges facing social and health-care, (ii) referring to the national healthcare strategies, connections are made with design thinking, social innovation, health and social care to facilitate a transition from applying design as a process to applying design as a strategy for cultural transformation. The viewpoint of persons with disabilities can be said to already include the viewpoint of "social design," which has three perspectives [316]. (1) Disability perspective; people with disabilities have already lived with their social needs. However, the needs vary according to individual obstacles. There are various obstacles, such as intellectual, physical, mental, visual, hearing, developmental disorders, and many other disorders. When considering technological development from the perspective of these various obstacles, it must be said that the quality of the product will increase even more. And the development of technology and products that can meet the needs of people with all disabilities will maximize AI/IoT technology. Taking it a step further, AI needs the perspective of people with intellectual disabilities, among others. Therefore, focusing on disabilities, especially difficult-to-see intellectual disabilities, is an essential viewpoint for the development of AI. (2) A bird's eye view of social design from the future; AI and IoT technologies are expected to have a great impact on society and the world as a whole. Therefore, we need to design in advance the social needs needed. Any technology depends on its use. However, what the society should be like is not so easy to draw. However, when focusing on the needs of people with disabilities from a perspective, it is possible to largely narrow down the technical development field to two fields. One is the field of "smart houses" and the other is the field of "mobility support systems." (3) Specific needs and technology perspective; then, how should we specifically clarify the needs? Let us look at the two areas of social design mentioned above. First of all, regarding "possibility of mobility support system" and "what technology is needed for mobility support?," the important keyword here is "advanced sensor & center function." The sensor function for grasping the state of a person with a disability or intellectual disability even if he/she is not aware of the state and cannot notify himself/herself is as follows. (i) sensing: (wearable) of physical condition sweating, palpitations, heart rate, to grasp the crisis situation of the person due to respiration number of times (→ automatic notification system), (ii) warning/notification in case of departure from planned route, (iii) warning and notification when the scheduled time has passed, (iv) confirmation program by voice to the person, (v) communication system with parents/supporters, (vi) sequential position grasp and movement grasp by position information, (vii) automatically notify parents of a meeting with a mobility supporter, contact between two persons, and contact, and (viii) emergency response contact system [316].

Published by the WHO in 1980, the International Classification of Impairments, Disabilities, and Handicaps (ICIDH) provides a conceptual framework for information by enabling classification, and hence description, of the four dimensions of the phenomenon of "disablement"; including (1) disease or disorder, (2) impairment, (3) disability, and (4) handicap [317]. Thus, the ICIDH is a model in which physical function is lost due to illness or other reasons, thereby causing disability and spreading to social disadvantage, but in order to more accurately capture the disorder, it has been no longer used in recent years. For example, a prosthetic runner who fills the record of a healthy person with a long jump is certified as a failure by the ICIDH definition. Since it is difficult to draw a line as to how far is a failure and how far it is not, it is a classification called ICF (international classification of functioning) that was revised in 2001. As a classification, ICF does not model the "process" of functioning and disability. It can be used, however, to describe the process by providing the means to map the different constructs and domains. It provides a multiperspective approach to the classification of functioning and disability as an interactive and evolutionary process. It provides the building blocks for users who wish to create models and study different aspects of this process. In this sense, ICF can be seen as a language: the texts that can be created with it depend on the users, their creativity and their scientific orientation [318].

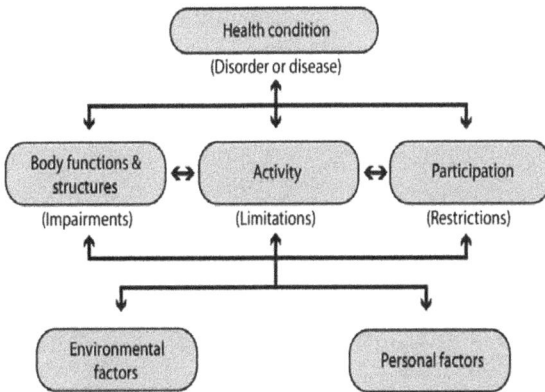

Figure 8.6: The International Classification of Functioning, Disability and Health for Children and Youth (ICF-CY) [318].

Figure 8.6 illustrates interactions between the components of ICF Health condition (disorder or disease), body functions and structures, environmental factors, activities, participation and personal factors. In this diagram, an individual's functioning in a specific domain is an interaction or complex relationship between the health condition and contextual factors (i.e., environmental and personal factors).

8.6.2 Precision medicine and Edge-AI

8.6.2.1 Precision medicine

President Obama announced that, in his 2015 State of the Union address, he's launching the Precision Medicine Initiative, which is a bold new research effort to revolutionize how we improve health and treat disease, with mission statement saying "to enable a new era of medicine through research, technology, and policies that empower patients, researchers, and providers to work together toward development of individualized care" [319]. Until now, most medical treatments have been designed for the average patient, such as "signs-and-symptoms" approach [320]. This approach will allow doctors and researchers to predict more accurately which treatment and prevention strategies for a particular disease will work in which groups of people. It is, in contrast to a one-size-fits-all approach, in which disease treatment and prevention strategies are developed for the average person, with less consideration for the differences between individuals [328]. Although the term precision medicine is relatively new, the concept has been a part of healthcare for many years. For example, a person who needs a blood transfusion is not given blood from a randomly selected donor; instead, the donor's blood type is matched to the recipient to reduce the risk of complications. Although examples can be found in several areas of medicine, the role of precision medicine in day-to-day healthcare is relatively limited. Researchers hope that this approach will expand to many areas of health and healthcare in the coming years [321]. Precision medicine refers to the tailoring of medical treatment to the individual characteristics of each patient. It does not literally mean the creation of drugs or medical devices that are unique to a patient, but rather the ability to classify individuals into subpopulations that differ in their susceptibility to a particular disease, in the biology or prognosis of those diseases they may develop, or in their response to a specific treatment. Preventive or therapeutic interventions can then be concentrated on those who will benefit, sparing expense and side effects for those who will not [322]. Although the term "personalized medicine" is also used to convey this meaning, that term is sometimes misinterpreted as implying that unique treatments can be designed for each individual [323].

NIH [324] considers that the precision medicine is an emerging approach for disease treatment and prevention that takes into account individual variability in genes, environment and lifestyle for each person. Precision medicine is also called personalized or individualized medicine. Pharmacogenomics is a component of precision medicine. By combining pharmacology and genomics, pharmacogenomics studies the impact a person's genomic fingerprint has on how they respond to a particular drug. This approach aims to improve the patient's drug response and decrease his or her treatment side effects by matching the right drug and drug dosage to an individual based on his or her genetic makeup [325]. FDA [326] sees that the precision medicine (or personalized medicine) is an innovative approach to tailoring disease prevention and treatment that takes into account differences in people's genes, environments, and lifestyles. The goal of precision medicine is to target the right treatments to the right patients at the right time. Advances in precision medicine have already led to powerful new discoveries and FDA-approved treatments that are tailored to specific characteristics of individuals, such as a person's genetic makeup, or the genetic profile of an individual's tumor. Patients with a variety of cancers routinely undergo molecular testing as part of patient care, enabling physicians to select treatments that improve chances of survival and reduce exposure to adverse effects. Precision care will only be as good as the tests that guide diagnosis and treatment. Next Generation Sequencing (NGS) tests are capable of rapidly identifying or "sequencing" large sections of a person's genome and are important advances in the clinical applications of precision medicine. Patients, physicians and researchers can use these tests to find genetic variants that help them diagnose, treat, and understand more about human disease. The NGS-based genetic tests are supported by three components, including clinical databases guidance, bioinformatics platform, and analytical validation guidance [326].

The idea of providing the right treatment for each patient, which is the basis of precision medicine, is different from the population-based evidence-based medicine (EBM). Doctors can use precision medicine to: (1) learn your disease risk. Testing your genes can reveal which conditions run in your family and how likely you are to get them; (2) prevent disease. Once you know you carry a certain gene, you may be able to make lifestyle changes or get medical treatment so you do not get sick. For example, women who carry the BRCA1 or BRCA2 gene mutation are at higher risk for breast cancer. To lower their risk, they may choose to have surgery to remove a breast, called a mastectomy; (3) find disease. If you know you're at risk for a certain disease, you can get tested for it. The earlier you find diseases like cancer, the easier they are to treat; (4) target treatments. Your genetic makeup can help guide your doctor to the drug that is most likely to work for you and cause the fewest side effects. Precision medicine can even help you decide what dose of a drug you should take; and (5) monitor your response. Doctors can use precision medicine techniques to see how well your condition responds to a treatment [327, 328].

AI/ML is generating a great deal of creativity in healthcare precision medicine. The healthcare organizations are implementing precision medicine initiatives as practitioners begin to incorporate genomic and molecular data into their decisions on clinical treatment. Precision medicine is an emerging approach to treatment and disease prevention that takes into account the individual variation in biology, climate, and lifestyle for each person. This method helps clinicians and researchers to help assess what groups of people are trying to treat and prevent those diseases. It requires a tremendous computing power (supercomputers), algorithms that can learn by themselves (DL) at an unparalleled rate, and an approach that uses the cognitive ability (AI) of newly developed doctors in general. Computer technology has become the frontline of countries that have shown their superiority in them. DL algorithms for the diagnosis of cardiology, dermatology for oncology have been demonstrated at least as well. Nevertheless, the importance of integrating these algorithms with physician expertise needs to be emphasized [329]. The essence of practicing medicine has been obtaining as much data about the patient's health or disease as possible and making decisions based on that. Physicians have had to rely on their experience, judgment, and problem-solving skills while using rudimentary tools and limited resources [330, 331]. With the cultural transformation called digital health, disruptive technologies have started to make advanced methods available not only to medical professionals but also to their patients. These technologies such as genomics, biotechnology, wearable sensors, or AI are gradually leading to three major directions: (1) making patients the point-of-care; (2) created a vast amount of data that require advanced analytics; and (3) made the foundation of precision medicine. It requires significant computing power (supercomputers); algorithms that can learn by themselves at an unprecedented rate (deep learning); and generally, an approach that uses the cognitive capabilities of physicians on a new scale (AI) [332, 333].

AI and nanotechnology are two fields that are instrumental in realizing the goal of precision medicine. Recent conversion between these two fields is enabling better patient data acquisition and improved design of nanomaterials for precision cancer medicine. Diagnostic nanomaterials are used to assemble a patient-specific disease profile, which is then leveraged, through a set of therapeutic nanotechnologies, to improve the treatment outcome. However, high intratumor and interpatient heterogeneities make the rational design of diagnostic and therapeutic platforms, and analysis of their output, extremely difficult. Integration of AI approaches can bridge this gap, using pattern analysis and classification algorithms for improved diagnostic and therapeutic accuracy. Nanomedicine design also benefits from the application of AI, by optimizing material properties according to predicted interactions with the target drug, biological fluids, immune system, vasculature, and cell membranes, all affecting therapeutic efficacy. Here, fundamental concepts in AI are described and the contributions and promise of nanotechnology coupled with AI to the future of precision cancer medicine are reviewed [334]. Using AI and NLP, a new project aims to accelerate health science and precision medicine. EHRs show the real way clinicians approach problems, subject to their working environments. These vital pieces of information are often written in unstructured text, which is difficult to gather information from accurately, at scale. The EU-supported SAVANA project, hosted by Savana Médica in Spain, developed a means of harnessing a specific branch of AI, known as clinical NLP, to capture the value from within this vast amount of text [335].

There are several clinical cases under treatments of precision medicine. In healthcare, the most common application of traditional ML is precision medicine – predicting what treatment protocols are likely to succeed on a patient based on various patient attributes and the treatment context [336]. The great majority of ML and precision medicine applications require a training dataset for which the outcome variable (e.g., onset of disease) is known; this is called supervised learning [13]. Cancers that appear pathologically similar often respond differently to the same drug regimens, so that methods to better match patients to drugs are in high demand. Lee et al. [336] demonstrated a promising approach to identify robust molecular markers for targeted treatment of acute myeloid leukemia (AML) by introducing: data from 30 AML patients including genome-wide gene expression profiles and in vitro sensitivity to 160 chemotherapy drugs, a computational method to identify reliable gene expression markers for drug sensitivity by incorporating multi-omic prior information relevant to each gene's potential to drive cancer. It was mentioned that (i) the method outperforms several state-of-the-art approaches in identifying molecular markers replicated in validation data and predicting drug sensitivity accurately, and (ii) SMARCA4 was identified as a marker and driver of sensitivity to topoisomerase II inhibitors, mitoxantrone, and etoposide, in AML by showing that cell lines transduced to have high SMARCA4 expression reveal dramatically increased sensitivity to these agents.

Healthcare is undergoing a transformation, and it is imperative to leverage new technologies to generate new data and support the advent of precision medicine. As a result, the big data revolution has provided an opportunity to apply AI/ML algorithms to this vast data set. The advancements in digital health opportunities have also arisen numerous questions and concerns on the future of healthcare practices in particular with what regards the reliability of AI diagnostic tools, the impact on clinical practice and vulnerability of algorithms. AI, ML algorithms, computational biology, and digital BMs (biomarkers) will offer an opportunity to translate new data into actionable information thus, allowing earlier diagnosis and precise treatment options. A better understanding and cohesiveness of the different components of the knowledge network is a must to fully exploit the potential of it [337]. AI-driven medicine may well be personalized medicine, and dependent on each individual's genetic code. That means that in order to benefit, our genomes will have to be out there in some sense. If my genome is out there, what if it contains potentially bad news, such as a likelihood of developing a debilitating disease in the future? Perhaps I want to know and want my doctors to know, in order to find the optimal preventative measures and, if necessary, treatments. Maybe I'd rather not know, in order not to waste the present worrying about something that may or may not happen in the future. Should insurers have access to such information? Or what about family members? These questions are likely to become more pressing in the coming years [338].

Panch et al. [339] mentioned about the situation for AI in clinical medicine that the age of big data and AI, each healthcare organization has built its own data infrastructure to support its individual needs, typically involving on-premises computing and storage. Data is balkanized along organizational boundaries, severely constraining the ability to provide services to patients across a care continuum within one organization or across organizations. This situation evolved as individual organizations had to buy and maintain the costly hardware and software required for healthcare, and has been reinforced by vendor lock-in, most notably in EMRs [340]. Within a practice or a hospital or even a small group of hospitals, the most detailed and most valuable store of data is in EMRs. To date, providers of EMR software have not been able to raise clinician satisfaction, which remains at a low point. As a result, completeness and availability of data lack the quality and governance that other enterprise applications possess. Most difficult of all, interoperability between different EMR providers is low, and even data extraction is challenging. To provide precision/personalized medicine, models cannot be trained with data from other sites that are not a match for local conditions [340].

8.6.2.2 Edge AI

There are three AI-related technologies: cloud AI, edge AI, and endpoint AI. (1) Cloud AI refers to AI processing within powerful cloud data centers. Cloud AI is undisputed in its ability to solve complex problems using ML. Yet as ML's use cases grow to include many mission-critical, real-time applications, these systems will live or die on how quickly decisions can be made. And when data has to travel thousands of miles from device to data center, there's no guarantee that by the time it has been received, computed and responded to it will still be useful. Applications such as safety-critical automation, vehicle autonomy, medical imaging and manufacturing all demand a near-instant response to data that is mere milliseconds old. The latency introduced in asking the cloud to process that weight of data would in many cases reduce its value to zero. (2) In a world where data's time to value or irrelevancy may be measured in milliseconds, the latency introduced in transferring data to the cloud threatens to undermine many of the IoTs most compelling use cases. Edge AI moves AI and ML processing from the cloud to powerful servers at the edge of the network such as offices, 5G base stations and other physical locations very near to their connected endpoint devices. Edge AI has the potential to benefit both the data and the network infrastructure itself. At a network level, it could be used to analyze the flow of data for network prediction and network function management, while enabling edge AI to make decisions over the data itself offers significantly reduced backhaul to the cloud, negligible latency and improved security, reliability and efficiency across the board. (3) Endpoint devices can be defined as physical devices connected to the network edge, from sensors to smartphones and beyond. This technology is now finding its way into smaller IoT devices. Endpoint AI also has its limitations: these devices are far more constrained in terms of performance, power, and storage than edge AI and cloud AI devices [341].

The combination of smart AI software paired with reliable, more powerful computer hardware allows for medical technology to incorporate deep learning, to ultimately lead to better patient care and medical outcomes of the patients [342]. Big data and the IoT have become major players in healthcare. For example, radiologists are tasked with analyzing ultrasounds to reach diagnoses for their patients. Whereas radiologists can review a few medical images at a time, AI software and it is computer hardware building blocks can analyze multiple images at a swift pace. Based on the vast amount of historical data that the AI server has stored and has access to, the system can accurately recognize any abnormalities and calculate results on its own. AI technology can take it a step further by offering customized treatment plans based on the patient's individual medical record. Once AI-based medical technology can be perfected, the healthcare industry can improve and save lives, while cutting costs. It can eliminate the risk of human error in diagnoses and offer reliable, accurate information for expedited care and treatments. It can also improve the clinical workflow and free up time for doctors and physicians to interact with and consult with their patients, lessening the amount of time patients would have to wait for treatment. To achieve that, highly reliable hardware that has compatibility with AI software is required. The hardware also requires high processing powers, multiple communication options and expandability to meet various needs of medical technology for easy deployment. It is mentioned that with expertise of R&D and engineering teams of Axiomtek developed certified medical all-in-one touch panel PCs, embedded motherboards, systems and gateway devices. They are feature-rich and designed for easy customization and upgrade. They are built to deliver flexibility – with multiple communication capabilities, compatibility with peripherals and interoperability with AI software [342]. Traditionally, these types of AI applications (for read/understand/interpret vast databases of biomedical knowledge in seconds, for enabling researchers to rapidly map epidemiology data, biomarker genes, molecular targets and for identifying potential treatment options) have been powered by data centers and cloud computing. Of course, big data will always be processed via the cloud. However, in time, AI has made its way closer to the user – into software and into IoMT endpoints and other medical devices. For example, wearable health monitors such as ECG monitors and blood pressure monitors can collect and analyze data locally, which a patient can share with their doctor for an instant health evaluation. As a result, more healthcare businesses involved in AI have started to realize the benefits of edge computing [343]. AI is undoubtedly a data-heavy and computing intensive technology. Concerns around bandwidth, latency, security and cost pose significant hurdles for the majority of healthcare businesses, particularly when a matter of seconds could determine the outcome for a patient. Edge benefits AI by helping overcome these technological challenges. Edge AI is the next wave of AI in healthcare and many medical technology companies are recognizing this. For example, GE Healthcare recently announced a new edge computing technology designed specifically for the needs of healthcare providers [358]. Ultimately, as healthcare enters a data-driven decade, there's a real need for data storage and data computation to be located on the device. Not forgetting other factors – speed, privacy and security – which enable clinicians to make faster, more informed decisions safely [343].

The aging population brings many challenges surrounding the QOL for older people and their caregivers, as well as impacts on the healthcare market. Several initiatives all over the world have focused on the problem of helping the aging population with AI technology, aiming at promoting a healthier society, which constitutes a main social and economic challenge. Pazienza et al. [344] introduced an ambient assisted living scenario in which a Smart Home Environment is carried out to assist elders at home, performing trustworthy automated complex decisions by means of IoT sensors, smart healthcare devices, and edge nodes. The core idea is to exploit the proximity between computing and information-generation sources. It was mentioned that taking automated complex decisions with the help AI-based techniques directly on the edge AI enables a faster, more private, and context-aware edge computing empowering, called edge intelligence (EI).

8.6.3 Pain management

8.6.3.1 Management

Pain management is a common focus and scientific studies indicate how virtual reality (VR) can help alleviate stress and anxiety for patients experiencing acute pain. Many healthcare institutions are already using this VR technology in some capacity. In fact, recent statistics show that more than 240 hospitals in the United States are using virtual reality to assist various health-related procedures and help patients visualize and understand their treatment plans. Although it might seem like an extravagant or expensive solution, VR is also surprisingly affordable and certainly accessible to public. Virtual reality creates an immersive experience by providing the user with images of 3D-environments. As part of this virtual experience, the user can sometimes change or alter the appearance of the environment [345, 346]. Medical VR is one of the most captivating emerging technologies the world has ever experienced. For example, VR was used to treat soldiers with PTSD to deactivate a deep-seated "flight or fight response," relieving fear and anxiety [347]. Some doctors are using VR to fight the opioid crisis by allowing people to escape to a virtual world in order to cope with acute pain [348]. All these examples prove the appeal of VR and its many applications.

Chronic pain is often related not only to qualitative factors but also to family relationships, psychological factors, social problems, etc., and these various factors make the disease state very complicated. The treatment of such chronic pain is often not improved by the usual symptomatic treatment. Hence, research of the interdisciplinary research with cooperation of medicine and engineering and enriched and augmented AI should be integrated to enhance the pain management and treatment. Back pain in young people is in a rise and in fact, lower back pain is the second most common cause of disability in the United States. While specific painkillers and injections can reduce some pain, some better ideas to cure with nothing more than an app and smartphone/tablets were developed. Recently, Kaia Health (a digital therapeutics company) uses AI and apps along with the guidance of experts in each medical field to develop an inter-disciplinary digital approach. Specifically, the app offers an interactive, multimodal therapy for two chronic conditions: back pain and chronic obstructive pulmonary disease (COPD). Kaia Health creates evidence-based treatments for a range of disorders including back pain, Parkinson's disease, osteoarthritis and chronic obstructive pulmonary disease [349].

Pain, while it has an important function of informing patients of physical abnormalities, is one of the factors that afflict patients the most. Chronic pain, especially when the pain continues even after the damage has healed, significantly reduces the patient's QOL over time, leading to significant social losses, such as reduced productivity and increased medical costs. The diagnosis is difficult to perform objective evaluation based on test values, and others, and it is currently dependent on the experience and subjectivity of the doctor in charge. The Interdisciplinary Pain Center of the Aichi Medical University (Japan) [350] has constructed a support system to enable AI to learn the diagnostic and treatment skills of the multidisciplinary chronic pain treatment team at the Pain Center, therefore, to make accurate diagnoses to patients and to transition to appropriate treatment that will lead to the improvement of pain as soon as possible. The "KIBIT" learns the characteristics of past medical record information of pain patients as teacher data, and the optimization of the learning algorithm is examined using the correlation and equivalence between the severity judgment of the pain by the analysis of KIBIT and the judgment result of the studying team as an index. By promoting the widespread use of the "Pain Care Support AI System" to clinics and hospitals in Japan, and by making it possible to provide efficient and accurate medical-care support for many pain patients, it was aimed to realize the economic benefits of medical care by reducing the burden on not only patients but also medical personnel [350].

Harada et al. [351, 352] proposed a method for humanoid robots to learn their own movable areas through interaction with themselves and the environment based on active movements. The arm of a humanoid robot, which is taken up as a control target, is a manipulator with a degree of freedom of redundancy, and generally has a problem of self-collision and collision with the external environment. In addition, when considering use in an unknown environment, a system that can adapt to changes in the environment is required. The proposed method protects the drive unit by detecting an overload on the actuator due to abnormal contact, based on the current flowing through the actuator of the joint drive unit, and can operate itself based on the posture information at that time. It is for learning various areas. In order to learn the movable region, we used Bayesian learning, which enables learning with high generalization ability even with a small amount of learning data. The learning data used for Bayesian learning is the joint angle of the arm, which is the control target, and the Bayesian identification surface for the range of motion of the robot is created in the joint angle space. This shows that it is possible to learn the movable area in the workspace based on its own operating load without preparing a model of the external environment in advance [351]. It is also mentioned that (i) it is possible to detect overload based on abnormal contact from the value of the current sensor and describe the proposed method for acquiring the operating range, (ii) it is possible to adapt to changes in the environment by planning the route based on the history of joint angles when overload occurs (iii) furthermore, using the joint angle when overload occurs during task execution, it is possible to acquire a movable area by Bayesian learning, and (iv) by linking the control and learning mechanism, the operating range can be sequentially learned online according to the operation, and it becomes possible to adapt to changes in the environment [352].

Babies' cry, which serves as the primary means of communication for infants, has not yet been extensively explored, because it is not a language that can be easily understood. Since cry signals carry information about a babies' well-being and can be understood by experienced parents and experts to an extent, recognition and analysis of an infant's cry is not only possible, but also has profound medical and societal applications. Liu et al. [353] analyzed audio features of infant cry signals in time and frequency domains. Based on the related features, they can classify given cry signals to specific cry meanings for cry language recognition. Features extracted from audio feature space include linear predictive coding (LPC), linear predictive cepstral coefficients (LPCC), Bark frequency cepstral coefficients (BFCC), and Mel frequency cepstral coefficients (MFCC). Compressed sensing technique was used for classification and practical data were used to design and verify the proposed approaches. It was concluded that (i) the proposed infant cry recognition approaches offer accurate and promising results and (ii) NLP can analyze babies' cry.

Although pain has a different path from the nervous system, such as tactile and somatosensory [354], it is not independent because the tactile system is excited by rubbing when it hurts, and the pain system is blocked [355]. Empathy, as originally defined, refers to an emotional experience that is shared among individuals. When discomfort or alarm is detected in another, a variety of behavioral responses can follow, including greater levels of nurturing, consolation or increased vigilance toward a threat. Moreover, changes in systemic physiology often accompany the recognition of distressed states in others. Employing a mouse model of cue-conditioned fear, we asked whether exposure to conspecific distress influences how a mouse subsequently responds to environmental cues that predict this distress. Chen et al. [356] found that (i) mice are responsive to environmental cues that predict social distress, (ii) their heart rate changes when distress vocalizations are emitted from conspecifics, and (iii) genetic background substantially influences the magnitude of these responses. Hence, the sense of sharing the pain is the source of empathy. As a result, Asada [355] suggested that embedding pain circuits separate from the tactile system into the nervous system of robots is not considered to be an excessive artificial bias from the viewpoint of living evolution.

8.6.3.2 Pain expression

Pain is a complex perceptual experience. When pain is acute, it is often psychologically inseparable from fear and general distress. When it is chronic, it can be the focal element in a complex network of suffering that involves depression, somatic preoccupation, physical limitation, sleep disturbance, and hopelessness [357].

When visiting a doctor, it is important to explain and convey the pain you are feeling. There are basically two methods for expressing your pain to a doctor; verbal expression and non-verbal expression. A combination of doctor and patient in the United States is not simple and easy, since the US society exhibits typical multilanguage and multiculture natures. It would not be difficult to observe how a patient whose mother tongue is, for example, Spanish is struggling his/her pain as much as close and precise to a doctor whose mother tongue is, for example, Korean using their mutual language of English. The adjective used to describe the nature of pain sensation may be of diagnostic value. These represent those attributes of pain apart from its intensity, locus, or temporal characteristics. Such terms as sharp, dull, aching, burning, tight, and others are of some importance. For example, pains associated with nerve damage are described as jabbing, shooting, or electric-shock-like; these are seen in some spinal cord injuries and neuralgias. Causalgia is usually described as burning. Vascular pain may be throbbing with vasodilation, or tight and cramping with vasoconstriction. Muscle pains are usually aching, and worsened by remaining too long in one position, but changes in position, and certain activities, are associated with sharp pain. Visceral pains are usually deep, with local tenderness to pressure, and have an aching quality with, occasionally, intermittent radiating sharp pain [358, 359]. Less educated patients do not always have a range of adjectives at their disposal. They may say, "I don't know how you describe it. It's just a hard pain, that's all." In such cases, when it is clear that the patient is not merely being vague, it is helpful to offer a list of commonest terms for the patient to choose from, or a series of pains, such as "Sharp or dull?," "Burning or icy?," "Swollen or tight?," and so forth. There are more adjectives, including stabbing pain, shooting pain, burning pain, stinging pain, searing pain, racking pain, tearing pain, linking pain, cold pain, pricking pain, and pinching pain.

There are several categories empirically developed to describe pain [360, 361]: (1) Sensory (e.g., throbbing, burning, cramping), which has the following characteristics; temporal (flickering, quivering, pulsing, throbbing, beating, pounding), spatial (jumping, flashing, shooting), punctuate pressure (pricking, boring, drilling, stabbing, lancinating), incisive pressure (sharp, cutting, lacerating), constrictive pressure (pinching, pressing, gnawing, camping, crushing), traction pressure (tugging, pulling, wrenching), thermal (hot, burning, scaling, searing), brightness (tingling, itchy, smarting, stinging), and dullness (dull, sore, hurting, acing, heavy). (2) Affective (e.g., sickening, terrifying, blinding), which has the following characteristics; tension (tiring, exhausting), autonomic (sickening, suffocating), fear (fearful, frightful, terrifying), and punishment (punishing, grueling, cruel, vicious, killing). (3)

Evaluative (e.g., annoying, miserable, unbearable), which has the following characteristics; annoying, troublesome, miserable, intense, and unbearable. (4) Affective-Evaluative-Sensory (miscellaneous), having wretched or blinding characteristics. (5) Supplementary sensory, which has the following characteristics; spatial pressure (spreading, radiating, piercing), pressure-dullness (tight, numb, drawing, squeezing, tearing), and thermal (cool, cold, freezing). (6) Supplementary affective with the following characteristics; nagging, nauseating, agonizing, dreadful, and torturing. These adjectives – mostly used in vividly description of pain – can communicate verbally and converted to the other persons (in most cases, patient–doctor relationship). This might work if both parties share with same cultural background, same linguistic capability, and same intelligential level. Hence, it immediately becomes a great doubt in a case, for example, for an infant/children–doctor relationship.

Non-verbal information may replace verbal report in some cases. Certain people are unable to provide self-report, although the need to understand whether pain is being experienced may be urgent. Those never able to communicate their distress verbally include very young children, the seriously mentally handicapped, and people with language deficits or speech disabilities [362]. For people who prefer the non-verbal means to communicate doctors with regard to their pain problem, the pain rating scale is available for simply pointing the scale of the pain they suffer from, as shown in Figure 8.7 [363, 364]. Explain to the person that each face is for a person who feels happy because he has no pain (hurt) or sad because he has some or a lot of pain and ask the person to choose the face that best describes how he is feeling. In figure, face 0 is very happy because s/he does not hurt at all, face 2 hurts just a little bit, face 4 hurts a little more, face 6 hurts even more, face 8 hurts a whole lot, and face 10 hurts as much as you can imagine, although you do not have to be crying to feel this bad.

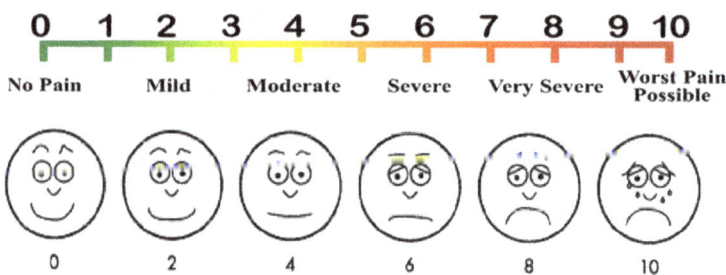

Figure 8.7: Pain rating scale [363, 364].

The use of behavioral and physiologic indicators is recommended for pain assessment in non-verbal patients. TBI can lead to neurologic changes and affect the way patients respond to pain. As such, commonly used indicators of pain may not apply to TBI patients. Arbour et al. [365] reviewed the literature about behavioral/physiologic indicators of pain in nonverbal TBI patients. An integrative review method was used. Medline (from 1948 to June 2011), Cinahl, and Cochrane databases were searched using any combination of the terms brain injury, behavioral indicators, behavioral scale, physiologic indicators, pain, pain assessment, and pain measurement. It was reported that (i) overall, TBI patients seemed to present a wider range of behavioral reactions to pain than other adult populations, (ii) in addition to the commonly observed grimace, agitation, and increased muscle tension, 14–72% of TBI patients showed raising eyebrows, opening eyes, weeping eyes, and absence of muscle tension when exposed to pain, (iii) those atypical reactions appeared to be present only in the acute phase of TBIs recovery, and (iv) Similarly to other populations, vital signs were identified as potential indicators of pain in TBI patients.

Facial expressions of virtually every common emotion, including pain, have been well characterized in humans, and can be reliably coded using the anatomically based "action units" of the facial action coding system. Similar scales have been adapted and have become useful tools in the assessment of pain and analgesia in clinical populations in which verbal communication is limited or nonexistent, such as infants and those with cognitive impairments. Despite evidence that nonhuman mammals including rats exhibit facial expressions of other emotional states, there has been no study of facial expressions of pain in any nonhuman species. Considering the pain field's heavy and continuing dependence on rodent models and the paucity of usable measures of spontaneous (as opposed to experimenter-evoked) pain in animals [366], the ability to reliably and accurately detect pain, in real time, using facial expression might offer a unique and powerful scientific tool in addition to having obvious benefits for veterinary medicine.

Facial expression is widely used as a measure of pain in infants; whether nonhuman animals display such pain expressions has never been systematically assessed. Lanford et al. [367] developed the mouse grimace scale (MGS), a standardized behavioral coding system with high accuracy and reliability; assays involving noxious stimuli of moderate duration are accompanied by facial expressions of pain. It was reported that the measure of spontaneously emitted pain may provide insight into the subjective pain experience of mice, indicating that mice show pain on their faces just like humans.

Medical treatments typically occur in the context of a social interaction between healthcare providers and patients. Although decades of research have demonstrated that patients' expectations can dramatically affect treatment outcomes, less is known about the influence of providers' expectations. Hence, Chen et al. [368] systematically manipulated provider's expectations in a simulated clinical interaction involving administration of thermal pain. It was found that (i) patient's subjective experiences of pain were directly modulated by provider's expectations of treatment success, as reflected in the patient's subjective ratings, skin conductance responses and facial expression behaviors, (ii) the belief manipulation also affected patients' perceptions of provider's empathy during the pain procedure and manifested as subtle changes in provider's facial expression behaviors during the clinical interaction, and (iii) importantly, these findings were replicated in two more independent samples, suggesting that these results provide evidence of a socially transmitted placebo effect, highlighting how healthcare provider's behavior and cognitive mindsets can affect clinical interactions. In the research [368], It has been confirmed that the doctor's fine facial expressions differ according to expectations, but it is not clear whether the patient is consciously aware of the change, and there is only a statistically slight tendency. However, there are studies that our bodies are unconsciously responding when facial expressions appear, even if they are not consciously recognizable [369]. Svetieva et al. [369] examined explicit recognition versus implicit reactions to spontaneous, brief expressions of emotion. Explicit recognition, where participants indicated which emotion they thought the stimulus individual was feeling, was compared to implicit reaction, based on (a) electrodermal response (skin conductance response amplitude) to the stimulus video, and (b) affiliative attitude toward the sender. Participants were exposed to a stimulus set of 20 spontaneously produced emotion expressions less than 500 ms in duration and that represented one of five universal emotions (anger, fear, sadness, disgust, and happiness), or a neutral comparison. It was reported that (i) in Experiment 1, individuals did not explicitly recognize emotion associated with the expressions, but had a higher electrodermal response, specifically for expressions of anger and sadness, and (ii) Experiment 2 (n = 80) replicated the electrodermal findings of Experiment 1 and also showed that indirect recognition effects can be seen in social behavior, with expressions of distress (fear and sadness) resulting in greater affiliative attitudes toward the sender than expressions of threat (anger and disgust) [369].

If the pain is prolonged, the nerves themselves will also change. If the same stimulus is given constantly, nerves perform a self-learning to enable to catch sensitively even a little stimulation. These changes are called plasticization. This process of plasticization is no exception in the process of transmitting pain. When neurons (neurons) involved in the sense of pain and its transmission become plasticized, even light stimuli can sense pain and become difficult to heal. We know that the pain is considered as the fifth vital signs (which include blood, pressure, pulse, respiration rate, and body temperature). Recently, Oshida had organized an interdisciplinary pain research team (composed of experts in computer science, cranial nerves, biology, bioscience and bioengineering) to study the pain with specific aims: (1) to find sensitive body signal(s) to respond pain which can be translated into digitized parameter, (2) to find differences expiratory airway pressure and inspiratory airway pressure and relate this difference to pain, (3) to find inside-body slow respiratory pressure and frequency, related to pain, and (4) to find any correlation between pain and respiratory noise through the noise analysis.

It would be interesting to know the diversity of etymology of pain: namely, Greek pain means "penalty," Latin pain means "endurance or bear," English pain means "stoicism," French pain means "complaint," German pain means "hurt," and Japanese pain implies "punishment."

8.6.3.3 Weather and health

According to the WHO report [370], over the last 50 years, human activities – particularly the burning of fossil fuels – have released sufficient quantities of carbon dioxide and other greenhouse gases to trap additional heat in the lower atmosphere and affect the global climate. In the last 130 years, the world has warmed by approximately 0.85 °C. Each of the last 3 decades has been successively warmer than any preceding decade since 1850. Sea levels are rising, glaciers are melting and precipitation patterns are changing. Extreme weather events are becoming more intense and frequent. There is a "+4 °C by 2030." Although global warming may bring some localized benefits, such as fewer winter deaths in temperate climates and increased food production in certain areas, the overall health effects of a changing climate are overwhelmingly negative. Climate change affects many of the social and environmental determinants of health – clean air, safe drinking water, sufficient food and secure shelter.

There are several weather-related index and alarm systems. By the arthritis index program [371], you enter your ZIP code or location to see how today's weather forecast and upcoming conditions could potentially affect your pain level. Weather-related arthritis ratings are ranged from 0 to 10; 0: there is a very low risk of weather-related arthritis pain, 1: there is a low risk of weather-related arthritis pain and less severe pain, 2: there is a low risk of weather-related arthritis pain and less severe pain, 3: the weather will not significantly impact the risk of arthritis pain, nor its severity, 4: the weather will not significantly impact the risk of arthritis pain, nor its severity, 5: there is some risk of weather-related arthritis pain, 6: there is some risk of weather-related arthritis pain, 7: there is a high risk of weather-related arthritis pain. It may likely be more severe and last longer, 8: there is an extremely high risk of weather-related arthritis and it is likely to be severe and last longer, and 9: there is an extremely high risk of weather-related arthritis and it is likely to be severe and last longer. The heat index (HI), aka the apparent temperature, is what the temperature feels like to the human body when relative humidity is combined with the air temperature. This has important considerations for the human body's comfort. When the body gets too hot, it begins to perspire or sweat to cool itself off. If the perspiration is not able to evaporate, the body cannot regulate its temperature. Evaporation is a cooling process. There is direct relationship between the air temperature and relative humidity and the heat index [372], meaning as the air temperature and relative humidity increase (decrease), the heat index increases (decreases). There are four classifications on the heat index: (1) caution (HI ranging from 26.7 to 32.2 °C), indicating fatigue possible with prolonged exposure and/or physical activity, (2) extreme caution (HI ranging from 32.2 to 39.4 °C), indicating heat stroke, heat cramps, or heat exhaustion possible with prolonged exposure and/or physical activity, (3) danger (HI ranging from 39.4 to 51.1 °C), indicating heat cramps or heat exhaustion likely, and heat stroke possible with prolonged exposure and/or physical activity, and (4) Extreme Danger (HI is 51.1 °C or higher), indicating heat stroke highly likely. There is still another weather-related index for heath management, that is, discomfort index (DI), which can be calculated by inputting air temperature and relative humidity [373]. There are six classifications on DI: (1) ~21, indicating no discomfort, (2) 21–24, indicating that under 50% population feels discomfort, (3) 24–27, indicating that most 50% population feels discomfort, (4) 27–29, indicating that most of population suffers discomfort, (5) 29–32, indicating that everyone feels severe stress, and (6) 32–, indicating that state of medical emergency [373].

A growing number of extreme climate events are occurring in the setting of ongoing climate change, with an increase in both the intensity and frequency. It has been shown that ambient temperature challenges have a direct and highly varied impact on cardiovascular health. Liu et al. [374] reviewed the recent publications regarding the impact of cold and heat on human populations with regard to CVD mortality/morbidity while also examining lag effects, vulnerable subgroups, and relevant mechanisms. It was reported that (i) although the relative risk of morbidity/mortality associated with extreme temperature varied greatly across different studies, both cold and hot temperatures were associated with a positive mean excess of cardiovascular deaths or hospital admissions, (ii) cause-specific study of CVD morbidity/mortality indicated that the sensitivity to temperature was disease-specific, with different patterns for acute and chronic ischemic heart disease, (iii) vulnerability to temperature-related mortality was associated with some characteristics of the populations, including sex, age, location, socioeconomic condition, and comorbidities such as cardiac diseases, kidney diseases, diabetes, and hypertension, and (iv) temperature-induced damage is thought to be related to enhanced sympathetic reactivity followed by activation of the sympathetic nervous system, renin-angiotensin system, as well as dehydration and a systemic inflammatory response. Based on these findings, it was recommended that (v) future research should focus on multidisciplinary adaptation strategies that incorporate epidemiology, climatology, indoor/building environments, energy usage, labor legislative perfection, and human thermal comfort models, and (vi) studies on the underlying mechanism by which temperature challenge induces pathophysiological response and CVD await profound and lasting investigation.

Closing this section, Oshida [375] proposed the face expression of today's weather which is also directly affecting to today's your health condition. As described in the above, there are various and valuable weather information including temperature, dry humidity, wet humidity, heat index, arthritis index, and DI. These individual data should be assigned to individual parts of face (such as eyebrows, eyes, nose, mouth, ears, and props). Then, each part of face element should be assigned by dimensions (length and width), location and angle, depending on the extent of each weather parameters. Hence, these combinations are large enough to cover and represent the above weather parameters. Figure 8.8 illustrates an example indicating a slight dangerous weather situation for people who has arthritis problem. This face can be animated so that weather face can be changing hourly due to ever-changing weather conditions. This program can be augmented by a full assistant by AI technologies. In addition, seasonal information of pollen scattering as well as yellow sand scattering (from Mainland China to Japan) should be added. These are directly related to air quality index, which will be discussed in the next section.

Figure 8.8: Face expression of weather-related health condition [375].

8.6.4 Allergy

The word allergy (originally a German word) has a combination of the Geek origins αλλος (allos: other) and εργο (ergo: action), implying that excessive immune responses cause harm to the body. As shown in Figure 8.9 [376], allergens and IgE antibodies are linked, and histamine and other substances are released from cells to cause allergic symptoms. There are different types of allergy, such as drug allergy, food allergy, latex allergy, mold allergy, pet allergy and pollen allergy. The typical symptoms of allergy include: sneezing, shortness of breath, wheezing, runny nose and eyes, pain over the sinuses (at the bridge of the nose, near the eyes, over cheeks and at the forehead), coughing, skin rashes (nettle rashes or hives), swelling of the lips or face, itching eyes, ears, lips, throat and roof of the mouth, nausea, vomiting, and/or abdominal cramps and diarrhea. American Academy of Allergy Asthma & Immunology (AAAAI) [377] reports as follows. As to an allergic rhinitis (i) roughly 7.8% of people 18 and over in the United States have hay fever, (ii) worldwide, allergic rhinitis affects between 10% and 30% of the population, (iii) worldwide, sensitization (IgE antibodies) to foreign proteins in the environment is present in up to 40% of the population. Besides, allergic rhinitis, there are drug allergy, food allergy, inset allergy, sinusitis and skin allergy.

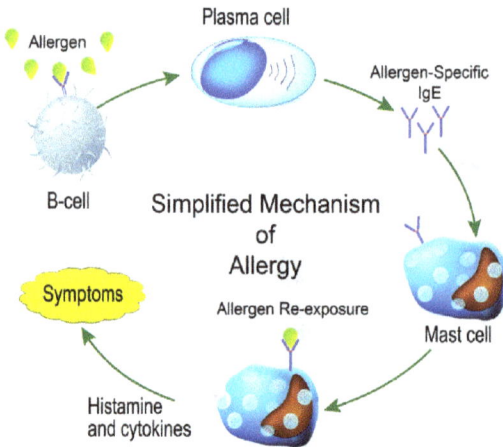

Figure 8.9: Simplified illustration for allergy mechanism [376].

Anaphylaxis [378] is the most severe of the symptoms of food allergies. Symptoms of allergies include skin, mucous membrane system, respiratory system, circulatory system, digestive system, and nervous system, and each organ has various symptoms from mild to heavy; including skin symptoms, mucosal symptoms, digestive symptoms, respiratory symptoms, systemic symptoms. Anaphylaxis is one of these with two or more severe symptoms occurring at the same time. For example, anaphylaxis occurs when the whole body has hives and palpitations, or when it repeatedly vomits and palpitations occur at the same time. When blood pressure drops and you have symptoms of unconsciousness, it is called anaphylactic shock, and if it does occur, it can be life-threatening if appropriate measures and treatment are not taken immediately. Anaphylaxis shock is sometimes abbreviated simply as anaphylaxis [378]. It is recently reported that a certain type of vaccine for COVID-19 might develop a severe anaphylaxis.

The "AI allergy" helps people explore new foods while avoiding allergens. A state-of-the-art NN is used to classify a picture of food, and a search algorithm is then employed to search a database of recipes for potential allergens. The AI/ML algorithm can classify food into 404 categories, including dishes from different countries (Hawaiian, Korean, Indonesian, Japanese, Italian, American, German, Chinese, French, Canadian, English, and Spanish cultures) and the recipe database contains more than 1,000 recipes per food category for a total of 400,000 recipes, all devoted to finding potential allergens [379]. There are various localized communication means showing allergy-related index. For example, the pollen scattering amount forecast map [380] indicates how the air quality would be, depending on your location. AAAAI [381] has also allergy forecasting. My Pollen Forecast [382] is also a common forecasting system. There are six classifications on air quality index: (1) good, (2) moderate, (3) unhealthy if sensitive, (4) unhealthy, (5) very unhealthy, and (6) hazardous. These indices can be reflected to facial presentation in Figure 8.8, as an asthma-face or allergy-face forecasting.

Pollinosis and allergies, which are said to be increasing, place a great burden on both patients and medical institutions. The Swiss Federal MeteoSwiss (MeteoSwiss) [383] is Europe's leader in real-time pollen surveillance network technology. The system placed all over the country creates a virtual hologram from the size and shape of pollen that has passed the laser beam, and it is possible to analyze the composition of pollen and identify the plant species with the AI algorithm. The bureau aims to utilize it for outdoor activity indicators and preventive treatment for hay fever patients. The American IT company "doc.ai" is developing a technology to analyze the health condition of a general user who provides a self-portrait photo, to obtain later suggestion on your health conditions [384]. Participants are asked to collect basic information about their overall health in the mobile app. Then, doc.ai worked to connect the dots between the biology, lifestyle, and environmental factors that may impact allergies. We call these pieces of health information, "personal omics." More specifically, doc.ai was looking at DNA, age, height, weight, and other important biological and health-related factors that can impact allergies. At the end, participants received an individualized report that included a visual snapshot of their symptoms, triggers, and the time of day they were more likely to experience allergies. The report was exportable so participants could share it with their clinician [385].

The team used IBM Watson ML for local model training of the data in order to help deliver a forecast that is able to assess the underlying conditions that cause allergens or symptoms and predict your seasonal allergy symptom risk in a given US location. It combines the world's most accurate weather data from The Weather Company – conditions that affect allergy triggers such as temperature, humidity, precipitation, wind and dew point – with anonymized health data and location data (used to understand the environment and local flora). These data inputs combine with AI technology to meet the scale of The Weather Channel digital properties, which see more than 300 million users monthly and live on the IBM Cloud. Pollen data and air quality data were excluded from the predictive model, since they proved unreliable indicators of seasonal allergy risk, but will continue to be reviewed. References to time of year were also removed, since the start of allergy season has changed over time due to climate change [386]. As seasonal allergy sufferers will attest, the concentration of allergens in the air varies every few paces. A nearby blossoming tree or sudden gust of pollen-tinged wind can easily set off sneezing and watery eyes. But concentrations of airborne allergens are reported city by city, at best. A network of DL-powered devices could change that, enabling scientists to track pollen density block by block [387].

Conventional bioaerosol sensing requires the sampled aerosols in the field to be transferred to a laboratory for manual inspection, which can be rather costly and slow, also requiring a professional for labeling and microscopic examination of the samples. Wu et al. [388] demonstrated label-free bioaerosol sensing using a field-portable and cost-effective device based on holographic microscopy and deep-learning, which screens bioaerosols at a throughput of 13 L/min. Two different deep NNs are designed to rapidly reconstruct the amplitude and phase images of the captured bioaerosols, and to classify the type of each bioaerosol that is imaged. As a proof-of-concept, we studied label-free sensing of common bioaerosol types, for example, Bermuda grass pollen, oak tree pollen, ragweed pollen, Aspergillus spore, and Alternaria spore and achieved > 94% classification accuracy, suggesting a 25% improvement over traditional ML methods. The presented label-free bioaerosol measurement device, with its mobility and cost-effectiveness, will find several applications in indoor and outdoor air quality monitoring.

8.6.5 Drug development

Drug discovery is a very time-consuming field that requires many processes, including target search, lead compound discovery and optimization, biological toxicity testing, preclinical trials, clinical trials, and approval and road. The process is long and normally takes more than 10 years. In addition, while drug R&D costs are increasing year by year, the success rate of drug development is decreasing year by year and is now 1/25,000. For this reason, streamlining the drug discovery process and reducing costs are urgent issues. In light of such environment surrounding drug development, famous pharmaceutical giants such as Pfizer, Novartis, and Merck are working on business alliances, investments, and acquisitions with drug discovery AI startups. Highly valued startups have their own rich databases as well as algorithms.

Since information on new drug development possesses a vast volume, details are discussed in Chapter 10.

8.6.6 Medical care operations

8.6.6.1 Drug-related death

Makary et al. [389] reported that medical errors are a major public health problem and a leading cause of mortality. With some 250,000 deaths per year in the United States, medical errors now rank after heart disease and cancer as the third leading cause of death. During hospitalization, the majority of adverse events are attributed to invasive procedures, hospital-acquired infections, and health products (drugs and medical devices) [390]. While at least 30% of adverse events are likely easily preventable, one study showed that adverse events associated with negligence (including medical errors) were more likely to be associated with injury to patients than other adverse events [391].

There are thousands of prescription drugs, with more and more new medications introduced every year, and hundreds of dietary supplements and herbal products are available in pharmacies and stores in the United States. Consumers should be aware of the dangers of combining certain drugs and other substances, including supplements, herbal products, foods, and beverages. Patients who take prescription medicine should know that in addition to potential side effects, it is important to be aware of how the drugs interact with other drugs, supplements, and certain foods and beverages [392]. According to the CDC report [393] after adjusting for age, about 34% of adults are taking one to four prescription drugs, and 11% are taking five or more. There are different types of drug interactions, including (i) drug–drug interactions, (ii) drug–food/beverage interactions, (iii) drug–condition interactions, and (iv) drug–supplement interactions. Possible side effects from interactions also vary widely and they should include drowsiness, rapid heartbeat, changes in blood pressure, bleeding, flushing, diarrhea, ruptured Achilles' tendon, coma, and even death [392]. The likelihood of interactions causing problems can depend on several factors, including genetics, age, lifestyle (such as the patient's diet and her exercise habits), other medical conditions, and the length of time both drugs are taken. Null et al. [243] reported that (i) the number of people having in-hospital, adverse reactions to prescribed drugs to be 2.2 million per year, (ii) the number of unnecessary antibiotics prescribed annually for viral infections is 20 million per year, (iii) the number of unnecessary medical and surgical procedures performed annually is 7.5 million per year, (iv) the number of people exposed to unnecessary hospitalization annually is 8.9 million per year, and (v) the total number of deaths caused by conventional medicine is an astounding 783,936 per year, suggesting a term "death by medicine."

8.6.6.2 Polypharmacy

Polypharmacy is the concurrent use of multiple medications. While there are almost always good intentions behind this – treating several chronic conditions at the same time, for example – it can cause side effects and even additional health concerns. Seniors, who are more likely to take many different prescriptions, are at the highest risk of adverse effects from polypharmacy. While you should never stop taking a medication because of this without your doctor's OK, you can play an active role in ensuring that you reduce any possible risks [394]. Various studies globally have shown that on an average two to nine medications per day are taken by the elderly people, and the prevalence of inappropriate medication used by the elderly people was found to be from 11.5% to 62.5% [395]. Polypharmacy poses several concerns: (1) excessive known side effects, particularly for the elderly, as sensitivities to medication increase with age, (2) additive side effects when several medications induce similar physical responses, (3) dangerous drug interactions, which may produce new side effects not typically associated with either medication, (4) diminished effectiveness of one medication due to the metabolic action another medication, and (5) confusion with regard to medication use, resulting in missed doses, extra doses, or other errors [394–398]. There are some countermeasures to prevent risks caused by polypharmacy. Polypharmacy can be managed as a team effort between healthcare providers, patients, and families. Stibich [394] listed what should be done. (1) Pick a point person: Be sure to select a primary care practitioner or a geriatrician who you are comfortable with and schedule regular appointments with this lead physician. (2) Keep track of medications: List out all of your medications and supplements and review it regularly for necessary updates. (3) Talk to your pharmacist: Pharmacists are trained to look for drug interactions and other problems, but they can only do that if they have all your information. (4) Do not stop your medications without approval from your doctor. You might be prescribed more than one heart or diabetes medication because one is not enough to control your condition. While polypharmacy can be harmful, medication adherence is important too. At the same time, healthcare systems are continuously looking for ways to play their part. Systemic solutions include (i) strict recording keeping and updating – tracking of prescriptions can allow your medical team to find your medication list more easily. This type of list may be generated within a medical clinic or hospital system, not between systems, (ii) "smart" systems that recognize and flag dangerous medication interactions and redundant prescriptions, and (iii) patient-friendly information describing the risks and benefits of each medication to be provided along with prescriptions [394].

To reduce the incidence and adverse effects of polypharmacy, medication regimes of elderly patients should be evaluated monthly. A single agent/drug should be prescribed instead of multiple drugs for the treatment of a single condition, if possible. Medications should be started with the lower drug dosage where clinically indicated and if required incremental increase can be done. Dagli et al. [395] mentioned that (i) identifying and avoiding the polypharmacy can lead to better outcomes in the elderly patients and also helps in improving the quality of life, and (ii) medication review is an essential part in the elderly patient to avoid adverse effects that can be caused due to polypharmacy. Farrell et al. [397] identified and examined the consensus on salient inter-professional roles, responsibilities, and competencies required in managing polypharmacy. Four focus groups with 35 team members working in geriatrics were generated to inform survey development. The sessions generated 63 competencies, roles or responsibilities, which were categorized into 4 domains defined by the Canadian Interprofessional Health Collaborative. The resulting survey was administered nationally to geriatric healthcare professionals who were asked to rate the importance of each item in managing polypharmacy; we sought agreement within and across professions using a confirmatory two-round Delphi method. It was found that (i) round 1 was completed by 98 survey respondents and round 2 by 72, (ii) there was high intraprofessional and interprofessional consensus regarding the importance of competencies among physicians, nurses and pharmacists; though pharmacists rated fewer competencies as important, and (iii) less consensus was observed among other healthcare professionals or they indicated the non-importance of competencies despite focus group discussion to the contrary. Based on the results, it was concluded that further exploration to understand the underutilization of professional expertise in managing polypharmacy will contribute to refining role clarity and translating competencies in practical settings, as well as guiding educators regarding curricular content [397].

Patients and caregivers can look for common symptoms of adverse reactions and drug interactions resulting from polypharmacy [398]. The common signs are loss of appetite, diarrhea, tiredness or reduced alertness, confusion and hallucinations, falls, weakness and dizziness, skin rashes, depression, anxiety, and excitability. Unfortunately, these signs and symptoms can also be related to the disease itself or be a consequence of aging. The most important measure that patients can take to prevent polypharmacy is to ensure that all of their healthcare providers have a complete and up-to-date list of their medications, including dosages and when they should be taken. This list should include OTC (over-the-counter) medications, supplements, and vitamins. It is excellent practice for patients to carry a list of their medications, updating it whenever there is a change. It is also suggested that before buying any OTC product or supplement, be sure to ask your pharmacist if it is safe to take with other medications or with your medical condition. Pharmacists can speak with your physician directly to better understand why a medication is prescribed and to discuss alternate therapies with a lower risk of drug interactions A complex medical system creates polypharmacy, in which older patients with multiple chronic conditions take several different medications with various dosages, prescribed by many separate physicians and perhaps filled at more than one pharmacy [398]. Polypharmacy is common among elderly persons because of the need to treat the various disease states that develop with age. Although the deprescribing of unnecessary medications is a way of limiting polypharmacy, the under-prescribing of effective therapies in older patients is a concern. Therefore, healthcare providers must evaluate each drug and balance its potential adverse effects against its potential benefits. Advances in information technologies such as electronic prescribing, EMRs, and electronic laboratory results will help prevent adverse drug effects and interactions. Medication management in nursing homes and outpatient settings is feasible because of alterations in administration and technology-driven prescribing systems. It is also important to mention that the drug is absorbed in the stomach and intestines, then broken down in the liver, and finally excreted as urine and feces from the kidneys and digestive tract. However, when the liver and kidneys function poorly with aging, it takes time to break down and excrete. Therefore, the remaining time in the body is extended, and the drug may be too effective and cause problems.

8.6.6.3 Risk detection

Corny et al. [390] improved patient safety and clinical outcomes by reducing the risk of prescribing errors by testing the accuracy of a hybrid clinical decision support system in prioritizing prescription checks. Data from EHRs were collated over a period of 18 months. Inferred scores at a patient level (probability of a patient's set of active orders to require a pharmacist review) were calculated using a hybrid approach (ML and a rule-based ES). A clinical pharmacist analyzed randomly selected prescription orders over a 2-week period to corroborate our findings. Predicted scores were compared with the pharmacist's review using the area under the receiving-operating characteristic curve and area under the precision-recall curve. These metrics were compared with existing tools: computerized alerts generated by a clinical decision support (CDS) system and a literature-based multicriteria query prioritization technique. Data from 10,716 individual patients (133,179 prescription orders) were used to train the algorithm on the basis of 25 features in a development dataset. It was found that while the pharmacist analyzed 412 individual patients (3,364 prescription orders) in an independent validation dataset, the areas under the receiving-operating characteristic and precision-recall curves of our digital system were 0.81 and 0.75, respectively, thus demonstrating greater accuracy than the CDS system (0.65 and 0.56, respectively) and multicriteria query techniques (0.68 and 0.56, respectively) and concluded that by primarily targeting high-risk patients, this novel hybrid decision support system improved the accuracy and reliability of prescription checks in a hospital setting.

8.6.6.4 Medical record and card

For use of AI technology effectively, data need to be digitized. Amazon HealthLake [399] aggregates an organization's complete data across various silos and disparate formats into a centralized AWS (Amazon Web Service) data lake and automatically normalizes this information using ML. The service identifies each piece of clinical information, tags, and indexes events in a timeline view with standardized labels so it can be easily searched, and structures all of the data into the Fast Healthcare Interoperability Resources (FHIR) industry standard format for a complete view of the health of individual patients and entire populations. As a result, Amazon Health-Lake makes it easier for customers to query, perform analytics, and run ML to derive meaningful value from the newly normalized data. Organizations such as healthcare systems, pharmaceutical companies, clinical researchers, health insurers, and more can use Amazon HealthLake to help spot trends and anomalies in health data so they can make much more precise predictions about the progression of disease, the efficacy of clinical trials, the accuracy of insurance premiums, and many other applications. Amazon HealthLake offers medical providers, health insurers, and pharmaceutical companies a service that brings together and makes sense of all their patient data, so healthcare organizations can make more precise predictions about the health of patients and populations. The new HIPAA-eligible service enables organizations to store, tag, index, standardize, query, and apply ML to analyze data at petabyte scale in the cloud.

Electronic medical services are becoming popular in the medical field, but recently EMR equipped with AI functions have appeared [400]. EMR, like paper medical record, have the advantage of streamlining operations because they do not require a large storage location such as a warehouse and are easy to manage. Currently, there are three types of EMR app. (1) "AI Inquiry Ubie" is a tool to support the initial consultation. "AI inquiry Ubie" allows patients to enter questionnaires on a tablet and reflect the contents in their medical record. As a result, the initial consultation time per person can be shortened to 1/3. In addition, AI can automatically generate questions based on patient input, so it is possible to listen to more detailed and deep content. (2) AI-powered cloud-based EMR "M3DigiKar." M3DigiKar is a cloud-based electronic medical service that reduces the burden of creating medical files by utilizing AI with unique automatic learning functions. Automatically learn data for all patients in your medical record to reduce input time and powerfully back up your medical record creation. (3) Completely free cloud-based EMR "Karte ZERO." There is also a completely free electronic medical service that is easy to consider even if you have a limited budget. "Karte ZERO" is a cloud-type electronic medical service that does not cost an initial cost or a monthly fee.

It can be predictable to put their health card into a computer, just as everyone puts our credit card into a gas pump, and that credit card's going to have all of medical data, all of medication history, all of health history, all record of anamnesis as well as current disease(s), all of DNA on that card, and the machine will read it and perhaps come up with a better diagnosis and treatment plan than a board-certified primary care provider can today. Just in an emergency incidence that you would encounter, your card might be a great assist for life-serving caregivers and medical staff members to determine the degree of triage and treatment plan, knowing your physical basic data, record of prescription medication(s), anamnesis as well.

8.6.7 Surgical robot technologies

AI-empowered surgical tools are becoming more and more common, utilizing their stability and accuracy to successfully perform a variety of surgical procedures, achieving a very high success rate and shortening the recovery time.

The da Vinci robotic surgical system (by the American company Intuitive Surgical) was FDA-approved in 2000. Thousands of surgeons around the world have been trained on da Vinci systems and have completed more than 7 million surgical procedures using da Vinci surgical systems. It is designed to facilitate surgery using a minimally invasive approach, and is controlled by a surgeon from a console. Over the past 25 years, surgical techniques have evolved from open surgery to minimally invasive laparoscopic procedures to robotic surgery. Robotic surgery can improve outcomes and potentially reduce complications, especially in urological surgery, gynecological surgery, and upper and lower abdominal surgery. The da Vinci systems were used in an estimated 200,000 surgeries in 2012 [401], in 2019–2020, there are an installed base of 4,986 units worldwide – 2,770 in the United States, 719 in Europe, 561 in Asia, and 221 in the rest of the world [402]. The da Vinci Surgical System provides the surgeon with (i) precision, dexterity, and control during surgery, and (ii) the ability to execute 1–2 cm incisions versus longer incisions. The system mainly consists of (1) an ergonomically designed surgeon's console, (2) a patient cart with four interactive robotic arms, and (3) a high-performance vision system and patented EndoWrist instruments [403]. The following surgeries are indicated: colorectal, gallbladder, gynecologic, hernia, kidney, and prostate [404]. Surgical robots, initially approved in the United States in 2000, provide "superpowers" to surgeons, improving their ability to see, create precise and minimally invasive incisions, stitch wounds, and so forth. Important decisions are still made by human surgeons, however. Common surgical procedures using robotic surgery include gynecologic surgery, prostate surgery and head and neck surgery [13, 405].

Over the next decade, AI is trying to change the scene of the operating room significantly with the development of robotics technology. More and more medical robotic companies are taking advantage of the new capabilities driven from AI and innovative computer vision algorithms, to provide solutions that add accuracy and stability, save time and improve supervision on many daily surgery procedures. Besides improving the surgery process, this AI revolution also shortens recovery time and reduces infection risks, by favoring minimally invasive methods, which expose only the minimal needed surface for the robotic arms operation. Hospitals are using robots to help with everything from minimally invasive (MI) procedures to open heart surgery. According to the Mayo Clinic, robots help doctors perform complex procedures with a precision, flexibility and control that goes beyond human capabilities. Robots equipped with cameras, mechanical arms and surgical instruments augment the experience, skill and knowledge of doctors to create a new kind of surgery. Surgeons control the mechanical arms while seated at a computer console while the robot gives the doctor a three dimensional, magnified view of the surgical site that surgeons could not get from relying on their eyes alone. The surgeon then leads other team members who work closely with the robot through the entire operation. Robot-assisted surgeries have led to fewer surgery-related complications, less pain and a quicker recovery time [14]. The Israel-based medical AI company (RSIP Vision) develops advanced AI and computer vision driven solutions for the medical field, with a strong lead in the Robotic Surgery thriving segment, to provide specific solutions for a variety of surgical operations, in the orthopedics, aesthetics and general surgery fields [406].

The interaction between humans and their animals has been an important part of people's lives since the beginning of time. The animal-assisted therapy (AAT) and pet visitation are making their way into hospitals, and long-term healthcare facilities are making room for companion pets. Whether you support or are unsure of the idea that animals have a role in healthcare settings, consider the following points. AAT is characterized by (1) for many people, pets are their most significant others, (2) interaction with animals has a calming effect, (3) interaction with animals can reduce pain, (4) there are three types of animal interactions in healthcare settings (AAT, pet visitation, and pets in long-term facilities), (5) infection risk from animals is very low, and (6) there are published guidelines for animals in healthcare [407, 408]. A recently developed communication robot possesses a similar function as AAT by animals as a non-human therapy means. In recent years, "communication robots" have been attracting attention in order to solve the problem of long working hours for medical professionals. A communication robot is a robot that not only works according to a fixed algorithm, but also recognizes and speaks human words. With the development of technologies such as voice recognition, it is spreading to daily life. With the declining birthrate and aging population, the issue of labor shortages has become apparent in all industries, and the use of robots that can perform work speedily without mistakes is expanding. Alternatively, using a communication robot to alleviate the feeling of loneliness of long-term patients is effective in improving the QOL of patients. In addition, even if it is difficult for the patient to ask the doctor directly, the robot can ask, and the hospital can also tell the patient directly through the robot, which is effective in preventing troubles and smooth communication. If you look not only at reducing the burden on the hospital side but also on the benefits on the patient side, you will find more hints on how to make effective use of communication robots [409].

The proliferation of anesthesia in the 1800s is America's greatest contribution to modern medicine and enabled far more complex, invasive, and humane surgical procedures. Now nearly 60,000 patients receive general anesthesia for surgery daily in the States [410]. Anesthesia is a reversible drug-induced state characterized by a combination of amnesia, immobility, anti-nociception, and loss of consciousness [410]. Reinforcement learning (RL) can be used to fit a mapping from patient state to a medication regimen. Prior studies have used deterministic and value-based tabular learning to learn a propofol dose from an observed anesthetic state. Deep RL replaces the table with a deep NN and has been used to learn medication regimens from registry databases. Schamberg et al. [411] performed the first application of deep RL to closed-loop control of anesthetic dosing in a simulated environment and employed the cross-entropy method to train a DNN to map an observed anesthetic state to a probability of infusing a fixed propofol dosage. The model was tested on simulated pharmacokinetic/pharmacodynamic model with randomized parameters to ensure robustness to patient variability. It was reported that the deep RL agent significantly outperformed a proportional integral-derivative controller (median absolute performance error 1.7% ± 0.6 and 3.4% ± 1.2), (ii) modeling continuous input variables instead of a table affords more robust pattern recognition and utilizes our prior domain knowledge, and (iii) deep RL learned a smooth policy with a natural interpretation to data scientists and anesthesia care providers alike [411].

8.7 Prognosis

8.7.1 Diagnosis and prognosis

Genetic programming (GP) is an evolutionary computing methodology capable of identifying complex, nonlinear patterns in large data sets. Bammister et al. [412] determined the utility of GP for the automatic development of clinical prediction models. It was concluded that (i) using empirical data, it was demonstrated that a prediction model developed automatically by GP has predictive ability comparable to that of manually tuned Cox regression, (ii) the GP model was more complex, but it was developed in a fully automated way and comprised fewer covariates, and (iii) it did not require the expertise normally needed for its derivation, thereby alleviating the knowledge elicitation bottleneck, indicating that GP demonstrated considerable potential as a method for the automated development of clinical prediction models for diagnostic and prognostic purposes. It was also demonstrated how AI can provide an equally accurate and reliable prognosis for patients with CVD, compared to traditional methods [413]. Cancer is an aggressive disease with a low median survival rate. Ironically, the treatment process is long and very costly due to its high recurrence and mortality rates. Accurate early diagnosis and prognosis prediction of cancer are essential to enhance the patient's survival rate. As AI, especially ML and DL, has found popular applications in clinical cancer research in recent years, cancer prediction performance has reached new heights [414]. Recently, there has been a momentous drive to apply advanced AI technologies to diagnostic medicine. The introduction of AI has provided vast new opportunities to improve healthcare and has introduced a new wave of heightened precision in oncologic pathology. Sultan et al. [415] mentioned that ML and DL methods designed to enhance prognostication of oral cancer have been proposed with much of the work focused on prediction models on patient survival and locoregional recurrences in patients with OSCC.

Glioblastoma is an aggressive brain tumor that is very difficult to treat successfully. Average survival after diagnosis is 12–18 months and fewer than 5% of patients are still alive after 5 years. Some patients do better than others, and so the ability to assess objectively patients' frailty and physical condition provides important information that can improve prognosis and help guide decisions on treatments, diet and exercise. If patients have sarcopenia – degenerative loss of skeletal muscle – they may be unable to tolerate surgery, chemotherapy or radiotherapy as well as patients without the condition. This can lead to adverse reactions to therapy, early discontinuation of treatment, accelerated progression of the disease and death. So indicators that can be assessed objectively, such as measures of sarcopenia, are needed. In the first study of its kind in cancer, Mi [416] had applied AI to measure the amount of muscle in patients with brain tumors to help improve prognosis and treatment. The usage of DL to evaluate MRI brain scans of a muscle in the head was as accurate and reliable as a trained person and was considerably quicker. Furthermore, it was shown that the amount of muscle measured in this way could be used to predict how long a patient might survive their disease as it was an indicator of a patient's overall condition.

Pancreatic cancer is a complex cancer of the digestive tract and has a high degree of malignancy. Given the difficulty of early diagnosis, pancreatic cancer often metastasizes after diagnosis. Despite significant progress in pancreatic cancer research in the past decade, treatment and prognosis still tend to be unsatisfactory. Diagnosis and treatment can be very difficult because of unclear early symptoms, the deep anatomical location of cancer tissues, and the high degree of cancer cell invasion. The prognosis is extremely poor; the 5-year survival rate of patients with pancreatic cancer is less than 1%. AI has great potential for application in the medical field. In addition to AI-based applications, such as disease data processing, imaging, and pathological image recognition, robotic surgery has revolutionized surgical procedures. To better understand the current role of AI in pancreatic cancer and predict future development trends, Lin et al. [417] reported the application of AI to the diagnosis, treatment, and prognosis of pancreatic cancer. Although there are systematic treatment plans, the effect of radiotherapy is poor because of the deep location of the pancreas and the tissue characteristics of the cancer. The special characteristics of pancreatic cancer also lead to drug resistance after chemotherapy, and surgical treatment is difficult because of the large number of important organs around the pancreas and its anatomical complexity. AI has the ability to replace or assist people in clinical work. It has great application prospects for the diagnosis, treatment, and prognosis of pancreatic cancer. Regarding molecular diagnosis, imaging diagnosis, and chemotherapy, ML can help researchers process data, perform analysis, and obtain experimental results. In radiotherapy, AI is mainly used for the automatic planning of radiation targets and radiation dose prediction. The development of robotic pancreatic surgery has increased the accuracy of pancreatic surgery and reduced complications, but automation cannot be fully achieved without continuous training and verification. Therefore, for a long time in the future, most AI applications for pancreatic cancer will continue to be used as practical auxiliary tools [417].

The prognosis for patients with brain metastases (BM) is known to be poor, as BM is one of the most deadly among various types of cancers [418, 419]. Ranging from early detection to intervention therapy, many innovative management models have been formulated with the goal of lowering the fatality rate of BM. Nevertheless, in the clinical practices, exploring the prognosis index markers of patients is often difficult and costly. Meanwhile, the challenging task of brain metastases survivability prediction could strongly benefit from the development of personalized and precise medicine. In this context, AI technology can be used to predict cancer as a means of inexpensive and practical research methodology [420]. Huang et al. [420] identified the optimum prognosis index for brain metastases by AI/ML. Seven hundred cancer patients with brain metastases were enrolled and divided into 446 training and 254 testing cohorts. Seven features and seven prediction methods were selected to evaluate the performance of cancer prognosis for each patient. The mutual information and rough set with particle swarm optimization (MIRSPSO) methods was used to predict patient's prognosis with the highest accuracy at AUC = 0.978 ± 0.06. The improvement by MIRSPSO in terms of AUC was at 1.72%, 1.29%, and 1.83% higher than that of the traditional statistical method, sequential feature selection (SFS), mutual information with particle swarm optimization (MIPSO), and mutual information with sequential feature selection (MISFS), respectively. It was concluded that (i) the clinical performance of the best prognosis was superior to conventional statistic method in accuracy, sensitivity, and specificity, and (ii) identifying optimal ML methods for the prediction of overall survival in brain metastases is essential for clinical applications. The accuracy rate by ML is far higher than that of conventional statistic methods [420].

There are prognostic monitoring devices and apps are developed for chronic patients. A new app, AlloCare from CareDx (NASDAQ: CDNA) uses AI/ML to help transplant patients track, monitor and manage their daily medicine doses, blood pressure, urine output, sleep, steps and weight. The AI models input from each of these factors and gives the patient a daily tracking score to monitor their progress with their transplant goals and take an active role in monitoring their organ health [421]. The Onduo virtual diabetes clinic (VDC) for people with type 2 diabetes (T2D) combines a mobile app, remote personalized lifestyle coaching, connected devices, and live video consultations with board-certified endocrinologists for medication management and prescription of real-time continuous glucose monitoring (RT-CGM) devices for intermittent use. Majithia et al. [422] evaluated glycemic outcomes associated with participation in the Onduo VDC for 4 months. It was reported that (i) participants in the Onduo VDC experienced significant improvement in HbA1c, increased TIR, decreased time in hyperglycemia, and no increase in hypoglycemia at 4 months, and (ii) improvements in other metabolic health parameters including weight and blood pressure were also observed, concluding that (iii) the Onduo VDC has potential to support people with T2D and their clinicians between office visits by increasing access to specialty care and advanced diabetes technology including the Onduo VDC has potential to support people with T2D and their clinicians between office visits by increasing access to specialty care and advanced diabetes technology including RT-CGM. Livongo is a data-based health coaching program that enables people with T2D to share blood glucose records with Certified Diabetes Educators (CDEs) [423]. People with diabetes receive feedback in real time – within minutes if there is an emergency. Livongo's goal is to reduce the burden of managing diabetes. They've created relationships with sponsors, such as health plans, health systems, or self-insured employers to empower people with diabetes to better control their blood sugar. The meter's over-the-air capability allows the system to provide people with diabetes, feedback throughout the day, allowing for actionable changes to their diabetes management right away. They do not have to wait to be seen by the doctor for changes to be made. When the Livongo system is used in a healthcare setting, the results can be integrated into EMRs, giving physicians access to a wealth of data and helping them to improve treatment plans. The Sword Health app is an on-line rehabilitation platform, by combining a team of real-life physical therapy and physicians with a digital physiotherapist using AI. By rehabilitating with a dedicated motion tracker, users can receive high-quality rehabilitation programs while at home, as digital PT provides real-time feedback and healthcare professionals intervene when needed [424].

8.7.2 Remaining life prediction

Predicting life expectancy is a challenging task. This is universal and not limited to human life, but also important engineering component such as engine parts for aircrafts. On the other hand, technology for predicting remaining life, based on knowing the extent of pre-damages, has been developed. For example, in engineering field, Oshida et al. [425] established a technique for assessment of remaining fatigue life of endodontic files. Similarly, remaining human life can be fairly well estimated. Improving the quality of end-of-life (Oshida prefers to call it as quality of dying – QOD, as opposing to QOL – quality of living) care for hospitalized patients is a priority for healthcare organizations. Studies have shown that physicians tend to over-estimate prognoses, which in combination with treatment inertia results in a mismatch between patients' wishes and actual care at the end of life. Avati et al. [426] addressed this problem using DL and HER data. The EHR data of admitted patients are automatically evaluated by an algorithm, which brings patients who are likely to benefit from palliative care services to the attention of the palliative care team. The algorithm is a DNN-trained on the EHR data from previous years, to predict all-cause 3- to 12-month mortality of patients as a proxy for patients that could benefit from palliative care. It was reported that the predictions enable the palliative care team to take a proactive approach in reaching out to such patients, rather than relying on referrals from treating physicians, or conduct time consuming chart reviews of all patients.

Of all the processes of medical care, it has been widely thought that the phase of human death should be with the utmost respect and always full of dignity. In recent years, AI has predicted human death and made brain death judgments with high accuracy, even foreshadowing the arrival of an era in which AI predicts and defines human death. Prognostic modeling using standard methods is well-established, particularly for predicting risk of single diseases. ML may offer potential to explore outcomes of even greater complexity, such as premature death. Weng et al. [427] developed novel prediction algorithms using ML, in addition to standard survival modeling, to predict premature all-cause mortality and reported that ML significantly improved accuracy of prediction of premature all-cause mortality in this middle-aged population, compared to standard methods, suggesting that the value of ML for risk prediction within a traditional epidemiological study design, and how this approach might be reported to assist scientific verification. Predictive modeling with EHR data is anticipated to drive personalized medicine and improve healthcare quality. Constructing predictive statistical models typically requires extraction of curated predictor variables from normalized EHR data, a labor-intensive process that discards the vast majority of information in each patient's record. Rajkomar et al. [428] proposed a representation of patients' entire raw EHR records based on the Fast Healthcare Interoperability Resources (FHIR) format. It was mentioned that (i) the models outperformed traditional, clinically used predictive models in all cases, and (ii) the approach can be used to create accurate and scalable predictions for a variety of clinical scenarios. In a case study of a particular prediction, demonstrating that (iii) NNs can be used to identify relevant information from the patient's chart. One goal of both studies [426, 427] was to assess how this information might help clinicians decide which patients might most benefit from intervention [429]. There is an issue of unconscious, or implicit, bias in healthcare, which has been studied extensively, both as it relates to physicians in academic medicine and toward patients. There are differences, for instance, in how patients of different ethnic groups are treated for pain, though the effect can vary based on the doctor's gender and cognitive load [429]. When it comes to death and end-of-life care, these biases may be particularly concerning, as they could perpetuate existing differences. We know that health providers can try to train themselves out of their implicit biases. The unconscious bias training that Stanford offers is one option, while other institutions have included training that focuses on introspection or mindfulness. But it is an entirely different challenge to imagine scrubbing biases from algorithms and the datasets they're trained on [429]. The algorithm is unbiased. However, the data that people put into algorithms can be biased. In an age where AI deals with even the death of people, there is a debate in front of us that can no longer be avoided.

Aging in humans is a process that affects all levels of biological organization: from molecular to systemic [430, 431]. Longitudinal studies of these processes have produced multiple "aging biomarkers," which are the most convenient and reliable features to determine the extent of the aging-related changes in the human body. The degree of such changes is usually expressed as "biological age" – a numeric value describing how typical the observed biomarker configuration is for healthy chronological age peers within a population [432]. Higher biological age values indicate the higher intensity of aging-related detrimental processes, while lower biological age – higher resilience to them. Traditionally, biological age metrics are designed to resemble the chronological age distribution within a cohort of healthy individuals, while being more predictive of a person's health status than chronological age itself [433]. Aging clocks that accurately predict human age based on various biodata types are among the most important recent advances in biogerontology. Since 2016 multiple DL solutions have been created to interpret facial photos, omics data, and clinical blood parameters in the context of aging. Some of them have been patented to be used in commercial settings. However, psychological changes occurring throughout the human lifespan have been overlooked in the field of "deep aging clocks." Zhavoronkov et al. [434] presented two DL predictors trained on social and behavioral data from Midlife in the United States (MIDUS) study: (a) PsychoAge, which predicts chronological age, and (b) SubjAge, which describes personal aging rate perception. The model is designed to accurately estimate subjective and psychological age from questionnaire question items. Using 50 distinct features from the MIDUS dataset these models have achieved a mean absolute error of 6.7 years for chronological age and 7.3 years for subjective age. It was also shown that both PsychoAge and SubjAge are predictive of all-cause mortality risk, with SubjAge being a more significant risk factor. More than 10,000 people aged 25–70 trained ML models, according to the study. In the verification study conducted on an independent data set, it was found that (i) the higher the psychological age, the higher the mortality rate, and (ii) if the psychological age is 5 years older than the actual age, the mortality rate is twice as high as that of those with normal age recognition. It was also mentioned that this psychological aging marker could help formulate effective interventions to make themselves feel youthful and live productive lives. Deep Longevity aims to build an integrated model for mental and physical health by extending its research scope to include differences in aging perceptions between men and women and the search for psychosocial markers related to mental health. Both clocks contain actionable features that can be modified using social and behavioral interventions, which enables a variety of aging-related psychology experiment designs. The features used in these clocks are interpretable by human experts and may prove to be useful in shifting personal perception of aging toward a mindset that promotes productive and healthy behaviors [445].

Aging is a biological process that causes physical and physiological deficits over time, culminating in organ failure and death. For species that experience aging, which includes nearly all animals, its presentation is not uniform; individuals age at different rates and in different ways. Biological age is an increasingly utilized concept that aims to more accurately reflect aging in an individual than the conventional chronological age. Biological measures that accurately predict health and longevity would greatly expedite studies aimed at identifying genetic and pharmacological disease and aging interventions. The identification of genes and interventions that slow or reverse aging is hampered by the lack of non-invasive metrics that can predict the life expectancy of pre-clinical models. Schultz et al. [435] conducted a study, attracting a lot of attention because it is a basic study to clarify the extent to which medical interventions affect healthy life expectancy extension in humans and is the first study to track frailty in mice in a entire life. Even mice usually need up to 3 years to see if a particular drug or diet slows the aging process. The research team tracked 60 mice to their deaths, continuously performed noninvasive tests such as walking ability, back flexion, and hearing, and built a model that detects the biological age of mice based on frailty and predicts the mouse's lifespan itself. Frailty Indices (FIs) in mice are composite measures of health that are cost-effective and non-invasive, but whether they can accurately predict health and lifespan is not known. Here, mouse FIs are scored longitudinally until death and ML is employed to develop two clocks. A random forest regression is trained on FI components for chronological age to generate the FRIGHT (Frailty Inferred Geriatric Health Timeline) clock, a strong predictor of chronological age. A second model is trained on remaining lifespan to generate the AFRAID (Analysis of Frailty and Death) clock, which accurately predicts life expectancy and the efficacy of a lifespan-extending intervention up to a year in advance. These enable accurate estimation of each outcome within two months, so it can be used as a tool to test how medical interventions work, and the speed of research can be expected to accelerate. It was mentioned that the factors that affect human health are much more diverse, like mice in the laboratory, so they cannot be applied immediately. On the other hand, based on the study, the researchers are looking to develop a system that can quickly and accurately predict medical interventions that contribute to human health span and life expectancy extension. Adoption of these clocks should accelerate the identification of longevity genes and aging interventions [435].

Molecular biologist says that age 120 years should be the longest healthy lifespan. Actually, we are witnessing that the world recoding living human is always close to 120 years old. Even the Genesis 1:1 of the Old Testament says that the God gave human a life of 120 years. Moreover, the *Tibetan Book of the Dead* (*Bardo Thödol*) and even the *Egyptian Book of the Dead* which is far older than the Tibetan book clearly mentioned that the longest lifespan for human can achieve is 120 years. Attia et al. [436] developed an AI-enabled ECG using a CNN to detect the electrocardiographic signature of atrial fibrillation present during normal sinus rhythm using standard 10-s, 12-lead ECGs. Included all patients were people of aged 18 years or older with at least one digital, normal sinus rhythm, standard 10-s, 12-lead ECG acquired in the supine position at the Mayo Clinic ECG laboratory between December 31, 1993, and July 21, 2017, with rhythm labels validated by trained personnel under cardiologist supervision. All subjects included 180,922 patients with 649,931 normal sinus rhythm ECGs for analysis: 454,789 ECGs recorded from 126,526 patients in the training dataset, 64,340 ECGs from 18,116 patients in the internal validation dataset, and 130,802 ECGs from 36,280 patients in the testing dataset. About 3,051 (8. 4%) patients in the testing dataset had verified atrial fibrillation before the normal sinus rhythm ECG tested by the model. It was found that (i) a single AI-enabled ECG identified atrial fibrillation with an AUC of 0.87 (95% CI 0.86–0.88), sensitivity of 79.0% (77.5–80.4), specificity of 79.5% (79.0–79.9), F1 score of 39.2% (38.1–40.3), and overall accuracy of 79.4% (79.0–79.9), (ii) including all ECGs acquired during the first month of each patient's window of interest (i.e., the study start date or 31 days before the first recorded atrial fibrillation ECG) increased the AUC to 0.90 (0.90–0.91), sensitivity to 82.3% (80.9–83.6), specificity to 83.4% (83.0–83.8), F1 score to 45.4% (44.2–46.5), and overall accuracy to 83.3% (83.0–83.7). It was, therefore, concluded that (iii) an AI-enabled ECG acquired during normal sinus rhythm permits identification at point of care of individuals with atrial fibrillation. Based on the thus established ECG technique, which is affected by sex and age, Attia et al. [437] hypothesized that a CNN could be trained through a process called deep learning to predict a person's age and self-reported sex using only 12-lead ECG signals. It was further hypothesized that discrepancies between CNN-predicted age and chronological age may serve as a physiological measure of health. Of 275,056 patients tested, 52% were males and mean age was 58.6 ± 16.2 years. It was found that (i) for sex classification, the model obtained 90.4% classification accuracy with an area under the curve of 0.97 in the independent test data, (ii) age was estimated as a continuous variable with an average error of 6.9 ± 5.6 years (R-squared = 0.7), (iii) among 100 patients with multiple ECGs over the course of at least 2 decades of life, most patients (51%) had an average error between real age and CNN-predicted age of <7 years, (iv) major factors seen among patients with a CNN-predicted age that exceeded chronologic age by >7 years included: low ejection fraction, hypertension, and coronary disease, and (v) in the 27% of patients where correlation was >0.8 between CNN-predicted and chronologic age, no incident events occurred over follow-up (33 ± 12 years). Based on these findings, it was concluded that (vi) applying AI to the ECG allows prediction of patient sex and estimation of age, and (vii) the ability of an AI algorithm to determine physiological age, with further validation, may serve as a measure of overall health [437].

8.8 Sports medicine

Sports medicine discipline covers relatively wide areas, including training/coaching, injury risk management, performance enhancement, or AI-empowered judgment [438]. The system, ISOTechne, evaluates a player's skill and consistency and who is passing or receiving with what frequency, as well as the structure of the team's defense. It even tracks the axis of spin and rate of rotation of the ball during soccer plays [439]. The majority of National Hockey League teams, including the last four Stanley Cup champions, used Sportlogiq's system to evaluate players [440]. Seattle Sports Sciences uses Labelbox (a training data platform) to create and manage the training data, people and processes in a single place [441]. It is anticipated that the entry of big data and AI in sports management, the process of recording and measuring these indicators of future success is becoming easier and more reliable [442].

The interdisciplinary nature of sports science introduces challenges such as multifaceted data collection, accuracy in knowledge formation, and equipment usability. AI of things (AIoT) technology presents a feasible solution adaptable to different sports. Taking weight training as an example, Chu et al. [443] applied AIoT technology to these challenges. Novatchkov et al. [444] demonstrated the potential of AI techniques in sports on the example of weight training, in particular on the implementation of pattern recognition methods for the evaluation of performed exercises on training machines. It was reported that the obtained modeling results showed good performance and prediction outcomes, indicating the feasibility and potency of AI techniques in assessing performances on weight training equipment automatically and providing sportsmen with prompt advice.

Wearable devices can help athletes or coaches to systematically manage athletic training and matches. Use of a commercially available wearable device to monitor jump load with elite volleyball players has become common practice. Skazalski et al. [445] evaluated the validity and reliability of this device, the Vert, to count jumps and measure jump height with professional volleyball players. The wearable devices can monitor functional movements, workloads, heart rate, etc., so they may be more widely used in sport medicine to maximize performance and minimize injury. Heat stroke can be potentially damaging for people while exercising in hot environments. To prevent this dangerous situation, Chen et al. [149] designed a wearable heat-stroke-detection device (WHDD) with early notification ability. Several physical sensors, such as galvanic skin response (GSR), heart-beat, and body temperature were used to acquire medical data from exercising people. In addition, risk evaluation functional components were designed, based on fuzzy theory to detect the features of heat stroke for users. It was reported that if a dangerous situation is detected, then the device will activate the alert function to remind the user to respond adequately to avoid heat stroke.

In various sports, not only AI but also IoT devices and digital products are being introduced. Takahashi [446] pointed out five important factors which will be further developed in sports: (1) AI-assisted simulation to build successful strategy, which might be helpful to players as well as coach and trainers staff members, (2) real-time display system during the race or games to provide better information to players and coaches, (3) camera relay while tracking competitors with automatic tracking drones, in particular cycling road races, cross-country races, athletics marathons and race walks, etc., (4) the congestion situation is grasped by base station and public transportation, and (5) AI utilization for judgment during the tennis, badminton, volleyball matches, and so on. Recently, a product that scores dance or ice dance techniques by video analysis has also been announced.

Some systems can detect and predict injuries. The application of AI opens an interesting perspective for predicting injury risk and performance in team sports. A better understanding of the techniques of AI employed and of the sports that are using AI is clearly warranted. Claudino et al. [447] identified which AI approaches have been applied to investigate sport performance and injury risk and to find out which AI techniques each sport has been using. Systematic searches through the PubMed, Scopus, and Web of Science online databases were conducted for articles reporting AI techniques or methods applied to team sports athletes. Fifty-eight studies were included in the review with 11 AI techniques or methods being applied in 12 team sports. It was found that (i) pooled sample consisted of 6,456 participants (97% male, 25 ± 8 years old; 3% female, 21 ± 10 years old) with 76% of them being professional athletes, (ii) the AI techniques or methods most frequently used were ANNs, decision tree classifier, SVM, and Markov process with good performance metrics for all of them, and (iii) soccer, basketball, handball, and volleyball were the team sports with more applications of AI. It was, accordingly, concluded that (iv) the results of this review suggest a prevalent application of AI methods in team sports based on the number of published studies, and (v) the current state of development in the area proposes a promising future with regard to AI use in team sports. Injuries have a great impact on professional soccer, due to their large influence on team performance and the considerable costs of rehabilitation for players. Existing studies in the literature provide just a preliminary understanding of which factors mostly affect injury risk, while an evaluation of the potential of statistical models in forecasting injuries is still missing. Rossi et al. [448] proposed a multidimensional approach to injury forecasting in professional soccer that is based on GPS measurements and ML. By using GPS tracking technology, data describing the training workload of players in a professional soccer club during a season was collected, followed by constructing an injury forecaster and show that it is both accurate and interpretable by providing a set of case studies of interest to soccer practitioners. An algorithm (with accompanied with parameters such as heart rate, sweating and psychophysical stress) was developed to enable to predict. In tests conducted on a professional team during an entire season, the algorithm has predicted more than 50% of muscle injuries. It has shown that a player is more likely to get hurt if he travels long distances with high running intensity and if he has already suffered injuries in recent times. According to the authors of the study, if the athletic trainers had had this data available during the championship, the team would have saved about 70% of the costs related to the recovery and rehabilitation of the players in the company [448].

8.9 Cosmetic surgery

The ML techniques can add another level of accuracy to computer-assisted 3D design models, which help plastic surgeons prepare for complex facial reconstruction surgery [14]. The human face is an extremely complex anatomical structure that fills crucial functional and aesthetic roles. A group of plastic surgeons at Yale is using virtual 3D planning technology to improve surgical outcomes. In the first part of the process, CT scans of patients' skulls are obtained, and these, in conjunction with 3D analysis tools, are used in pre-operative planning for reconstructive surgery. Three-dimensional planning then allows the surgeon to virtually simulate the surgery using the composite images. Other crucial components of 3D planning include construction of cutting guides, which assist the surgeon in performing the procedure, 3D printing of splints and other prosthetics and the ability to review the relevant anatomical landscape in advance of the procedure. Three-dimensional planning allows surgeons to use computer-aided design and computer-aided manufacturing technology to bring industrial standards of precision to the operating room [449,450].

One of the biggest fears for any patient who is considering a surgical cosmetic procedure is the outcome of the whole process [450]. Three-dimensional imaging allows the surgeon to show a simulation of what the given procedure will like on the patient's actual body. Three-dimensional imaging removes the fear of the unknown. With the software, a surgeon can take a picture of whatever body part that you are looking to improve such as the nose, face or breasts. Then, in real time, the surgeon can show a patient an extremely realistic simulation of predicted outcomes. In fact, with something like breast implants, it allows you to see what each different implant would look like on your body. Just as 3D imaging is a valuable tool for the patient, it is equally valuable for the surgeon. Every surgeon wants to do an exemplary job, to deliver an excellent surgical result. Part of this comes down to technical skill and experience, but another part of it is delivering a result that meets the patent's expectations. In order to do this, there must be a clear channel of communication between the surgeon and the patient. Three-dimensional imaging allows the surgeon to point out things on a body that may limit or affect the final results. One of the biggest things that 3D imaging helps to predict is asymmetries. All human beings are naturally asymmetric [451]. No one has a right side that is exactly the same as the left side. In many cases, these differences are nominal. But because a cosmetic surgical procedure is such an important decision for the patient to make, if the patient is not aware that these asymmetries exist before surgery, s/he may be disappointed afterward if they see it in their final results [451]. With 3D imaging, there are no surprises. Three-dimensional imaging has increased patient satisfaction exponentially [452].

Mental health can have an impact on patient satisfaction with rhinoplasty and a major hurdle of rhinoplasty is predicting outcomes. However, the association between mental health and patient satisfaction with functional outcomes of rhinoplasty is poorly understood. Strazdins et al. [453] determined whether pre-operative mental health is associated with satisfaction with functional outcomes of rhinoplasty. It was concluded that (i) rhinoplasty imparts similar benefits to nasal function assessed by patient-reported outcome measures and objective airflow measures regardless of preoperative mental health status and (ii) the study provides some evidence that patients perceive improvement in nasal function after surgery, regardless of their mental status.

8.10 Concerns and future

As we have seen in the above, AI technologies have been utilized in four major area of medicine (i.e., prevention, diagnosis, treatment, and prognosis). If mechanism of medical action, major cause of diseases, and symptoms can be digitized, all information can be fed to AI brain. Accordingly, as Wagner et al. [454] pointed out, the

AI in healthcare is over-hyped due to the advances in the DL technologies, accompanied with several critical issues.

8.10.1 Concerns

We have already discussed the reliability and bias of medical information to be input to AI systems. There is a concern about the lack of AI validation [455]. As with any other medical devices or technologies, the importance of thorough clinical validation of AI algorithms before their adoption in clinical practice through adequately designed studies to ensure patient benefit and safety while avoiding any inadvertent harms cannot be overstated [456]. Clinical validation of AI technologies can be performed at different levels of efficacy: diagnostic performance, effects on patient outcome, and societal efficacy that considers cost-benefit and cost-effectiveness [457]. Kim et al. [456] evaluated the design characteristics of studies that evaluated the performance of AI algorithms for the diagnostic analysis of medical images. Of 516 eligible published studies, only 6% (31 studies) performed external validation. It was reported that (i) none of the 31 studies adopted all three design features: diagnostic cohort design, the inclusion of multiple institutions, and prospective data collection for external validation, and (ii) no significant difference was found between medical and non-medical journals. It was then concluded that (iii) nearly all of the studies published in the study period that evaluated the performance of AI algorithms for diagnostic analysis of medical images were designed as proof-of-concept technical feasibility studies and did not have the design features that are recommended for robust validation of the real-world clinical performance of AI algorithms [456].

In the last several years or so, the use of AI in medical care has progressed so rapidly that the required legal development has been delayed. As a matter of fact, there has not been enough discussion on how to handle the diagnostic results shown by medical AI. There is a minimum need to clarify the guidelines for building algorithms that are essentially effective, the requirements for AI systems to be approved as medical devices, and certain restrictions on what is not approved. It is very risky and dangerous for medical AI with unknown effectiveness to go on the market without any restrictions. Jiang et al. [58] pointed out that, although the AI technologies are attracting substantial attentions in medical research, the real-life implementation is still facing obstacles. The first hurdle comes from the regulations. Current regulations lack of standards to assess the safety and efficacy of AI systems. To overcome the difficulty, the US FDA made the first attempt to provide guidance for assessing AI systems [458]. Lack of standardization will be another concern. Even if health records were open to the public, it would not solve the quality and standardization issues. Medical information is collected in many formats, with standards varying greatly across organizations. So, it takes scientists significant time to clean and label data before feeding it to models. Typically, DL algorithms work as black boxes. They do not explain why they jump to certain conclusions. While for many areas the lack of interpretability is not a problem, it certainly matters in healthcare, where people's lives are at stake. Clinicians and their patients need to know what makes the machine generate its verdicts and if there is evidence behind them. Otherwise, they can hardly rely on diagnoses suggested by IT systems [52].

The data exchange would be another issue [58]. In order to work well, AI systems need to be trained (continuously) by data from clinical studies. However, once an AI system gets deployed after initial training with historical data, continuation of the data supply becomes a crucial issue for further development and improvement of the system. Current healthcare environment does not provide incentives for sharing data on the system. Nevertheless, a healthcare revolution is under way to stimulate data sharing in the United States [459]. The reform starts with changing the health service payment scheme. Many payers, mostly insurance companies, have shifted from rewarding the physicians by shifting the treatment volume to the treatment outcome. Furthermore, the payers also reimburse for a medication or a treatment procedure by its efficiency. Under this new environment, all the parties in the healthcare system, the physicians, the pharmaceutical companies and the patients, have greater incentives to compile and exchange information.

There are also a variety of ethical implications around the use of AI in healthcare. Healthcare decisions have been made almost exclusively by humans in the past, and the use of smart machines to make or assist with them raises issues of accountability, transparency, permission and privacy [13, 454]. Perhaps the most difficult issue to address given today's technologies is transparency. Many AI algorithms, particularly DL algorithms used for image analysis, are virtually impossible to interpret or explain. If a patient is informed that an image has led to a diagnosis of cancer, the patient will likely want to know why. DL algorithms, and even physicians who are generally familiar with their operation, may be unable to provide an explanation [13]. Mistakes will undoubtedly be made by AI systems in patient diagnosis and treatment, and it may be difficult to establish accountability for them. There are also likely to be incidents in which patients receive medical information from AI systems that they would prefer to receive from an empathetic clinician. ML systems in healthcare may also be subject to algorithmic bias, perhaps predicting greater likelihood of disease on the basis of gender or race when those are not actually causal factors [460]. We are likely to encounter many ethical, medical, occupational and technological changes with AI in healthcare. It is important that healthcare institutions, as well as governmental and regulatory bodies, establish structures to monitor key issues, react in a responsible manner and establish governance mechanisms to limit negative implications. This is one of the more powerful and consequential technologies to impact human societies, so it will require continuous attention and thoughtful policy for many years [13].

8.10.2 Future

Healthcare needs the transformative power of digital – DX technology. The report [236] overviewed the future of digital health, describing the skepticism and hype and what is needed for the medical community to embrace a world where data, machines, and analytics are employed to deliver higher quality, more efficient care. It was reported that we have entered the next era of medicine, and an era where we focus on health and wellness rather than disease. An era where technology not only makes care safer, more efficient, and higher quality, but also where it improves access to everyone. Today, the technology exists to analyze the quadrillions of bits of data previously siloed and ignored. Torn between the skepticism and the hype, it is incumbent on all of us to realize that the potential of digital health is already being realized. We are in the midst of the digitally enabled health revolution, where the cloud connects people, data, and machines, and in which analytics has become a critical tool to delivering higher quality, more efficient care [236]. It seems increasingly clear that AI systems will not replace human clinicians on a large scale, but rather will augment their efforts to care for patients. Over time, human clinicians may move toward tasks and job designs that draw on uniquely human skills like empathy, persuasion and big-picture integration. Perhaps the only healthcare providers who will lose their jobs over time may be those who refuse to work alongside AI [13].

References

[1]   Ahuja AS. The impact of artificial intelligence in medicine on the future role of the physician. Peer Journal. 2019, 7, e7702, doi: 10.7717/peerj.7702.
[2]   Okamoto M The latest use cases of medical AI. 2019; https://aitimes.media/medicalai_use.
[3]   Greenhill AT, Edmunds BR. A primer of artificial intelligence in medicine. Techniques and Innovations in Gastrointestinal Endoscopy. 2020, 22, 85–9.
[4]   Miller DD. The medical AI insurgency: what physicians must know about data to practice with intelligent machines. npj Digital Medicine. 2019, 2, 62, https://doi.org/10.1038/s41746-019-0138-5.
[5]   Sreedharan S, Mian M, Robertson RA, Yang N. The top 100 most cited articles in medical artificial intelligence: a bibliometric analysis. Journal of Medical Artificial Intelligence. 2020, 3, https://jmai.amegroups.com/article/view/5216/html.
[6]   Frost LA, Sullivan DL Transforming healthcare through artificial intelligence systems. 2016; https://docplayer.net/36848717-Transforming-healthcare-through-artificial-intelligence-systems.html.
[7]   Daley S 32 Examples of AI in Healthcare that will make you feel better about the future. 2020; https://builtin.com/artificial-intelligence/artificial-intelligence-healthcare.
[8]   Spatharou A, Hieronimus S, Jonathan Jenkins J Transforming healthcare with AI: The impact on the workforce and organizations. 2021; https://www.mckinsey.com/industries/health care-systems-and-services/our-insights/transforming-healthcare-with-ai.
[9]   Buch VH, Ahmed I, Maruthappu M. Artificial intelligence in medicine: current trends and future possibilities. British Journal of General Practice. 2018, 68, 143–4.

[10] Krittanawong C. The rise of artificial intelligence and the uncertain future for physicians. European Journal of Internal Medicine. 2018, 48, e13–e14.

[11] Pearson D Artificial intelligence in radiology: the game-changer on everyone's mind. 2017; https://www.radiologybusiness.com/topics/technology-management/artificial-intelligence-radiology-game-changer-everyones-mindRadiologybusiness.

[12] Recht M, Bryan RN. Artificial intelligence: threat or boon to radiologists?. Journal of American College of Radiologists. 2017, 14, 1476–80.

[13] Davenport T, Kalakota R. The potential for artificial intelligence in healthcare. Future Healthcare JOURNAL. 2019, 6, 94–8.

[14] Raven K. Artificial Intelligence in Medicine: Getting Smarter One Patient at a Time. 2020; https://www.yalemedicine.org/stories/ai-medicine/.

[15] Khillar S. Difference Between AI and Expert System. 2020; http://www.differencebetween.net/technology/difference-between-ai-and-expert-system/.

[16] Fieschi M. Artificial Intelligence in Medicine: Expert Systems. Springer US. 1990.

[17] Krishnamoorthy CS, Rajeev S. Artificial Intelligence and Expert Systems for Engineers, CRC Press, Boca Raton, Florida, 1996.

[18] Difference Between Artificial Intelligence And Expert System In 2020. 2020; https://azati.ai/the-return-of-expert-systems/.

[19] The difference between machine learning and expert systems. 2020; http://tooljp.com/windows/chigai/html/AI/ExpertSystem-MachineLearning-chigai.html.

[20] Difference between AI and Expert System. 2020; https://www.geeksforgeeks.org/difference-between-ai-and-expert-system/.

[21] Vihinen M, Samarghian C. Medical Expert Systems. Current Bioinformatics. 2008, 3, 56–65.

[22] Nath P. AI & expert system in medical field: A study by survey method. AITHUN. 2015; https://www.researchgate.net/publication/272163982_AI_Expert_System_in_Medical_Field_A_study_by_survey_method.

[23] Mycin; https://en.wikipedia.org/wiki/Mycin#:~:text=MYCIN%20was%20an%20early%20backward,antibiotics%20themselves%2C%20as%20many%20antibiotics.

[24] Aikins JS, Kunz JC, Shortliffe EH, Fallat RJ. PUFF: An expert system for interpretation of pulmonary function data. Computers and Biomedical Research. 1983, 16, 199–208.

[25] CADUCEUS: An Experimental Expert System for Medical Diagnosis. In: Winston PH. et al., ed., The AI Business: Commercial Uses of Artificial Intelligence, MIT Press, 1986.

[26] Lauritzen SL, Spiegelhalter DJ. Local computations with probabilities on graphical structures and their application to expert systems. Journal of the Royal Statistical Society. Series B (Methodological). 1988, 50, 157–224.

[27] Kannan A. The science of assisting medical diagnosis: From Expert systems to Machine-learned models. Curai Health Technology. 2019, https://medium.com/curai-tech/the-science-of-assisting-medical-diagnosis-from-expert-systems-to-machine-learned-models-cc2ef0b03098.

[28] Diamandis PH, Kotler S. The Future Is Faster Than You Think, Simon & Schuster, New York NY, 2020.

[29] Arndt RZ The slow upgrade to artificial intelligence. https://www.modernhealthcare.com/indepth/artificial-intelligence-in-healthcare-makes-slow-impact/#firststory.

[30] IoT Standards and Protocols; https://networklessons.com/cisco/evolving-technologies/iot-standards-and-protocols.

[31] Sheldon. LAN vs MAN vs WAN: What's the Difference? 2020; https://community.fs.com/blog/lan-vs-man-vs-wan-whats-the-difference.html.

[32] Wu M, Luo J. Wearable technology applications in healthcare: A literature review. Journal of Nursing Informatics Contributors. 2019, 23, http://www.himss.org/ojni.

[33] From Telehealth To Infection Control, How 5G, Edge Computing, & The Internet Of Medical Things Are Transforming Healthcare. 2020; https://www.cbinsights.com/research/internet-of-medical-things-5g-edge-computing-changing-healthcare/.

[34] Steger A How the Internet of Medical Things Is Impacting Healthcare. 2020; https://health techmagazine.net/article/2020/01/how-internet-medical-things-impacting-healthcare-perfcon.

[35] Hurt A. What the internet of medical things means to your practice. Medical Econimics Journal. 2019, 96, 17, https://www.medicaleconomics.com/view/what-internet-medical-things-means-your-practice.

[36] Franklin R What is the Internet of Medical Things (IoMT)? 2019; https://www.mobius.md/2019/03/06/what-is-the-iomt/.

[37] Dimitrov DV. Medical internet of things and big data in healthcare. Healthcare Informatics Research. 2016, 22, 156–63.

[38] Attanà S What is intrabody communication? 2020; https://www.sintec-project.eu/what-is-in trabody-communication/.

[39] Naranjo-Hernández D, Callejón-Leblic A, Lučev Vasić ŽL, Seyedi MH, Yue-Ming Gao Y-M. Past results, present trends, and future challenges in intrabody communication. Wireless Communications and Mobile Computing. 2018, https://doi.org/10.1155/2018/9026847.

[40] Artificial Intelligence in Healthcare – A Comprehensive Account. https://marutitech.com/arti ficial-intelligence-in-healthcare/.

[41] Andreoni G, Standoli CE, Perego P. Defining requirements and related methods for designing sensorized garments. Sensors. 2016, 16, 769, DOI: 10.3390/s16060769.

[42] AI predicts risk of lifestyle-related diseases. 2019; https://ledge.ai/sompo-ai/.

[43] Medical care can predict from lifestyle-related diseases to infectious diseases with AI. 2020; https://medicaldx-jp.com/prevention/9.

[44] Fall prevention by gait analysis. 2020; https://ledge.ai/care-exawizards-ai/.

[45] AI in Healthcare: 4 Examples in Health Informatics. https://healthinformatics.uic.edu/blog/ai-in-healthcare-4-examples-in-health-informatics/.

[46] Emc IDC. The Digital Universe: Driving Data Growth in Healthcare. 2014; https://www.cyclo neinteractive.com/cyclone/assets/File/digital-universe-healthcare-vertical-report-ar.pdf.

[47] Azcoitia SS Expert systems in the world of medicine. Luca. 2018; https://business.blogthink big.com/expert-systems-in-world-of-medicine/.

[48] Miller RA. Medical diagnostic decision support systems – past, present, and future: a threaded bibliography and commentary. Journal of the American Medical Informatics Association. 1994, 1, 8–27.

[49] Barnett ML, Boddupalli D, Nundy S, Bates DW. Comparative accuracy of diagnosis by collective intelligence of multiple physicians vs individual physicians. JAMA Network Open. 2019, 2, e190096, doi:10.1001/jamanetworkopen.2019.0096.

[50] Saposnik G, Redelmeier D, Ruff CC, Philippe N, Tobler PN. Cognitive biases associated with medical decisions: a systematic review. MC Medical Informatics and Decision Making. 2016, 16, 138, doi: 10.1186/s12911-016-0377-1.

[51] Xue Q, Chuh MC. Explainable deep learning based medical diagnostic system. Smart Health. 2019, 13, 100068, https://doi.org/10.1016/j.smhl.2019.03.002.

[52] Deep Learning in Medical Diagnosis: How AI Saves Lives and Cuts Treatment Costs. 2020; https://www.altexsoft.com/blog/deep-learning-medical-diagnosis/.

[53] Millard M. How AI and machine learning are transforming clinical decision support. 2020; https://www.healthcareitnews.com/news/how-ai-and-machine-learning-are-transforming-clinical-decision-support.

[54]   Kent J How artificial intelligence is changing radiology, pathology. Health IT Anaytics. 2018. https://healthitanalytics.com/news/how-artificial-intelligence-is-changing-radiology-pathology.

[55]   Liew C. The future of radiology augmented with Artificial Intelligence: a strategy for success. European Journal of Radiology. 2018, 102, 152–6.

[56]   Mishra S, Dash A, Jena L. Use of Deep Learning for Disease Detection and Diagnosis. In: Bhoi A, et al. ed, Bio-inspired Neurocomputing. Studies in Computational Intelligence, Vol. 903, 2021, Springer, Singapore, https://doi.org/10.1007/978-981-15-5495-7_10.

[57]   Liang H, Tsui BY, Ni H. 70 co-authors. Evaluation and accurate diagnoses of pediatric diseases using artificial intelligence. Nature Medicine. 2019, 25, 433–8.

[58]   Jiang F, Jiang Y, Zhi H, Dong Y, Li H, Ma S, Wang Y, Dong Q, Shen H, Wang Y. Artificial intelligence in healthcare: past, present and future. Stroke and Vascular Neurology. 2017, 2, 230–43.

[59]   Pastorello N, Berry K Deep Learning in Medical Imaging. https://www.asianhhm.com/diagnostics/artificial-intelligence-deep-learning-medical-imaging.

[60]   Yoshizawa M The forefront of healthcare and medical startups in the world where AI application is advancing. 2019; https://coralcap.co/2019/08/healthtech-report-one/.

[61]   Souquet J AI is transforming diagnostic imaging. 2018; https://www.beckershospitalreview.com/healthcare-information-technology/ai-is-transforming-diagnostic-imaging.html.

[62]   Ardila D, Kiraly AP, Bharadwaj S, Choi B, Reicher JJ, Peng L, Tse D, Etemadi M, Ye W, Corrado G, Naidich DP, Shetty S. End-to-end lung cancer screening with three-dimensional deep learning on low-dose chest computed tomography. Nature Medicine. 2019, 25, 954–61.

[63]   Conner-Simons A, Gordon R. Using AI to predict breast cancer and personalize care. MIT News. 2019; https://news.mit.edu/2019/using-ai-predict-breast-cancer-and-personalize-care-0507.

[64]   Press G AI Startup Zebra Medical Vision Enlists Deep Learning To Save Lives. 2019; https://www.forbes.com/sites/gilpress/2019/04/22/ai-startup-zebra-medical-vision-enlists-deep-learning-to-save-lives/?sh=611696f61035.

[65]   Mahler T BGU Researchers Propose New Technique to Prevent Medical Imaging Cyberthreats. 2020; https://in.bgu.ac.il/en/pages/news/imaging_cyberthreats.aspx.

[66]   Fisher C Google researchers taught an AI to recognize smells. 2019; https://www.engadget.com/2019-10-24-google-researchers-train-ai-smells.html?fbclid=IwAR2m-4tOjD2s2A49ckitIkrK2hRRUefAbQ5rD7YbSuNitoJzA84NdcwPOic.

[67]   Imam N, Cleland TA. Rapid online learning and robust recall in a neuromorphic olfactory circuit. Nature Machine Intelligence. 2020, 2, 181–91.

[68]   To a new era of disease diagnosis – Detect the signs of disease with an odor sensor https://project.nikkeibp.co.jp/mirakoto/atcl/wellness/h_vol19/.

[69]   WHO. Preventing cancer. https://www.who.int/activities/preventing-cancer.

[70]   WHO. Cancer. 2018; https://www.who.int/news-room/fact-sheets/detail/cancer.

[71]   Lemjabbar-Alaoui H, Hassan O, Yang Y-W, Buchanan P. Lung cancer: biology and treatment options. Biochim Biophys Acta. 2015, 1856, 189–210.

[72]   Johnson K Google's lung cancer detection AI outperforms 6 human radiologists. The Machine. 2019; https://venturebeat.com/2019/05/20/googles-lung-cancer-detection-ai-outperforms-6-human-radiologists/.

[73]   Lu MT, Raghu VK, Mayrhofer T, Aerts HJWL. Deep learning using chest radiographs to identify high-risk smokers for lung cancer screening computed tomography: development and validation of a prediction model. Annals of Internal medicine. 2020, 173, 704–13.

[74] Sinagra E. 10 co-authors. Use of artificial intelligence in improving adenoma detection rate during colonoscopy: Might both endoscopists and pathologists be further helped. World Journal of Gastroenterology : WJG. 2020, 26, 5911–8.

[75] Wang K-W, Dong M. Potential applications of artificial intelligence in colorectal polyps and cancer: Recent advances and prospects. World Journal of Gastroenterol. 2020, 26, 5090–100.

[76] Wang P. 14 co-authors. Effect of a deep-learning computer-aided detection system on adenoma detection during colonoscopy (CADe-DB trial): a double-blind randomised study. Lancet Gastroenterol Hepatol. 2020, 5, 343–51.

[77] Wang P. 13 co-authors. Real-time automatic detection system increases colonoscopic polyp and adenoma detection rates: a prospective randomised controlled study. Gut. 2019, 68, 1813, http://dx.doi.org/10.1136/gutjnl-2018-317500.

[78] Niehous K, Wan N, White B, Kannan A, Gafni E, Liu T-Y, Haque I, Putcha G. Early stage colorectal cancer detection using artificial intelligence and whole-genome sequencing of cell-free dna in a retrospective cohort of 1,040 patients. American Journal of Gastroenterology. 2018, 113, S169. https://journals.lww.com/ajg/Fulltext/2018/10001/Early_Stage_Colorectal_Cancer_Detection_Using.307.aspx.

[79] Jiang D. 11 co-authors. A machine learning-based prognostic predictor for stage III colon cancer. Scientific Reports. 2020, 10, 10333, DOI:10.1038/s41598-020-67178-0.

[80] Armitage H AI tool created to guide colorectal cancer care with more precision. 2020; https://scopeblog.stanford.edu/2020/10/01/ai-tool-created-to-guide-colorectal-cancer-care-with-more-precision/.

[81] Bibault J-E, Chang DT, Xing L. Development and validation of a model to predict survival in colorectal cancer using a gradient-boosted machine. Gut. 2020, doi: 10.1136/gutjnl-2020-321799.

[82] Ferlay J, Soerjomataram I, Dikshit R, Eser S, Mathers C, Rebelo M, Parkin DM, Forman D, Bray F. Cancer incidence and mortality worldwide: sources, methods and major patterns in glOBOcan 2012. International Journal of Cancer. 2015, 136, e359–86.

[83] Karim-Kos HE, De Vries E, Soerjomataram I, Lemmens V, Siesling S, Coebergh JW. Recent trends of cancer in Europe: a combined approach of incidence, survival and mortality for 17 cancer sites since the 1990s. European Journal of Cancer. 2008, 44, 1345–89.

[84] Lee KJ, Inoue M, Otani T, Iwasaki M, Sasazuki S, Tsugano S. Gastric cancer screening and subsequent risk of gastric cancer: a large-scale population-based cohort study, with a 13-year follow-up in Japan. International Journal of Cancer. 2006, 118, 2315–21.

[85] Hosokawa O, Hattori M, Douden K, Hayashi H, Ohta K, Kaizaki Y. Difference in accuracy between gastroscopy and colonoscopy for detection of cancer. Hepatogastroenterology. 2007, 54, 442–4.

[86] Ikenoyama Y. 13 co-authors. Detecting early gastric cancer: Comparison between the diagnostic ability of convolutional neural networks and endoscopists. Digestive Endoscopy. 2020, 33, 141–50, doi: 10.1111/den.13688.

[87] Hirasawa T. 10 co-authors. Application of artificial intelligence using a convolutional neural network for detecting gastric cancer in endoscopic images. Gastric Cancer. 2018, 21, 653–60.

[88] Japanese Researchers Build An AI That Identifies Early-Stage Of Stomach Cancer. 2019; https://techgrabyte.com/japanese-researchers-ai-identify-stomach-cancer/.

[89] Nakahira H. 20 co-authors. Stratification of gastric cancer risk using a deep neural network. Journal of Gastroenterology and Hepatology Open. 2020, 4, 466–71.

[90] Jin P, Ji X, Kang W, Li Y, Liu H, Ma F, Ma S, Hu H, Li W, Tian Y. Artificial intelligence in gastric cancer: a systematic review. Journal of Cancer Research and Clinical Oncology. 2020, 146, 2339–50.

[91] Mori Y, Berzin TM, Kudo S-E. Artificial intelligence for early gastric cancer: early promise and the path ahead. Gastrointestinal Endoscopy. 2019, 89, 816–7.

[92] Forner A. 11 co-authors. Diagnosis of hepatic nodules 20 mm or smaller in cirrhosis: prospective validation of the noninvasive diagnostic criteria for hepatocellular carcinoma. Hepatology. 2008, 47, 97–104.

[93] DeSantis CE, Siegel RL, Sauer AG, Miller KD, Fedewa SA, Alcaraz KI, Jemal A. Cancer statistics for African Americans, 2016: progress and opportunities in reducing racial disparities. CA Cancer Journal for Clinician. 2016, 66, 290–308.

[94] Preisler SR Liver Cancer: Fast Facts. 2019; https://ezra.com/liver-cancer-fast-facts/.

[95] Kather JN, Calderaro J. Development of AI-based pathology biomarkers in gastrointestinal and liver cancer. Nature Reviews. Gastroenterology & Hepatology. 2020, 17, 591–2.

[96] Zhen S-H. 14 co-authors. Deep learning for accurate diagnosis of liver tumor based on magnetic resonance imaging and clinical data. Frontiers in Oncology. 2020, https://doi.org/10.3389/fonc.2020.00680.

[97] Nishida N, Kudo M. Artificial intelligence in medical imaging and its application in sonography for the management of liver tumor. Frontiers in Oncology. 2020, https://doi.org/10.3389/fonc.2020.594580.

[98] Pérez MJ, Grande RG. Application of artificial intelligence in the diagnosis and treatment of hepatocellular carcinoma: A review. World Journal of Gastroenterol. 2020, 26, 5617–28.

[99] Hu W, Yang H, Xu H, Mao Y. Radiomics based on artificial intelligence in liver diseases: where are we?. Gastroenterology Report. 2020, 8, 90–7.

[100] Lucence Diagnostics to Develop AI Tools for Liver Cancer Treatment. 2019; https://www.itnonline.com/content/lucence-diagnostics-develop-ai-tools-liver-cancer-treatment.

[101] Siegel RL, Miller KD, Jemal A. Cancer statistics, 2019, A Cancer Journal for Clinicians, CA, 2019, DOI: 10.3322/caac.21551.

[102] McKinney SM, Sieniek M. 32 co-authors. International evaluation of an AI system for breast cancer screening. Nature. 2020, 577, 89–94.

[103] Koh D AI helps radiologists improve accuracy in breast cancer detection with lesser recalls. 2020; https://www.healthcareitnews.com/news/asia-pacific/ai-helps-radiologists-improve-accuracy-breast-cancer-detection-lesser-recalls.

[104] AAMC (Association of American Medical Colleges). New Findings Confirm Predictions on Physician Shortage. 2019; https://www.aamc.org/news-insights/press-releases/new-findings-confirm-predictions-physician-shortage.

[105] Basen-Engquist K, Scruggs S, Jhingran A, Bodurka DC, Lu K, Ramondetta L, Hughes D, Taylor CC. Physical activity and obesity in endometrial cancer survivors: associations with pain, fatigue, and physical functioning. American Journal of Obstetrics and Gynecology. 2009, 2009, 200, e281–8.

[106] Rossi A, Frechette L, Miller D, Miller E, Friel C, Van Arsdale A, Lin J, Shankar V, Kuo DYS, Nevadunsky NS. Acceptability and feasibility of a Fitbit physical activity monitor for endometrial cancer survivors. Gynecologic Oncology. 2018, 149, 470–5.

[107] Nguyen NH, Hadgraft NT, Moore MM, Rosenberg DE, Lynch C, Reeves MM, Lynch BM. A qualitative evaluation of breast cancer survivors' acceptance of and preferences for consumer wearable technology activity trackers. Supportive Care in Cancer. 2017, 25, 3375–84.

[108] Robust AI tools to predict future cancer. 2021; https://www.csail.mit.edu/news/robust-ai-tools-predict-future-cancer.

[109] Six O, Veldhuis W, Akin O The ultimate guide to AI in prostate cancer. https://www.quantib.com/the-ultimate-guide-to-ai-in-prostate-cancer.

[110] Bardid MD, Houshyar R, Changd PD, Ushinsky A, Glavis-Bloom J, Chahine C, Bui T-L, Rupasinghe M, Filippi CG, Chow DS. Applications of artificial intelligence to prostate

multiparametric mri (mpmri): current and emerging trends. Cancers (Basel). 2020, 12, 1204, doi: 10.3390/cancers12051204.

[111] Harmon SA, Tuncer S, Sanford T, Cholyke PL, Türkbey B. Artificial intelligence at the intersection of pathology and radiology in prostate cancer. Diagnostic and Interventional Radiology. 2019, 25, 183–8.

[112] Broderick JM AI tool shows promise in predicting biochemical recurrence in prostate cancer. Urology Times. 2021; https://www.urologytimes.com/view/ai-tool-shows-promise-in-predicting-biochemical-recurrence-in-prostate-cancer.

[113] Liu Q, Li J, Liu F, Yang W, Ding J, Chen W, Wei Y, Li B, Zheng L. A radiomics nomogram for the prediction of overall survival in patients with hepatocellular carcinoma after hepatectomy. Cancer Imaging. 2020, 20, 82, https://doi.org/10.1186/s40644-020-00360-9.

[114] Palmer WJ MRI AI tool could lead to more targeted prostate cancer treatment. Diagnostic Imaging. 2021; https://www.diagnosticimaging.com/view/mri-ai-tool-could-lead-to-more-targeted-prostate-cancer-treatment.

[115] Jiang Y, Chen C, Xie J, Wang W, Zha X, Lv W, Chen H, Hu Y, Li T, Yu J, Zhou Z, Xu Y, Li G. Radiomics signature of computed tomography imaging for prediction of survival and chemotherapeutic benefits in gastric cancer. The LANCET EBio Medicine. 2018, 36, 171–82.

[116] Esteva A, Kuprel B, Novoa RA, Ko J, Swetter SM, Blau HM, Thrun S. Dermatologist-level classification of skin cancer with deep neural networks. Nature. 2017, 542, 115–18.

[117] Haenssle HA. 58 co-authors. Man against machine: diagnostic performance of a deep learning convolutional neural network for dermoscopic melanoma recognition in comparison to 58 dermatologists. Annals of Oncology. 2018, 29, 1836–42.

[118] Han SS, Park I, Chang SE, Lim W, Kim MS, Park GH, Chae JB, Huh CH, Na J-I. Augmented intelligence dermatology: Deep neural networks empower medical professionals in diagnosing skin cancer and predicting treatment options for 134 skin disorders. Journal of Investigative Dermatology. 2020, 140, 1753–61.

[119] Diabetic Complications. US National Library of Medicine. https://medlineplus.gov/diabetescomplications.html.

[120] Heintzman ND. A digital ecosystem of diabetes data and technology: services, systems, and tools enabled by wearables, sensors, and apps. Journal of Diabetes Science and Technology. 2016, 10, 35–41.

[121] Dudde R, Vering T, Piechotta G, Hintsche R. Computer-aided continuous drug infusion: setup and test of a mobile closed-loop system for the continuous automated infusion of insulin. IEEE Transactions on Information Technology in Biomedicine. 2006, 10, 395–402.

[122] Brown SA. 22 co-authors. Overnight closed-loop control improves glycemic control in a multicenter study of adults with type 1 diabetes. The Journal of Clinical Endocrinology and Metabolism. 2017, 102, 3674–82.

[123] Hetterich C, Pobiruchin M, Wiesner M, Pfeifer D. How Google Glass could support patients with diabetes mellitus in daily life. Studies in Health Technology and Informatics. 2014, 205, 298–302.

[124] Lee H, Song C, Hong YS, Kim MS, Cho HR, Kang T, Shin K, Choi SH, Hyeon T, Kim D-H. Wearable/disposable sweat-based glucose monitoring device with multistage transdermal drug delivery module. Science Advances. 2017, 3, e1601314, DOI: 10.1126/sciadv.1601314.

[125] Mishra SG, Takke AK, Auti ST, Suryavanshi SV, Oza MJ. Role of artificial intelligence in health care. Biochemical Journal. 2017, 15, https://www.tsijournals.com/articles/role-of-artificial-intelligence-in-health-care-13471.html.

[126] Marateb HR, Mansourian M, Faghihimani E, Amini M, Farina D. A hybrid intelligent system for diagnosing microalbuminuria in type 2 diabetes patients without having to measure urinary albumin. Computers in Biology and Medicine. 2014, 45, 34–42.

[127] Cho BH, Yu H, Kim K-W, Kim TH, Kim IY, Kim SI. Application of irregular and unbalanced data to predict diabetic nephropathy using visualization and feature selection methods. Artificial Intelligence in Medicine. 2008, 42, 37–53.

[128] Sánchez CI, Niemeijer M, Dumitrescu AV, Suttorp-Schulten MSA, Abràmoff MD, Van Ginneken B. Evaluation of a computer-aided diagnosis system for diabetic retinopathy screening on public data. Invest Ophthalmol Visual Science. 2011, 52, 4866–71.

[129] Mestman J, Umpierrez G. Gestational diabetes. The Journal of Clinical Endocrinology and Metabolism. 2007, 92, E1, https://doi.org/10.1210/jcem.92.6.9997.

[130] Wu Y-T. 10 co-authors. Early prediction of gestational diabetes mellitus in the chinese population via advanced machine learning. The Journal of Clinical Endocrinology and Metabolism. 2020, dgaa899, https://doi.org/10.1210/clinem/dgaa899.

[131] Shen J. 18 co-authors. An innovative artificial intelligence–based app for the diagnosis of gestational diabetes mellitus (gdm-ai): development study. Journal of Medical Internet Research. 2020, 22, e21573, doi:10.2196/21573.

[132] Liu B. 14 co-authors. Early diagnosis of gestational diabetes mellitus. The Journal of Clinical Endocrinology and Metabolism. 2020, 105, e4264–74.

[133] Huang H-F AI model predicts gestational diabetes risk in first trimester. Endocrinology. 2020; https://www.healio.com/news/endocrinology/20201222/ai-model-predicts-gestational-diabetes-risk-in-first-trimester.

[134] Atzi NS. 9 co-authors. Prediction of Gestational Diabetes based on Nationwide Electronic Health Records. Nature Medicine. 2020, 26, 71–6.

[135] Yan Y, Zhang J-W, Zang G-Y, Pu J. The primary use of artificial intelligence in cardiovascular diseases: what kind of potential role does artificial intelligence play in future medicine?. Journal of Geriatric Cardiology. 2019, 16, 585–91.

[136] Mordecai A. Heart Attack Risk Prediction Using Machine Learning. 2020; https://towardsdatascience.com/heart-disease-risk-assessment-using-machine-learning-83335d077dad.

[137] CDC. Heart Disease Facts. 2019; https://www.cdc.gov/heartdisease/facts.htm.

[138] Felman A What to know about cardiovascular disease. 2019; https://www.medicalnewstoday.com/articles/257484#lifestyle-tips.

[139] Winokur ES, Delano MK, Sodini CG. A wearable cardiac monitor for long-term data acquisition and analysis. IEEE Transactions on Biomedical Engineering. 2013, 60, 189–92.

[140] Yang H-K, Lee J-W, Lee K-H, Lee Y-J, Kim K-S, Choi H-J, Kim D-J Application for the wearable heart activity monitoring system: analysis of the autonomic function of HRV. Engineering in Medicine and Biology Society, 2008. EMBS 2008. 30th Annual International Conference of the IEEE. DOI: 10.1109/IEMBS.2008.4649392.

[141] He DD, Winokur ES, Sodini CG An Ear-worn Continuous Ballistocardiogram (BCG) SENSOR FOR CARDIOVASCULAR MONITORING. Engineering in Medicine and Biology Society (EMBC), 2012 Annual International Conference of the IEEE. DOI: 10.1109/EMBC.2012.6347123.

[142] Kelin HU, Meltendorf U, Reek S, Smid J, Kuss S, Cygankiewicz I, Jons C, Szymkiewicz S, Buhtz F, Wollbrueck A, Zareba W, Moss AJ. Bridging a temporary high risk of sudden arrhythmic death. Experience with the wearable cardioverter defibrillator (WCD). Pacing and Clinical Electrophysiology. 2010, 33, 353–67.

[143] Weng SF, Reps J, Kai J, Garibaldi JM, Qureshi N. Can machine-learning improve cardiovascular risk prediction using routine clinical data?, PLOS ONE, 2017, 12, 2017, https://doi.org/10.1371/journal.pone.0174944.

[144] Dawes TJW, De Marvao A, Shi W, Fletcher T, Watson GMJ, Wharton J, Rhodes CJ, Howard LSGE, Gibbs JSR, Rueckert D, Cook SA, Wilkins MR, O'Regan DP. Machine learning of three-dimensional right ventricular motion enables outcome prediction in pulmonary hypertension: a cardiac MR imaging study. Radiology. 2017, 283, 381–90.

[145] Poplin R, Varadarajan AV, Blumer K, Liu Y, McConnel MV, Corrado GS, Peng L, Webster DR. Prediction of cardiovascular risk factors from retinal fundus photographs via deep learning. Nature Biomedical Engineering. 2018, 2, 158–64.

[146] Fan R, Zhang N, Yang L, Ke J, Zhao D, Cui Q. AI-based prediction for the risk of coronary heart disease among patients with type 2 diabetes mellitus. Scientific reports. 2020, 10, 14457, https://doi.org/10.1038/s41598-020-71321-2.

[147] Hernandez-Silveira M, Ahmed K, Ang -S-S, Zandari F, Mehta T, Weir R, Burdett A, Toumazou C, Brett SJ. Assessment of the feasibility of an ultra-low power, wireless digital patch for the continuous ambulatory monitoring of vital signs. BMJ Open. 2015, 5, e006606, doi: 10.1136/bmjopen-2014-006606.

[148] Kroll RR, Boyd JG, Maslove DM. Accuracy of a wrist-worn wearable device for monitoring heart rates in hospital inpatients: a prospective observational study. Journal of Medical Internet Research. 2016, 18, e253, doi: 10.2196/jmir.6025.

[149] Chen S-T, Lin -S-S, Lan C-W, Hsu H-Y. Design and Development of a Wearable Device for Heat Stroke Detection. Sensors. 2018, 18, 17, doi: 10.3390/s18010017.

[150] Carfagno J AI System Predicts Risk of Cardiovascular Death from ECG Data. 2019; https://www.docwirenews.com/docwire-pick/ai-system-predicts-risk-of-cardiovascular-death-from-ecg-data/.

[151] Kuo P-C, Tsai -C-C, Celi LA. AI Model Aids in TB Detection via Smartphone. Radiological Society of North America. 2020, https://press.rsna.org/timssnet/media/rsna/pdf/2020-snapshot-11-29TB.pdf.

[152] Tey C-K, An J, Chung W-Y. A novel remote rehabilitation system with the fusion of noninvasive wearable device and motion sensing for pulmonary patients. Computational and Mathematical Methods in Medicine. 2017, 5823740, doi: 10.1155/2017/5823740.

[153] Le S, Pellegrini E, Green-Saxena A, Summers C, Hoffman J, Calvert J, Das R. Supervised machine learning for the early prediction of acute respiratory distress syndrome (ARDS). Journal of Critical Care. 2020, 60, 96–102.

[154] Whooley S. Aidoc, Imbio partner on AI to detect blood clots. 2020; https://www.medicaldesignandoutsourcing.com/aidoc-imbio-partner-on-ai-to-detect-blood-clots/#:~:text=Aidoc%20announced%20today%20that%20it,according%20to%20a%20news%20release.

[155] CDC. Heart Failure. 2020; https://www.cdc.gov/hreatdisease/heart_failure.htm.

[156] Conner-Simons A Anticipating heart failure with machine learning. https://news.mit.edu/2020/anticipating-heart-failure-machine-learning–1001.

[157] Burns A, Adeli H. 2017. Wearable technology for patients with brain and spinal cord injuries. Reviews in the Neurosciences. 2017, 28, 913–20.

[158] Burridge JH, Lee ACW, Turk R, Stokes M, Whitall J, Vaidyanathan R, Clatworthy P, Hughes A-M, Meagher C, Franco E, Yardley L. Telehealth, wearable sensors, and the internet: will they improve stroke outcomes through increased intensity of therapy, motivation, and adherence to rehabilitation programs?. Journal of Neurologic Physical Therapy. 2017, 41, S32–8.

[159] CDC. Stroke Facts. https://www.cdc.gov/stroke/facts.htm.

[160] Arbabshirani MR, Fornwalt BK, Mongelluzzo GJ, Suever JD, Geise BD, Patel AA, Moore GJ. Advanced machine learning in action: identification of intracranial hemorrhage on computed tomography scans of the head with clinical workflow integration. npj Digital Medicine. 2018, 1, https://doi.org/10.1038/s41746-017-0015-z.

[161] Murray NM, Unberath M, Hager GD, Hui FK. Artificial intelligence to diagnose ischemic stroke and identify large vessel occlusions: a systematic review. Journal of Neurointerventional Surgery. 2019, 12, http://dx.doi.org/10.1136/neurintsurg-2019-015135.

[162] Kummer S. Can Artificial Intelligence Tell the Difference Between Ischemic and Hemorrhagic Stroke? 2019; https://ysjournal.com/can-artificial-intelligence-tell-the-difference-between-ischemic-and-hemorrhagic-stroke/.

[163] Yu M, Cai T, Huang X, Wong K, Volpi J, Wang JZ, Wong STC. Toward Rapid Stroke Diagnosis with Multimodal Deep Learning. In: Martel AL, et al. ed., Medical Image Computing and Computer Assisted Intervention – MICCAI 2020. MICCAI 2020. Lecture Notes in Computer Science, 2020, 12263, Springer, Cham, https://doi.org/10.1007/978-3-030-59716-0_59.

[164] Avicenna.AI Partners With Canon Medical on AI Stroke Detection. 2020; https://avicenna.ai/partnership-avicenna-ai-canon-medical/.

[165] Marmar CR and 11 co-authors. Speech-based markers for posttraumatic stress disorder in US veterans. Depression and Anxiety. 2019; https://doi.org/10.1002/da.22890.

[166] Czeisler MÉ and 12 co-authors. Mental Health, Substance Use, and Suicidal Ideation During the COVID-19 Pandemic – United States, June 24–30, 2020. CDC. Weekly. 2020, 69, 1049–57.

[167] Ellipsis Health Announces Rising Higher, New Voice-Based App, to Help Healthcare Providers Identify and Manage Behavioral Health at Scale. 2020; https://www.prnewswire.com/news-releases/ellipsis-health-announces-rising-higher-new-voice-based-app-to-help-healthcare-providers-identify-and-manage-behavioral-health-at-scale-301121040.html.

[168] Diagnosis of neurological diseases to benefit from AI. 2020; https://www.ncl.ac.uk/press/articles/latest/2020/09/aiforparkinsonsandalzheimers/

[169] Myszczynska MA. 8 co-authors. Applications of machine learning to diagnosis and treatment of neurodegenerative. Nature reviews. Neurology. 2020, 16, 440–56.

[170] Marquez JR Georgia State Researchers Developing Deep Learning Framework to Map Brain Disorders. 2020; https://news.gsu.edu/2020/08/26/georgia-state-researchers-developing-deep-learning-framework-to-map-brain-disorders/.

[171] Parkinson's Foundation. Statistics. https://www.parkinson.org/Understanding-Parkinsons/Statistics.

[172] Belić M, Bobić V, Badža M, Šolaja N, Đurić-Jovičić M, Kostić VS. Artificial intelligence for assisting diagnostics and assessment of Parkinson's disease-A review. Clinical Neurology and Neurosurgery. 2019, 184, 105442, doi: 10.1016/j.clineuro.2019.105442.

[173] Types of Parkinson's Disease. Parkinson's News Today. https://parkinsonsnewstoday.com/types-of-parkinsons-disease/.

[174] Kim DH, Wit H, Thurston M. Artificial intelligence in the diagnosis of Parkinson's disease from ioflupane-123 single-photon emission computed tomography dopamine transporter scans using transfer learning. Nuclear Medicine Communications. 2018, 39, 887–93.

[175] Lin Z, Dai H, Xiong Y, Xia X, Horng S-J. Quantification assessment of bradykinesia in Parkinson's disease based on a wearable device. Engineering in Medicine and Biology Society (EMBC). Annual International Conference of the IEEE. 2017, 2017, doi: 10.1109/EMBC.2017.8036946.

[176] Delrobaei M, Baktash N, Gilmore G, McIsaac K, Jog M. Using wearable technology to generate objective Parkinson's disease dyskinesia severity score: Possibilities for home monitoring. IEEE Transactions on Neural Systems and Rehabilitation Engineering. 2017, 25, 1853–63.

[177] SpeechVive Efficacy Studies. https://www.purdue.edu/hhs/motorspeechlab/speechvive-efficacy-studies/.

[178] Neubert AP Innovative research helping people with Parkinson's disease communicate better. 2020; https://www.purdue.edu/newsroom/releases/2020/Q4/innovative-research-helping-people-with-parkinsons-disease-communicate-better.html.

[179] Brooks L. Eye Exam Could Lead to Early Parkinson's Disease Diagnosis. Radiological Society of North America. 2020; https://press.rsna.org/timssnet/media/pressreleases/14_pr_target.cfm?ID=2229.

[180] Diaz M, Tian J, Ramirez-Zamora A, Fang R Machine learning for Parkinson Disease diagnosis using fundus eye images. Presented at RSNA 2020; November 29–December 5, 2020. Abstract: IN-1A–07.

[181] What Is Alzheimer's Disease? 2017; https://www.nia.nih.gov/health/what-alzheimers-disease.

[182] What is Alzheimer's Disease? Alzheimer's Association. https://www.alz.org/alzheimers-dementia/what-is-alzheimers.

[183] CDC. Alzheimer's Disease. https://www.cdc.gov/aging/aginginfo/alzheimers.htm.

[184] WHO. Dementia: A Public Health Priority. 2012; http://www.who.int/mental_health/publications/dementia_report_2012/en/.

[185] Rudzicz F. Toward dementia diagnosis via artificial intelligence. Today's Geriatric Medicine. 2016, 9, https://www.todaysgeriatricmedicine.com/archive/MA16p8.shtml.

[186] Astell AJ, Bouranis N, Hoey J, Lindauer A, Mihailidis A, Nugent C, Robillard JM. Technology and dementia: The future is now. Dementia & Geriatric Disorders. 2019, 47, 131–9.

[187] Dallora AL, Eivazzadeh S, Mendes E, Berglund J, Anderberg P. Machine learning and microsimulation techniques on the prognosis of dementia: A systematic literature review. PLOS ONE. 2017, 12, e0179804. https://pubmed.ncbi.nlm.nih.gov/28662070/.

[188] Deetjen UM, Schroeder R. Big data for advancing dementia research: an evaluation of data sharing practices in research. OECD Digit Econ. 2015, https://doi.org/10.1787/20716826.

[189] Chen PH, Lee DD, Yang MH. Data mining the comorbid associations between dementia and various kinds of illnesses using a medicine database. Computers & Electrical Engineering. 2018, 70, 12–20.

[190] Alipour AP, Khademi M. Alzheimer's disease detection using data mining techniques, MRI imaging, Blood-based biomarkeres and neuropsychological tests. Research Journal of Recent Sciences. 2015, 4, 1–5.

[191] Zhang M, Schmitt-Ulms G, Sato C, Xi Z, Zhang Y, Zhou Y, St George-Hyslop P, Rogaeva E. Drug repositioning for Alzheimer's disease based on systematic "omics" data mining. PLOS ONE. 2016, 11, e0168812, https://doi.org/10.1371/journal.pone.0168812.

[192] Mathotaarachchi S, Pascoal TA, Shin M, Benedet AL, Kang MS, Beaudry T, Fonov VS, Gauthier S, Rosa-Neto P. Alzheimer's Disease Neuroimaging Initiative. Alzheimer's Disease Neuroimaging Initiative Identifying incipient dementia individuals using machine learning and amyloid imaging. Neurobiology Aging. 2017, 59, 80–90.

[193] Cheng B, Liu M, Zhang D, Munsell BC, Shen D. Domain transfer learning for MCI conversion prediction. IEEE Transactions on Bio-medical Engineering. 2015, 62, 1805–17.

[194] Korcovelos EAFK, Meltzer J, Hirst G, Rudzicz F. Studying neurodegeneration with automated linguistic analysis of speech data. Alzheimer's and Dementia. Alzheimer's Association. 2017, 13, P164–161P165, https://doi.org/10.1016/j.jalz.2017.06.2612.

[195] Dodge HH, Mattek NC, Austin D, Hayes TL, Kaye JA. In-home walking speeds and variability trajectories associated with mild cognitive impairment. Neurology. 2012, 78, 1946–52.

[196] Schlegl T, Waldstein SM, Bogunovic H, Endstraßer F, Sadeghipour A, Philip AM, Podkowinski D, Gerendas BS, Langs G, Schmidt-Erfurth U. Fully automated detection and quantification of macular fluid in OCT using deep learning. Ophthalmology. 2018, 125, 549–58.

[197] Liu X, Jiang J, Zhang K, Long E, Cui J, Zhu M, An Y, Zhang J, Liu Z, Lin Z, Li X, Chen J, Cao Q, Li J, Wu X, Wang D, Lin H. Localization and diagnosis framework for pediatric cataracts based on slit-lamp images using deep features of a convolutional neural network. PLOS ONE. 2017, 12, e0168606, 10.1371/journal.pone.0168606.

[198] Kim SJ, Cho KJ, Oh S. Development of machine learning models for diagnosis of glaucoma. PLOS ONE. 2017, 12,, e0177726, 10.1371/journal.pone.0177726.

[199] Liu H. 29 co-authors, Development and validation of a deep learning system to detect glaucomatous optic neuropathy using fundus photographs. JAMA Ophthalmology. 2019, 137, 1353–60.

[200] Ruiz HI, Rozema JJ, Saad A, Gatinel D, Rodriguez P, Zakaria N, Koppen C. Validation of an objective keratoconus detection system implemented in a Scheimpflug tomographer and comparison with other methods. Cornea. 2017, 36, 689–95.

[201] Brown JM. 11 co-authors. Automated diagnosis of plus disease in retinopathy of prematurity using deep convolutional neural networks. JAMA Ophthalmology. 2018, 136, 803–10.

[202] Bur AM, Shew M, New J. Artificial Intelligence for the Otolaryngologist: A State of the Art Review. Otolaryngology--head and Neck Surgery : Official journal of American Academy of Otolaryngology-Head and Neck Surgery. 2019, 160, 603–11.

[203] Hughes C, Agrawal S. Machine learning and the future of otolaryngology. ENT & Audiology News. 2020, https://www.entandaudiologynews.com/features/ent-features/post/machine-learning-and-the-future-of-otolaryngology.

[204] Senaras C, Moberly AC, Teknos T, Essig G, Elmaraghy C, Taj-Schaal N, Yu L, Gurcan M. Autoscope: automated otoscopy image analysis to diagnose ear pathology and use of clinically motivated eardrum features. Proc. SPIE. Medical Imaging 2017: Computer-Aided Diagnosis. 101341X, https://doi.org/10.1117/12.2250592.

[205] Ghosh A, Torres JMM, Danieli M, Riccardi G Detection of essential hypertension with physiological signals from wearable devices. Paper presented at the Engineering in Medicine and Biology Society (EMBC), Annual International Conference of the IEEE. 2015; DOI:10.1109/EMBC.2015.7320272.

[206] Goldberg EM, Levy PD. New approaches to evaluating and monitoring blood pressure. Current Hypertension Reports. 2016, 18, 49, 10.1007/s11906-016-0650-9.

[207] Fujikawa T, Tochikubo O, Kura N, Kiyokura T, Shimada J, Umemura S. Measurement of hemodynamics during postural changes using a new wearable cephalic laser blood flowmeter. Circulation Journal. 2009, 73, 1950–5.

[208] Iwasaki W, Nogami H, Takeuchi S, Furue M, Higurashi E, Sawada R. Detection of site-specific blood flow variation in humans during running by a wearable laser Doppler flowmeter. Sensors. 2015, 15, 25507–19.

[209] Lomas N Sight Diagnostics starts selling an AI-based diagnostics device for faster blood tests. 2018; https://techcrunch.com/2018/07/12/sight-diagnostics-launches-an-ai-based-diagnostics-device-for-faster-blood-tests/.

[210] Buescher AV, Cidav Z, Knapp M, Mandell DS. Costs of autism spectrum disorders in the United Kingdom and the United States. JAMA pediatrics. 2014, 168, 721–8.

[211] Daniels J, Haber N, Voss C, Schwartz J, Tamura S, Fazel A, Kine A, Washington P, Phillips J, Winograd T, Feinstein C, Wall DP. Feasibility Testing of a Wearable Behavioral Aid for Social Learning in Children with Autism. Applied Clinical Informatics. 2018, 9, 129–40.

[212] Ramirez-Celis A, Becker M, Nuño M, Schauer J, Aghaeepour N. Van de Water J. Risk assessment analysis for maternal autoantibody-related autism (MAR-ASD): a subtype of autism. Molecular Psychiatry. 2021, https://doi.org/10.1038/s41380-020-00998-8.

[213] Valenza G, Citi L, Gentili C, Lanata A, Scilingo EP, Barbieri R. Characterization of depressive states in bipolar patients using wearable textile technology and instantaneous heart rate variability assessment. IEEE Journal of Biomedical and Health Informatics. 2015, 19, 263–74.

[214] Roh T, Hong S, Yoo H-J Wearable depression monitoring system with heart-rate variability. Engineering in Medicine and Biology Society (EMBC), Annual International Conference of the IEEE. 2014;2014, 562–5; doi: 10.1109/EMBC.2014.6943653.

[215] Luo Y, Eran A, Palmer N, Avillach P, Levy-Moonshine A, Szolovits P, Kohane IS. A multidimensional precision medicine approach identifies an autism subtype characterized by dyslipidemia. Nature Medicine. 2020, 26, 1375–9.

[216] European College of Neuropsychopharmacology. Depression risk detected by measuring heart rate changes. 2020; https://medicalxpress.com/news/2020-09-depression-heart.html.

[217] The American Psychiatric Association. What Is ADHD? 2017; https://www.psychiatry.org/pa tients-families/adhd/what-is-adhd.

[218] Scientists use AI to develop better predictions of why children struggle at school. ADHD in the News. 2018-10-04; https://chadd.org/adhd-in-the-news/scientists-use-ai-to-develop-bet ter-predictions-of-why-children-struggle-at-school/.

[219] Chen M, Li H, Wang J, Dillman JR, Parikh NA, He L. A Multichannel Deep Neural Network Model Analyzing Multiscale Functional Brain Connectome Data for Attention Deficit Hyperactivity Disorder Detection. Radiology: Artificial Intelligence. 2019, https://doi.org/10.1148/ryai.2019190012.

[220] American Psychiatric Association. What Is Schizophrenia? https://www.psychiatry.org/pa tients-families/schizophrenia/what-is-schizophrenia.

[221] Kalmady SV, Greiner R, Agrawal R, Shivakumar V, Narayanaswamy JC, Brown MRG, Greenshaw AJ, Dursun SM, Venkatasubramanian G. Towards artificial intelligence in mental health by improving schizophrenia prediction with multiple brain parcellation ensemble-learning. npj Schizophrenia. 2019, 5, https://doi.org/10.1038/s41537-018-0070-8.

[222] Gheiratmand M, Rish I, Cecchi GA, Brown MRG, Greiner R, Polosecki PI, Bashivan P, Greenshaw AJ, Ramasubbu R, Dursun SM. Learning stable and predictive network-based patterns of schizophrenia and its clinical symptoms. npj Schizophrenia. 2017, 3, 22, https://doi.org/10.1038/s41537-017-0022-8.

[223] Rezaii N, Walker E, Wolff P. A machine learning approach to predicting psychosis using semantic density and latent content analysis. npj Schizophrenia. 2019, 5, https://doi.org/10.1038/s41537-019-0077-9.

[224] Liu X, Faes L, Kale AU, Wagner SK, Jack FJ, Bruynseels A, Mahendiran T, Moraes G, Shamdas M, Kern C, Ledsam JR, Schmid MK, Balaskas K, Topol EJ, Bachmann LM, Keane PA, Denniston AK. A comparison of deep learning performance against health-care professionals in detecting diseases from medical imaging: a systematic review and meta-analysis. The LANCET Digital Health. 2019, https://doi.org/10.1016/S2589-7500(19, 30123–2.

[225] Martin N Artificial Intelligence Is Being Used To Diagnose Disease And Design New Drugs. 2019; https://www.forbes.com/sites/nicolemartin1/2019/09/30/artificial-intelligence-is-being-used-to-diagnose-disease-and-design-new-drugs/#3fe1dae444db.

[226] Caption Health Presents New Research on Caption Guidance Performance at ESC Congress 2020; https://oltnews.com/caption-health-presents-new-research-on-caption-guidance-per formance-at-esc-2020.

[227] Recht MP. 21 co-authors. Using Deep Learning to Accelerate Knee MRI at 3 T: Results of an Interchangeability Study. American Journal of Roentgenology. 2020, 215, 1421–9.

[228] Kooi T, Litijens G, Ginneken B, Gubern-Mérida A, Sánchez CI, Mann R, Heeten A, Karssemeijer N. Large scale deep learning for computer aided detection of mammographic lesions. Medical Image Analysis. 2017, 35, 303–12.

[229] Serag A, Ion-Margineanu A, Qureshi H, McMillan R, Saint Marin M-J, Diamond J, O'Reilly P, Hamilton P. Translational AI and Deep Learning in Diagnostic Pathology. Fronters in Medicine. 2019, https://doi.org/10.3389/fmed.2019.00185.

[230] Kolachalama VB, Singh P, Lin CQ, Mun D, Belghasem ME, Henderson JM, Francis JM, Salant DJ, Chitalia VC. Association of Pathological Fibrosis With Renal Survival Using Deep Neural Networks. Kidney Int'l Reports. 2018, 3, 464–75.

[231] Bresnick J Machine learning, imaging analytics predict kidney function. 2018; https://health itanalytics.com/news/machine-learning-imaging-analytics-predict-kidney-function.

[232] Dhir A, Lee CM Integrating Overlapping Datasets Using Bivariate Causal Discovery. Thirty-Fourth AAAI Conference on Artificial Intelligence. 2020; https://arxiv.org/abs/1910.11356.

[233] Artificial intelligence can spot when correlation means causation. 2020; https://www.ucl.ac.uk/news/2020/feb/artificial-intelligence-can-spot-when-correlation-means-causation.

[234] López-Cortés A, Cabrera-Andrade A, Vázquez-Naya JM, Pazos A, Gonzáles-Díaz H, Paz-y-miño C, Guerrero S, Pérez-Castillo Y, Tejera E, Cristian R, Munteanu CR. Prediction of breast cancer proteins involved in immunotherapy, metastasis, and RNA-binding using molecular descriptors and artificial neural networks. Scientific Reports. 2020, 10, 8515, https://doi.org/10.1038/s41598-020-65584-y.

[235] Kent J. Artificial Intelligence Bests Clinicians in Flagging Prostate Cancer. Health IT Analytics. 2020, https://healthitanalytics.com/news/artificial-intelligence-bests-clinicians-in-flagging-prostate-cancer.

[236] Big Data, Analytics & Artificial Intelligence: The Future of Health Care is Here. UCSF https://www.gehealthcare.com/static/pulse/uploads/2016/12/GE-Healthcare-White-Paper_FINAL.pdf.

[237] Bejnordi BE, Veta M, Van Diest PJ, Van Ginneken B, Karssemeijer N, Litjens G, Van Der Laak JAWM. the CAMELYON16 Consortium, Diagnostic Assessment of Deep Learning Algorithms for Detection of Lymph Node Metastases in Women With Breast Cancer. JAMA. 2017, 318, 2199–210.

[238] Packin NG Consumer Finance and AI: The Death of Second Opinions? 2020; https://papers.ssrn.com/sol3/papers.cfm?abstract_id=3361639.

[239] I think of AI as an automated second opinion. 2019; https://medium.com/@Philips/i-think-of-ai-as-an-automated-second-opinion-785ed3f3e79d.

[240] For a Second Medical Opinion, Concilio Provides the Best of Online and Face-To-Face Consultations. AIT News Desk. 2020; https://aithority.com/technology/life-sciences/for-a-second-medical-opinion-concilio-provides-the-best-of-online-and-face-to-face-consultations/

[241] Iatrogenesis. https://en.wikipedia.org/wiki/Iatrogenesis#:~:text=Iatrogenesis%20is%20the%20causation%20of,intervention%2C%20error%2C%20or%20negligence.

[242] Ivan II. Ivan Illich, Limits to Medicine: Medical Nemesis, the Expropriation of Health, Marion Boyars, New York, NY, 2010.

[243] Null G, Dean C, Feldman M, Rasio D, Smith D Death by Medicine. http://www.webdc.com/pdfs/deathbymedicine.pdf.

[244] Global Burden of Disease Study 2013 Collaborators. Global, regional, and national incidence, prevalence, and years lived with disability for 301 acute and chronic diseases and injuries in 188 countries, 1990–2013: a systematic analysis for the Global Burden of Disease Study 2013. The Lancet. 2015, https://doi.org/10.1016/S0140-6736(15)60692-4.

[245] GBD 2013 Mortality and Causes of Death Collaborators. Global, regional, and national age–sex specific all-cause and cause-specific mortality for 240 causes of death, 1990–2013: a systematic analysis for the Global Burden of Disease Study 2013. The Lancet. 2014, https://doi.org/10.1016/S0140-6736(14)61682-2.

[246] Weber EJ. Triage: Making the simple complex?. Emergency Medicine Journal : EMJ. 2018, 36, 64–5.

[247] Ohshita J AI triages emergency patients and supports destination determination. 2017; https://xtech.nikkei.com/dm/atcl/feature/15/011000049/011600009/.

[248] Miles J, Turner J, Jacques R, Williams J, Mason S. Using machine-learning risk prediction models to triage the acuity of undifferentiated patients entering the emergency care system: a systematic review. Diagnostic and Prognostic Research. 2020, 4, 16, https://doi.org/10.1186/s41512-020-00084-1.

[249] Kang D-Y, Cho K-J, Kwon O, Kwon J-M, Jeon K-H, Park H, Lee Y, Park J, Oh B-H. Artificial intelligence algorithm to predict the need for critical care in prehospital emergency medical services. Scandinavian Journal of Trauma, Resuscitation and Emergency Medicine. 2020, 28, 17, https://doi.org/10.1186/s13049-020-0713-4.

[250] Farahmand S, Shabestari O, Pakrah M, Hossein-Nejad H, Arbab M, Artificial Intelligence-Based B-HS. Triage for Patients with Acute Abdominal Pain in Emergency Department; a Diagnostic Accuracy Study. Advanced Journal of Emergency Medicine. 2017, 1, e5, 10.22114/AJEM.v1i1.11.

[251] Katz C Emergency room triage with AI. 2020; https://www.aidoc.com/blog/emergency-room-triage-with-ai/.

[252] American College of Surgeons. AI outperforms clinicians' judgment in triaging postoperative patients for intensive care. Science daily. 2019; https://www.sciencedaily.com/releases/2019/10/191029182456.htm.

[253] Jha S. Value of Triage by Artificial Intelligence. Academic Radiology. 2020, 27, 153–5.

[254] Sweeney E FDA approves AI stroke application, signaling a shift in triage software oversight. 2018; https://www.fiercehealthcare.com/regulatory/fda-approval-clinical-decision-support-ai-triage-app.

[255] Etherington D FDA-cleared AI-based medical triage tool goes free for customers to help busy radiology diagnostics departments. 2020; https://techcrunch.com/2020/04/30/fda-cleared-ai-based-medical-triage-tool-goes-free-to-help-busy-radiology-diagnostics-departments/.

[256] O'Hear S Infermedica scores $10M Series A for its "AI-driven" diagnosis and triage platform. 2020; https://techcrunch.com/2020/08/06/infermedica/.

[257] Drees J Mayo Clinic to implement AI-powered predictive patient triage platform. 2020; https://www.beckershospitalreview.com/artificial-intelligence/mayo-clinic-to-implement-ai-powered-predictive-patient-triage-platform.html.

[258] Can artificial intelligence provide real value in triaging urgent care patients?. Journal of Urgent care Medicine. 2021, https://www.jucm.com/can-artificial-intelligence-provide-real-value-in-triaging-urgent-care-patients/.

[259] Baker A, Perov Y, Middleton K, Baxter J, Mullarkey D, Sangar D, Butt M, DoRosario A, Johri S. A Comparison of Artificial Intelligence and Human Doctors for the Purpose of Triage and Diagnosis. Frontiers in Artificial Intelligence. 2020, https://doi.org/10.3389/frai.2020.543405.

[260] Rowland SP, Fitzgerald JE, Holme T, Powell J, McGregor A. What is the clinical value of mHealth for patients?. npj Digital Medicine. 2020, 3, 4, 10.1038/s41746-019-0206-x.

[261] Millenson ML, Baldwin J, Zipperer L, Singh H. Beyond dr. google: the evidence on consumer-facing digital tools for diagnosis. Diagnosis. 2018, 5, 105–95.

[262] Medvedovsky V Artificial intelligence-based triage. Using AI to triage patients in a healthcare facility. https://proxet.com/blog/artificial-intelligence-based-triage-using-ai-to-triage-patients-in-a-healthcare-facility/.

[263] Levin S, Toerper M, Hamrock E, Hinson JS, Barnes S, Gardner H, Dugas A, Linton B, Kirsch T, Kelen G. Machine-Learning-Based Electronic Triage More Accurately Differentiates Patients With Respect to Clinical Outcomes Compared With the Emergency Severity Index. Annals of Emergency Medicine. 2018, 71, 565–74.

[264] Yan L and 28 co-authors. A machine learning-based model for survival prediction in patients with severe COVID-19 infection. medRxiv. 2020; doi: https://doi.org/10.1101/2020.02.27.20028027.

[265] Wang M, Xia C, Hunag L, Xu S, Qin C, Liu J. Deep learning-based triage and analysis of lesion burden for COVID-19: a retrospective study with external validation. The Lancet Digital Health. 2020, 2, E605–15.

[266] Lai L, Wittbold KA, Dadabhoy FZ, Sato R, Landman AB, Schwamm LH, He S, Patel R, Wei N, Zuccotti G, Lennes IT, Medina D, Sequist TD, Bomba G, Keschner YG, Zhang H. Digital triage: Novel strategies for population health management in response to the COVID-19 pandemic. Health (Amst). 2020, 8, 100493, 10.1016/j.hjdsi.2020.100493.

[267] Ryynänen O-P, Leppänen T, Kekolahti P, Mervaala E, Töyräs J. Bayesian Network Model to Evaluate the Effectiveness of Continuous Positive Airway Pressure Treatment of Sleep Apnea. Healthcare Informatics Research. 2018, 24, 346, 10.4258/hir.2018.24.4.346.

[268] About cancer treatment methods. 2020; https://translate.google.com/translate?hl=en&sl=ja&u=https://www.ganchiryo.com/prevention/cure.php&prev=search&pto=aue.

[269] Gershgorn D Google is teaming up with a London hospital to inject AI into cancer treatment. 2016; https://qz.com/769974/google-deepmind-cancer-artificial-intelligence-deep-learning-university-college-london-hospital/.

[270] Supporting cancer research and treatment. https://www.ibm.com/watson-health/solutions/cancer-research-treatment.

[271] Zou F-W, Tang Y-F, Liu C-Y, Ma J-A, Hu C-H. Concordance Study Between IBM Watson for Oncology and Real Clinical Practice for Cervical Cancer Patients in China: A Retrospective Analysis. Frontiers in Genetics. 2020, 11, 200, 10.3389/fgene.2020.00200.

[272] Beck JT. 19 co-authors. Cognitive technology addressing optimal cancer clinical trial matching and protocol feasibility in a community cancer practice. Journal of Clinical Oncology. 2017, 35, 10.1200/JCO.2017.35.15_suppl.6501.

[273] Researchers aim artificial intelligence at rising oral cancers with $3.3 million grant from National Cancer Institute. 2021; https://thedaily.case.edu/researchers-aim-artificial-intelligence-at-rising-oral-cancers-with-3-3-million-grant-from-national-cancer-institute/.

[274] Volker N. Whole Genome Sequencing. http://www.genesinlife.org/testing-services/testing-genetic-conditions/whole-genome-sequencing.

[275] Wheeler T, Den Bakker H, Wiedmann M. Whole Genome Sequencing (WGS) 101. https://www.uspoultry.org/foodsafety/docs/WGS_lay_summary072916-01.pdf.

[276] A high-resolution view of the entire genome. https://www.illumina.com/techniques/sequencing/dna-sequencing/whole-genome-sequencing.html.

[277] Staaf J. 20 co-authors. Whole-genome sequencing of triple-negative breast cancers in a population-based clinical study. Nature Medicine. 2019, 25, 1526–33.

[278] G-TECH: General window for advanced cancer treatment. https://gtech.ne.jp/column-5g-ai-change/.

[279] French healthtech startup using AI to treat cardiac arrhythmias raises €23M. 2021; https://siliconcanals.com/news/startups/french-healthtech-startup-volta-medical-funding/.

[280] VOLTA MEDICAL's VX1 AI software for use in atrial fibrillation mapping now FDA cleared. 2020; https://www.prnewswire.com/news-releases/volta-medicals-vx1-ai-software-for-use-in-atrial-fibrillation-mapping-now-fda-cleared-301189550.html.

[281] Arsene C. Artificial Intelligence in Healthcare: the future is amazing. Healthcare Weekly. 2020; https://healthcareweekly.com/artificial-intelligence-in-healthcare/.

[282] Hutson M The Future of the Brain: Implants, Ethics, and AI. 2019; https://www.psychologytoday.com/us/blog/psyched/201912/the-future-the-brain-implants-ethics-and-ai.

[283] Reardon S AI-controlled brain implants for mood disorders tested in people. 2017; https://www.nature.com/news/ai-controlled-brain-implants-for-mood-disorders-tested-in-people-1.23031.

[284] Scientists Want to Use Brain Implants to Tune the Mind. https://www.kavlifoundation.org/science-spotlights/scientists-want-use-brain-implants-tune-mind#.YB3C8OhKhyw.

[285] Pineau J, Montemerlo M, Pollack M, Roy N, Thrun S. Towards robotic assistants in nursing homes: challenges and results. Robotics and Autonomous Systems. 2002, 1048, 1–11.

[286] Coradeschi SC and 14 co-authors. GiraffPlus: Combining social interaction and long term monitoring for promoting independent living. 6th International Conference on Human System Interactions. 2013, 578–85; 10.1109/HSI.2013.6577883.

[287] Louie W-YG, Li J, Mohamed C, Despond F, Lee V, Nejat G. Tangy the Robot Bingo Facilitator: A Performance Review. Journal of Medical Devices. 2015, 9, 020936, https://doi.org/10.1115/1.4030145.

[288] Dimirioglou N, Kardaras D, Barbounaki S Multicriteria Evaluation of the Internet of Things Potential in Health Care: The Case of Dementia Care: IEEE 19th Conference on Business Informatics. 2017, DOI: 10.1109/CBI.2017.34.

[289] Addlesee A. How Dementia Affects Conversation: Building a More Accessible Conversational AI. 2019; https://heartbeat.fritz.ai/how-dementia-effects-conversation-f538d2d9507a.

[290] Artificial Intelligence to help look after patients with dementia. 2020; https://www.maastrichtuniversity.nl/news/artificial-intelligence-help-look-after-patients-dementia.

[291] What If AI Could Advance The Science Surrounding Dementia? [Science And Technology Podcast]. Scientific Foresight. 2020; https://epthinktank.eu/2020/06/11/what-if-ai-could-advance-the-science-surrounding-dementia-science-and-technology-podcast/.

[292] Wathour J, Govaerts PJ, Deggouj N. From manual to artificial intelligence fitting: Two cochlear implant case studies. Cochlear Implants International. 2020, 21, 299–305.

[293] Lakhani M. Why Schizophrenia Is SUCH a Big problem- and how AI can help. 2019; https://medium.com/datadriveninvestor/why-schizophrenia-is-such-a-big-problem-and-how-ai-can-help-472029614e2b.

[294] Fernández-Caballero A, Navarro E, Fernández-Sotos P, González P, Ricarte JJ, Latorre JM, Rodriguez-Jimenez R. Human-Avatar Symbiosis for the Treatment of Auditory Verbal Hallucinations in Schizophrenia through Virtual/Augmented Reality and Brain-Computer Interfaces. Frontiers in Neuroinformatics. 2017, 11, 64, 10.3389/fninf.2017.00064.

[295] Craig TKJ, Rus-Calafell M, Ward T, Leff JP, Huckvale M, Howarth E, Emsley R, Garety PA. AVATAR therapy for auditory verbal hallucinations in people with psychosis: a single-blind, randomised controlled trial. The Lancet Psychiatry. 2017, 2017, 10.1016/S2215-0366(17, 30427-3.

[296] Dellazizzo L, Potvin S, Phraxayayong K, Lalonde P, Dumais A. Avatar Therapy for Persistent Auditory Verbal Hallucinations in an Ultra-Resistant Schizophrenia Patient: A Case Report. Frontiers in Psychiatry. 2018, 9, 131, 10.3389/fpsyt.2018.00131.

[297] More deceased-donor organ transplants than ever. UNOS. 2020; https://unos.org/data/transplant-trends/.

[298] FACTS: DID YOU KNOW? American Transplant Foundation. https://www.americantransplantfoundation.org/about-transplant/facts-and-myths/.

[299] WHO. GKT1 Activity and Practices. https://www.who.int/transplantation/gkt/statistics/en/.

[300] WHO. WHO Task Force on Donation and Transplantation of Human Organs and Tissues. https://www.who.int/transplantation/donation/taskforce-transplantation/en/.

[301] Heinrich J Organ donation – a new frontier for AI? 2017; https://phys.org/news/2017-04-donationa-frontier-ai.html.

[302] Machine learning to improve organ donation rates and make better matches. Innovation, Science and Economic Development Canada. 2020; https://www.ic.gc.ca/eic/site/101.nsf/eng/00070.html.

[303] Kohli T Change of heart: how algorithms could revolutionise organ donations. The New Economy. 2020; https://www.theneweconomy.com/technology/change-of-heart-how-algorithms-could-revolutionise-organ-donations.

[304] Briceño J. Artificial intelligence and organ transplantation: challenges and expectations. Organ Transplantation. 2020, 25, 393–8.

[305] Díez-Sanmartín C, Cabezuelo AS. Application of Artificial Intelligence Techniques to Predict Survival in Kidney Transplantation: A Review. Journal of Clinical Medicine. 2020, 9, 572, https://doi.org/10.3390/jcm9020572.

[306] Organ Donation and Transplantation Statistics. National Kidney Foundation. 2014. https://www.kidney.org/news/newsroom/factsheets/Organ-Donation-and-Transplantation-Stats.

[307] Organ Procurement and Transplantation Network. http://optn.transplant.hrsa.gov/.

[308] Progress Through Research. http://www.usrds.org/2015/view/v2_07.aspx.

[309] Kidney transplant. https://www.mayoclinic.org/tests-procedures/kidney-transplant/about/pac-20384777.

[310] Pahl E. Predicting Deceased Donor Kidney Transplant Outcomes: Comparing KDRI/KDPI with Machine Learning. 2019 American Transplant Congress, B167; https://atcmeetingabstracts.com/abstract/predicting-deceased-donor-kidney-transplant-outcomes-comparing-kdri-kdpi-with-machine-learning/.

[311] Burlacu A, Iftene A, Jugrin D, Popa IV, Lupu PM, Vlad C, Covic A. Using Artificial Intelligence Resources in Dialysis and Kidney Transplant Patients: A Literature Review. BioMed Research International. 2020, https://doi.org/10.1155/2020/9867872.

[312] Sanchez-Madhur R. One Drop Launches New Digital Membership, Bringing Its Multi-Condition Platform To Consumers Worldwide. 2020; https://onedrop.today/blogs/press-releases/digital-membership.

[313] Nusem E, Straker K, Wrigley C. Design Innovation for Health and Medicine, Palgrave MacMillan Pub., 2020.

[314] Day TM, Guillot R. Social Design, Creative Spaces, and the Future of Healthcare. https://www.gensler.com/research-insight/publications/dialogue/33/social-design-creative-spaces-and-the-future-of-healthcare.

[315] Valentine L, Kroll T, Bruce F, Lim C, Moutain R. Design Thinking for Social Innovation in Health Care. An International Journal for All Aspects of Design. 2017, 20, 755–74.

[316] Does AI (Artificial Intelligence) dream of supporting people with disabilities? 2016; https://translate.google.com/translate?hl=en&sl=ja&u=https://www.zaikei.co.jp/article/20160712/316526.html&prev=search&pto=aue.

[317] Shuntermann MF. International Classification of Impairments, Disabilities and Handicaps ICIDH–results and problems. Die Rehabilitation. 1996, 35, 6–13.

[318] WHO. International Classification of Functioning, Disability and Health for Children and Youth (ICFCY). Geneva; 2007. https://apps.who.int/iris/bitstream/handle/10665/43737/9789241547321_eng.pdf?sequence=1.

[319] The Precision Medicine Initiative. https://obamawhitehouse.archives.gov/precision-medicine.

[320] König IR, Fuchs O, Hansen G, Von Mutius E, Kopp MV. What is precision medicine?. European Respiratory Journal. 2017, 50, 1700391, 10.1183/13993003.00391-2017.

[321] What is precision medicine? https://medlineplus.gov/genetics/understanding/precisionmedicine/definition/.

[322] Precision Medicine. https://en.wikipedia.org/wiki/Precision_medicine.

[323] Timmerman L What's in a Name? A Lot, When It Comes to "Precision Medicine". 2013; https://xconomy.com/national/2013/02/04/whats-in-a-name-a-lot-when-it-comes-to-precision-medicine/.

[324] The Precision Medicine Initiative Cohort Program – Building a Research Foundation for twenty-first Century Medicine. 2015; https://www.nih.gov/sites/default/files/research-training/initiatives/pmi/pmi-working-group-report-20150917-2.pdf

[325] What is Precision Medicine? https://www.healio.com/hematology-oncology/learn-genomics/course-introduction/what-is-precision-medicine.

[326] FDA. Precision Medicine. https://www.fda.gov/medical-devices/vitro-diagnostics/precision-medicine.

[327] Traditional vs. Precision Medicine: How They Differ. 2019; https://www.webmd.com/cancer/precision-vs-traditional-medicine#:~:text=Precision%20medicine%20can%20predict%20which,makes%20side%20effects%20less%20likely.

[328] What is the difference between precision medicine and personalized medicine? What about pharmacogenomics? https://medlineplus.gov/genetics/understanding/precisionmedicine/precisionvspersonalized/.

[329] How Artificial Intelligence and Machine Learning are Advancing Precision Medicine? The CE VIEWS; https://theceoviews.com/how-artificial-intelligence-and-machine-learning-are-advancing-precision-medicine/.

[330] Mesko B. The role of artificial intelligence in precision medicine. Journal of Expert Review of Precision Medicine and Drug Development. 2017, 2, https://doi.org/10.1080/23808993.2017.1380516.

[331] Martin N. How Artificial Intelligence Is Advancing Precision Medicine. AI & Big Data. 2019; https://www.forbes.com/sites/nicolemartin1/2019/10/18/how-artificial-intelligence-is-advancing-precision-medicine/#67b1dcaa4d57.

[332] Filipp FV. Opportunities for Artificial Intelligence in Advancing Precision Medicine. Current Genetic Medicine Reports. 2019, 7, 208–13.

[333] Schork NJ. Artificial Intelligence and Personalized Medicine. Cancer Treatment and Research. 2019, 178, 265–83.

[334] Adir O, Poley M, Chen G, Froim S, Krinsky N, Shklover J, Shainsky-Roitman J, Lammers T, Schroeder A. Integrating Artificial Intelligence and Nanotechnology for Precision Cancer Medicine. Advanced Materials: Materials for Precision Medicine. 2020, 32, https://onlinelibrary.wiley.com/doi/10.1002/adma.201901989.

[335] Massive reutilization of Electronic Health Records (EHRs) through AI to enhance clinical research and precision medicine. 2020; https://cordis.europa.eu/article/id/422215-unlocking-electronic-medical-records-to-revolutionise-health-research.

[336] Lee SI. 12 co-authors. A machine learning approach to integrate big data for precision medicine in acute myeloid leukemia. Nature Communications. 2018, 9, 42, https://doi.org/10.1038/s41467-017-02465-5.

[337] Seyhan AA, Carini C. Are innovation and new technologies in precision medicine paving a new era in patients centric care?. Journal of Translational Medicine. 2019, 17, 114. https://translational-medicine.biomedcentral.com/articles/10.1186/s12967-019-1864-9.

[338] Eureka. Ethical dilemmas that artificial intelligence raise in the lab (2019). https://eureka.criver.com/ethical-dilemmas-that-artificial-intelligence-raise-in-the-lab/.

[339] Panch T, Mattie H, Celi LAG. The "inconvenient truth" about AI in healthcare. Npj Digital Medicine, 2019, 2, 10.1038/s41746-019-0155-4.

[340] Raden N Precision medicine and AI – data problems ahead. Diginomica. 2019; https://diginomica.com/precision-medicine-and-ai-data-problems-ahead.

[341] Cloud AI, Edge AI, Endpoint AI What's the Difference? 2020; https://www.arm.com/blogs/blueprint/cloud-edge-endpoint-ai.

[342] AI and Edge Computing Healthcare Technology. http://axiomtek-medical.com/discovering-ai-edge-computing/.

[343] Offin N The future of healthcare: a marriage of AI and edge. Med Tech Onnovation News. 2020; https://www.med-technews.com/medtech-insights/ai-in-healthcare-insights/the-future-of-healthcare-a-marriage-of-ai-and-edge/.

[344] Pazienza A, Mallarfi G, Fasciano C, Vitulano F Artificial Intelligence on Edge Computing: a Healthcare Scenario in Ambient Assisted Living. http://ceur-ws.org/Vol-2559/paper2.pdf.

[345] Bulgaru I. Virtual Reality Innovations Revolutionizing Healthcare in 2021. Healthcare Weekly. 2019; https://healthcareweekly.com/virtual-reality-in-healthcare/.

[346] Arsene C J&J's revolutionary Virtual Reality program for doctors: all you need to know. Healthcare Weekly. 2018; https://healthcareweekly.com/johnson-and-johnson-virtual-reality/.

[347] Fink C VR Brings Dramatic Change To Mental Health Care. 2017; https://www.forbes.com/sites/charliefink/2017/10/25/vr-brings-dramatic-change-to-mental-health-care/?sh=551e34844361.

[348] Kite-Powell J See How This Company Uses Virtual Reality To Change Patient Healthcare. 2018; https://www.forbes.com/sites/jenniferhicks/2018/09/30/see-how-this-company-uses-virtual-reality-to-change-patient-healthcare/?sh=17ed1bdb455e.

[349] Germany-based digital therapeutics startup Kaia Health raises €8.8M to heal your back pain with AI app. 2019; https://aitimes.media/2019/01/31/1980/.

[350] Aichi Medical University, Nippon Organ Pharmaceutical, and FRONTEO start joint research on "AI system for pain treatment support" – joint research of Aichi Medical University, Nippon Organ Pharmaceutical, and FRONTEO; https://robotstart.info/2017/05/09/kibit-health.html.

[351] Harada A, Suzuki K Action oriented bayesian learning of the operating space for a humanoid robot. 2009 IEEE Intl. Conf. On Robotics and Biomimetics (ROBIO). 2009, 633-8, 11153327; DOI: 10.1109/ROBIO.2009.5420600.

[352] Harada A, Suzuki K A robot that perceives pain: Action oriented probabilistic self-modeling for a humanoid robot. http://www.ai.iit.tsukuba.ac.jp/research/004-j.html.

[353] Liu L, Li W, Wu X, Zhou BX Action oriented bayesian learning of the operating space for a humanoid robot. 2009 IEEE Intl Conf on Robotics and Biomimetics (ROBIO). 2009, 778-88, 18669777; DOI: 10.1109/JAS.2019.1911435.

[354] Purves D, Augustine G, Fitzpatrick D, Hall W, Lamantia A-S WL. Neuroscience, 5th, Sunderland, MA, 2012.

[355] Asada M. Consciousness, Ethics, and Legal System for AI and Robots Who Feel Pain. Artificial Intelligence. 2018, http://www.er.ams.eng.osaka-u.ac.jp/Paper/2018/Asada2018a.pdf.

[356] Chen QL, Panksepp JB, Lahvis GP. Empathy Is Moderated by Genetic Background in Mice. PLOS ONE. 2009, https://doi.org/10.1371/journal.pone.0004387.

[357] Chapman CR, Casey KL, Dubner R, Foley KM, Gracely RH, Reading AE. Pain measurement: an overview. Pain. 1985, 22, 1–31.

[358] Sternbach RA. Clinical Aspect of Pain. In: The Psychology of Pain, Ra S, ed. The Psychology of Pain Raven Press, New York, 1978, 223–7.

[359] Tursky B, Jamner LD. Behavioral Assessment of Pain. In: Surwit RS. Et al., ed. Behavioral Treatment of Disease. NATO Conference Series (III Human Factors), 1982, Vol. 19, 329–51. Springer, Boston, MA, https://doi.org/10.1007/978-1-4613-3548-1_23.

[360] Melzak R, Torgerson WS. On the language of pain. Anesthesiology. 1971, 34, 50–9.

[361] Lascaratou C The Language of Pain: Expression or description? https://benjamins.com/catalog/celcr.9; 2007. John Benjamins Publishing Company.

[362] Craig KD. The facial expression of pain. APS Journal. 1992, 1, 153–62.

[363] Wong-Baker Pain Rating Scale. https://wongbakerfaces.org/wp-content/uploads/2016/05/FACES_English_Blue_w-instructions.pdf.

[364] Pain Scale Chart: 1 to 10 Levels. https://www.disabled-world.com/health/pain/scale.php.

[365] Arbour C, Gélinas C. Behavioral and physiologic indicators of pain in nonverbal patients with a traumatic brain injury: an integrative review. Pain Managing Nursing. 2014, 15, 506–18.

[366] Mogil JS. Animal models of pain: progress and challenges. Nature reviews. Neuroscience. 2009, 10, 283–94.

[367] Lanford DJ. 18 co-authors. Coding of facial expressions of pain in the laboratory mouse. Nature Methods. 2010, 7, 447–9.

[368] Chen P-H, Cheong JH, Jolly E, Elhence H, Wager TD, Chang LJ. Socially transmitted placebo effects. Nature Human Behaviour. 2019, 3, 1295–305.

[369] Svetieva E, Frank MG Seeing the Unseen: Evidence for Indirect Recognition of Brief, Concealed Emotion. 2016; https://ssrn.com/abstract=2882197 or http://dx.doi.org/10.2139/ssrn.2882197.

[370] WHO. Climate change and health. 2018; https://www.who.int/news-room/fact-sheets/detail/climate-change-and-health.

[371] Weather & Arthritis Index. https://www.arthritis.org/weather.

[372] National Weather Service. What is the heat index? https://www.weather.gov/ama/heatindex.

[373] Discomfort Index Calculator. https://keisan.casio.com/exec/system/1351058230.

[374] Liu C, Yavar Z, Sun Q. Cardiovascular response to thermoregulatory challenges. American journal of physiology. Heart and circulatory physiology. 2015, 309, H1793–812, 10.1152/ajpheart.00199.2015.

[375] Oshida Y. Putting a New Face on Weather Reports. Weatherwise. 1987, 40, 330–2, https://doi.org/10.1080/00431672.1987.9932079.

[376] Does your pet suffer with allergic skin disease? https://leicesterskinvet.co.uk/allergy.

[377] AAAAI. Allergy Statistics. https://www.aaaai.org/about-aaaai/newsroom/allergy-statistics.

[378] Anaphylaxis. https://www.mayoclinic.org/diseases-conditions/anaphylaxis/symptoms-causes/syc-20351468.

[379] Machine learning allergen detection. https://devpost.com/software/ai-allergy.

[380] Pollen scattering amount forecast map; https://www.pollen.com/map.

[381] National Allergy Bureau. American Academy of Allergy Asthma & Immunology. https://www.aaaai.org/.

[382] My Pollen Forecast – Allergies. https://apps.apple.com/us/app/my-pollen-forecast-allergies/id1244428929.

[383] Switzerland to create world's first real-time pollen monitoring system. 2018; https://translate.googleusercontent.com/translate_c?depth=1&hl=en&prev=search&pto=aue&rurl=translate.google.com&sl=janmt4&u=https://www.swissinfo.ch/eng/society/meteoswiss_switzerland-to-create-world-s-first-real-time-pollen-monitoring-system/44403062&usg=ALkJrhjOWkVtzTXrkBIEb6LMtOZx5fbjkQ.

[384] Selfie shows your health !? New data mining spreads in the United States. 2019; https://translate.googleusercontent.com/translate_c?depth=1&hl=en&prev=search&pto=aue&rurl=translate.google.com&sl=janmt4&u=https://aitimes.media/2019/01/31/1954/&usg=ALkJrhjbj_nQ_PXC1rUECVsoMFsw3gygOg.

[385] Allergies. https://doc.ai/research-trials/allergies.

[386] Chukura R, Sulpovar M. How AI and Weather Data Can Help You Plan for Allergy Season. 2020; https://www.ibm.com/blogs/think/2020/04/the-question-this-spring-is-it-covid-or-allergies/.

[387] Salian I Springing into Deep Learning: How AI Could Track Allergens on Every Block. 2019; https://blogs.nvidia.com/blog/2019/05/01/deep-learning-allergens-ucla/#:~:text=Researchers%20at%20the%20University%20of,over%20traditional%20machine%20learning%20methods.

[388] Wu Y. 11 co-authors. Label-Free Bioaerosol Sensing Using Mobile Microscopy and Deep Learning. American Chemical Society Photonics. 2018, 5, 4617–27.

[389] Makary MA, Daniel M. Medical error–the third leading cause of death in the US. British Medical Journal. 2016, 353, https://doi.org/10.1136/bmj.i2139.

[390] Corny J, Rajkumar A, Martin O, Dode X, Lajonchère J-P, Billuart O, Bézie Y, Buronfosse A. A machine learning–based clinical decision support system to identify prescriptions with a

high risk of medication error. Journal of the American Medical Informatics Association. 2020, 27, 1688–94.

[391] Brennan TA. 9 co-authors. Incidence of Adverse Events and Negligence in Hospitalized Patients – Results of the Harvard Medical Practice Study I. The New England Journal of Medicine. 1991, 324, 370–6.

[392] Drug Interactions. https://www.drugwatch.com/health/drug-interactions/.

[393] CDC. Health, United States 2018. https://www.cdc.gov/nchs/data/hus/hus18.pdf.

[394] Stibich M. Polypharmacy Concerns and Risks. 2020; https://www.verywellhealth.com/what-is-polypharmacy-2223450.

[395] Dagli RJ, Sharma A. Polypharmacy: A Global Risk Factor for Elderly People. Journal of International Oral Health. 2014, 6, i–ii.

[396] Parulekar MS, Rogers CK. Polypharmacy and Mobility. Geriatric Rehabilitation. 2018. 121–9.

[397] Farrell B. 9 co-authors. Health care providers' roles and responsibilities in management of polypharmacy: Results of a modified Delphi. Can Pahram J (Ott). 2018, 151, 396–407.

[398] Polypharmacy. U. S. pharmacist. 2017, 42, 13–14.

[399] AWS Announces Amazon HealthLake. 2020; https://www.businesswire.com/news/home/20201208005491/en/AWS-Announces-Amazon-HealthLake.

[400] Electronic Medical Record. https://www.digital-transformation-real.com/blog/what-changes-will-bring-to-ai-medical-electronic-medical-records.

[401] The kindness of strangers. 2012; https://www.economist.com/babbage/2012/01/18/the-kindness-of-strangers.

[402] The Slow Rise of the Robot Surgeon. MIT Technology Review. https://www.technologyreview.com/video/the-slow-rise-of-the-robot-surgeon/.

[403] About the da Vinci Surgical System. https://www.uchealth.com/services/robotic-surgery/patient-information/davinci-surgical-system/.

[404] Da Vinci Robotic Surgery System. https://www.rwhs.org/services/surgical-clinics/da-vinci-robotic-surgery-system.

[405] Davenport TH, Glaser J. Just-in-time delivery comes to knowledge management. Harvard Business Review. 2002, 80, 107–11. https://europepmc.org/article/med/12140850.

[406] RSIP Vision CEO: AI in medical devices is reducing dependence on human skills and improving surgical procedures and outcome. https://www.prnewswire.com/news-releases/rsip-vision-ceo-ai-in-medical-devices-is-reducing-dependence-on-human-skills-and-improving-surgical-procedures-and-outcome-301109031.html.

[407] Paton F Animals in Health Care Facilities? Consider These 6 Points. 2020; https://nurseslabs.com/animals-health-care-facilities-consider-6-points/.

[408] Alpert JS. Animals in the Hospital. The American Journal of Medicine. 2019, 132, 779–80.

[409] About the medical revolution and communication robots in the medical field, Digital Transformation Channel. 2020; https://translate.google.com/translate?hl=en&sl=ja&u=https://www.digital-transformation-real.com/blog/the-medical-revolution-and-communication-robots-in-the-medical-field&prev=search&pto=aue.

[410] Brown EN, Lydic R, Schiff ND. General anesthesia, sleep, and coma. New England Journal of Medicine. 2010, 363, 27, 2638–50.

[411] Schamberg G, Badgeley M, Brown EN Controlling Level of Unconsciousness by Titrating Propofol with Deep Reinforcement Learning. International Conference on Artificial Intelligence in Medicine AIME. 2020: Artificial Intelligence in Medicine, 26–36; https://arxiv.org/pdf/2008.12333v2.pdf.

[412] Bannister CA, Halcox JP, Currie CJ, Preece A, Irena Spasić I. A genetic programming approach to development of clinical prediction models: A case study in symptomatic cardiovascular disease. PLOS ONE. 2018, https://doi.org/10.1371/journal.pone.0202685.

[413] Ives J. Adopting AI in diagnosis and prognosis of disease could help extend people's lives. 2019; https://www.news-medical.net/news/20190130/Adopting-AI-in-diagnosis-and-prognosis-of-disease-could-help-extend-peoples-lives.aspx.

[414] Huanh S, Yang J, Fong S, Zhao Q. Artificial intelligence in cancer diagnosis and prognosis: Opportunities and challenges. Cancer Letters. 2020, 471, 61–71.

[415] Sultan AS, Elgharib MA, Tavares T, Jessri M, Basile JR. The use of artificial intelligence, machine learning and deep learning in oncologic histopathology. Journal of Oral Pathology & Medicine : Official Publication of the International Association of Oral Pathologists and the American Academy of Oral Pathology. 2020, 49, 849–56.

[416] Mi E Artificial intelligence can help to improve prognosis and treatment for glioblastoma. National Cancer Research Institute. 2020; https://medicalxpress.com/news/2020-10-artificial-intelligence-prognosis-treatment-glioblastoma.html.

[417] Lin HM, Xue XF, Wang XG, Dang SC, Gu M. Application of artificial intelligence for the diagnosis, treatment, and prognosis of pancreatic cancer. Artificial Intelligence in Gastroenterology. 2020, 1, 19–29.

[418] Alexander M. 10 co-authors. Lung cancer prognostic index: A risk score to predict overall survival after the diagnosis of non-small-cell lung cancer. British Journal of Cancer. 2017, 117, 744–51.

[419] Fujima N. 10 co-authors. Machine-Learning-Based Prediction of Treatment Outcomes Using MR Imaging-Derived Quantitative Tumor Information in Patients with Sinonasal Squamous Cell Carcinomas: A Preliminary Study. Cancers. 2019, 11, 800, https://doi.org/10.3390/cancers11060800.

[420] Huang S, Yang J, Fong S, Zhao Q. Mining Prognosis Index of Brain Metastases Using Artificial Intelligence. Cancers. 2019, 11, 1140, https://doi.org/10.3390/cancers11081140.

[421] Kite-Powell J. This App Uses AI To Help Transplant Patients Monitor Organ Health. 2020; https://www.forbes.com/sites/jenniferhicks/2020/08/17/this-app-uses-ai-to-help-transplant-patients-monitor-organ-health/#59956dcf4dd3.

[422] Majithia A. 10 co-authors. Improved Glycemic Outcomes in Adults with Type 2 Diabetes Participating in Continuous Glucose Monitor Driven Virtual Diabetes Clinic: A Prospective Trial. Journal of Medical Internet Research. 2020, 22, e21778, 10.2196/21778.

[423] Cervoni B. Livongo: An Interactive Meter and Coaching Service. 2019; https://www.verywellhealth.com/livongo-meter-and-coaching-for-diabetes-management-4175919.

[424] Lovett L Digital physical therapy company Sword Health raises additional $9M. 2020; https://www.mobihealthnews.com/news/digital-physical-therapy-company-sword-health-raises-additional-9m.

[425] Oshida Y, Farzin-Nia F. Progressive Damage Assessment of TiNi Endodontic Files. In: Yahia L, ed. Shape Memory Implants, Springer, New York NY, 2000, 236–49.

[426] Avati A, Jung K, Harman S, Downing L, Ng A, Shah NH Improving Palliative Care with Deep Learning. IEEE International Conference on Bioinformatics and Biomedicine. 2017; arXiv:1711.06402.

[427] Weng SF, Vaz L, Qureshi N, Kai J. Prediction of premature all-cause mortality: A prospective general population cohort study comparing machine-learning and standard epidemiological approaches. PLOS ONE. 2019, 14, e0214365, 10.1371/journal.pone.0214365.

[428] Rajkomar A. 34 co-authors. Oren E, Scalable and accurate deep learning with electronic health records. npj Digital Medicine. 2018, 1, 18, https://doi.org/10.1038/s41746-018-0029-1.

[429] AI Could Predict Death. But What If the Algorithm Is Biased? 2019; https://www.wired.com/story/ai-bias-predict-death/.

[430] Butler RN. 17 co-authors. Biomarkers of aging: from primitive organisms to humans. The Journals of Gerontology. Series A, Biological Sciences and Medical Sciences. 2004, 59, B560–7.

[431] Johnson TE. Recent results: biomarkers of aging. Experimental Gerontology. 2006, 41, 1243–6.

[432] Galkin F, Mamoshina P, Aliper A, De Magalhães JP, Gladyshev VN, Zhavoronkov A. Biohorology and biomarkers of aging: current state-of-the-art, challenges and opportunities. Ageing Research Reviews. 2020, 60, 101050, 10.1016/j.arr.2020.101050.

[433] Zhavoronkov A, Li R, Ma C, Mamoshina P. Deep biomarkers of aging and longevity: from research to applications. Aging (Albany NY). 2019, 11, 1077–80.

[434] Zhavoronkov A, Kochetov K, Diamandis P, Mitina M. PsychoAge and SubjAge: development of deep markers of psychological and subjective age using artificial intelligence. Aging. 2020, 12, 23548–77.

[435] Schultz MB. 9 co-authors. Age and life expectancy clocks based on machine learning analysis of mouse frailty. Nature Communications. 2020, 11, 4618, https://doi.org/10.1038/s41467-020-18446-0.

[436] Attia ZI. 11 co-authors. An artificial intelligence-enabled ECG algorithm for the identification of patients with atrial fibrillation during sinus rhythm: a retrospective analysis of outcome prediction. The Lancet. 2019, 394, 10201, 861–7, https://doi.org/10.1016/S0140-6736(19)31721-0.

[437] Attia ZI. 11 co-authors. Age and Sex Estimation Using Artificial Intelligence From Standard 12-Lead ECGs. Circulation. Arrhythmia and Electrophysiology. 2019, 12, https://doi.org/10.1161/CIRCEP.119.007284.

[438] Smith CS What to be better at sports? Listen to the machines. 2020; https://www.nytimes.com/2020/04/08/technology/ai-sports-athletes-machine-learning.html.

[439] Make Better Decisions and Improve Performance. https://www.isotechne.com/.

[440] Sport's Sharpest Edge. https://sportlogiq.com/en/.

[441] Get to production AI faster. https://labelbox.com/.

[442] Joshi N. Here's How AI Will Change The World Of Sports! 2019; https://www.forbes.com/sites/cognitiveworld/2019/03/15/heres-how-ai-will-change-the-world-of-sports/#5f36e321556b.

[443] Chu W-C-C, Shih C, Chou W-Y, Ahamed SI, Hsiung P-A. Artificial Intelligence of Things in Sports Science: Weight Training as an Example. Computer. 2019, 52, 52–61, 10.1109/MC.2019.2933772.

[444] Novatchkov H, Baca A. Artificial Intelligence in Sports on the Example of Weight Training. Journal of Sports Science & Medicine. 2013, 12, 27–37.

[445] Skazalski CM, Whiteley R, Hansen C, Bahr R. A valid and reliable method to measure jump-specific training and competition load in elite volleyball players. Scandinavian Journal of Medicine & Science in Sports. 2018, 28, 1579–85.

[446] Takahashi T Five factors that AI will change sports in 2020, from strategy planning to watching. 2020; https://ledge.ai/appier-sports-ai/.

[447] Claudino JG, De Oliveira Capanema D, De Souza TV, Serrão JC, Pereira ACM, Nassis GP. Current Approaches to the Use of Artificial Intelligence for Injury Risk Assessment and Performance Prediction in Team Sports: a Systematic Review. Sports Medicine-Open. 2019, 5, 28, https://doi.org/10.1186/s40798-019-0202-3.

[448] Rossi A, Pappalardo L, Cintia P, Iaia FM, Fernàndez J, Medina D. Effective injury forecasting in soccer with GPS training data and machine learning. PLOS One. 2018, 13, e0201264, 10.1371/journal.pone.0201264.

[449] Veeramani A 3-D technology improves surgery at Yale. 2016; https://yaledailynews.com/blog/2016/12/06/3d-technology-improves-surgery-at-yale/.

[450] 3-D Planning for Facial Plastic Surgery: A New Frontier. 2020; https://www.yalemedicine.org/news/3d-planning-facial-plastic-surgery.

[451] Oshida Y. Aesthetics of the Face – from Face to be seen to Face to show -, Ishiyaku Pub., Tokyo, 2019.

[452] Stuart K, Diaz J Predict Your Results with 3D Imaging. https://www.theplasticsurgerychannel.com/2018/06/28/predict-results-3d-imaging/.

[453] Strazdins E and 7 co-authors. Association Between Mental Health Status and Patient Satisfaction With the Functional Outcomes of Rhinoplasty. JAMA Facial Plastic Surgery. 2018, 20; https://doi.org/10.1001/jamafacial.2018.0001.

[454] Wagner T, Banja J, Hoff T Episode: Managing Health Care AI "Megarisks". AMA Journal of Ethics. 2021; https://journalofethics.ama-assn.org/ethics-talk-podcast-transcript-managing-health-care-ai-megarisks.

[455] O'Connor M A majority of AI studies don't adequately validate. Artificial Intelligence. 2019, https://www.healthimaging.com/topics/artificial-intelligence/majority-ai-radiology-studies-dont-validate-methods.

[456] Kim DW, Jang HY, Kim KW, Shin Y, Seong Ho Park SH. Design Characteristics of Studies Reporting the Performance of Artificial Intelligence Algorithms for Diagnostic Analysis of Medical Images: Results from Recently Published Papers. Korean Journal of Radiology : Official Journal of the Korean Radiological Society. 2019, 20, 405–10.

[457] Park SH, Kressel HY. Connecting technological innovation in artificial intelligence to real-world medical practice through rigorous clinical validation: what peer-reviewed medical journals could do. Journal of Korean Medical Science. 2018, 33, e152, https://doi.org/10.3346/jkms.2018.33.e152.

[458] Graham J. Artificial Intelligence, Machine Learning, And The FDA. 2016; https://www.forbes.com/sites/theapothecary/2016/08/19/artificial-intelligence-machine-learning-and-the-fda/#4aca26121aa1.

[459] Kayyali B, Knott D, Kuiken SV The big-data revolution in US health care: Accelerating value and innovation. 2013. https://www.mckinsey.com/industries/healthcare-systems-and-services/our-insights/the-big-data-revolution-in-us-health-care.

[460] Davenport TH, Dreyer K AI will change radiology, but it won't replace radiologists. Harvard Business Review 2018. https://hbr.org/2018/03/ai-will-change-radiology-but-it-wont-replace-radiologists.

Chapter 9
AI in practice of dentistry

9.1 Dentistry versus medicine

Before getting into the main topics, we need to explain why dentistry and medicine are separated from each other. Beck [1] described the situation by saying that doctors are doctors, and dentists are dentists, and never the twain shall meet and whether you have health insurance is one thing, whether you have dental insurance is another. The current separation between dentistry and medicine needs to visit a historical event. In the barber-surgeon days, dentist skills were among one of the many personal services that barber surgeons provided, like leeching and cupping and tooth extractions. They approached it as a mechanical challenge, to repair and extract teeth. Barber surgery was a denturist. But the dental profession really became a profession in 1840 in Baltimore. When the Maryland University School of Medicine (United States) was founded, Horace Hayden and Chapin Harris approached the physicians at the college of medicine with the idea of adding dental instruction to the medical course there, because they really believed that dentistry was more than a mechanical challenge, that it deserved status as a profession, and a course of study, and licensing, and peer-reviewed scientific consideration. But the physicians rejected their proposal and said that the dentistry is art but not science. Since then, medical and dental educations have been still provided separately almost everywhere in this country and our two systems have grown up to provide care separately, too [1, 2].

For becoming an MD, students study internal medicine, obstetrics, gynecology, surgery, and other medicines. Hence, even if a man with abdominal pain rushes in obstetricians, the obstetrician cannot refuse to perform medical examinations on male patient. Dentists only study dentistry, so you cannot see people with abdominal pain. As a result of separation of dentistry from medicine, there are several distinct characters associated with dentistry, including that (1) Dentistry is a specialty. A general physician (GP) oversees the health of the whole body. If there is an issue with a certain part of it that needs some "expertise," the doctor will refer their patient to a specialist. Some medical specialties include eye doctors, neurologists, ENT (ear, nose, and throat) doctors, chiropractors, gastroenterologists, gynecologists, proctologists, podiatrists, dentists, and more. (2) Dentists take care of something GPs do not have time for. Although a holistic diagnosis and treatment are ideal, unfortunately, there simply is not enough time for one person to learn about all the intricacies of the human body. There is not enough time in the day for a GP to take care of all of their patients' needs, including oral health. That's why dentists are important. We focus on treating periodontal disease and dental caries, while the

https://doi.org/10.1515/9783110717853-010

doctor focuses on heart disease, diabetes, obesity, and treating infections. (3) Aesthetics is a large part of dentistry. In dentistry, while we get people coming in for toothaches, dental caries, and periodontal disease, we also get a lot of people who come in for cosmetic purposes. Aesthetics is a very important part of dentistry [3]. More clearly, dentistry, unlike medicine, requires a special technique. This is because the object of dentistry is not soft tissues, but mainly hard tissues (such as bones and teeth), which are not just hard tissue; they are very special hard tissues that are exposed to the surface of the human body. None of the bones are on the surface of the human body, and if they are exposed on surface, they are very abnormal. However, this abnormal condition is normal for tooth to function properly. Moreover, bones and soft tissues are regenerated and repaired by cells even if they are damaged somewhat, so if the wound is small, it will heal cleanly, while if the tooth is damaged even a little, it will not return to the original. In that respect, we have to think from a different point of view from general medicine, and the technology of treatment becomes special.

But tooth structure belongs still to a same body which might be subjected to examine or diagnose if necessary. The former surgeon general mentioned that we must recognize that (i) oral health and general health are inseparable, and (ii) getting corporation it seems like things such as incorporation between two disciplines are changing, very slowly though [4]. In recent years, the relationship between the oral cavity and the whole body has been reviewed due to the aging of society and the progress of research on medical care. Accordingly, current dental care is changing to the original dental care for diseases in a wide range of oral areas, not just lesions of teeth and surrounding tissues. Such holistic approaches will be discussed in the last portion of this chapter.

As we may notice on the street, there are variety of medical advertisements such as ENT or eye doctors, orthodontics, and implant dentistry. ENT represents ear, nose, and throat specialists and professionally they are called otolaryngologists. Eye doctors can include ophthalmologist (medical and surgical eye care), optometrist (vision care and eye care services), and optician (eyeglasses and contact lenses specialist). Similarly, in dental clinic office, there are various professions working, including dentist(s), dental hygienist(s), dental assistant(s), and dental laboratory technician, who work in the same office or in separate laboratory. Separate and distinct work tasks are assigned and allowed to exercise to each of these professions. Furthermore, we have various dental specialists and disciplines, including periodontics, endodontics, prosthodontics, operative dentistry, orthodontics, implant dentistry, pediatric dentistry, geriatric dentistry, oral surgery, and dental radiology.

In the following sections, we will be looking at AI-powered or AI-assisted dentistry in general, and each specified discipline.

9.2 AI dentistry

9.2.1 Digital dentistry

Digital dentistry refers to the use of dental technologies or devices that incorporates digital or computer-controlled components to carry out dental procedures rather than using mechanical or electrical tools. Since AI dentistry and digital dentistry share same (or similar) digitized raw data, we will look at the digital dentistry first. We need here to clearly differentiate between digitization and digitalization and beyond there is a term DX (digital transformation). Digitization is foundational and refers to creating a digital representation (either 1s or 0s) of physical objects or attributes, serving the connection between the physical world and software. Then computerized systems can then use it for various use cases. On the other hand, digitalization refers to enabling or improving processes by leveraging digital technologies and digitized data and increases productivity and efficiency while reducing costs. DX is really business transformation enabled by digitalization [5, 6]. Figure 9.1 illustrates pyramid relationship among these three techniques [5].

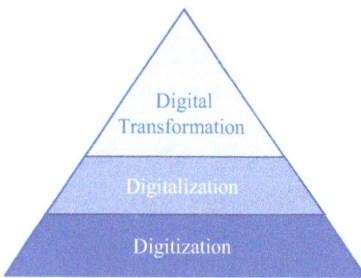

Figure 9.1: Hierarchy of digitized data handling [5].

There can be found various examples of advanced dental technologies based on digitized raw data. Digital dentistry may be defined in a broad scope as any dental technology or device that incorporates digital or computer-controlled components in contrast to that of mechanical or electrical alone. This broad definition can range from the most commonly identified area of digital dentistry – CAD/CAM (computer-aided design/computer-aided manufacturing) – to those that may not even be recognized, such as computer-controlled delivery of nitrous oxide. The goal of digital dentistry is to enhance the efficiency and accuracy of a dental procedure, while producing predictable clinical outcomes. Advances often improve patient comfort, safety, and overall satisfaction. As more practitioners begin to adopt digital technology, innovations in tools, materials, and educational services also increase to support the shift in methods [7, 8]. The main areas of digital dentistry extend from the diagnosis and procedures all the way to virtual networks and training, as listed in the following as typical applications in dental area and all are assumed to incorporate some type of digital components [7–10]. They include (1) CAD/CAM and intraoral imaging for both laboratory-controlled and clinician-controlled; (2) caries diagnosis, (3) computer-aided implant dentistry, including design and fabrication of surgical guides; (4) digital radiography – intraoral and extraoral, including cone beam computed tomography (CBCT); (5) electric and surgical/implant handpieces; (6) lasers (advances in lasers include the expanded use in almost every field of dentistry, such as periodontics, endodontics, surgery, prosthodontics, and general practice); (7) occlusion and temporomandibular joint (TMJ) analysis and diagnosis; (8) photography for both extraoral and intraoral situations; (9) practice and patient record management, including digital patient education; and (10) shade matching. There are many other areas of digital dentistry available. Clear advantages which are common in the above applications must include three things: (i) improved efficiency in both cost and time, (ii) improved accuracy in comparison to previous methods, and (iii) a high level of predictability of outcomes [7].

Digital dentistry has caused disruption on many fronts, bringing new techniques, systems, and interactions that have improved dentistry. Innovation has spurred opportunities for material scientists' future research [10]. Digital dentistry also promises to improve communication at different levels. On enlarged screen displaying the images in an operatory room, patients see the problems their dentist is referring to, so it is easier for those patients to accept the treatment recommendations of the dentist. In addition to that, digital dentistry also allows dental offices to send patients online reminders or notifications about appointments and other related issues. In the future, the applications of digital dentistry will expand even further as the dental practice management software is further exploited in the wake of all the technological developments in the industry [11]. Table 9.1 lists typical software or device applications in digital dentistry [12–23].

9.2.2 Health data management

Biomedical research has recently moved through three stages in digital healthcare: data collection, data sharing, and data analytics. With the explosion of stored health data, dental medicine is edging into its fourth stage of digitization using AI [24]. Health data availability can be viewed from four major perspectives in dentistry: (1) patients – provision of tools and/or data to engage patients and make them part of the decision-making process about their own care, (2) healthcare providers – provision of all relevant data to allow clinicians to make the right decisions about patient care and reflect on their practice, (3) research and academia – provision of oversight of data that enables the health sector to identify hypotheses that may otherwise remain undetected, and (4) policy makers – access to data to make decisions regarding value of care, and to support system efficiency and resource allocation as well as improve safety and quality [24].

Table 9.1: Currently available AI-powered software and devices in dentistry [12–23].

	ORCA Dental AI	Denti.AI	VideaHealth	Pearl	Glidewell.io	Smilecloud	Dental monitoring	Dentistry AI
Easier access to oral/dental care								
Augmented reality dental simulation								
Checking proper tooth brushing								
Accurate margin marking								
Provide analytics across dental offices								
Identify common pathology								
Detection of dental anomalies								
Dental assistant, in general								
Orthodontics monitoring							O	
Digital smile design						O		
Prosthesis design					O			
Patient data analysis				O				
Crown preparation margin analysis				O	O			
Treatment plan suggestion			O	O	O			
Caries detection			O	O	O			O
Anatomy annotation	O			O				
Image enhancement	O			O				
Automated analysis	O	O	O	O				

D Assistant				O		O
DEXvoice				O		O
ChairFill			O			
Kapanu		O				
DX Vision					O	
Second Opinion					O	
Prac. Intelligence					O	
Smart Margin					O	
Smart Toothbrush					O	
DentSim Simulator					O	
TeleDent						O
Intraoral Camera	O				O	O
SureSmile		O				

AI is predicted as the further development of the digital revolution. AI tools in healthcare are rapidly maturing and are expected to have a profound impact in the near future, including, as some argue, the replacement of whole professions [25, 26]. AI in healthcare uses algorithms and software applications to approximate human cognition in the analysis of complex data, including health data [24]. Joda et al. [24] listed up advantages of digital health data in dental research: (1) improved ability to find patient-level information through deliberate use of standardized data, (2) development of a regulatory framework permitting timely access by trusted users, (3) collabo-ration with health information technology (HIT) ecosystem partners to succeed inter-operability using standardized platforms for electronic health records (EHR), (4) promotion of re-use of data already obtained and stored for the benefit of the community, (5) systematic filtering of patients for prospective research applying trial-specific inclusion criteria, (6) high-quality clinical research with large sample sizes and accelerated recruitment, (7) epidemiological surveys for public health-related statistics, (8) early identification of merging health issues and future research needs, (9) simplified translation from dentistry to medicine and laboratory to clinical science, (10) assessment of safety of treatment protocols and pharmaceutical therapy on an ongoing basis, and (12) evaluate the effectiveness and efficiency of healthcare services and policy.

9.2.3 AI dentistry

The evolution of AI makes the analysis of big data possible, which provides reliable information and improves the decision-making process [27]. The following three AI and ML opportunities play important roles in digital dentistry as well as AI dentistry. In fact, most of software or device listed in Table 9.1 has been utilizing one or more elements thereof [28]. (1) Predictive analytics using large volumes of data. Technology supported by ML allows dental practices to not only capture and analyze that data but also quickly compare it with larger data sets (i.e., data points from hundreds or even thousands of past campaigns and consumer interactions) to determine what's most likely to achieve the best marketing outcomes in the future. By deriving these predictive analytics through ML, dental practices can use what they already know about their target patient base and their past campaign performance to create campaigns that are more likely to drive engagement and bring new patients through the door. And they can do it with greater speed and accuracy than ever before because a machine, not a human, is processing these predictive analytics. (2) Real-time engagement with prospective patients through chatbots. Dentists and their office staff cannot possibly be available 24/7. Dental offices are open during only certain hours of the day, and even during those hours, phone lines become busy and current and prospective patients may have to wait to get their questions answered. Consider that 51% of people believe professional service businesses (e.g., a dental practices) should be accessible 24 h a day, and 73% of people say timeliness is critical to good service. When dental practices pair chatbot technology and live chat services with smart patient intake practices, they can more easily provide seamless, around-the-clock support for their website visitors and capture more leads, allowing them to grow their patient volume. (3) Compatibility with voice search. A survey shows that nearly half (48%) of consumers are using voice for general web searches. Hence, dentists should take voice search into account when creating marketing plans for their practices [28].

As AI/ML becomes more integral to how dental offices find, connect with, and attract new patients, dentists should consider how they can leverage these resources for their own practices. With the right technologies and tactics in place, dentists have the potential to increase their patient volume, maximize their revenue, and become the go-to dental care providers in their communities [28]. The dental industry will be keeping its eyes on technology that enables AI, including (1) graphical processing units, which provide heavy computer power, (2) IoT, which collects data from connected devices, (3) advanced algorithms, which could one day predict rare events, and (4) application processing interfaces (API), which is code that enables the adding of AI functionality to existing products [12, 29]. Although it is generally believed that AI-based applications will streamline care, relieving the dental workforce from laborious routine tasks, increasing health at lower costs for a broader population, and eventually facilitate personalized, predictive, preventive, and participatory dentistry. Schwendicke et al. [30] concern that AI solutions have not by large entered routine dental practice, mainly due to (1) limited data availability, accessibility, structure, and comprehensiveness, (2) lacking methodological rigor and standards in their development, and (3) practical questions around the value and usefulness of these solutions, but also ethics and responsibility.

It is estimated that the human misdiagnosis rate of caries from X-rays may be higher than 20% [31]. Such is not the case with AI – machines are not hindered by the inherent human bias and error, nor do they suffer from fatigue. As long as they are provided with the proper data set, and correctly trained on how to recognize patterns, machines can facilitate faster, more efficient outcomes [32]. AI-based diagnosis is gaining importance due to its ability to detect and diagnose lesions which may go unnoticed to the human eye, thereby paving way for a holistic practice. AI can be a useful modality in diagnosis and treatment of lesions of oral cavity and can be employed in screening and classifying suspicious altered mucosa undergoing premalignant and malignant changes. The latest advancement in ANNs can detect and diagnose lesions which can go undetected to human eye. These systems help connect dental healthcare professionals all over the world. With their personal smart devices, the patients can enter the symptoms that they are experiencing and can be made conscious of the most probable diagnosis of the illness. AI is growing tremendously but it can never replace human knowledge, skill, and decision ability. As biological systems are much more complex, it is not only diagnosing but relating it to clinical findings and personalized care is also important in comprehensive care. Although AI can assist in numerous ways, final call should be made by a dentist as dentistry is a multidisciplinary approach [33]. Dentistry has witnessed tremendous advances in all its branches over the past three decades. With these advances, the need for more precise diagnostic tools, specially imaging methods, has become mandatory. From the simple intraoral periapical X-rays, advanced imaging techniques like computed tomography, CBCT, MRI, and US have also found place in modern dentistry. Changing from analogue to digital radiography has not only made the process simpler and faster but also made image storage, manipulation (brightness/contrast, image cropping, etc.), and retrieval easier. The 3D imaging has made the complex cranio-facial structures more accessible for examination and early and accurate diagnosis of deeply seated lesions [34].

Conventional way for machines to perform reading radiographs, first of all, machines must be trained on huge data sets to recognize meaningful patterns, and to understand new information in the form of spoken language, written text or images with proper context and nuance. Then, machines must be able to make intelligent decisions regarding that new information and then learn from mistakes to improve the decision-making process. In order for an AI system to have a practical benefit in the real world, all of this must happen in about the same time that a human being can perform the same task [35]. Prados-Privado et al. [36] conducted systematic review to identify the state of the art of neural networks in caries detection and diagnosis using sources from PubMed, Institute of Electrical and Electronics Engineers (IEEE) Xplore, and ScienceDirect. Unfortunately, many variability complicate the conclusions that can be made about the reliability or not of a neural network to detect and diagnose caries. A comparison between neural network and dentist results is also necessary.

Suppose I have a toothache and visit local dental clinic. Whether it is an advanced X-ray or a laser that diagnoses cavities, some state-of-the-line medical device provides accurate examinations on me. AI presents dentists and patients with appropriate diagnostics and treatments derived from big data. If AI tells me that either tooth extraction or taking a nerve should do, I cannot help but be convinced. Who is the boss here? AI or real human dentist? If the necessary treatment is decided, while AI is judging the part of the infected tooth cavities and the part of the tooth that is not drilled, the therapeutic robot is drilling infected tooth portion accurately, according to the data read by X-ray or others. After the necessary cutting (infected lesion removal), the cut shape is read with a dedicated camera to produce the data of the filling. The data is made by another robot placed next to the interconnecting robot, and it creates a filling with good fit by 3D printing technology. Regarding the material for the filling, AI comprehensively determines the most appropriate material, based on patient's demand, age, aesthetics, and cost, so that any unwanted mistakes can be avoided.

It was mentioned that an innovative inter-professional coordination among clinicians, researchers, and engineers will be the key to AI development in the field of dentistry. Despite the potential misinterpretations and the concern of patient privacy, AI will continue to connect with dentistry from a comprehensive perspective due to the need for precise treatment procedures and instant information exchange. Moreover, such developments will enable professionals to share health-related big data and deliver insights that improve patient care through hospitals, providers, researchers, and patients [13].

9.3 AI dental disciplines

There are advantages and disadvantages associated with AI dentistry. As to advantages, AI dentistry shows (1) performing tasks in almost no time, (2) logical and feasible decisions which results in an accurate diagnosis, and (3) procedures can be standardized; while AI dentistry possesses following shortcomings; (1) mechanism/system complexity, (2) costly setup, (3) adequate training is required, (4) data is often used for both training and testing, leading to "data snooping bias," and (5) the outcomes of AI in dentistry are not readily applicable [37].

9.3.1 Implantology

More than one million dental patients suffer from dental implant failure, which often leads to further oral health concerns, such as damage to surrounding bone, nerves, sinus cavities, and adjacent teeth. These treatment failures cause considerable patient frustration and anguish as well as huge financial losses for practitioners and insurance companies [38]. One of the most critical decisions dentists and patients must make is when they should use fixed denture prosthetics or dental implants. AI can help with evidence-based decision-making. AI can quickly analyze data including underlying conditions, radiological scans including CBCT, and any other patient-related information [39]. When all necessary personal data on a patient's personality, financial situation, systemic health condition, bone quality and quantity, and other factors are input to AI system, AI will be able to reply the long-term prediction and reliability; say 10-year survival rate will be xx% or 20-year survival rate will be yy% for 10 years, it would be much better to explain the implant treatment. Such consultation can avoid unwanted lawsuits and other contingencies. For example, DTX Studio, Blue Sky Plan, NeoGuide System, DIOnavi, Simplant, or Novadontics software are some of the software available for digital implant planning, and some of them can be downloaded without any costs [40]. Sadat et al. [41], using model classifiers such as W-J48, SVM, NNs, K-NN, and Naïve Bayes, have developed a combined predictive model to evaluate the success of dental implants, enabling help surgeons to make a more reliable decision on level of success of implant operation prior to surgery. Peri-implantitis is the main reason of dental implant failure. Papantonopoulos et al. [42] developed an AI model to predict severe bone loss around implants that functioned for at least 2 years, by applying the principal component analysis (PCA) for feature reduction and an artificial neural network of the kind of multilayer perceptron (MLP) regressor. It was concluded that (i) AI models can be the basis for clinical decision-making in the beginning of dental implant therapy by indicating possible future peri-implantitis complications and (ii) a software application with the insertion of a few parameters can provide a risk profile for a scheduled implant [42].

Jaskari et al. [43] based the model on training and DNNs. The model was trained by using a dataset consisting of 3D CBCT, have developed a new model for locating mandibular canals, reducing the likelihood of nerve damage in dental implant operations. Ferro et al. [40] valued the AI technique as an impacting tool for education and training purposes. Education in implant dentistry will evolve quickly over the next decade as technologies already being used in other industries are incorporated into new and innovative learning models. The merging of technological innovations culminating in "digital dentistry" makes the "digital education of digital dentistry" inevitable [40]. Augmentation reality technology is also widely used in medical education and training. In dentistry, oral and maxillofacial surgery is the primary area of use, where dental implant placement and orthognathic surgery are the most frequent applications [29]. The combination of augmented reality (AR) and virtual reality (VR) has been termed mixed reality (MR), such as Microsoft Hololens 2 is an example software, was designed for educational application [40, 44]. Xiaojun et al. [45] introduced a modular software package named Computer Assisted Preoperative Planning for Oral Implant Surgery (CAPPOIS) used in conjunction with a VR-enabled haptic feedback system for teaching implant treatment planning and subsequent surgery, providing the operator with tactile feedback during virtual surgery based on the treatment plan and relevant anatomy at the surgical site.

Due to advanced AI technology and digital processes, the dental digital implant process can be achieved by the workflow; step 1 (accurate and fast digital scan, utilizing the latest advances in AI is statistically more accurate than conventional impressions, can image dynamic occlusion, and can even reproduce realistic colors) → step 2 (the scanning data is imported into a digital CAD software where real-time 3D modeling and manipulation can occur. Dental prosthetics are manipulated in real-time 3D restorations, shaped, positioned, and finalized. This takes all the guesswork out of the process and guarantees a high degree of accuracy) → step 3 (upon completion of the 3D model is complete, 3D data is imported into a 3D printer which prints custom, prosthetic component with tolerances as small as approximately 35 microns when printing with dental grade resin) → step 4 (the printed component is finished, by washing, drying, post-curing, and touching up the final prosthetic form) → step 5 (once the finishing touches are complete, the newly crafted replacement tooth can be fitted intraorally) [46]. This entire process can be done while the patient is sitting on the dental chair.

A robot dentist has carried out the first successful autonomous implant surgery by fitting two new teeth into a female patient's mouth. Two implants were implanted under local anesthesia. The error compared to the pre-operation simulation was within 0.2 to 0.3 mm. One-hour successful procedure raises hopes technology could avoid problems caused by human error and help overcome shortage of qualified dentists [47]. It was reported that Neocis Inc has received US Food and Drug Administration (FDA) clearance to market Yomi®, a robotic guidance system for dental implant procedures. Yomi is a computerized navigational system intended to provide assistance in both the planning (pre-operative) and the surgical (intra-operative) phases of dental implantation surgery. The system provides software to preoperatively plan dental implantation procedures and provides navigational guidance of the surgical instruments [48]. One of key purposes of these autonomous medical operation robots is to eliminate human errors during the operation. There are softwares for monitoring and checking the post-operative condition.

9.3.2 Prosthodontics

Lerner et al. [49] demonstrated the use of AI to fabricate implant-supported monolithic zirconia crowns (MZCs) cemented on customized hybrid abutments to evaluate variables including quality of the marginal adaptation, quality of interproximal contact points, quality of occlusal contacts, chromatic integration, survival and success of MZCs. Total 90 patients (35 males, 55 females; mean age 53.3 ± 13.7 years) restored with 106 implant-supported MZCs were included in the study. The follow-up varied from 6 months to 3 years. It was found that (i) the quality of the fabrication of individual hybrid abutments revealed a mean deviation of $44 \, \mu m$ (± 6.3) between the original CAD design of the zirconia abutment, and the mesh of the zirconia abutment captured intraorally at the end of the provisionalization, (ii) at the delivery of the MZCs, the marginal adaptation, quality of interproximal and occlusal contacts, and aesthetic integration were excellent and (iii) the 3-year cumulative survival and success of the MZCs were 99.0% and 91.3%, respectively. Based on these findings, it was concluded that AI seems to represent a reliable tool for the restoration of single implants with MZCs cemented on customized hybrid abutments via a full digital workflow. Martin et al. [50] developed an AI-based method to assess service quality in the dental prosthesis sector, using strategic options development and analysis (SODA), which is grounded on cognitive mapping, and the measuring attractiveness by a categorical based evaluation technique (MACBETH), a constructivist decision support system was designed to facilitate the assessment of service quality in the dental prosthesis sector. The system was tested, and the results were validated both by the members of an expert panel and by the vice-president of the Portuguese association of dental prosthesis technicians. It was observed that the process developed in this study was extremely versatile and its practical application facilitated the development of an empirically robust evaluation model in this study context; specifically, the profile analyses carried out in actual clinics allowed the cases in which improvements are needed to be identified [50].

CAD/CAM application in dentistry is the process by which is attained finished dental restoration through fine milling process of ready ceramic blocks. CAD/CAM technology is developed to design manufacture of inlays, onlays, crowns, and bridges. In order to operate more efficiently, reduce costs, increase user/patient satisfaction and ultimately achieve profits, many dental offices in the world have their attention focused on implementation of modern IT solutions in everyday practice. In addition to the specialized clinic management software, inventory control, or hardware such as the use of lasers in cosmetic dentistry or intraoral scanning, recently the importance is given to the application of CAD/CAM technology in the field of prosthetic. After the removal of pathologically altered tooth structure, it is necessary to achieve restoration that will be most similar to the anatomy of a natural tooth. Applying CAD/CAM technology on applicable ceramic blocks it can be obtained very quick, but also very accurate restoration, in the forms of inlays, onlays, bridges, and crowns [51]. In order to provide perfect prosthesis to the patient, there are several factors which a dentist has to keep in his mind like anthropological calculations, facial measurements, aesthetics, and patient preferences. The use of computer aided technology for precise fit of prosthesis is yet another breakthrough of AI in dentistry. Also, CAD/CAM based systems are used in dentistry to attain finished dental restorations with great precision. Furthermore, AI based systems are used to design inlays, onlays, crowns, and bridges. This system has replaced the conventional method of casting the prosthesis, reduces time and errors [37].

9.3.3 Orthodontics

Basically, there are two major methods to rearrange dentition: one is a traditional orthodontic mechanotherapy and the other is rather modern aligner-supported tooth movement.

9.3.3.1 Mechanotherapy

Early attempts to use AI in dentistry and orthodontics were in the form of knowledge-based ES. These systems were mainly aimed at helping non-specialist dentists develop diagnoses and treatment plans [52, 53]. For last decades, AI/ML application in orthodontics has progressed slowly, despite promising results in the areas of orthodontic diagnosis, treatment planning, growth evaluations, and in the prediction of treatment outcomes [54]. Currently, general dentists have more advanced ML systems available to them that can diagnose a broader range of orthodontic cases and determine treatment needs and several advanced systems have been developed to help orthodontists diagnose and treatment plan and evaluate treatment outcomes and growth [54, 55].

Faber et al. [56] introduced an interesting story of a 25-year-old male patient with chief complaint of discomfort regarding the appearance of his teeth and chin. He visits an orthodontist and is diagnosed with an Angle class III malocclusion, for which the only treatment is combined orthodontic-orthognathic surgery. From another opinion from a second orthodontist, who confirms the diagnosis as Class III, understanding, nonetheless, that the problem is of intermediate severity and suggests a treatment using aligners. Furthermore, this patient got the third orthodontist says that he would not recommend surgery if the young man was his son but asserts that the treatment is too complex to be performed using aligners and, hence, proposes a treatment using fixed appliances and lower premolars extractions. His journey among orthodontists never ended and he got fourth orthodontist who agreed with the third orthodontist on the use of braces without requiring orthognathic surgery; however, he proposes to extract the third molars and use a skeletal anchorage system to distalize all the lower teeth. Four orthodontists and four different treatment plans. Thus far, no tools exist to lead patients and clinicians out of the decision-making uncertainty, in which they are trapped when they face a condition that has several possible correct treatment options – though some better than others. It is in this context that AI can make a significant contribution [56], since AI has been seen uncountable more orthodontic patients than all these four orthodontists. There could be useful data sources to be fed to AI such as Dental Monitoring^{TM} [57], which is the world's largest collections of dental pictures depicting more than 170 different clinical situations, from hygiene insufficiency to debonded brackets, unseated aligners, tooth movement, or tooth wear. AI/ML is an excellent tool to help orthodontists to choose the best way to move, for instance, a tooth or group of teeth from point A to point B, once the orthodontist instructs the machine where the final position should be. This is useful because orthodontics performed in a totally traditional way – with brackets only – require high manual skill, and many professionals do not have or have not received proper training to develop it [56].

Park et al. [58] compared the accuracy and computational efficiency of two of the latest deep-learning algorithms for automatic identification of cephalometric landmarks, using cephalometric radiographs. Studied DL algorithms were YOLOv3 [59] and SSD [60]. It was found that (i) the YOLOv3 algorithm outperformed SSD in accuracy for 38 of 80 landmarks; while the other 42 of 80 landmarks did not show a statistically significant difference between YOLOv3 and SSD, (ii) error plots of YOLOv3 showed not only a smaller error range but also a more isotropic tendency, (iii) the mean computational time spent per image was 0.05 s and 2.89 s for YOLOv3 and SSD, respectively and (iv) YOLOv3 showed approximately 5% higher accuracy compared with the top benchmarks in the literature, concluding that (v) between the two latest DL methods applied, YOLOv3 seemed to be more promising as a fully automated cephalometric landmark identification system for use in clinical practice.

Studies conducted by Kunz et al. [61] and Hwang et al. [62] showed excellent accuracy in identifying the landmarks similar to the trained human examiners using a specialized AI algorithm and DL-based automated identification system respectively. Moreover, excellent results with automated skeletal classification with lateral cephalometry based on the AI Model were reported [63]. The results of the above-mentioned studies indicate that, these systems prove to be a viable option for repeatedly identifying multiple cephalometric landmarks [64]. Kunz et al. [61] created an automated cephalometric X-ray analysis using a specialized AI algorithm to compare the accuracy of this analysis to the current gold standard (analyses performed by human experts) and to evaluate precision and clinical application of such an approach in orthodontic routine. It was reported that there were almost no statistically significant differences between humans' gold standard and the AI's predictions, while differences between the two analyses do not seem to be clinically relevant. It was then concluded that an AI algorithm able to analyze unknown cephalometric X-rays at almost the same quality level as experienced human examiners (current gold standard), claiming that their study is one of the first to successfully enable implementation of AI into dentistry, in particular orthodontics, satisfying medical requirements [61].

AI has been used to diagnose [65] and establish accurate treatment planning in osteoarthritis in the TMJ [66] or in orthognathic surgery [64]. Choi et al. [67] developed a new AI model for surgery/non-surgery decision and extraction determination and evaluated the performance of this model. The sample for the study consisted of 316 patients in total. Of the total sample, 160 were planned with surgical treatment and 156 were planned with non-surgical treatment. The AI model of ML consisted of 2-layer NN with one hidden layer. The learning was carried out in three stages, and four best performing models were adopted. Using these models, decision-making success rates of surgery/non-surgery, surgery type, and extraction/non-extraction were calculated. It was reported that the success rate of the model showed 96% for the diagnosis of surgery/non-surgery decision and showed 91% for the detailed diagnosis of surgery type and extraction decision, suggesting that the AI model using neural network ML could be applied for the diagnosis of orthognathic surgery cases. KöK et al. [68] determined cervical vertebrae stages (CVS) for growth and development periods by the frequently used seven AI classifiers, and compared the performance of these algorithms with each other. It was reported that a mean accuracy of 77.02%, using AI algorithms for determining the growth and development by CVS when applied on the cephalometric radiographs.

In some cases, prior to orthodontic treatment, tooth needs to be extracted. The decision for tooth extraction can be made by AI technology. It was reported that ANN model was applied for deciding if extractions are necessary using lateral cephalometric radiographs and the results were promising [69]. Jung et al. [70] showed 92% accuracy using AI expert system for deciding on permanent tooth extraction, using lateral cephalometric radiographs. Khanagar et al. [64] evaluated these results suggestive that the AI modes were effective and accurate in predicting the need for extraction. Li et al. [71] employed MLP ANNs to predict orthodontic treatment plans, including the determination of extraction–nonextraction, extraction patterns, and anchorage patterns. It was reported that (i) the neural network can output the feasibilities of several applicable treatment plans, offering orthodontists flexibility in making decisions, (ii) the neural network models show an accuracy of 94.0% for extraction–nonextraction prediction, and (iii) the accuracies of the extraction patterns and anchorage patterns are 84.2% and 92.8%, respectively, indicating that the proposed method based on artificial neural networks can provide good guidance for orthodontic treatment planning for less experienced orthodontists. Jung et al. [70] constructed an AI/ES for the diagnosis of extractions using neural network ML and evaluated the performance of this model. The subjects included 156 patients. Input data consisted of 12 cephalometric variables and an additional 6 indexes. Output data consisted of 3 bits to divide the extraction patterns. Four neural network ML models for the diagnosis of extractions were constructed using a back-propagation (BP) algorithm and were evaluated. It was obtained that the success rates of the models were 93% for the diagnosis of extraction versus nonextraction and 84% for the detailed diagnosis of the extraction patterns, suggesting that (i) AI/ES with neural network ML could be useful in orthodontics and (ii) improved performance was achieved by components such as proper selection of the input data, appropriate organization of the modeling, and preferable generalization.

It is important to note that AI models are limited and have drawbacks. They should be used only after careful considerations. Like any statistical model, the ML algorithms are based on assumptions and have limitations. If used incorrectly, they can give misleading information [54]. Given the large-scale deployment of EHR, secondary use of EHR data will be increasingly needed in all kinds of health services or clinical research. Botsis et al. [72] reported some data quality issues encountered in a survival analysis of pancreatic cancer patients. Incompleteness was the leading data quality issue; many study variables had missing values to various degrees. Inaccuracy and inconsistency were the next common problems. Hence, the quality of data is very important [72]. Data with a lot of noise, missing information, and more variables than observations can result in poor models [54]. Moreover, the phenomena called overfitting occurs when a model is trained too many times on too few observations [73]. Such models perform poorly when introduced to new data. Keeping that in mind, orthodontists should understand that these AI models are meant to assist with the clinical judgment and not to substitute for the knowledge and expertise of humans [54]. Faber et al. [56] pointed out that AI today completely ignores the existence of oral diseases [74, 75] and possible previous health treatments that may affect the prescription of orthodontic corrections, either with aligners or fixed appliances [76]. Patients with periodontitis seem to be more interested in correcting the alignment of their teeth, as pathological tooth migration is a common consequence of periodontitis. However, performing orthodontic movement with active disease is contraindicated [56]. Thus, it is essential that an orthodontist prepares a proper anamnesis, examines the patient, makes a diagnosis, and only then prescribes the appropriate treatment before performing it. More often than not, orthodontics is performed after essential endodontic, periodontal, restorative, and other treatments. Another limitation of AI algorithms being implemented today is that they do not incorporate patients' facial analysis, their proportions, and aesthetics. There is a direct interaction between orthodontic dental movements and facial aesthetics. Only a qualified orthodontist can perform these analyses because tooth movement in any direction of the space is commonly connected with facial and smile aesthetics. In addition, facial analysis is the first step toward determining whether dentofacial deformities are present and thereby the possibility of surgical orthodontic corrections [56].

In the future, information such as the shape of the root, occlusal force, jaw movement, muscle strength of the lips and cheek muscles, jawbone bone density, and bone metabolism activity will be included in the tooth movement simulation, and a new phase will finally be developed in orthodontic practice by further improving the predictive intelligence [77]. The greatest goal of digitalization technology in orthodontic treatment at present is to introduce biomechanical methods that integrate function and morphology, that is, to increase the science of diagnosis. In addition, it is necessary to advance the digitization of a lot of biological information (motion, power, quality, etc.) and to advance the simulation itself, but biomechanical analysis is absolutely necessary for the integration and interpretation of morphological information and motion information. Currently, motion analysis, which is not often used due to the complexity of calculations, and new analytical models that change property values over time, should also be developed for predicting bone and cartilage metabolism and estimating optimal occlusal conditions. Moreover, advanced simulations required in the orthodontic field are very useful tools in other medical areas such as orthopedics, plastic surgery, rehabilitation departments, and industrial areas including medical support robots. Digital imaging and communications in medicine (DICOM) is the standard for the communication and management of medical imaging information and related data. DICOM is most commonly used for storing and transmitting medical images enabling the integration of medical imaging devices such as scanners, servers, workstations, printers, network hardware, and picture archiving and communication systems (PACS) from multiple manufacturers. It has been widely adopted by hospitals and is making inroads into smaller applications like dentists' and doctors' offices [78]. Standard triangle language or standard tessellation language (STL) is a file format native to the stereo-lithography CAD software created by 3D Systems. This file format is supported by many other software packages; it is widely used for rapid prototyping, 3D printing and CAM [79]. Hence, in particular, if the connection between DICOM data and STL files can be easily performed, and various simulations are possible, it will be directly linked from diagnosis to the production of treatment device or equipment [77].

There are several 3D wire bending tabletop-type machines have been developed such as DIWEIRE [80] or KWC-082F [81]. Both computer numerical control (CNC) wire bender systems provide desktop production. The wire is freely machined according to the curve drawn by scalable vector graphic (SVG) or other equivalent programs. There could be some amount of unavoidable springbacks, depending on the bending material, thickness, and angle of bending. Such springbacks are measured by a prior calibration and adjusted for the final bending procedure. As well known, NiTi wire as a typical orthodontic archwire exhibits a relatively large springback [82]. These techniques can be applied to time-consuming archwire bending procedures.

9.3.3.2 Aligner

AI-driven customized orthodontic treatment is becoming popular among young (in particular, female) patients. With the help of these 3D scans, aligners can be printed, and treatment can be customized. After these printed aligners a data algorithm is created that intelligently decides how the teeth or tooth of the patients should be moved, how much pressure should be applied and even also recognize the pressure points for that specific teeth. The AI-conjugated aligners not only provide precise treatment but also reduce the chances of error and time for treatment [37].

There are several types of clear aligner systems commercially available. The Invisalign system [83] is one of advanced clear aligner systems. The Invisalign system is used for straightening teeth with a series of custom-made aligners for each patient. It is a combination of proprietary virtual modeling software, rapid manufacturing processes and mass customization, and custom-made aligners with patented aligner material. This is a custom-made aligner that is interchanged roughly every two weeks for a period of six to eighteen months, or longer depending on the severity of misalignment. Similar to a mouth or dental retainer (which is designed to keep teeth from shifting out of place), this clear aligner is used for orthodontic treatment as a technique to move and properly align teeth for a beautiful smile. This clear aligner is usually computer generated from a mold (or impression) of the patient's teeth – taken by either a dentist or an orthodontist – and the fitting is unique to each patient only. Comparing to the conventional braces, there can be several advantages; the clear aligner system is one of the most convenient methods to straighten teeth, requiring patients to make minor life changes to accommodate the orthodontic treatment; for instance, patients can take out the aligners when brushing, flossing, eating, or drinking – a significant advantage over cumbersome braces that not only complicate oral care, but also are especially prone to damage when eating hard or chewy foods (i.e., corn on the cob, crunchy taco shells, and beef jerky) [84]. In general, clear aligner systems may correct some common dental problems including gapped teeth, mild bite problems (including overbite, under bite, crossbite, or open bite), and crowded teeth. One serious drawback of this technology is that it currently cannot fix severe malocclusion or a badly aligned bite and not always effective in moving and aligning tooth roots in the back of the mouth (molars), especially when gaps are present. Unlike fixed braces, many patients take out their aligners frequently, increasing the risk of misplacing the plastic tray and delaying orthodontic progress [84]. It is the automated extension of AI-assisted teledentistry which has been used in orthodontics since the 1990s [85]. Estai et al. [85] evaluated benefits of integrating teledentistry into routine health services, by presenting an overview of the evidence for the effectiveness and economic impact of teledentistry and reported that the most convincing published evidence regarding the efficacy of teledentistry was provided by studies on pediatric dentistry, orthodontics, and oral medicine. Treatment times depend on a number of factors and all virtual treatment monitoring can dectect progress, breakages and aligner fit (tracking) against the proposed treatment plan. If all other factors are favorable, we should see a reduction in aligner unseats and be able to assess breakages quickly. Katyal [86] showed that aligner patients are more compliant with prescribed wear time when AI technologies are used – the compliance rose from 50% to 90%, indicating that we should be finishing treatments earlier and more predictably [86].

Besides the "Invisalign" system, there are other teledentistry-originated orthodontic treatments such as "Smile Mate" [87] or "Oh my teeth" [88]. Usually, orthodontics had to be continued for at least 1–2 years, and it was a burden to come to the clinic by replacing the mouthpiece. There may be a lot of people who hesitated thinking about going to the dental office for a long time though it was a problem of the tooth line. These applications are designed to realize orthodontic treatments while receiving support on a smartphone at home. Only one time clinic visit is required for taking tooth mold with a 3D scanner. The scanned tooth pattern is immediately 3Ded, allowing to know tooth condition in real-time base. Later, while looking at the 3D images sent to smartphone at home, the patient will hold a meeting with the dentist about treatment plans during Zoom's video session. Choose a correction plan that is suitable for the patient and complete the online purchase agreement. Start straightening using the mouthpiece sent to realign dentition. New mouthpieces will be mailed sequentially according to the plan, so there is no need to come to the clinic to replace the mouthpiece. In addition, a dedicated assistant will support the patient online every day (if necessary). In order to realize high-quality orthodontics only online method, the timing of replacing the mouthpiece is determined by the dentist or dental hygienist by the user taking a photo of the mouthpiece and submitting it online. In the traditional way, taking photos with a standard mouthpiece is difficult, and submitted photos are not able to determine whether to replace the mouthpiece by camera stabilizing or different angles, which has been a major challenge for online orthodontics. Hence, tooth-lined shooting application equipped with AI was developed [87] and, by using AI to detect image stabilizes and identify the position of teeth, anyone can easily shoot with high accuracy according to the standard.

9.3.3.3 Outcome prediction

The prediction of orthognathic treatment outcome is an important part of orthognathic planning and the process of patient's informed consent (IC). The predicted results must be presented to the patients prior to treatment in order to assess the treatment's feasibility, optimize case management, and increase patient understanding and acceptance of the recommended treatment [89]. Smith et al. [90] investigated perceived differences in the ability of current software to simulate the actual outcome of orthognathic surgery. Ten difficult test cases with vertical discrepancies and "retreated" them using the actual surgical changes were chosen. Five programs, including Dentofacial Planner Plus, Dolphin Imaging, Orthoplan, Quick Ceph Image, and Vistadent were evaluated, by using both the default result and a refined result created with each program's enhancement tools. Three panels (orthodontists, oral-maxillofacial surgeons, and laypersons) judged the default images and the retouched simulations by ranking the simulations in side-by-side comparisons and by rating each simulation relative to the actual outcome on a 6-point scale. The results were (i) for the default and retouched images, Dentofacial Planner Plus was judged the best default simulation 79% and 59% of the time, respectively, and its default images received the best (lowest) mean score (2.46) on the 6-point scale. It also scored best (2.26) when the retouched images were compared, but the scores for Dolphin Imaging (2.83) and Quick Ceph (3.03) improved, (ii) retouching had little impact on the scores for the other programs, and (iii) although the results show differences in simulation ability, selecting a software package depends on many factors. It was also mentioned that (iv) performance and ease of use, cost, compatibility, and other features such as image and practice management tools are all important considerations [90].

Fudalej et al. [91] reviewed systematically the orthodontic literature to assess the effectiveness of a prediction of outcome of orthodontic treatment in subjects with a Class III malocclusion. A structured search of electronic databases, as well as hand searching, retrieved 232 publications concerning the topic. Following application of inclusion and exclusion criteria, 14 studies remained. Among other data, sample ethnicity, treatment method, age at the start and completion of treatment, age at follow-up, outcome measures, and identified predictors were extracted from the relevant studies. Thirty-eight different predictors of treatment outcome were identified: 35 cephalometric and three derived from analysis of study casts. Prediction models comprising three to four predictors were reported in most studies. However, only two shared more than one predictor. Gonial angle was identified most frequently in five publications. The studies were of low or medium quality. It was concluded that, due to the large variety of predictors and differences among developed prediction models, the existence of a universal predictor of outcome of treatment of Class III malocclusions is questionable [91]. Cephalometrics is a routine part of the diagnosis and treatment planning process and also allows the clinician to evaluate changes following orthognathic surgery. Traditionally cephalometry has been employed manually; nowadays, computerized cephalometric systems are very popular. Cephalometric prediction in orthognathic surgery can be done manually or by computers, using several currently available software programs, alone or in combination with video images. Both manual and computerized cephalometric prediction methods are two-dimensional and cannot fully describe three-dimensional phenomena. Today, three-dimensional prediction methods are available, such as three-dimensional computerized tomography (3DCT), 3D magnetic resonance imaging (3DMRI), and surface scan/cone-beam CT. Based on this background, Kolokitha et al. [89] discussed the different methods of cephalometric prediction of the orthognathic surgery outcome. It was mentioned that (i) the manual methods of cephalometric prediction of the orthognathic outcome are time consuming; while computerized methods facilitate and speed the performance of the visualized treatment objective, (ii) both manual and computerized cephalometric prediction methods are two-dimensional and will always have limitations, because they are based on correlations between single cephalometric variables and cannot fully describe a three-dimensional biological phenomenon, and (iii) despite the promising capabilities of 3D technology there is not yet a reliable technique for orthognathic prediction, although, the different method of prediction are useful tools for orthognathic surgery planning and facilitates patient communication [89].

9.3.4 Periodontics

Periodontal diseases are one of the most common oral diseases affecting the mankind, being mainly responsible for the early loss of teeth [64]. Various studies have been done to ascertain AI technology application to diagnose and predict periodontal diseases. Lee et al. [92] developed a computer-assisted detection system based on a deep convolutional neural network (CNN) algorithm and evaluated the potential usefulness and accuracy of this system for the diagnosis and prediction of periodontally compromised teeth (PCT). It was reported that (i) with the DL algorithm, the diagnostic accuracy for PCT was 81.0% for premolars and 76.7% for molars, and (ii) using 64 premolars and 64 molars that were clinically diagnosed as severe PCT, the accuracy of predicting extraction was 82.8% (95% CI, 70.1–91.2%) for premolars and 73.4% (95% CI, 59.9–84.0%) for molars. It was then concluded that (iii) the deep CNN algorithm was useful for assessing the diagnosis and predictability of PCT, and (iv) with further optimization of the PCT dataset and improvements in the algorithm, a computer-aided detection system can be expected to become an effective and efficient method of diagnosing and predicting PCT. Krois et al. [93] applied deep CNNs to detect periodontal bone loss (PBL) on panoramic dental radiographs, by synthesizing a set of 2001 image segments from panoramic radiographs. For comparison, six dentists assessed the image segments for PBL. It was found that (i) averaged over 10 validation folds the mean (SD) classification accuracy of the CNN was 0.81 (0.02), (ii) mean sensitivity and specificity were 0.81 (0.04), 0.81 (0.05), respectively, (iii) the mean accuracy of the dentists was 0.76 (0.06), but the CNN was not statistically significant superior compared to the examiners and (iv) mean sensitivity and specificity of the dentists was 0.92 (0.02) and 0.63 (0.14), respectively. It was concluded that (v) a CNN trained on a limited amount of radiographic image segments showed at least similar discrimination ability as dentists for assessing PBL on panoramic radiographs, and (vi) dentists' diagnostic efforts when using radiographs may be reduced by applying ML-based technologies.

9.3.5 Endodontics

The success of root canal treatment mainly depends on accuracy of working length determination [64]. The issue of final apical preparation size remains controversial despite considerable clinical and in vitro research. The astute clinician must be aware of the research before choosing any instrumentation system because the informed clinician's decision must be guided by the best available evidenced-based information. Baugh et al. [94] reviewed article generated a Medline-based search strategy to disclose these studies and provides a critique and summary of the results and mentioned that the prognosis of the treatment can only be ensured when instrumentation terminates at the apical constriction. Saghiri et al. [95] developed a new approach for locating the minor apical foramen (AF) using feature-extracting procedures from radiographs and then processing data using an ANN as a decision-making system. It was found that analysis of the images from radiographs (test samples) by ANN showed that in 93% of the samples, the location of the AF had been determined correctly by false rejection and acceptation error methods, concluding that (i) ANNs can act as a second opinion to locate the AF on radiographs to enhance the accuracy of working length determination by radiography, and (ii) ANNs can function as a decision-making system in various similar clinical situations. Similarly, Saghiri et al. [96] evaluated the accuracy of the ANN in a human cadaver model in an attempt to simulate the clinical situation of working length determination. Obtained results showed that (i) there were significant differences between data obtained from endodontists and ANN and data obtained from endodontists and real measurements by stereomicroscope after extraction, (ii) the correct assessment by the endodontists was accurate in 76% of the teeth and (iii) ANN determined the anatomic position correctly 96% of the time, indicating that (iii) ANN was more accurate than endodontists' determinations when compared with real working length measurements by using the stereomicroscope as a gold standard after tooth extraction, and (iv) the ANN is an accurate method for determining the working length.

Detection of vertical root fractures (VRFs) in their initial stages is a crucial issue, which prevents the propagation of injury to the adjacent supporting structures, so that designing a suitable NN-based model could be a useful method to diagnose the VRFs. Johari et al. [97] designed a probabilistic neural network (PNN) to diagnose the VRFs in intact and endodontically treated teeth of periapical and CBCT radiographs and compared the efficacy of these two imaging techniques in the detection of VRFs. It was found that (i) in the periapical radiographs, the maximum accuracy, sensitivity and specificity values in the three groups were 70.00%, 97.78%, and 67.7%, respectively and (ii) these values in the CBCT images were 96.6%, 93.3%, and 100%, respectively, at the variance change range of 0.1–0.65. Based on these findings, it was concluded that (iii) the designed NN can be used as a proper model for the diagnosis of VRFs on CBCT images of endodontically treated and intact teeth; in this context, CBCT images are more effective than similar periapical radiographs, and (ii) limitations of the study are the use of sound one-rooted premolar teeth without carious lesions and dental fillings and not simulating the adjacent anatomic structures. Fukuda et al. [98] evaluated the use of a CNN system for detecting VRF on panoramic radiography. It was mentioned that (i) of the 330 VRFs, 267 were detected. Twenty teeth without fractures were falsely detected and (ii) recall was 0.75, precision 0.93, and F measure 0.83, concluding that the CNN learning model has shown promise as a tool to detect VRFs on panoramic images and to function as a CAD tool.

9.3.6 Pediatric dentistry

Reviewing the history, concept, and the latest application of AI in the field of pediatric dentistry, Zakirulla et al. [99] mentioned that the various applications of AI in pediatric dentistry include ANNs, genetic algorithms (GA), and fuzzy logic. Over the years, AI has changed how dentistry is practiced with computer-based diagnosis and radiological applications gaining enormous momentum. In pediatric dentistry AI has many potential applications which would change the face of behavioral pediatric practice in future [100]. Pain control with AI enabled devices is the new, smarter way toward injection-free pedodontic practice [101]. The various 4D goggles, movies, animations, and VR-based games can be used as a behavior modification aid effectively for pediatric patients. Thus, a dramatic improvisation of AI-based pediatric dentistry would change the way we practice as well as teach [100]. Although MLCs have already demonstrated strong performance in image-based diagnoses, analysis of diverse and massive EHR data remains challenging. Liang et al. [102] applies an automated NLP system using DL techniques to extract clinically relevant information from EHRs. In total, 101.6 million data points from 1,362,559 pediatric patient visits presenting to a major referral center were analyzed to train and validate the framework. It was demonstrated that (i) high diagnostic accuracy across multiple organ systems and is comparable to experienced pediatricians in diagnosing common childhood diseases, and (ii) the study provides a proof of concept for implementing an AI-based system as a means to aid physicians in tackling large amounts of data, augmenting diagnostic evaluations, and to provide clinical decision support in cases of diagnostic uncertainty or complexity, indicating that although this impact may be most evident in areas where healthcare providers are in relative shortage, the benefits of such an AI system are likely to be universal [102].

9.3.7 Oral cancers

The WHO [103] reported that (i) there are an estimated 657,000 new cases of cancers of the oral cavity and pharynx each year, and more than 330,000 deaths, (ii) oral cancers include the main subsites of lip, oral cavity, nasopharynx, and pharynx and have a particularly high burden in South Central Asia due to risk factor exposures, and (iii) a comprehensive approach is needed for oral cancer to include health education and literacy, risk factor reduction, and early diagnosis. AI technology has been used for detecting cancers [104, 105]. Oral squamous cell carcinoma (OSCC) is a common type of cancer of the oral epithelium. Early detection and accurate outline estimation of OSCCs would lead to a better curative outcome and a reduction in recurrence rates after surgical treatment. Confocal laser endomicroscopy (CLE) records sub-surface micro-anatomical images for in vivo cell structure analysis and recent CLE studies showed great prospects for a reliable, real-time ultrastructural imaging of OSCC in situ. Based on this background, Aubreville et al. [106] evaluated a novel automatic approach for OSCC diagnosis using DL technologies on CLE images and compared to textural feature-based ML approaches that represent the current state of the art. It was mentioned that the present approach is found to outperform the state of the art in CLE image recognition with an AUC of 0.96 and a mean accuracy of 88.3% (sensitivity 86.6% and specificity 90%). AI technology has also been used for predicting postoperative facial swelling after extraction of teeth [64]. Patients' postoperative facial swelling following third molars extraction may have both biological impacts and social impacts. Zhang et al. [107] evaluated the accuracy of ANNs in the prediction of the postoperative facial swelling following the impacted mandibular third molars extraction. The improved conjugate grads BP algorithm combining with adaptive BP algorithm and conjugate gradient BP algorithm together was used. It was reported that the accuracy of this model was 98.00% for the prediction of facial swelling following impacted mandibular third molars extraction, indicating that the present AI model is approved as an accurate method for prediction of the facial swelling following impacted mandibular third molars extraction.

9.3.8 Radiology

Advanced breakthroughs in image recognition techniques using AI and media researcher statements had led to the impression of AI being equated with the demise of radiologists. However, the work performed by radiologists includes many other tasks apart from image recognition that require common sense and general intelligence for solving tasks that cannot be achieved through AI. A thorough knowledge regarding the adaptation of technology will not only help in better and precise patient care but also reducing the work burden of the radiologist [108, 109]. Based on this current status, Yaji et al. [108] reviewed to focus on AI and its application in dentomaxillofacial radiology perspective. As to advantages of AI, there should include (i) it is a powerful tool to identify patterns, predict behavior or events, or categorize objects, (ii) improve radiology departmental workflow through precision scheduling, identify patients most at risk of missing appointments, and empower individually tailored exam protocols, and (iii) ML directly with medical data can help in preventing the errors due to cognitive bias. At the same time, as many other technologies, AI possesses some limitations as many other technologies, including (i) it requires a very huge and sound data base of knowledge, if not may result in inappropriate answers when presented with images outside of their knowledge set; for example, when image techniques not appropriate or if there are any artifacts may result in faulty interpretation of image, (ii) it may not adapt with new imaging software or new machine immediately, (iii) not all the algorithms used are apt for clinical application, and (iv) more trials to recommend the analytic programs for different scenarios [108, 109]. Based on these features associated with AI technology in general, types of AI applications can be divided into (1) detection: to identify an anomaly within an image (e.g., a nodule), (2) segmentation: to isolate a structure from the remainder of the study (e.g., defining the boundary of an organ); and (3) classification: to assign an image or lesion within an image is assigned to a category (e.g., is presence or absence of pulmonary embolism on a CT scan).

DL technique was applied to medical field for the teeth detection and classification of dental periapical radiographs, which is important for the medical curing and postmortem identification [110]. CNN is one of the most popular models used today. This neural network computational model uses a variation of MLPs and contains one or more convolutional layers that can be either entirely connected or pooled. These convolutional layers create feature maps that record a region of image which is ultimately broken into rectangles and sent out for nonlinear processing. CNN is a method to perform a multi-class detection and classification, but it needs a lot of training data to get a good result if used directly. It was reported [111] that, for small training dataset, compared to the precision and recall by training a 33-class (32 teeth and background) state-of-the-art CNN directly, the proposed approach reaches a high precision and recall of 95.8% and 96.1% in total, which is a big improvement in such a complex task. A novel solution based on CNNs was applied to panoramic radiographs. A separate testing set of 222 images was used to evaluate the performance of the system and to compare it to the expert level. It was obtained that (i) for the teeth detection task, the system achieves the following performance metrics: a sensitivity of 0.9941 and a precision of 0.9945, (ii) for teeth numbering, its sensitivity is 0.9800 and specificity is 0.9994, (iii) experts detect teeth with a sensitivity of 0.9980 and a precision of 0.9998 and their sensitivity for tooth numbering is

0.9893 and specificity is 0.999, and (iv) the detailed error analysis showed that the developed software system makes errors caused by similar factors as those for experts. Based on these findings, it was concluded that (v) the method has the potential for practical application and further evaluation for automated dental radiograph analysis, (vi) computer-aided teeth detection and numbering simplify the process of filling out digital dental charts, and (vii) automation could help to save time and improve the completeness of electronic dental records [111]. Chen et al. [112] proposed an application of faster regions with convolutional neural network features (faster R-CNN) in the TensorFlow tool package to detect and number teeth in dental periapical films. It was reported that (i) both precisions and recalls exceed 90% and the mean value of the Intersection-Over-Union (IOU) between detected boxes and ground truths also reaches 91%, (ii) three dentists are also invited to manually annotate the test dataset (independently), which are then compared to labels obtained by our proposed algorithms, indicating that machines already perform close to the level of a junior dentist. Application of CNN algorithms for detection and diagnosis of dental caries on periapical radiographs demonstrated considerably good performance [113] and similar results were reported using deep learning model designed for the detection and localization of dental lesions in near-infrared transillumination (NIT) images [114] and using of near-infrared-light transillumination (NILT) images for diagnosing dental caries and showed that the performance of this AI-based models was satisfactory [115].

There are still more successful implementations, including diagnosing proximal caries using the bitewing radiographs with AI-based ANN model [116], predicting root caries [117], detecting apical lesion with CNN model [118], detection of Sjögren's syndrome (SjS) on CT images with CNN model [119], diagnosing maxillary sinusitis on panoramic radiography [120, 121], diagnosing lymph node metastasis on CT images [122], diagnosing extra nodal extension of cervical lymph node metastases in CT images [117], diagnosis and detection of osteoporosis using the deep convolutional neural network (DCNN) based computer-assisted diagnosis (CAD) systems [123, 124], diagnosing and classifying the root morphology of mandibular first molars on panoramic radiographs and dental cone-beam CT (CBCT) [125], and dental radiology in general [37, 64, 126].

Closing this section, it should be worthy to pay attention conclusive remarks by Tandon et al. [37] and Khanagar et al. [64], saying as follows. Applications of AI technology in every area are growing day by day. These systems bring terrific value to the table by improving the accuracy of diagnosis, enhancing clinical decision-making, and predicting the treatment prognosis which can help clinicians in rendering best quality care to their patients. While in no way it can replace the role of dentist as dental practice is not about diagnosis of disease, but it also includes correlation with various clinical findings and provides treatment to the patient. Nevertheless, a clear understanding of the techniques and concepts of AI surely have an advantage in the coming future. We soon hope to see AI to be completely implied in orthodontics, endodontics and restorative dentistry (reconstructive surgeries). Although AI is widely used in various fields of dentistry, some specialties such as pedodontics and oral pathology still lack the development and application of AI technology. The only limitations to use of AI presently are the availability of insufficient and inaccurate data. Hence, it is the responsibility of dentists and clinicians to focus on collecting and entering the authentic data in their database that will be fully utilized for AI in dentistry in near future [37, 64].

9.4 AI dental laboratory and technicians

The dental industry has experienced unprecedented change in the last decade with rise of digital technologies-based AI methods and significant developments in restorative materials (such as monolithic zirconia: zirconium oxide ceramics) [127]. Until recently, the utilization of AI in dentistry was largely untapped because of the difficulty of transferring this amazing technology to real-life solutions. Now, however, AI is enabling the delivery of precision-fit crowns – directly from the innovative dental lab or the next-generation in-office mill. This is accomplished through generative adversarial networks, or GANs [128]. A GAN is a class of ML frameworks [129, 130]. In the past, to design a CAD/CAM crown, the technician would begin by using one standard template chosen from a limited library. But because there are many variables in teeth, the challenge was in producing each new crown to suit the needs of the individual patient, with a narrow scope of information available to create the design. Researchers recognized that if they could capture all the important data that is implicitly in each doctor-accepted crown – and store the information and constantly add to it, so that a growing number of successful crowns are the basis for design – it could improve the precision, functionality, and accuracy of restorations. The concept for AI-produced restoration design was born [128].

Amid advancing AI technologies employed in dental laboratory, there are still important roles of human dental technicians. According to Budny [131], there are unique three professions possessed by dental technicians, including art, science, and technology. First of all, an art is needed because each restoration is unique to each patient. The restoration must imitate or improve the beauty and the function of the patient's natural dentition and be in harmony with the rest of the system. The technician's greatest challenge is to create a restoration that looks and feels completely natural in the patient's mouth. Next, a science that is advancing rapidly. In order to fabricate the fixed or removable dental prostheses, dental technicians must have a keen knowledge and understanding of tooth anatomy, masticatory functions and the materials and processes utilized in the creation of such devices. A variety of high-tech materials, such as zirconia, ceramics (i.e., lithium disilicate, feldspatic porcelains), plastics (i.e., PMMA, acrylics, composite resins), and metal alloys (i.e., metal substructures, implants, attachments, wires) are utilized in dental laboratories. And finally, the last recognized profession is driven by technology today more than ever. In the past decade, technology has taken over dentistry. In fact, the biggest dental advancements came from the field of dental laboratory technology in the form of CAD/CAM technologies. Today, CAD/CAM is utilized as an integral part of the laboratory's everyday practice. With these technological advances and those to come, there is a very high demand for dental designers who can design and assist in manufacturing of CAD/CAM restorations. Because of these advances, aspiring dental lab technicians should take courses in computer skills and programming [131]. After receiving proper training, dental lab technicians can be certified in next specialty areas: crowns and bridges, ceramics, partial dentures, complete dentures, implants, and orthodontic appliances.

There are several AI softwares available to fabricate crowns and other appliances. Glidewell Dental is using AI to automatically generate advanced dental restorations, designed to provide perfect fit and ideal function, while exceeding aesthetic expectations. This is accomplished through GANs. The approach uses sophisticated algorithms to reduce the amount of manual CAD adjustments required to produce restorations, saving CAD technicians a significant amount of time in the laboratory. In addition, when doctors choose to mill crowns in their practice with the glidewell.io™ In-Office Solution (Glidewell Direct; Irvine, CA, USA), the design is AI-enabled, which allows dentists to spend more time treating patients and less time at the computer [132]. Dental Support Co., Ltd. with cooperation from 9DW Co., Ltd. has developed a method that can complete the design of single tooth in 20 s by utilizing AI technology, in particular, which employs IYO as a core system for AI engine that has the strength of being able to perform various identifications such as speech recognition, natural language processing, and image recognition in one engine [133]. Responding to the digitization needs of dentists and dental laboratories, Kulzer offers more than just digital-compatible materials, also deploying digital tools and software such as 3D scanners, CAD software, 3D printers, and milling machines to contribute to further growth in the dental industry. The development of CAD software with incorporated AI makes it possible to output nearly perfectly designed bridge data after scanning an oral cavity. A laboratory would only need to put the finishing touches on the design. The goal is to drastically cut the time needed for bridge design work from 15 min to just 30 s. This innovation will help optimize operations at laboratories struggling with labor shortages and make it possible to provide patients with speedier treatment [134]. Conventionally, artificial teeth (especially ceramics) created during tooth treatment have been produced in an analog process in which dentists take photos of the patient's oral cavity and dental technicians look at the photo data and faithfully reproduce it in a way that adjusts the color tone with other teeth in the oral cavity. With this method, the aesthetics of artificial teeth depended greatly on the experience and sensibility of dental technicians, and quality instability was a major issue. Using AI, development of a color determination system to reproduce the original tooth tone in denture production has been initiated. When color matching task becomes possible with AI, the burden on dental technicians will be reduced, leading to a productivity improvement [135, 136].

9.5 Clinical recording and data security

According to Sir William Osler, "Variability is the law of life, and as no two faces are the same, so no two bodies are alike, and no two individuals react alike and behave alike under the abnormal conditions which we know as disease" [137]. Sir William Osler, MD (1849–1919), was one of the founding professors in Johns Hopkins School of Medicine and has been called the "Father of Modern Medicine." As stated clearly, each patient visits clinic to see discuss with MD or DDS about his/her chief compliant(s). Following each patient visit, physicians must draft a detailed clinical summary. Taking and keeping patient's record is important, particularly with EHRs, these records must be digitized, so that these valuable medical data can be fed to AI system.

There are two SOAPs sharing same acronym: one is SOAP (Simple Object Access Protocol) which is a message protocol allowing distributed elements of an application to communicate through such as an API [138] and the other SOAP is the one that we are interested in here.

9.5.1 SOAP

In order to facilitate a standard method for providing patient information, clinicians use the SOAP (Subjective–Objective–Assessment–Plan) format for both writing notes and presenting patients on rounds. A SOAP note is information about the patient, which is written or presented in a specific order, which includes certain components. SOAP notes are used for admission notes, medical histories, and other documents in a patient's chart. Many hospitals use electronic medical records, which often have templates that plug information into a SOAP note format. Most healthcare clinicians including nurses, physical and occupational therapists, and doctors use SOAP notes [139–141].

A SOAP note consists of four sections including subjective, objective, assessment, and plan.

9.5.1.1 Subjective
Subjective section of your documentation refers to subjective observations that are verbally expressed by the patient, such as information about feeling and symptoms and how they've been since the last review in their own words. If the patient mentions multiple symptoms you should explore each of them, having the patient

describe them in their own words. When considering what to include in the subjective section of your SOAP notes remember the mnemonic OLDCHARTS. Each letter stands for a question to consider when documenting symptoms. Consider the following: Onset (determine from the patient when the symptoms first started), Location (if pain is present, location refers to what area of the body hurts), Duration (how long has the pain or symptom been experienced for?), Character (character refers to the type of pain, such as stabbing, dull or aching), Alleviating factors (determine if anything reduces or eliminates symptoms and if anything makes them worse), Radiation (in addition to the main source of pain, does it radiate anywhere else?), Temporal patterns (temporal pattern refers to whether symptoms have a set pattern, such as occurring every evening), and Symptoms associated (in addition to the chief complaint, determine if there are other symptoms) [141, 142].

9.5.1.2 Objective
The second section of a SOAP note involves objective observations, which means factors you can measure, see, hear, feel, or smell. This is the section where you should include vital signs, such as pulse, respiration, and temperature. Information from a physical exam including color and any deformities felt should also be included. Results of diagnostic tests, such as lab work and x-rays can also be reported in the objective section of the SOAP notes. Objective observations should include (i) appearance (e.g., the patient appeared to be very pale and in significant discomfort); (ii) vital signs (blood pressure, pulse rate, respiratory rate, SpO_2, and temperature); (iii) fluid balance as an assessment of the patient's fluid intake and output (including oral fluids, nasogastric fluids/feed, intravenous fluids, urine output, vomiting, and drain output/stoma output); (iv) clinical examination findings; and (v) investigation results including recent lab results (e.g., blood tests/microbiology) and imaging results (e.g., chest X-ray/CT abdomen) [141].

9.5.1.3 Assessment
An assessment is the diagnosis or condition the patient has. In some instances, there may be one clear diagnosis (or differential diagnosis), which will be based on the information collected in the previous two sections. In other cases, a patient may have several things wrong. There may also be other times where a definitive diagnosis is not yet made, and more than one possible diagnosis is included in the assessment.

9.5.1.4 Plan
The last section of a SOAP note is the plan, which refers to how you are going to address the patient's problem raised during the review. It may involve ordering additional tests to rule out or confirm a diagnosis. It may also include treatment that is prescribed, such as medication or surgery. The plan may also include

further investigations (e.g., laboratory tests, imaging), treatments (e.g., medications, intravenous fluids, oxygen, and nutrition), referrals to specific specialties, review date/time, frequency of observations and monitoring of fluid balance, and planned discharge date (if relevant).

Analyzing and decomposing the problems into extractive and abstractive subtasks, exploring a spectrum of approaches according to how much they demand from each component, Krishna et al. [142] observed that the performance improves constantly as the extractive subtask is made more complex and suggested that the best performing methods are (i) extracts noteworthy utterances via multi-label classification, assigning each to summary section(s), (ii) clusters noteworthy utterances on a per-section basis, and (iii) generates the summary sentences by conditioning on the corresponding cluster and the subsection of the SOAP sentence to be generated [143].

9.5.2 POS/POMR

Problem-oriented system (POS) focuses on a patient's medical problems and a series of work systems that strive for the best patients care. This system is not just about making medical records in new ways, but is also to correct it as a complete scientific medical record of the patient to provide a useful mechanism. POS consists of the following three stages: Stage I: Production of the problem-oriented medical record (POMR), Stage II: Auditing POMR, and Stage III: Revision of records in POMR [143, 144]. POMR is a way of recording patient health information in a way that is easy for physicians to read and revise. The basic idea of POMR is to equip doctors with the ability to understand the patient's medical history. Since its introduction to the medical world, POMR has been an important resource for supporting patients with chronic illnesses and other complicated medical problems. POMR was also the world's first electronic health record. The patient will start by telling the doctor or nurse of their symptoms and other issues [145]. POMR has five important components: (1) database (the database contains the patient's medical history, including their lab results, x-rays, and physical exam results); (2) problem list (this is a complete problem list outlining the patient's medical issues after the hospital or clinic admitted them; it will also include information from the database); (3) initial plans (based on the problem list, the physician will then write out a complete plan of action for the patient's care); (4) daily progress notes (the clinic will then update the POMR with the patient's progress as well as their medical problems, which could be one or multiple problems); and (5) discharge summary (finally, the discharge summaries will outline the patient's care over time, i.e., from the point where you admitted them to their stay at the clinic/hospital) [145–147].

9.5.3 PKC

Aiming that the ultimate goal with patients should be to couple the knowledge of the unique patient to the knowledge in the literature and get the best possible match, Weed et al. [148, 149] developed a concept of problem-knowledge coupling (PKC). The PKC system was developed to help overcome the inherent limitations of the human mind in decision-making when faced with a complex set of data, the norm in most medical situations. The coupler principle is simple: gather a large number of variables (medical history findings, physical examination findings, laboratory data) and use a computer to sort them into all the diagnostic or treatment possibilities for that patient's unique clinical situation. The logic is combinatorial rather than probabilistic or algorithmic. Probabilistic logic would cause us to miss the rare possibility, and algorithmic logic forces an either–or decision, but in fact, there may be two simultaneous choices (migraine and muscle contraction headache) [150]. When the coupler session is saved, the user is presented with a comment dialogue box. The final document is then saved in the patient coupler record. The subjective and objective findings with the diagnostic or management options selected for consideration, plus my plan summary, may then be displayed as an encounter report. It is a simple step to copy this report into the EMR under the history of present illness and vital signs. There is no dictation, and typing is minimal. The patient leaves with a copy of that note and the printed options information. Traditionally, the term "medical record" was used, indicating a record of diagnosis and treatment which are designed and conducted for physicians' convenience. And it is characterized by the event-oriented record with chronological progresses of what is happening on the patient. In contrast, the modern record recognized as "health record" is a record regarding the patient's health conditions and characterized by the problem-oriented nature. The description purposes of the health records should include (i) for conducting evidence-based medical care (EBM), (ii) for sharing information with other medical personal(s), (iii) for conducting sufficient explanation to and obtaining IC from the patient, and (iv) for proving medical and legal justification of the physician.

Getting the right diagnosis is a key aspect of healthcare – it provides an explanation of a patient's health problem and informs subsequent healthcare decisions. The diagnostic process is a complex, collaborative activity that involves clinical reasoning and information gathering to determine a patient's health problem. As shown in Figure 9.2, several important elements are involved toward the reliable healthcare decisions, in which the health information technology (health IT) and tools are playing an important role [151]. Health IT covers a broad range of technologies used in healthcare, including EHRs, clinical decision support, patient engagement tools, computerized provider order entry, laboratory and medical imaging information systems, health information exchanges, and medical devices. Health IT plays key roles in various aspects of the diagnostic process: capturing information about a patient that informs the diagnostic process, including the clinical history and interview, physical exam, and diagnostic testing results; shaping a clinician's workflow and decision-making in the diagnostic process; and facilitating information exchange [151].

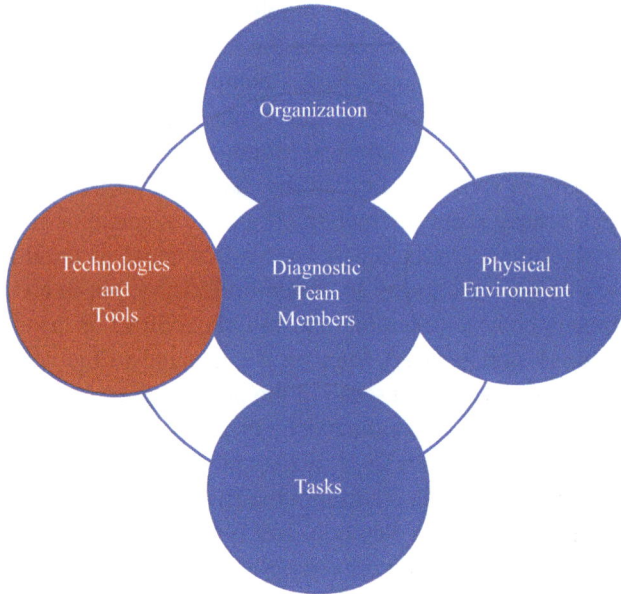

Figure 9.2: Technologies and tools are playing important roles in the work system for successful diagnostic process [151].

9.5.4 HIPAA

Protected health information (PHI) is broadly defined as health information, including demographic information, relating to an individual's mental or physical condition, treatment, and payment for healthcare services. Health Insurance Portability and Accountability Act (HIPAA) establishes a federal floor for safeguarding the PHI. HIPAA establishes (i) requirements on the use and disclosure of PHI, (ii) rights of individuals with respect to PHI, (iii) administrative, physical, and technical security safeguards to be implemented to protect PHI, (iv) notification obligations in the event of a breach of unsecure PHI, and (v) (being a federal regulation) administrative requirements [152].

What if AI were deployed within a health system to apply ML to patient information, in part, to allow patients to download information and wellness numbers (such as steps, blood pressure, and blood glucose levels) and to check on their own well-being without coming in for a professional visit? These activities could bring the AI developer under federal and state laws protecting health information privacy, particularly the HIPAA and its implementing regulations [152–155]. With regard to how to implement AI in accordance with HIPAA compliance, there are several issues to be mentioned: (i) access to stored data: HIPAA law requires access management to safeguard PHI. Access should only be granted to those that need it as part of their job function, (ii) data encryption: when your data is processed it passes through a server. Sending data outside an organization means that it passes through a third-party server. Although data sent within your organization does not need to be encrypted it is recommended to do so. Data sent externally, however, must be encrypted, (iii) de-identifying data: when conducting research, HIPAA law does not require patient permission if the data is adequately deidentified. This means that the data used cannot be tied to an individual in any way. If it is even slightly possible that the data can be tied to a specific individual, than it is not in accordance with HIPAA regulations, (iv) updated policies and procedures: as stated previously, HIPAA law is constantly changing. When implementing new technology an organization must look to their internal policies to ensure that their procedures are HIPAA compliant, and (v) business associate agreement (BAA): a BAA must be in place before any PHI can be transmitted. Since AI solutions have contact with PHI, an organization must have a signed BAA with the technology company before they can use any new technologies [154]. According to a 2017 report [156], the comprehensive healthcare damages from cybercrime will amount to greater than $1 trillion in 2018. Since healthcare firms want to avoid the huge costs of a breach (and all the ramifications related to HIPAA compliance and reputation), there is a huge incentive to adopt more intuitive and adaptive security protections [157]. Many healthcare companies are approaching the AI field through incorporation of semantic data lakes. The semantic approach is an upgrade from relational databases, which have very exact structural requirements. The relational method makes it necessary to have a pre-established, organized schematic strategy. Isolated data silos result. Data lakes are collections of data in a wide variety of formats. Insights are derived by bringing together data without any known relationship by applying unique standardized identifiers to each. This capability is permitted by the incorporation of natural language processing within data lakes. All of these applications will require a HIPAA compliant database [155].

Concern over health data privacy is growing primarily because data breaches at covered entities are becoming more and more common. It is not just that hackers are going after large medical companies known to have thousands of patient records – they are also going after just about any entity with large repositories of healthcare data, including the US Department of Health and Human Services (HHS). In the era of AI, what is important is the raw data. When that data can be correlated and compared against other sets of health data – that is when unique identities can be discovered [155]. Before AI, all of that information was protected by HIPAA. The problem, quite frankly, is that AI is what can be called a black box technology. Researchers can guarantee what goes into that black box, but they no longer know for certain how or why an AI system comes up with a decision. If a human makes a decision, you can ask him or her how they came up with the decision. It is a lot harder to ask a machine how it came up with a decision. Thus, AI researchers talk a lot about "poisonous" biases, or the fact that feeding the wrong data to the wrong AI system can lead to some pretty negative outcomes. Based on the above, it is easy to see how both state laws and federal law will need to be re-thought and re-imagined for the AI era. For the American medical and healthcare establishment, there will need to be a modern AI version of the HIPAA Privacy Rule that will require covered entities to ensure health data privacy, no matter how intelligent AI systems become. The reality for now is that AI has opened health data privacy to attack, and something needs to be done to deter and defend against those attacks [155].

9.6 Advanced dentistry in future

9.6.1 Dentistry in general

Most AI deploys some combination of computer vision (the processing and understanding of digital imagery), ML (data-driven algorithms that enable computers to learn underlying patterns about the data they process), and predictive analytics (statistical modeling used to find or forecast patterns and outcomes) [16]. AI technology continues to influence healthcare, daily life on QOL level, and its evolution makes the analysis of big data possible, providing reliable information and assisting the decision-making process in both medicine and dentistry [16, 155–157]. Taking a brief look at clinical applications of AI technologies, they can include (i) in orthodontics, diagnosis for need of extraction using ANN or treatment outcome analysis using CNN, (ii) in periodontics, diagnosis for PCT using CNN, (iii) in radiology, caries diagnosis using DL, and (iv) in oral medicine, oral cancer risk assessment using ANN [13]. The AI comprehensive dental care (ACDC) system was proposed [13], composed of pre-appointment (involved by AI patient manager, AI patient history analyzer, and AI scientific data library) → inter-appointment (supported by AI problem detector, AI treatment proposals, and AI instant feedback mechanism) → post-appointment (assisted with AI laboratory work designers, AI patient data library, and AI clinician evaluation). This patient care flow will collect enough reliable data for AI, utilizing for improving patient care. AI will continue to connect with dentistry from a comprehensive perspective due to the need of precise treatment procedures and instant information exchange. Such developments will enable professionals to share health-related big data and deliver insights that improve patient care through hospital, providers, researchers, and patients. AI-powered predictive analytic technology enables to analyze unstructured data within a patient's medical record and tie it back to more massive datasets to discover other health concerns. This plays into the trend toward an integrative practice methodology that will give dentists the chance to take a more significant role in their patient's overall health. Hence,

Stanley foresees that an AI "second opinion" will be particularly valuable for dentists in solo practices. It will help us avoid diagnostic mistakes while making patient education easier. With an AI-backed report justifying our diagnoses, we will be able to focus on clearly communicating treatment plans with compassion and empathy [18]. AI in dentistry is going to revolutionize the dental industry. Patients will be fundamentally different in the future. According to current analyses, patients aged 65 and over have a close relationship with their doctors, while baby boomers (born until 1970) often want a second opinion. Due to the growing Internet penetration, young patients are looking for specialized doctors, and at the same time – contradictively – want "holistic" care. The latter has resulted in a trend toward closer interprofessional collaboration, e.g., in dentistry and dental technology.

AI is unlikely to replace the dentist for the foreseeable future, but it will be able to relieve it of increasingly complex tasks. But, we've already witnessed the potential of AI/ML in dentistry [158].

9.6.2 Digital dentistry

Digital technology has changed the landscape of every industry, altering manual tasks and independent processes into integrated, streamlined workflows [159]. Digital dentistry refers to technology or devices that use digital, computer-based components instead of mechanical or electrical methods. The goal of digital dentistry is to enhance the efficiency and accuracy of a dental procedure, while producing predictable clinical outcomes. Advances often improve patient comfort, safety, and overall satisfaction. As more practitioners begin to adopt digital technology, innovations in tools, materials, and educational services also increase to support the shift in methods [159–161]. Basically, if elements involved in the workflows can be translated into digitized form, they can be subjected to the digitized technique. Accordingly, the main areas of digital dentistry extend from the diagnosis and procedures all the way to virtual networks and training. Different levels of digital integration allow for easier adoption of new technology, as case studies and returns on investment become more clearly defined. As digital dentistry continues to shape the future of the industry, significant advances have been made in several key areas [160]. Digital-supported dental technologies can include intraoral and extraoral scanning, CBCT technology, the CAD/CAM milling of restorations, guided implant surgery, complete denture methodologies, removable partial denture techniques, and applications to maxillofacial prosthetics.

The cone beam computerized tomography (CBCT) provides dentists with a quick 3D image of a patient's oral or maxillofacial anatomy. It is the basis for implant surgical guides used by oral surgeons and periodontists when placing dental implants. Such pre-surgical imaging techniques have made implant placement easier and more predictable, which helps ensure greater treatment success [162]. The array of digital equipment available to cosmetic dentists, general dentists, and implant dentists, etc., has increased significantly. Such advancements in dental technology enable patients to receive modern solutions to traditional dental problems.

Intraoral scanners have raised the bar for accuracy and speed in obtaining 3D impressions. For many practices, this is the starting point for incorporating digital dentistry into an established workflow. From orthodontics to implants, today's intraoral scanners/cameras and digital sensors offer high-precision results and intelligent processing to quickly assess potential treatment options. Another benefit is that they can share what dental team see with you, allowing you to better understand and improve your oral hygiene. Images can also be shared with lab technicians to match crowns and bridges to the shade of your actual teeth [160–162]. Digital impressions offer patients the convenience of not having to suffer through traditional impressions involving unpleasant tasting materials, bulky and cumbersome trays and possible gagging [162].

There are improvements in diagnostic x-ray imagining procedures. X-rays have been used in dental offices for a very long time, but the traditional x-ray process required film processing, which took time, was expensive, and the prints needed to be filed away in cabinets and physically delivered to other offices and specialists if necessary. Digital radiography is faster. Digital X-rays provide greater comfort than traditional X-rays and reduce radiation exposure (four digital radiographs equal one "film" X-ray). The images are immediately available to see on a computer screen. The files are stored on a server or in the cloud, and the images can be shared easily with specialists if necessary, with an internet connection [160–162]. Additionally, digital radiographs allow dentists to magnify images for greater diagnostic accuracy, ensuring more timely and appropriate treatments.

CAD/CAM technology enables dental prosthesis (crowns, veneers, inlays and onlays, bridges, implant-supported restorations or dentures) to be fabricated using computerized milling technology. Your dentist may work with in-office CAD/CAM to complete same-day tooth restorations that would otherwise require two or more visits to complete. Alternatively, if your case is more extensive, your dentist may work with a dental laboratory that uses CAD/CAM technology to create your restorations and the prosthesis is then either milled out of a solid block of material or 3D printed [161, 162].

Digitized dental bio-information can be easily managed. Not too long ago, dental records were all physical copies stored away in filing cabinets. A mix of technological innovation and federal regulation has motivated dental offices around the country to update their systems. This has improved scheduling, made it easier for your dental professional to access records right when they need them, improved workflows, and simplified patient information sharing between offices when necessary. Digital technology in dentistry is accelerating rapidly, and dental health professionals who harness scientifically tested and proven advances in their field may be able to offer the best care. Digital dentistry is constantly providing dental professionals with ways in which they can help you in faster, safer, more comfortable, and more reliable ways than they ever have before [161]. With a variety of appointment scheduling programs available, dentists today are making it easier for patients to make and keep their oral hygiene and treatment appointments via the Internet. Additionally, communication programs make it easy for dentists to securely share information about a patient's case with their laboratories and specialists to ensure proper care and eliminate unnecessary patient office visits. Real-time computer consultations also are possible while patients are in the chair, so any aesthetic or functional issues can be discussed and resolved [162]. There are still digitization-supported technologies available in dentistry; TekScan (T-Scan) which is a computer using an ultra-thin electronic sensor to digitally evaluate a patient's bite relationships, and the Wand which is a computerized tool that can deliver anesthesia in a slow and methodic manner. The sensation of pain often associated with an injection is caused as a result of the pressure of the liquid being injected rather than the needle itself. The slow and gentle delivery associated with The Wand often makes injections painless. The delivery holder is small and easy for the dentist to use. Digital technologies continue to aid dentists' efforts to ensure patients receive the best possible treatment under the most comfortable of circumstances. Talk to your dentist about the technologies they use to discover how they may impact your oral health treatments [162].

The digital dentistry has been developed in a win-win situation, brining advantages to both patients and dentists as well. Although patients may not directly identify the elements of integrated technology in the practice, they recognize the overall experience is efficient, smooth, and comfortable. Patient satisfaction increases with confidence and trust in the practice. With more accurate diagnostics, treatments can be more comprehensive and effective. Early detection improves clinical outcomes and reduces recovery time. Instruments enhanced with smart technology simplify surgical procedures, reducing chair time or successive visits. An integrated operatory allows the practitioner to seamlessly adjust to the needs of the patient, accommodating for changes in diagnostics, treatment, and scheduling without leaving the room. Technology becomes more than just a piece of equipment. Digital dentistry is a feature of the dental practice that now touches every part of the patient's visit. At the same time, digital dentistry empowers practitioners to perform at a higher level. From the way the office is managed to the way a treatment plan is developed, computer-aided technology allows practices to do more and do it better than ever before. Advancements in diagnostics and patient-specific treatment options promote positive clinical outcomes. Technical integration improves workflows in daily care and instantly connects partners and specialists to create a reliable source of information and guidance [161]. Rekow [163] mentioned that fresh approaches are bringing greater efficiency and accuracy, capitalizing on the interest, capabilities, and skills of those involved. New ways for effective and efficient inter-professional and clinician-patient interactions have evolved.

9.6.3 Teledentistry

Teledentistry is a combination of telecommunications and dentistry involving the use of electronic information, imaging and communication technologies, including interactive audio, video, data communications as well as store and forward technologies, to provide and support dental care delivery, diagnosis, consultation, treatment, transfer of dental information and education, and to improve access to oral healthcare [164–167]. It can include virtual consultations and high-tech monitoring of patients which offer less expensive and more convenient care options for patients. Teledentistry has been proven to (i) improve the dental hygiene of patients, (ii) be more affordable than in-office dentistry, (iii) be a more innovative solution for the mainstream healthcare industry, (iv) align with today's patients' needs for modern forms of communication, (v) improve access to care for patients, (vi) reduce the amount of time employees spend away from the office, (vii) make in-office appointment times more accessible to patients who really need them, and (viii) provide the same level of care to patients as in-office visits [165].

American Dental Association (ADA) stated Teledentistry can include patient care and education delivery using, but not limited to, the following modalities: namely, synchronous, asynchronous, remote patient monitoring (RPM), and mobile health (mHealth) [168, 169]. It continued by the general considerations. While in-person (face to face) direct examination has been historically the most direct way to provide care, advances in technology have expanded the options for dentists to communicate with patients and with remotely located licensed dental team members. The ADA believes that examinations performed using teledentisty can be an effective way to extend the reach of dental professionals, increasing access to care by reducing the effect of distance barriers to care. Teledentistry has the capability to expand the reach of a dental home to provide needed dental care to a population within reasonable geographic distances and varied locations where the services are rendered. In order to achieve this goal, services delivered via teledentistry must be consistent with how they would be delivered in-person. Examinations and subsequent interventions performed using teledentistry must be based on the same level of information that would be available in an in-person environment, and it is the legal responsibility of the dentist to ensure that all records collected are sufficient for the dentist to make a diagnosis and treatment plan. The treatment of patients who receive services via teledentistry must be properly documented and should include providing the patient with a summary of services. A dentist who uses teledentistry shall have adequate knowledge of the nature and availability of local dental resources to provide appropriate follow-up care to a patient following a teledentistry encounter. A dentist shall refer a patient to an acute care facility or an emergency department when referral is necessary for the safety of the patient or in case of emergency. As the care provided is equivalent to in person care, insurer reimbursement of services provided must be made at the same rate that it would be made for the services when provided in person, including reimbursement for the teledentistry codes as appropriate [168].

It was reported that, according to the Centers for Disease Control and Prevention (CDC), half of all adults in the US have a chronic condition, resulting in approximately 86% of health-care costs. In addition, more than 64 million Americans over the age of 30 have gum disease [170]. Given that oral health is directly linked to systemic health, it is imperative that the two are viewed as a mirror image of overall health and that dental care is considered a medical necessity rather than a luxury [171]. Oral health is a key indicator of a healthy person, as good dental care leads to reduced risk of diabetes, heart disease, stroke, premature or low-birth-weight babies, and chronic oral pain [172]. In order to facilitate total systemic health and reduce these adverse risks, barriers to dental care must be addressed. All things considered, access to dental care is still an impractical problem in the US Barriers to care limit or prevent people from receiving adequate healthcare. In the case of dental care, the most common barriers are financial hardship, geographic location, pressing health needs, and poor oral health literacy. Language, education, and cultural and ethnic barriers may compound the problem. Essentially, addressing barriers of access to care lie within two realms: the ability to access oral healthcare associated with socio-economic factors, and the overwhelming shortage of dental providers across the nation [171].

AI allows teledental, telehealth, and telemedicine professionals to compare their work more securely with their peers. AI Teledentistry offers dentists an arm of support by means of algorithm and computational statistical assuredness. AI teledentistry works best when it is virtually connecting doctors to patients. It is easy to connect doctors and patients when you are using an intuitive application [173]. A virtual platform that allows dentists to provide dental care to patients comes with an endless list of benefits that will last much longer than the impact of COVID-19 on the dental community. Due to typical characteristics of decreasing overhead costs and increasing revenue, overcoming geographical barriers, and better patient experience, teledentistry has the potential to completely revolutionize the field of dentistry in the near future [174]. Teledentistry is a rapidly forming subset of telehealth, a field that already has considerable impact on the healthcare industry. Recent advances have created new opportunities for teledentistry, and changes in diverse technologies have created new tools for the practitioner. Technologies currently available are beginning to change the dynamics of dental care delivery. As teledentistry evolves, it will offer new opportunities to improve the level of patient care and reshape current business models [175].

Teledentistry is already impacting dental care and creating exciting opportunities. Teledentistry services are designed differently and programed to meet the specific needs of the populations, including traditional dental practices. DiGangi [176] mentioned that teledentistry can be used as a catalyst for collaborative care and increased referrals. Teledentistry connections create a very strong communication/referral system inter-professionally and intra-professionally. The inter-professional system is a network that connects a dental setting with a medical setting, such as (i) a private medical practitioner (i.e., oncologist, cardiologist, pediatrician, and obstetrician), (ii) an emergency room or acute care facility, and (iii) any institution that provides residential accommodations. On the other hand, the intra-professional system is a network that connects different entities within the dental industry together, such as (i) specialists to general practitioners, (ii) provider-owned remote clinics with hygienists/assistant providers, (iii) dental service organizations with resource sharing, and (iv) dental lab to dentists.

Early intraoral cancer can be detected and accurately diagnosed by comprehensive diagnosis system using any one of currently commercially available device (OralID [177], Vizilite [178], VELscope [176] and Identafi [177]) through the teledental mechanism. There are currently numerous types of dental apps available to both potential patients and dentists as well. The most common practices are taking intraoral images by patients and send images to dentist to have an opinion with regard to outcomes of brushing, whitening, or prognosis. Moreover, this teledental communication can be done for obtaining an appropriate suggestive diagnosis, too. Recently, selfied intraoral images are put on a specified auction to find and select the most suitable orthodontist after negotiating treatment plan (duration and cost), if uncertain protectability and security of such publicly released personal biodata can be ignored. If someone could use this app to have a second opinion from public domain.

The generally accepted services from the teledentistry can include online prescriptions, one-on-one voice or video calls with a dentist, remote monitoring of patients, forwarding images/video to a dental practitioner to aid in diagnosis or treatment, online dental education, and remote treatment (such as guidance for in-home orthodontics). There are pros and cons associated with the teledentistry. Pros of telehealth dental care can include (i) increased access to patient care, (ii) opportunity for patient education, and (iii) prescription of antibiotics for dangerous infections; while cons of telehealth dental care should include (i) not available in every state, (ii) certain treatments need in-office visits, (iii) no x-rays or in-person evaluation limits the accuracy of diagnosis, and (iv) accuracy, treatment limitations, and assurance of the advice you've been given, as ADA's guideline [168, 179, 180].

9.6.4 Systemic treatment

We started this chapter with the historical separation between medicine and dentistry. We well know that almost all (relatively large scale) general hospitals and specified hospitals assign a certain region for dental treatment purpose for in-house patients who cannot visit dental clinic due to another medical reasons. The time is here for us to assist to re-unite these two disciplines. As mentioned previously, oral health is directly linked to systemic health, it is imperative that the two are viewed as a mirror image of overall health and that dental care is considered a medical necessity rather than a luxury [181] and oral health is a key indicator of a healthy person, as good dental care leads to reduced risk of diabetes, heart disease, stroke, premature or low-birth-weight babies, and chronic oral pain [182].

There is a growing need to fully understand the link between periodontal disease and systemic diseases such as other respiratory diseases, dementia, or diabetes. Figure 9.3 [183] is a typical illustration, depicting how systemic factors and intraoral factors are affecting adversely the periodontal disease.

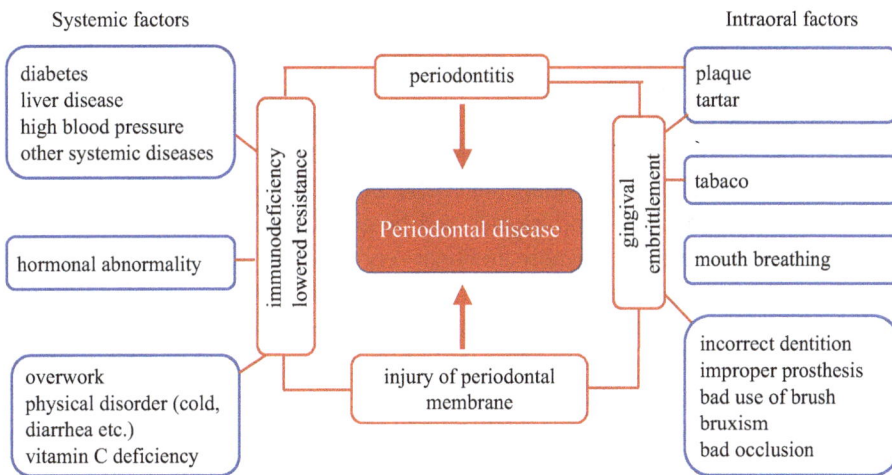

Figure 9.3: Systemic factors affecting periodontal disease [183].

At the same time, periodontic disease might trigger other types of diseases, including angina pectoris/myocardial infarction (toxins in bacteria causing periodontic disease enter the bloodstream, leading to arteriosclerosis and other heart diseases), endocarditis (periodontitis may be transmitted to the valve of the heart. Be careful of people with basic diseases such as cardiac peritoneal disease), underweight and premature birth of the fetus (during pregnancy, inflammation of gums is likely to occur, and many people have periodontitis; it is thought that substances caused by periodontitis inflammation adversely affect the placenta); osteoporosis (menopausal women need to be careful. The bone density decreases, and the alveolar bone becomes easy to lose weight. The decrease in female hormones after menopause is the main cause. It is the interrelationship with periodontic disease); dementia (periodontic disease is deeply related to cerebrovascular dementia because it is related to arteriosclerosis and stroke. In addition, since the brain is activated by chewing, it is also deeply related to Alzheimer's-type dementia), pneumonia (bacteria in the mouth, including periodontic disease bacteria, enter the trachea, and cause pneumonia. Elderly, the bedridden or those with reduced swallowing ability might be affected to develop pneumonia), diabetes (the relationship between diabetes mellitus and periodontic disease is close. Treating diabetes improves periodontic disease, and treating periodontic disease improves diabetes), cancer (if inflammation continues due to periodontic disease, normal cells may become abnormal, leading to carcinogenesis), and obesity (chewing is involved in the mechanism to prevent obesity. A decrease in the ability to chew leads to metabolic syndrome. If you keep in mind the prevention of periodontic disease in your diet, it will lead to obesity prevention). Therefore, intraoral condition and systemic conditions are strongly interacted [181–183].

The main purpose of the holistic dentistry is to promote the overall health of patients through systemic dental care. Hence, this vision of future dentistry can be result in naming the biological dentistry. Since all dental problems are systemic problems in patients, the holistic dentistry deals with problem foreseen from systemic observation and also emphasizes to prevent dental diseases, by paying close attention to correct nutritional knowledge, emphasis on the prevention of gums and gum disease, etc. To this end, dental education should include basic medical education and vice versa. In the future, these educated professionals will be able to introduce patients to each other, opening up a way to diagnose and treat very systemic diseases in both medical and dental terms. All problems related to the oral cavity will be discarded from the old sense of medicine, and a system that can share data on patients with medical and dental care (especially risk factors like NCDs), such as carding medical files (like a credit card), will become essential in the future [183].

While utilizing the advanced frailty studies, medical care and care to support the culmination of life are required by respecting the value of the person in question and the view of life and death in clinical practice and avoiding overtreatment or undertreatment. To achieve this aim, it is necessary to perform the Advanced Care Planning (ACP) [184]. Key point of the ACP is the communication between clinicians and patients and is a process in which the person or family members make decisions regarding medical care in consultation with the medical person. In the process of ACP, it is required for medical and care professionals, including dental professionals, to think together with the person himself and his family the question of how to live (QOL) and how to die (QOD). Accordingly, collaboration between family doctors and family dentists is also required. Not only integrating collaboration between medicine and dentistry, there is other type of integration of medical disciplines, i.e., the Integrative Medicine (or Integrative Health), which is the combination of regular and alternative medicine by Western medicine to treat patients. The alternative medicine refers to "medicine, which is usually used instead of medicine." Specifically, Chinese medicine (Chinese drugs, acupuncture, shiatsu, qigong), Indian medicine, yoga breathing methods, immunotherapy (lymphocyte therapy, etc.), medicinal foods and health functional foods (antioxidant foods, immunosuppressed foods, various preventive and supplemental foods, etc.), herbal medicine, aromatherapy, vitamin therapy, diet, psychotherapy, hot spring therapy, and oxygen therapy are all included in alternative medicine.

In today's fast-paced and increasingly complex world, medical science and the evolution of unique models for healthcare are bringing us closer to truly personalized medicine. Given these evolving trends, the need for a comprehensive and dynamically responsive plan for health is even more important. For this reason, a creation of personalized health plans that are as responsive to current health needs as they are adaptable to future health has been introduced, as shown in Figure 9.4 [184].

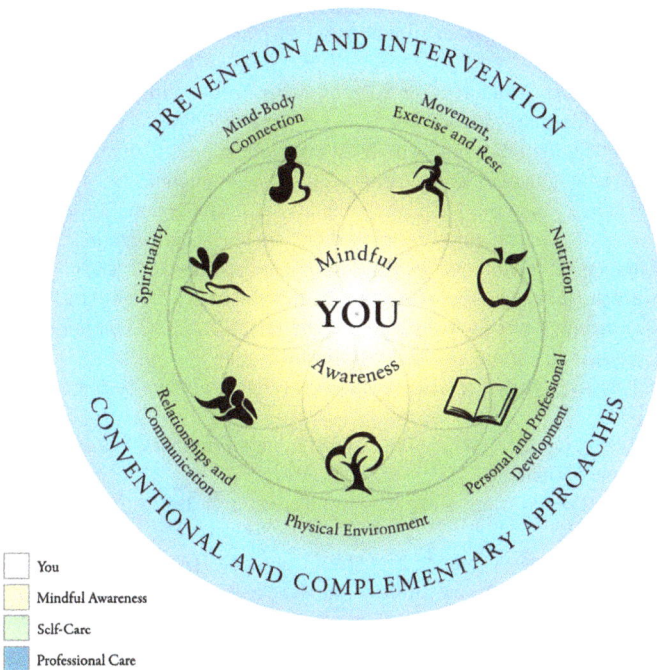

Figure 9.4: Wheel of health [184].

In figure, core of the health circle is yourself, who should be responsible to physical, mental, social, and spiritual well-being. The core is surrounded by the self-care, resonating deeply in all aspects of your health and well-being, and is reflected in lifestyle, priorities, motivation, and habits. Finally these essential portions are supported and protected by outer shell of professional care, including therapies that are aimed at staying healthy (prevention) as well as treating conditions that arise (intervention) [184]. Integrative medicine combines evidence-based therapies from both conventional (western or allopathic medicine) and complementary therapies. Providers include physicians, nurses, nutritionists, and other conventional healthcare workers, as well as integrative health coaches, acupuncturists, massage therapists, psychologists, mind–body therapists, and a wide range of other complementary health care providers in the alternative medicine, as mentioned above [184].

References

[1]   Beck J. Why Dentistry Is Separate From Medicine. 2017; https://www.theatlantic.com/health/archive/2017/03/why-dentistry-is-separated-from-medicine/518979/.

[2]   Otto M. Teeth: The Story of Beauty, Inequality, and the Struggle for Oral Health in America. New Press,2017.

[3]   Dentistry versus Medicine: 5 Reasons why they are different. 2017; https://youfirstdental.ca/dentistry-versus-medicine-blog/.

[4]   Satcher D. Oral Health in America: A Report of the Surgeon General. 2000; https://www.nidcr.nih.gov/sites/default/files/2017-10/hck1ocv.%40www.surgeon.fullrpt.pdf.

[5]   Gupta BS. What is Digitization, Digitalization, and Digital Transformation? 2020; https://www.arcweb.com/blog/what-digitization-digitalization-digital-transformation.

[6]   Hapon M. What Is the Difference Between Digitization, Digitalization and Digital Transformation. 2020; https://www.netguru.com/blog/digitization-and-digitalization.

[7]   Child PL. Digital dentistry: Is this the future of dentistry? 2011; https://www.dentaleconomics.com/science-tech/article/16394539/digital-dentistry-is-this-the-future-of-dentistry.

[8]   Digital Dentistry 2020 Overview. Dentsply Sirona. https://www.dentsplysirona.com/en-us/discover-by-topic/digital-dentistry.html.

[9]   Maguire RM. Rebooting dental hygiene department. 2021; www.dentaleconomics.com.www.dentaleconomics.com.

[10] Sedler A. The Future of Digital Dentistry. LIFE AFTER COVID-19. https://burbankdental.com/the-future-of-digital-dentistry-teledentistry/.

[11] Rekow ED. Digital dentistry: The new state of the art – Is it disruptive or destructive? Dental Materials. 2020, 36, 9–24.

[12] Shuman L. How artificial intelligence is shaping dentistry. 2019; https://www.dentaleconomics.com/macro-op-ed/article/16386252/how-artificial-intelligence-is-shaping-dentistry.

[13] Chen Y-W, Stanley K, Att W. Artificial intelligence in dentistry: current applications and future perspectives. Quintessence International. 2020, 51, 248–57, http://www.quintpub.com/userhome/qi/qi_51_3_chen_p248.pdf.

[14] 'D Assistant' – Virtual Assistant with a human interface. 2016; https://medium.com/@dentem/d-assistant-virtual-assistant-with-a-human-interface-d307948413a5.

[15] Artificial Intelligence In Dentistry. 2019; http://drkantor.com/wordpress_1691305916/?p=1256.

[16] Stanley K. Artificial Intelligence: The Future of Dentistry. 2019; https://www.dentistrytoday.com/news/todays-dental-news/item/4631-artificial-intelligence-the-future-of-dentistry#:~:text=AI%20can%20identify%20caries%20and,while%20making%20patient%20education%20easier.

[17] Marr B. Artificial Intelligence And Other Tech Innovations Are Transforming Dentistry. 2019; https://www.forbes.com/sites/bernardmarr/2019/05/31/artificial-intelligence-and-other-tech-innovations-are-transforming-dentistry/#688f87023abc.

[18] Mallon S. How AI Technology Is Shaping The Dental Industry. https://www.healthworkscollective.com/how-ai-technology-is-shaping-the-dental-industry/.

[19] DiamondClean Smart; https://www.usa.philips.com/c-m-pe/electric-toothbrushes#trigger name=color_white.
[20] DentSim Simulator. https://image-navigation.com/home-page/dentsim/.
[21] The benefits of using the teledentists. https://www.theteledentists.com/.
[22] 9 Technologies That Will Shape The Future Of Dentistry. The Medical Futurist. 2020; https://medicalfuturist.com/the-amazing-future-of-dentistry-and-oral-health/.
[23] SureSmile® Clear AlignersAll-new option for clinicians. https://www.dentsplysirona.com/en-us/categories/orthodontics/suresmile-aligner.html.
[24] Joda T, Waltimo T, Probst-Hensch N, Pauli-Magnus C, Zitzmann NU. Health data in dentistry: an attempt to master the digital challenge. Public Health Genomics. 2019, 22, 1–7.
[25] Lange K, Chi EC, Zhou H. A brief survey of modern optimization for statisticians. International Statistical Review. 2014, 82, 46–70.
[26] Susskind R, Susskind D. Technology will replace many doctors, lawyers, and other professionals. Harvard Business Review. 2016, https://hbr.org/2016/10/robots-will-replace-doctors-lawyers-and-other-professionals.
[27] How Artificial Intelligence will Affect In dentistry! 2019; https://kckdirect.com/how-artificial-intelligence-will-affect-in-dentistry/.
[28] Farkas D. Machine learning and AI are poised to transform dental practices in 2020. https://www.dentalproductsreport.com/view/machine-learning-and-ai-are-poised-transform-dental-practices–2020.
[29] Kwon HB, Park YS, Han JS. Augmented reality in dentistry: a current perspective. Acta Odontologica Scandinavica. 2018, 76, 497–503.
[30] Schwendicke F, Samek W, Krois J. Artificial intelligence in dentistry: chances and challenges. Journal of Dental Research. 2020, 99, 769–74.
[31] Dillard R. Artificial Intelligence: The Future of Dentistry. 2018; https://www.thedentalgeek.com/2018/11/artificial-intelligence-the-future-of-dentistry/.
[32] Gupta S. The Future of Artificial Intelligence in Dentistry. 2018; https://dzone.com/articles/the-future-of-artificial-intelligence-in-dentistry.
[33] Bhatia AP, Tiwari S. Artificial intelligence: an advancing front of dentistry. Acta Scientific Dental Sciences. 2019, 3, 135–8.
[34] Shah N, Bansal N, Logani A. Recent advances in imaging technologies in dentistry. World Journal of Radiology. 2014, 6, 794–807.
[35] Jablow M, Parashar A. The future of artificial intelligence in dentistry. 2018; https://www.dentalproductsreport.com/view/future-artificial-intelligence-dentistry.
[36] Prados-Privado M, Villalón JG, Martínez-Martínez CH, Ivorra C, Prados-Frutos JC. Dental caries diagnosis and detection using neural networks: a systematic review. Journal of Clinical Medicine. 2020, 9. doi: 10.3390/jcm9113579.
[37] Tandon D, Rajawat J. Present and future of artificial intelligence in dentistry. Journal of Oral Biology and Craniofacial Research. 2020, 10, 391–6, https://www.ncbi.nlm.nih.gov/pmc/articles/PMC7394756/.
[38] Solutions. 2020; https://www.orca-ai.com/solutions/prosthetics/.
[39] Phillips A. AI-Powered Innovations That Are Transforming Dentistry. 2018; https://becominghuman.ai/ai-innovations-transforming-dentistry-aef03479664d.
[40] Ferro AS, Nicholson K, Koka S. Innovative trends in implant dentistry training and education: a narrative review. Journal of Clinical Medicine. 2019, 8, 1618, doi: 10.3390/jcm8101618.
[41] Sadat R, Khalili M, Nazari M. A Hybrid. Method to Predict success of dental implants. International Journal of Advanced Computer Science and Applications. 2016, 7. doi: 10.14569/IJACSA.2016.070501.

[42] Papantonopoulos G, Takahashi K, Loos BG. PR613: artificial intelligence modelling delivers actionable insights about personalized implant dentISTRY. Journal of Clinical Periodontology. 2018, doi: https://doi.org/10.1111/jcpe.617_12915.

[43] Jaskari J, Sahlsten J, Järnstedt J, Mehtonen H, Karhu K, Sundqvist O, Hietanen A, Varjonen V, Vesa Mattila V, Kaski K. Deep learning method for mandibular canal segmentation in dental cone beam computed tomography volumes. Scientific Reports. 2020, 10, 5842, doi: https://doi.org/10.1038/s41598-020-62321-3.

[44] Roy E, Bakr MM, George R. The need for virtual reality simulators in dental education: A review. Saudi Dental Journal. 2017, 29, 41–7.

[45] Xiaojun C, Yanping L, Yiqun W, Chengtao W. Computer-aided oral implantology: methods and applications. Journal of Medical Engineering & Technology. 2007, 31, 459–67.

[46] Haghighat K. How Digital Technology is Changing the Dental Implant Process. https://portlandperioimplantcenter.com/digital-technology-dental-implant-process/.

[47] Yan A. World's first implant surgery by dentist robot takes place in China. 2017; https://www.scmp.com/news/china/article/2112197/chinese-robot-dentist-first-fit-implants-patients-mouth-without-any-human.

[48] Volchok J. Neocis Inc. Announces FDA Clearance for First Robotic System for Dental Implant Procedures. 2017; https://www.businesswire.com/news/home/20170302005444/en/Neocis-Announces-FDA-Clearance-Robotic-System-Dental.

[49] Lerner H, Mouhyi J, Admakin O, Mangano F. Artificial intelligence in fixed implant prosthodontics: a retrospective study of 106 implant-supported monolithic zirconia crowns inserted in the posterior jaws of 90 patients. BMC Oral Health. 2020, 20, https://bmcoralhealth.biomedcentral.com/articles/10.1186/s12903-020-1062-4.

[50] Martins SM, Ferreira FAF, Ferreira JJM, Marques CSE. An artificial-intelligence-based method for assessing service quality: insights from the prosthodontics sector. Journal of Service Management. 2020, https://www.emerald.com/insight/content/doi/10.1108/JOSM-03-2019-0084/full/html.

[51] Susic I, Travar M, Susic M. The application of CAD/CAM technology in Dentistry. IOP Conference Series: Materials Science. 2016, 200, doi: 10.1088/1757-899X/200/1/012020. 2017 Engineering.

[52] Sims-Williams J, Brown I, Matthewman A, Stephens C. A computer-controlled expert system for orthodontic advice. British Dental Journal. 1987, 163, 161–6.

[53] Williams JS, Matthewman A, Brown D. An orthodontic expert system. Fuzzy Sets and Systems. 1989, 30, 121–33.

[54] Asiri SN, Tadlock LP, Schneiderman E, Buschang PH. Applications of artificial intelligence and machine learning in orthodontics. APOS Trends Orthod. 2020, 10, 17–24, https://apospublications.com/applications-of-artificial-intelligence-and-machine-learning-in-orthodontics/.

[55] Brown I, Adams S, Stephens C, Erritt S, Sims-Williams J. The initial use of a computer-controlled expert system in the treatment planning of class II division 1 malocclusion. British Dental Journal. 1991, 18, 1–7.

[56] Faber J, Faber C, Faber P. Artificial intelligence in orthodontics. APOS Trends in Orthodontics. 2019, 9, 201–5, doi: 10.25259/APOS_123_2019.

[57] Dabney CW. Artificial Intelligence (AI) and orthodontics – can you imagine where we will be in the very near future? Orthodontics. https://orthopracticeus.com/artificial-intelligence-ai-orthodontics-can-imagine-will-near-future/.

[58] Park JH, Hwang HW, J.H. Moon, Yu Y, Kim H, Her S-B, Srinivasan G, Aljanabi MNA, Donatelli RE, Lee S-J. Automated identification of cephalometric landmarks: part 1-comparisons between the latest deep-learning methods YOLOV3 and SSD. The Angle Orthodontist. 2019, 89, 903–9.

[59]   Redmon J, Farhadi A. Yolov3: An Incremental Improvement. 2018; https://arxiv.org/pdf/1804.02767.pdf.

[60]   Liu W, Anguelov D, Erhan D, Szegedy C, Reed S, Fu C-Y, Berg AC. SSD: Single Shot MultiBox Detector. In: Proceedings of the European Conference on Computer Vision. New York: Springer; 2016:21–37.

[61]   Kunz F, Stellzig-Eisenhauer A, Zeman F, Boldt J. Artificial intelligence in orthodontics: evaluation of a fully automated cephalometric analysis using a customized convolutional neural network. Journal of Orofacial Orthopedics. 2020, 81, 52–68.

[62]   Hwang HW, Park J-H, Moon J-H, Yu Y, Kim H, Her S-B, Srinivasan G, Alijanabi MNA, Donatelli RE, Lee S-J. Automated identification of cephalometric landmarks: Part 2-Might it be better than human? The Angle Orthodontist. 2020, 90, 69–76.

[63]   Yu HJ, Cho SR, Kim MJ, Kim WH, Kim JW, Choi J. Automated skeletal classification with lateral cephalometry based on artificial intelligence. Journal of Dental Research. 2020, 99, 249–56.

[64]   Khanagar SB, Al-ehaideb A, Maganur PC, Vishwanathaiah S, Patil S, Baeshen HA, Sarode SC, Bhandi S. Developments, application, and performance of artificial intelligence in dentistry – A systematic review. Journal of Dental Sciences. 2021, 16, 508–22.

[65]   Arnett GW, Bergman RT. Facial keys to orthodontic diagnosis and treatment planning. Part I. American Journal of Orthodontics and Dentofacial Orthopedics. 1993, 103, 299–312.

[66]   Ribera NT, De Dumast P, Yatabe M, Ruellas A, Ioshida M, Paniagua B, Styner M, Gonçalves JR, Bianchi J, Cevidanes L, Prieto J-C. Shape variation analyzer: A classifier for temporomandibular joint damaged by osteoarthritis. Proc SPIE Int Soc Opt Eng 10950, Medical Imaging 2019: Computer-Aided Diagnosis, 10950212019;10950:1095021; https://doi.org/10.1117/12.2506018.

[67]   Choi HI, Jung S-K, Baek S-H, Lim WH, Ahn S-J, Yang I-H, Kim T-W. Artificial intelligent model with neural network machine learning for the diagnosis of orthognathic surgery. Journal of Craniofacial Surgery. 2019, 30, 1986–9.

[68]   Kök H, Acilar AM, İzgi MS. Usage and comparison of artificial intelligence algorithms for determination of growth and development by cervical vertebrae stages in orthodontics. Progress in Orthodontics. 2019, 20. doi: https://doi.org/10.1186/s40510-019-0295-8.

[69]   Xie X, Wang L, Wang A. Artificial neural network modeling for deciding if extractions are necessary prior to orthodontic treatment. The Angle Orthodontist. 2010, 80, 262–6.

[70]   Jung SK, Kim TW. New approach for the diagnosis of extractions with neural network machine learning. American Journal of Orthodontics and Dentofacial Orthopedics. 2016, 149, 127–33.

[71]   Li P, Kong D, Tang T, Su D, Yang P, Wang H, Zhao Z, Liu Y. Orthodontic Treatment Planning based on Artificial Neural Networks. Scientific Reports. 2019, 9, 2037, doi: https://doi.org/10.1038/s41598-018-38439-w.

[72]   Botsis T, Hartvigsen G, Chen F, Weng C. Secondary use of EHR: Data quality issues and informatics opportunities. Summit on Translational Bioinformatics. 2010, 2010, 1–5.

[73]   Rodvold DM, McLeod DG, Brandt JM, Snow PB, Murphy GP. Introduction to artificial neural networks for physicians: Taking the lid off the black box. Prostate. 2001, 46, 39–44.

[74]   Tanimoto Y, Miyawaki S, Imai M, Takeda R, Takano-Yamamoto T. Orthodontic treatment of a patient with an impacted maxillary second premolar and odontogenic keratocyst in the maxillary sinus. The Angle Orthodontist. 2005, 75, 1077–83.

[75   Elhaddaoui R, Bahije L, Chbicheb S, Zaoui F. Cervico-facial irradiation and orthodontic treatment. International Orthodontics. 2015, 13, 139–48.

[76]   Hirschfeld J, Reichardt E, Sharma P, Hilber A, Meyer-Marcotty P, Stellzig-Eisenhauer A, Schlagenhauf U, Sickel FE. Interest in orthodontic tooth alignment in adult patients affected by periodontitis: A questionnaire-based cross-sectional pilot study. Journal of Periodontology. 2019, 90, 957–65.

[77] Maki K. Digitalization technology in orthodontics. Showa Gakusikai. 2015, 75, 21–9.

[78] DICOM PS3.1 2020e – Introduction and Overview. http://dicom.nema.org/medical/dicom/cur rent/output/chtml/part01/chapter_1.html.

[79] Introduction to the Standard Template Library. https://www.boost.org/sgi/stl/stl_introduc tion.html#:~:text=The%20Standard%20Template%20Library%2C%20or,the%20STL%20is% 20a%20template.

[80] Tabletop CNC Wire Bender; http://www.nihonbinary.co.jp/Products/3DModeling/diwire. html.

[81] 3D CNC 08 type Wire Bender KWC-082F; https://www.ipros.jp/product/detail/2000182672/.

[82] Oshida Y, Tominaga T. Nickel-Titanium Materials – Biomedical Applications. De Gruyter, Berlin, 2020.

[83] The Invisalign system; https://www.aligntech.com/solutions.

[84] Lewis GM. What is Invisalign®? 2018; https://lovethatsmile.com/dental-articles/what-is-invisalign.

[85] Estai M, Kanagasingam Y, Tennant M, Bunt SM. A systematic review of the research evidence for the benefits of teledentistry. Journal of Telemedicine and Telecare. 2018, 24, 147–56.

[86] Katyal V. Artificial Intelligence (AI) driven customised orthodontics is the future. 2018; https://www.breakthroughwithboss.com/948-2/.

[87] Smile Mate; https://dental-monitoring.com/smilemate/.

[88] Oh my teeth; https://ledge.ai/stayhome-teeth-ai/.

[89] Kolokitha O-E, Topouzelis N. Cephalometric methods of prediction in orthognathic surgery. Journal of Oral and Maxillofacial Surgery. 2011, 10, 236–45.

[90] Smith JD, Thomas PM, Proffit WR. A comparison of current prediction imaging programs. American Journal of Orthodontics and Dentofacial Orthopedics. 2004, 125, 527–36.

[91] Fudalej P, Dragan M, Wedrychowska-Szulc B. Prediction of the outcome of orthodontic treatment of Class III malocclusions–a systematic review. European Journal of Orthodontics. 2011, 33, 197–7.

[92] Lee JH, Kim DH, Jeong SN, Choi SH. Diagnosis and prediction of periodontally compromised teeth using a deep learning-based convolutional neural network algorithm. Journal of Periodontal & Implant Science. 2018, 48, 114–23.

[93] Krois J, Ekert T, Meinhold L, Golla T, Kharbot B, Wittemeir A, Dörfer C, Schewendicke F. Deep learning for the radiographic detection of periodontal bone loss. Scientific Reports. 2019, 9, 8495, doi: https://doi.org/10.1038/s41598-019-44839-3.

[94] Baugh D, Wallace J. The role of apical instrumentation in root canal treatment: a review of the literature. Journal of Endodontics. 2005, 31, 333–40.

[95] Saghiri MA, Asgar K, Boukani KK, Lotfi M, Aghili H, Delvarani A, Karamiafr K, Saghiri AM, Mehrvarzfar P. A new approach for locating the minor apical foramen using an artificial neural network. International Endodontic Journal. 2012, 45, 257–65.

[96] Saghiri MA, Garcia-Godoy F, Gutmann JL, Lotfi M, Asgar K. The reliability of artificial neural network in locating minor apical foramen: a cadaver study. Journal of Endodontics. 2012, 38, 1130–4.

[97] Johari M, Esmaeili F, Andalib A, Garjani S, Saberkari H. Detection of vertical root fractures in intact and endodontically treated premolar teeth by designing a probabilistic neural network: an ex vivo study. Dentomaxillofacial Radiology. 2017, 46. doi: https://doi.org/10.1259/ dmfr.20160107.

[98] Fukuda M, Inamoto K, Shibata N, Ariji O, Yanashita Y, Kutsuna S, Nakata K, Katsumata A, Fujita H, Ariji E. Evaluation of an artificial intelligence system for detecting vertical root fracture on panoramic radiography. Oral Radiology. 2020, 36, 337–43.

[99]   Zakirulla M, Javed S, Assiri NEH, Alqahtani AM, Alzahrani RH, Laheq MT, Alamri AA, Alqahtani AM, Al-Qahatani AHK, Alghozi AA, Alawwad SM, Almoammar S. An Overview of Artificial Neural Network in the Field of Pediatric Dentistry. Journal of Dental and Orofacial Research. 2020, 16, 20–5.

[100]   Baliga MS. Artificial intelligence – The next frontier in pediatric dentistry. Journal of Indian Society of Pedodontics and Preventive Dentistry. 2019, 37, 315, doi: 10.4103/JISPPD. JISPPD_319_19.

[101]   McGrath H, Flanagan C, Zeng L, Lei Y. Future of artificial intelligence in anesthetics and pain management. Journal of Biosciences and Medicines. 2019, 7, 111–18.

[102]   Liang H and 70 co-authors. Evaluation and accurate diagnoses of pediatric diseases using artificial intelligence. Nature Medicine. 2019, 25, 433–8.

[103]   WHO. Oral cancer. 2020; https://www.who.int/cancer/prevention/diagnosis-screening/oral-cancer/en/.

[104]   Svoboda E. Artificial intelligence is improving the detection of lung cancer. Nature. 2020, https://www.nature.com/articles/d41586-020-03157–9.

[105]   Dlamini Z, Francies FZ, Hull R, Marima R. Artificial intelligence (AI) and big data in cancer and precision oncology. Computational and Structural Biotechnology Journal. 2020, 18, 2300–11.

[106]   Aubreville M, Knipfer C, Oetter N, Jaremenko C, Rodner E, Denzler J, Bohr C, Neumann H, Stelzle F, Maier A. Automatic classification of cancerous tissue in laserendomicroscopy images of the oral cavity using deep learning. Scientific Reports. 2017, 7(119), 79, https://doi.org/10.1038/s41598-017-12320-8.

[107]   Zhang W, Li J, Li Z-B, Li Z. Predicting postoperative facial swelling following impacted mandibular third molars extraction by using artificial neural networks evaluation. Scientific Reports. 2018, 8, 12281, doi: https://doi.org/10.1038/s41598-018-29934-1.

[108]   Yaji A, Prasad S, Pai A. Artificial Intelligence in Dento-Maxillofacial Radiology. Acta Scientific Dental Sciences. 2019, 3, 116–21.

[109]   Heo M-S, Kim J-E, Hang J-J, Han S-S, Kim J-S, Yi W-J, Park I-W. Artificial intelligence in oral and maxillofacial radiology: what is currently possible? DentoMaxilloFacial Radiology. 2020, 2020(03), 75, 10.1259/dmfr.20200375.

[110]   Zhang K, Wu J, Chen H, Lyu P. An effective teeth recognition method using label tree with cascade network structure. Computerized Medical Imaging and Graphics. 2018, 68, 61–70.

[111]   Tuzoff DV, Tuzova LN, Bornstein MM, Krasnov AS, Kharchenko MA, Nikolenko SI, Sveshnikov MM, Bednenko GB. Tooth detection and numbering in panoramic radiographs using convolutional neural networks. Dentomaxillofacial Radiol. 2019, 48, 20180051, doi: 10.1259/dmfr.20180051.

[112]   Chen H, Zhang K, Lyu P, Li H, Zhang L, Wu J, Lee C-H. A deep learning approach to automatic teeth detection and numbering based on object detection in dental periapical films. Scientific Reports. 2019, 9, 3840, doi: https://doi.org/10.1038/s41598-019-40414-y.

[113]   Lee JH, Kim DH, Jeong SN, Choi SH. Detection and diagnosis of dental caries using a deep learning-based convolutional neural network algorithm. 2018, 77, 106–11.

[114]   Casalegno F, Newton T, Daher R, Abdelaziz M, Lodi-Rizzini A, Schürmann F, Krejci I, Markam H. Caries detection with near-infrared transillumination using deep learning. Journal of Dental Research. 2019, 98, 1227–33.

[115]   Schwendicke F, Elhennawy K, Paris S, Friebertshäuser P, Krois J. Deep learning for caries lesion detection in near-infrared light transillumination images: a pilot study. Journal of Dentistry. 2020, 92, 103260, doi: https://doi.org/10.1016/j.jdent.2019.103260.

[116]   Devito KL, De Souza Barbosa F, Felippe Filho WN. An artificial multilayer perceptron neural network for diagnosis of proximal dental caries. Oral Surgery, Oral Medicine, Oral Pathology, Oral Radiology, and Endodontology. 2008, 106, 879–84.

[117] Hung M, Voss MW, Li W, Su W, Xu J, Bounsanga J, Ruiz-Negrón B, Lauren E, Licari FW. Application of machine learning for diagnostic prediction of root caries. Gerodontology. 2019, 36, 395–404.

[118] Ekert T, Krois J, Meinhold L, Elhennawy K, Emara R, Golla T, Schwendicke F. Deep learning for the radiographic detection of apical lesions. Journal of Endodontics. 2019, 45, 917–22.

[119] Kise Y, Ikeda H, Fujii T, Fukuda M, Ariji Y, Fujita H, Katsumata A, Ariji E. Preliminary study on the application of deep learning system to diagnosis of sjögren's syndrome on CT images. Dentomaxillofacial Radiology. 2019, 48, 20190019, doi: https://doi.org/10.1259/dmfr.20190019.

[120] Murata M, Ariji Y, Ohashi Y, Kawai T, Fukuda M, Funakoshi T, Kise Y, Nozawa M, Katsumata A, Fijita H, Ariji E. Deep-learning classification using convolutional neural network for evaluation of maxillary sinusitis on panoramic radiography. Oral Radiology. 2019, 35, 301–7.

[121] Kim Y, Lee KJ, Sunwoo L, Choi D, Nam C, Cho J, Kim J, Bae YJ, Yoo R-E, Choi BS, Jung C, Kim JH. Deep Learning in diagnosis of maxillary sinusitis using conventional radiography. Investigative Radiology. 2019, 54, 7–15.

[122] Ariji Y, Sugita Y, Nagao T, Nakayama A, Fukuda M, Kise Y, Nozawa M, Nishiyama M, Katumata A, Ariji E. CT evaluation of extranodal extension of cervical lymph node metastases in patients with oral squamous cell carcinoma using deep learning classification. Oral Radiology. 2020, 36, 148–55.

[123] Lee JS, Adhikari S, Liu L, Jeong HG, Kim H, Yoon SJ. Osteoporosis detection in panoramic radiographs using a deep convolutional neural network-based computer-assisted diagnosis system: a preliminary study. Dentomaxillofacial Radiology. 2019, 48, 20170344, doi: https://doi.org/10.1259/dmfr.20170344.

[124] Lee KS, Jung SK, Ryu JJ, Shin SW, Choi J. Evaluation of transfer learning with deep convolutional neural networks for screening osteoporosis in dental panoramic radiographs. Clinical Medicine. 2020, 9, 392, doi: https://doi.org/10.3390/jcm9020392.

[125] Hiraiwa T, Ariji Y, Fukuda M, Kise Y, Nakata K, Katsumata A, Fujita H, Ariji E. A deep-learning artificial intelligence system for assessment of root morphology of the mandibular first molar on panoramic radiography. Dentomaxillofac Radiol. 2019, 48, 20180218, doi: 10.1259/dmfr.20180218.

[126] Mudrak J. Artificial Intelligence and Deep Learning in Dental Radiology: A Way Forward in Point of Care Radiology. 2019, https://www.oralhealthgroup.com/features/artificial-intelligence-and-deep-learning-in-dental-radiology-a-way-forward-in-point-of-care-radiology/.

[127] Leeson D. The digital factory in both the modern dental lab and clinic. Dental Materials. 2020, 36, 43–52.

[128] Rella A. Making better Dental Crowns using Artificial Intelligence in dentistry. 2019; https://www.dentidesk.com/en/dental-crowns-artificial-intelligence/.

[129] Goodfellow IJ, Pouget-Abadie J, Mirza M, Xu B, Warde-Farley D, Ozair S, Courville A, Bengio Y. Generative Adversarial Networks. Proceedings of the International Conference on Neural Information Processing Systems (NIPS 2014), 2672–80; arXiv:1406.2661.

[130] Adaloglou N. GANs in computer vision: Introduction to generative learning. AI Summer. 2020; https://theaisummer.com/gan-computer-vision/.

[131] Budny R. Dental Laboratory Technician. https://explorehealthcareers.org/career/dentistry/dental-laboratory-technician/.

[132] Azernikov S. Making Better Crowns Using Artificial Intelligence (AI). 2018; https://glidewelldental.com/education/chairside-dental-magazine/volume-13-issue-2/crowns-artifical-intelligence.

[133] Developed a program to design dental work in 20 seconds using artificial intelligence. 2017; https://ai-biblio.com/articles/570/.

[134] Kulzer Sets out to Develop New Dental CAD Software with Artificial Intelligence (AI). 2018; https://jp.mitsuichemicals.com/en/release/2018/2018_0530_01.htm.

[135] Tooth color judgment by AI. 2019; https://www.kanaloco.jp/article/entry-195357.html.

[136] AI technology application to dental treatment. 2019; https://www.meisei-u.ac.jp/2019/2019082901.html.

[137] Cushing H. The Life of Sir William Osler. Clarendon Press, Oxford, 1925.

[138] Rouse M. SOAP (Simple Object Access Protocol). 2019; https://searchapparchitecture.techtarget.com/definition/SOAP-Simple-Object-Access-Protocol.

[139] Understanding SOAP format for Clinical Rounds. 2015; https://www.globalpremeds.com/blog/2015/01/02/understanding-soap-format-for-clinical-rounds/.

[140] Potter L. How to Document a Patient Assessment (SOAP). 2021; https://geekymedics.com/document-patient-assessment-soap/.

[141] Andales J. SOAP Note Templates. 2020; https://safetyculture.com/checklists/soap-note-template/.

[142] Krishna K, Khosla S, Bigham JP, Lipton ZC. Generating SOAP Notes from Doctor-Patient Conversations. 2020; arXiv:2005.01795v2.

[143] Takabayashi K. POMR (problem-oriented medical record). 2017; https://www.jstage.jst.go.jp/article/naika/106/12/106_2529/_pdf/-char/en.

[144] POS(Problem-Oriented System). 2016; https://blog.goo.ne.jp/yasuharutokuda/e/82ac104e06ba40c9928801dbaeb70aeb.

[145] Guide to Problem Oriented Medical Records. https://www.truenorthitg.com/problem-oriented-medical-records.

[146] Salmon P, Rappaport A, Bainbridge M, Hayes G, Williams J. Taking the problem oriented medical record forward. Proc AMIA Annu Fall Symp. 1996, 463–7; https://www.ncbi.nlm.nih.gov/pmc/articles/PMC2233232/.

[147] Jain N. Takling the problem-oriented medical records approach (POMR), 2016; https://www.m-scribe.com/blog/taking-the-problem-oriented-medical-records-approachpomr-is-it-worth-the-effort.

[148] Weed LL. Knowledge coupling, medical education and patient care. Critical Reviews in Medical Informatics. 1986, 1, 55–79.

[149] Weed LL, Zimmy NJ. The problem-oriented system, problem-knowledge coupling, and clinical decision making. Physical Therapy. 1989, 69, 565–8.

[150] Burger C. The. Use of Problem-Knowledge Couplers in a Primary Care Practice. Permanente Journal. 2010, 14, 47–50.

[151] Technology and Tools in the Diagnostic Process. In: Balogh EP, et al. ed. Improving Diagnosis in Health Care, National Academies Press, Washington (DC), 2015, https://www.nap.edu/read/21794/chapter/7#230.

[152] Williams RL. Privacy Please: HIPAA and Artificial Intelligence – Part I. 2018; https://www.dwt.com/blogs/privacy–security-law-blog/2018/03/privacy-please-hipaa-and-artificial-intelligence.

[153] Schneider J. How HIPAA Is Undermining IT and AI's Potential To Make Healthcare Better. 2019; https://electronichealthreporter.com/how-hipaa-is-undermining-it-and-ais-potential-to-make-healthcare-better/.

[154] HIPAA Compliance and AI Solutions. https://compliancy-group.com/hipaa-compliance-and-ai-solutions/.

[155] Lindsey N. New Advances in AI Could Have a Significant Impact on Health Data Privacy. 2019; https://www.cpomagazine.com/data-privacy/new-advances-in-ai-could-have-a-significant-impact-on-health-data-privacy/.

[156] Gavel M. Machine learning in cybersecurity to boost big data, intelligence, and analytics spending to $96 billion by 2021. ABI Research. 2017, https://www.prnewswire.com/news-re leases/machine-learning-in-cybersecurity-to-boost-big-data-intelligence-and-analytics-spend ing-to-96-billion-by-2021-300398664.html.

[157] Roberts K. The Future of Machine Learning and AI in Healthcare Security. 2018; https://www. atlantic.net/hipaa-compliant-hosting/the-future-of-machine-learning-and-ai-in-healthcare-se curity/.

[158] Navarro P. AI Assisted Dentist: How Artificial Intelligence Will Reform Dentistry by 2050. 2020; https://datafloq.com/read/artificial-intelligence-dentistry-going-revolutionize-dental-indus try/8705.

[159] Dentsply Sirona. Digital Dentistry 2021 Overview; https://www.dentsplysirona.com/en-us/ discover-by-topic/digital-dentistry.html.

[160] Digital dentistry; https://en.wikipedia.org/wiki/Digital_dentistry.

[161] Colgate. The Changing World of Digital Dentistry; https://www.colgate.com/en-us/oral-health/dental-visits/the-changing-world-of-digital-dentistry.

[162] Digital Dentistry: Overview of Digital Dental Technologies; https://www.yourdentistryguide. com/digital-dentistry/.

[163] Rekow ED. Digital dentistry: The new state of the art – Is it disruptive or destructive? Dental Materials. 2020, 36, 9–24.

[164] Janpani ND, Nutalapati R, Dontula BSK, Boyapati R. Applications of teledentistry: A literature review and update. Journal of International Society of Preventive and Community Dentistry. 2011, 1, 37–44.

[165] Facts About Teledentistry; https://www.americanteledentistry.org/facts-about-teledentistry/.

[166] McClure D. What Is Teledentistry Anyway? 2019; https://virtudent.com/2019/12/10/what-is-teledentistry-anyway/.

[167] Burke E. How did we get here? A brief history of Teledentistry. 2020; https://medium.com/ oraleye-network-news/how-did-we-get-here-a-brief-history-of-teledentistry-f985b8e0b7b3.

[168] ADA. ADA Policy on Teledentistry. 2020; https://www.ada.org/en/about-the-ada/ada-posi tions-policies-and-statements/statement-on-teledentistry.

[169] Bayba M. Dental Telemedicine (Teledentistry). 2019; https://www.newmouth.com/blog/tele dentistry/.

[170] Towers S. Why connect dental data to population health? 2017; https://www.ahip.org/why-connect-dental-data-to-population-health/.

[171] Webb A, Sommers S. AI technology and teledentistry: Breaking barriers to dental care. 2020; https://www.dentistryiq.com/practice-management/systems/article/14188148/ai-technol ogy-and-teledentistry-breaking-barriers-to-dental-care.

[172] Bresnick J. Dental care is the missing piece of population health management. 2016; https:// healthitanalytics.com/news/dental-care-is-the-missing-piece-of-population-health-management.

[173] Artificial Intelligence Teledentistry is going to Lead Dentistry by 2025. 2020; https://ai-tele dentistry.com/.

[174] Montisano V. The Future of Teledentistry. 2020; https://www.oralhealthgroup.com/blogs/ the-future-of-teledentistry/.

[175] Brinbach JM. The future of teledentistry. Journal of the California Dental Association. 2000, 28, 141–3.

[176] DiGangi P. Teledentistry for a Prosperous Future. Dentistry Today. 2018; https://www.dentist rytoday.com/viewpoint/10487-teledentistry-for-a-prosperous-future.

[177] OralID. 2021; https://forwardscience.com/oralid.

[178] ViziLite PRO® Oral Lesion Screening System. 2021, https://www.denmat.com/vizilite-pro-oral-lesion-screening-system.html.

[179] What is Teledentistry and Does It Work for My Dental Care Needs? https://www.dentistry.com/articles/what-is-teledentistry.

[180] Romanban VR. Teledentistry: What are its benefits and drawbacks? 2019; https://onemedall.com/teledentistry-what-are-its-benefits-and-drawbacks/.

[181] VELscope; https://velscope.com/velscope/.

[182] Dentafi; https://www.dentalez.com/product/identafi/.

[183] Systemic and introoral factors on periodontal disease; http://www.seiei-dental.com/blog/2013/11/post-4-175397.html.

[184] Wheel of Health; https://dukeintegrativemedicine.org/patient-care/wheel-of-health/.

Chapter 10
AI in drug development

The pharmaceutical industry is characterized by being (1) closely related to life science, (2) large variety of products with small quality production, (3) research oriented, and (4) knowledge integrated with high-added value. The pharmaceutical industry plays an important role in promoting disease treatments, strengthening life supporting systems, or enhancing the quality of life. To be more specific, the pharmaceutical industry has advanced to (1) establish treatment strategies for intractable diseases, (2) prevent disease, (3) develop life supporting systems and/or complete nutrition feeding systems, or (4) control progressive disease [1]. Historically, a drug production has been characterized for searching a main component of the new drug by analyzing species such as organic compounds or fermented substances, or those extracted from animals and plants. Since this conventional drug-developing manner has found its limitation, new drugs are currently researched and developed by applying the biotechnology, based on information obtained from human genetic information by the genome analysis. A principal concept of pharmaceuticals is based on a simple equation of "drug = substance + information." During new drug development process, chemical compounds or extracted natural products are just drug candidate material, to which necessary information with regard to three crucial elements of quality, efficacy, and safety should be added before claiming a real pharmaceutical candidate. This is just a beginning of the time-consuming and costly procedure before we can see and use it.

Developing and bringing a new single drug from original idea to launch of a finished product is a long time-consuming (in a range from 10 to 17 years) and expensive process (in excess of 1 billion dollars) [2, 3]. Main reason for such a long and costly process involved in a new drug development is due to its complicated stages and check-points toward safe pharmaceutics. The sequential process is basically as follows: Basic (idea) research → Target research (or identification) → Lead discovery → Candidate selection → Principal development → Non-clinical animal tests → IND (Investigation New Drug Application) filing → Clinical developments (phases I, II, III, including theranostic tests) → NDA (New Drug Application) filing → Approval (by, e.g., FDA in the USA, EMA in EU, MHRA in England, or PMDA in Japan) → Quality Management and Practice → Shipping and Marketing → (phase IV studies). Besides the pharmacological effects, there is one important non-negligible factor associated with drugs. It is the drug delivery system.

In this chapter, we will be discussing AI-assisted drug development, AI-assisted drug delivery system, adverse drug reaction, the role of pharmaceutical chemists, and effects of placebo and nocebo.

https://doi.org/10.1515/9783110717853-011

10.1 AI-assisted new drug development

Pharmaceutics can be simply defined as a synergy of substance and information. Chemically synthesized substances or drug candidate materials extracted from natural substances are just substances. To these substances, medical/chemical (or biopharmaceutical) information should be added to form a pharmaceutics. Since three major elements (quality, efficacy, and safety) should be satisfied at higher than a certain required level, such information should possess these elements. Traditionally, substances are relied on organic syntheses or fermented substances or extracted material from plants or animals; however there found a limitation by utilizing these substances. Accordingly, new pharmaceutics, based on human genetic information alayzed by the genomic project, have been R/Ded by fully employing biotechnology [1].

Phase I studies are usually the first tests of a drug under development in healthy (from 20 to 80) volunteers and conducted to determine a drug's safety profile, including the safe dosage range, plus how the drug is absorbed, distributed, metabolized, and excreted, and the duration of its action. Phase I trials take, on the average, 1 year. Phase II clinical trials (take normally 2 years) are slightly larger studies that are done in patients with the disease for which the drug is intended. This phase is usually designed to identify what are the minimum and maximum dosages. The trials generally involve 100–300 volunteer patients and are controlled in design. They are done to assess the drug's effectiveness. Phase III clinical trials are the definitive, largely randomized trials that are submitted to the FDA in order to obtain approval of a drug. This phase examines the effectiveness as well as the safety (adverse events) of the new drug. Phase III trials usually involve 1,000–3,000 patients in clinics and hospitals. Patients are usually asked a list of possible side effects, often derived from what was observed in phase II studies. Patients are also free to report any other side effects that occur while they are on the new drug or the placebo (the "sugar pill" that is given to a percentage of patients in a trial study). Phase III takes on the average 3 years. Phase IV studies are any organized collection of data from patients who are taking a drug that has already received approval from the FDA. New Drug Application (NDA): Following the phase III clinical trials, the drug manufacturer analyzes all the data from the studies and files an NDA with the FDA (provided the data appear to demonstrate the safety and effectiveness of the drug). The NDA contains all data gathered to date about the drug. An NDA typically consists of at least 100,000 pages. The average NDA review time for new drugs approved in 1992 was close to 30 months (2.5 years). In phase IV studies, patients may check boxes on a list (as in phase III studies) or they may just report other symptoms. Phase IV studies are commonly called "post-marketing studies" [2].

Success rates of new drug development are varied in a wide range, depending on which stage they are considered and type of disease (such as allergy, autoimmune, cardiovascular, chronic high prevalence diseases, endocrine, gastroenterology, hematology, infectious disease, metabolic, neurology, oncology, ophthalmology, psychiatry, rare diseases, respiratory, or urology [4]) [1, 4–9]. During 2006–2015, the success rate was 9.6% [4]. The high failure rates associated with pharmaceutical development are referred to as the "attrition rate" problem. Careful decision-making during drug development is essential to avoid costly failures [8]. In many cases, intelligent program and clinical trial design can prevent false-negative results. Well-designed, dose-finding studies and comparisons against both a placebo and a gold-standard treatment arm play a major role in achieving reliable data [9].

During the lengthy process toward the final approval of newly designed drugs, in almost each phase (mentioned above) AI technology has been recognized for its effectiveness and power.

In early stage of drug development, in order to enhance the success rate, the application of in silico (in computer simulation) modeling has been employed. A computer simulation model is a computer program or algorithm which simulates changes of a modeled system in response to input signals and useful adjunct to purely mathematical models in promoting the research and investigation, science, mathematics, and technology. Utilizing AI program, since a broad data source can be handled, any potentially highly effective methods for clinical test designs and/or selection of clinical trial subjects, based on gene information or evidence for treatments, can be determined. Actually, AI's deployment in the drug development stages is recognized for its versatility, involving target identification, drug designing, big data analysis, prediction of drug risk, and patient matching.

10.1.1 Drug design

A large variety of computer-aided methods and tools were developed to improve the drug development process. There has also been a remarkable advancement in computational power coupled with big data analysis that enables AI to revolutionize the drug discovery and development process. While some protagonists point to vast opportunities potentially offered by such tools, others remain skeptical, waiting for a clear impact to be shown in drug discovery projects. The reality is probably somewhere in between these extremes, yet it is clear that AI is providing new challenges not only for the scientists involved but also for the biopharma industry and its established processes for discovering and developing new medicines [10]. The top pharmaceutical companies have already implemented AI methods to improve their development success rates. The research paradigm is clearly changing, presenting great challenges and opportunities for scientists.

The discovery of molecular structures with desired properties for applications in drug discovery or chemical biology is among the most impactful scientific challenges. However, given the complexity of biological systems and the associated cost for experiments and trials, molecular design is also scientifically very challenging, prone to failure, inherently expensive, and time consuming [11]. The number of publications is increasing, and a good collaboration between scientific experts across disciplines is required to fully evaluate the potential of a hypothesis. Such evidence-based learning (EBL) should be carefully conducted in terms of its degree of reliability. There is an established protocol for ranking the reliability of information sources – that is an especially useful tool in the medical and dental fields. The ranking, from the most to least reliable evidence source, follows: (1) clinical reports using placebo and double blind studies, (2) clinical reports not using placebo, but conduced according to well-prepared statistical test plans, (3) study reports on time effect on one group of patients during predetermined period of time, (4) study/comparison reports, at one limited time, on many groups of patients, (5) case reports on a new technique or idea, or both, and (6) retrospective reports on clinical evidence [12]. Unfortunately, the number of published articles increases in this descending reliability order. The greatest advantage of the EBL reviews is that it helps identify common phenomena in diversity of fields and literature. It is then possible to create synthetic, inclusive hypotheses that deepen and further understanding. The theoretical space of chemistry, even when limited by molecular size, is huge and dramatically exceeds what we can assess experimentally and even computationally. Accordingly, data mining and statistics on reliable sources have been integrated into molecular discovery and design pipelines to provide computational support in the prioritization of molecular hypotheses [11].

ML algorithms have been part of the routine toolbox of computational and medicinal chemists for decades. The recent increase in applications and coverage of these methodologies has been attributed to advances in computational power, the growing amount of digitized research data, and an increasing theoretical understanding of the algorithms and their shortcomings. However, given the gradual character of these evolutions, it might be counterintuitive to expect a dramatic revolution of molecular design. Nevertheless, extravagant claims have been made for the ability of AI to accelerate the design process [11, 13, 14].

Over the last few decades, computer-aided drug design has emerged as a powerful technique playing a crucial role in the development of new drug molecules. AI/DL provides opportunities for the discovery and development of innovative drugs. Various ML approaches have recently (re)emerged, some of which may be considered instances of domain-specific AI that have been successfully employed for drug discovery and design [15]. Structure-based drug design (SBDD) and ligand-based drug design (LBDD) are two methods commonly used in computer-aided drug design [16, 17]. SBDD method attempts to use the structure of proteins as a basis for designing new ligands by applying the principles of molecular recognition by analyzing macromolecular target three-dimensional structural information, typically of proteins or RNA, to identify key sites and interactions that are important for their respective biological functions. Such information can then be utilized to design antibiotic drugs that can compete with essential interactions involving the target and thus interrupt the biological pathways essential for survival of the microorganism(s) [17]. Selective high-affinity binding to the target is generally desirable since it leads to more efficacious drugs with fewer side effects [16, 17]. LBDD methods focus on known antibiotic ligands for a target to establish a relationship between their physiochemical properties and antibiotic activities, referred to as a structure–activity relationship (SAR), information that can be used for optimization of known drugs or guide the design of new drugs with improved activity. Moreover, LBDD is an approach used in the absence of the receptor 3D information, and it relies on knowledge of molecules that bind to the biological target of interest [16, 17].

Long term, AI offers the hope of a streamlined and automated approach across these various stages. A future AI may hold within its databases the sum of all knowledge about biology, genes, and chemical interactions. It will be able to identify new targets and find candidate molecules for a particular target in silico from vast libraries and develop and refine molecules to home in on the best ones. It will be able to specify how to synthesize the candidate molecules, gather test data, and refine further. Analytics and statistical models have long been used to reduce trial and error in drug discovery. AI has the potential to remove much more, and home in on better answers much quicker than is currently possible. Even short term, AI could conceivably shave a year off the development of many drugs, which would be worth billions. To benefit from AI short term, companies are looking at how it can deliver across different parts of the discovery process: some in a piecemeal way, some with a view to building toward complete AI driven digitalization as the technology develops [18].

10.1.2 Target identification

There are two paradigms in traditional drug discovery: physiology-based and target-based. Physiology-based designs have an unknown target and phenotypic read-outs, meaning that researchers may not know the target on which a compound that is mitigating a disease is acting.

Target-based designs have a known target and read-outs based on the activity or expression of the target. After a target molecule is chosen and validated, high-throughput screening is conducted to identify hits with the desired activity, from which a lead compound is selected and optimized [19]. Target identification is the process of identifying the direct molecular target – for example, protein or nucleic acid – of a small molecule and involved early validation of disease-modifying targets, is an essential first step in the drug discovery pipeline. Indeed, the drive to determine protein function has been stimulated, both in industry and in academia, by the completion of the human genome project. In clinical pharmacology, target identification is aimed at finding the efficacy target of a drug/pharmaceutical or other xenobiotic. The techniques used may be based on principles of biochemistry, biophysics, genetics, chemical biology, or other disciplines. With the target identified,

AI can be used for molecular generation from scratch, identifying proposed new molecules and performing virtual test cycles. While AI is some way from creating drugs without human guidance, this is in invaluable in bringing focus to the idea generation process [18, 20].

Target validation is the first step in discovering a new drug and can typically take 2–6 months. The process involves the application of a range of techniques that aim to demonstrate that drug effects on the target can provide a therapeutic benefit with an acceptable safety window. Early in-depth target validation increases understanding between target manipulation and disease efficacy, leading to increased likelihood of success in the clinic. Once a target has reached an acceptable level of validation and disease linkage, the project moves into the hit identification phase. Figure 10.1 shows the spectrum of target validation techniques [21].

Figure 10.1: The multidisciplinary influences on target validation and identification [21].

Since without foreknowledge of the complete drug target information, development of promising and affordable approaches for effective treatment of human diseases is challenging, Zeng et al. [22] developed a deepDTnet, a DL methodology for new target identification and drug repurposing in a heterogeneous drug–gene–disease network embedding 15 types of chemical, genomic, phenotypic, and cellular network profiles. It was reported that (i) after trained on 732 US FDA-approved small-molecule drugs, deepDTnet shows high accuracy (AU-ROC: 0.963) in identifying novel molecular targets for known drugs, outperforming previously published state-of-the-art methodologies; (ii) experimentally validate that deepDTnet-predicted topotecan (an approved topoisomerase inhibitor) is a new, direct inhibitor (IC50: 0.43 μM) of human retinoic-acid-receptor-related orphan receptor-gamma t (ROR-γt); and (iii) by specifically targeting ROR-γt, topotecan reveals a potential therapeutic effect in a mouse model of multiple sclerosis; summarizing that the deepDTnet offers a powerful network-based DL methodology for target identification to accelerate drug repurposing and minimize the translational gap in drug development [22]. Using AI in the target identification process allows scientists to really explore all the available evidence to better understand a disease and its underlying biology. The technology can synthesize data, and then make inferences with regard to the best targets, in a way that would be impossible for an individual human, or even a team of people, to do. This allows the scientist to make more informed and therefore hopefully better decisions about which targets are most likely to succeed and in which patients they are most likely to be important. Hunter et al. [23] reported that the use AI/ML for the discovery of potential new drug targets for chronic kidney disease and idiopathic pulmonary fibrosis had accomplished.

10.1.3 Lead compound identification

A lead compound in drug discovery is a chemical compound that has pharmacological or biological activity likely to be therapeutically useful, but may nevertheless have suboptimal structure that requires modification to fit better to the target; lead drugs offer the prospect of being followed by backup compounds. Its chemical structure serves as a starting point for chemical modifications to improve potency, selectivity, or pharmacokinetic parameters. Furthermore, newly invented pharmacologically active moieties may have poor druglikeness and may require chemical modification to become drug-like enough to be tested biologically or clinically [2]. The objective of a lead optimization, as shown in Figure 10.2 [24], campaign is to deliver one or more clinical candidate for evaluation in good laboratory practices safety studies prior to clinical trials by monitoring and manipulating multiple parameters pertaining various properties (including absorption, distribution, metabolism, excretion, and toxicity (ADMET)) as well as potency and selectivity while remaining within a patentable chemical space.

Lead compound identification is a complex process, with many decision-making points converging upon the selection of promising compounds to compete in the race for lead compound optimization and, ultimately, clinical candidate development. Initially, the focus was on having a large, chemically diverse library or collection of small molecules for testing to maximize the chances of finding a lead. Issues such as purity, chemically reactive functional groups, and drug-like properties have been addressed over recent years to enhance the chances of finding a successful lead. Other issues, including physical properties such as solubility and aggregation, are also critical to the filtering of hits versus non-hits. Depending on the decision-making strategy, lowering the concentration of lead compounds used during screening may reduce the impact of physical properties such as poor solubility or high propensity for aggregation. However, lowering the concentration also increases the risk of missing new

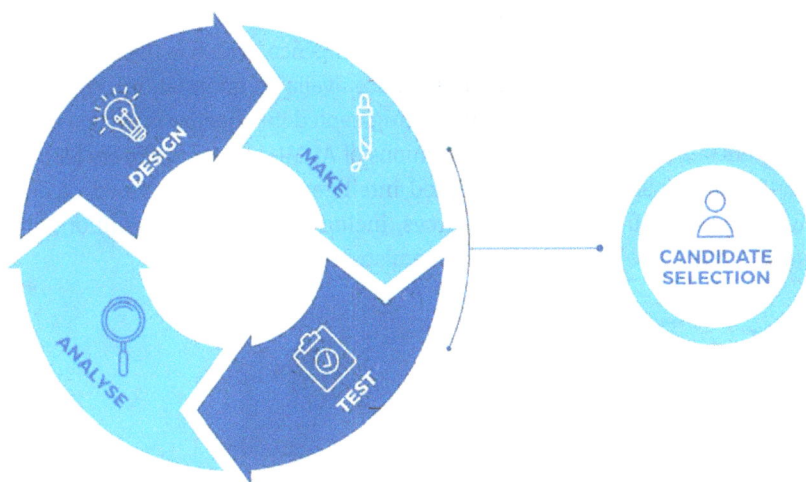

Figure 10.2: Multidisciplinary tasks toward lead optimization of drug candidate [24].

small molecules that have the potential to be optimized by iterative drug design and biological screening [25, 26].

In drug design and discovery, what is required is expertise in chemistry and biology, to curate and analyze the data set. Understanding the questions being asked and the relevance of these to the available data is of crucial importance. It is critical to avoid discarding or ignoring information from prior analysis and modeling. The dimensionality of drug space is simply too big even when big data is available. Computational drug design is finally achieving industry recognition and is moving toward acceptance as indicated by numerous high profile deals in 2018. However, exact methods and applications are still part of active research. It was hence shown that the methods had successfully employed on numerous targets [27, 28]. Rouse [29] defined the augmented intelligence. AI is an alternative conceptualization of AI that focuses on AI's assistive role, emphasizing the fact that cognitive technology is designed to enhance human intelligence rather than replace it. The choice of the word "augmented," which means "to improve," reinforces the role human intelligence plays when using ML and DL algorithms to discover relationships and solve problems [29]. Hence, the augmented intelligence is a design pattern for a human-centered partnership model of people and AI working together to enhance cognitive performance, including learning, decision-making, and new experiences. Specifically, the use of augmented intelligence, instead of black box AI, provides solutions to the challenges of traditional drug design. We have demonstrated it by leveraging a mixture of ML and other computational tools in drug design to successfully fully utilize of prior data and domain expertise. AI can greatly benefit the drug discovery field by lowering the high failure rate, high cost, and finding novel intellectual property. However, the hype surrounding the use of AI/ML can be found in almost every field today. When AI/ML is applied to drug design, many problems hinder progress. An exciting option is the use of Augmented Intelligence, which is the application of AI methods (such as big data and ML to enhance available information), computational chemistry, and other non-AI algorithms. Augmented intelligence overcomes several problems encountered in common applications of AI/ML in drug discovery [28].

Figure 10.3 demonstrates the augmented intelligence being combined of data processing tools driven from multiple sources, including AI, human intuition, and knowledge and traditional computational chemistry tools [28].

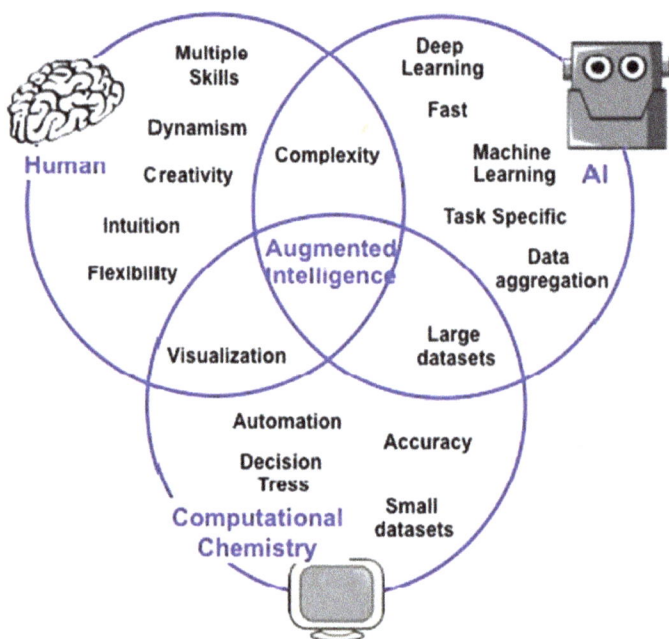

Figure 10.3: General conceptual scheme of the augmented intelligence [28].

Green et al. [30] recognized the augmented intelligence in a course of drug development as shown in Table 10.1.

10.1.4 Clinical research

The clinical research refers to studies or trials that are done in human. As the developers design the clinical study plan (or a protocol), they will consider what they want to accomplish for each of the different clinical research phases I, II, and III [31], as mentioned previously. Then, the clinical studies are further followed by the IND application and FDA application for the final approval. Data acquired by the

Table 10.1: Five levels of artificial intelligence involvements in drug discovery process [30].

Level	Name	Definition	Target object
0	No assistance	Manual design, synthesis, and analysis	Human
1	Analytical assistance	Input from computational analysis	Human
2	Partial design assistance	Human design and synthesis with occasional input from AI design methods	Human
3	Augmented design and partial synthesis assistance	AI design and partial automated synthesis with significant input from humans in both design and synthesis	Human
4	Conditional automation	AI design and automated synthesis with occasional input from humans	Human and system
5	High automation	AI design and automated synthesis with minimal or no input from humans	System

clinical studies are so important that in various stage of the entire drug development process they are effectively utilized. In the target research stage, clinical data are used along with published patents, pathway data, compound data, literature-based genome, omics, and images. For the lead compound research stage, clinical data are also crucial along with genome and protein information of patients and healthy individuals. Of course, clinical data are very important in the clinical research stage [32]. While protocols and standards in clinical research have become increasingly complex, slowing progress and increasing costs, companies from start-ups to Big Pharma (which is a popular naming for the world's pharmaceutical industry) are identifying opportunities to apply AI to enhance trial efficiency, patient enrollment, and outcomes targeting. The fact that we are at the dawn of AI technology indicates that its role in clinical research could grow exponentially in the years ahead [33]. AI-powered capabilities, including data integration and interpretation, pattern recognition and evolutionary modeling, are essential to gather, normalize, analyze, and harness the growing masses of data that fuel modern therapy development. AI has many potential applications in clinical trials both near- and long-term. AI technologies make possible innovations that are fundamental for transforming clinical trials, such as seamlessly combining phases I and II of clinical trials, developing novel patient-centered endpoints, and collecting and analyzing data. The AI transformation of clinical trials starts with protocol development, reducing, or replacing outcome assessments that may be more responsive to change than traditional methods and utilizing remote connected technologies that reduce the need for patients to travel long distances for sites visits. Data-driven protocols and strategies powered by advanced AI algorithms processing data collected from mobile sensors and applications, electronic medical and administrative records, and other sources have the potential to reduce trial costs. They achieve this by improving data quality, increasing patient compliance and retention, and identifying treatment efficacy more efficiently and reliably than ever before [34]. The traditional linear and sequential clinical trials process remains the accepted way to ensure the efficacy and safety of new medicines. However, suboptimal patient selection, recruitment, and retention, together with difficulties managing and monitoring patients effectively, are extending the length of trials and contributing to high trial failure rates. AI can improve clinical cycle times while reducing the cost and burden of clinical development [35].

Clinical trials are an area with great potential for optimization; only 12% of drug development programs ended in success in a 2000–2019 study according to recent research. Inability to demonstrate efficacy or safety, flawed study design, participant dropouts, or unsuccessful recruitment all contributed to the low success rate of clinical trials. Vendors active in this field are therefore focusing on the use of AI-based software in three main areas: information engines, patient stratification, and clinical trial operations. AI is being applied in several aspects for the clinical trial process today, from analyzing real-world data and scientific information to providing improved patient stratification and predictive outcomes and assisting with different aspects of clinical trial operations [36]. Phase III clinical trial outcomes used to evaluate new drugs, therapies and vaccines are among the most complex experiments performed in medicine. Around 50% of phase III trials fail, due to the difficulty of predicting clinical results in a wide patient base, even with the backing of solid data [37]. More importantly, these barriers are costing healthcare industries, governments, and academic research hospitals billions of dollars, and further drive up costs and delay lifesaving treatments to patients and in some cases lead to adverse events. The crucial roadblock is the limited knowledge of key parameters, which need to be considered to test candidate molecules, eliminate adverse events and select optimal half maximal inhibitory concentrations (IC_{50}). New ML architectures were developed with automated learning and predictions that were gleaned from past experimental successes and failures of drugs leading to the design of faster, safer, and more efficacious clinical trials [38].

10.1.5 Overall

To develop a new drug, since human intervention is inevitable, a team of medicinal chemists and structural biologists can work to set the direction of research (de novo drug designing), guide AI tools, and make decisions along the way. The next step is to discover the protein in the body that is the target of the drug (target identification), followed by the search for substances that are candidates for the drug (lead compound identification). This involves huge numbers of experiments, predictive models, and expertise, applied across several rounds of optimization, each with modifications to the best set of potential molecules. Such identified molecules also need a whole host of other properties prediction – such as ADMET. For identified candidates, AI can perform in silico property prediction, allowing poor candidates to be eliminated early, and increasing throughput of good quality leads. While in silico modeling is nothing new, it is getting better with more data and better algorithms. Then, if the compound that is the active ingredient is optimized and the clinical trials applied to humans are cleared through animal experiments, the application filings and final governmental approval are followed. In practice, however, there are tens of thousands of drug candidates in the library of a pharmaceutical company like Big Pharma, and there are more than 100,000 proteins in vivo, so it is not an easy task to discover the most optimal combination from them. It is said that development of drugs having a relatively simple structure with low-molecular-weight compounds have already been exhausted, and the probability of successful development of new drugs in modern times is less than 1/25,000, the development period is more than 10 years and it will cost more than 1 billion dollars [39].

AI drug discovery utilizing AI is expected to be effective in developing new drugs. By letting AI perform in silico calculations such as comparing compounds in the vast library and virtual compounds that do not yet exist with proteins in the living body and data from the three billion human genomes one by one and predicting the binding of the compounds, it has become enable to discover and optimize compounds and to specify how to synthesize the candidate molecules. Analytics and statistical models have long been used to reduce trial and error in drug discovery and AI has the potential to complete these processes much quicker with less cost than is currently possible. The key technologies for drug discovery using AI should include the ML to realize AI and the DL which is one of the methods. AI learns a large number of cases in advance and collates the optimum one from the learned patterns, but the ML is a mechanism that makes a machine learn a large amount of data and memorize the characteristics and recognize pattern characteristics of the data. Among the ML, the DL has been developed by imitating the function of human brain nerve cells. DL, in which the computer itself finds problems, makes it possible to discover small differences and features that human cannot notice, and to open up fields that could not be realized by human hands. This is the reason why AI-powered drug discovery that utilizes DL technology is attracting attention. AI-powered drug discovery efforts are enabling Big Pharma and biotechnology companies to streamline R&D efforts, including calculating vast patient data sets into digestible, tangible information, identifying personalized/precision medicine opportunities, or forecasting potential responses to new drugs. With a well-developed AI platform, precision medicine is creating and executing a new promise to cut drug costs and development time significantly [18, 40, 41].

There are several issues associated with the AI-assisted drug development. Although a big data is nicely synchronized and AI revolutionary changes were made in

drug development, a number of identified target substances and identified lead compounds are ever-increasing. Taking an example of cancer, its type, and progression
can be varied in wide range, indicating that massive data are required. In addition,
there is the issue of how long it takes to collect data. In the field of mental health
medication, where you need a variety of treatment options for different types of people, this type of efficient drug discovery assisted by AI can lead to increased effectiveness. The pharmaceutical industry is also embracing the trendy technology for its
abilities in effectively advancing and/or addressing the ever-changing drug or therapeutic needs from those who suffer from everyday viruses to complex diseases, like
pancreatic cancer or Alzheimer's. The demand for treatments for Alzheimer's disease,
which causes dementia, is very high, but since it is a disease that often develops only
in old age, it is specified to collect basic comparative data. It can take decades to
track a population of people for a period of time. Currently, AI tools can be used to
screen compounds, identify them for use, suggest modifications while humans focus
on assessing the decisions of AI and how these decisions useful. AI tools can also be
used to minimize drug interactions that may cause an issue down the line. This is a
concern for many elderlies in the world that take a cocktail of medication each day
[42].

Besides the fact that effectiveness and productivity are needed to be improved,
data privacy protection and security are another challenge. AI-driven personalized
medicine where you can use each person's genetic code to produce drugs will need
your personal information. There are risks to using or manipulating this information
without the proper security measures taken. Since medical data as private information is the most sensitive, their handing should be carefully performed. Hence, it is
a key point to mask the information specifying an individual. It is urgently required
to establish technology as well as legislation to protect the privacy using, for example, the blockchain, which is a combined technique of cryptogram and authentication function [43, 44].

10.1.6 COVID-19 vaccine development

With the spread of pandemic of COVID-19 showing no signs of slowing and its relatively high mortality rate when compared to other viral bases illnesses such as influenza, it is essential to develop vaccines and antiviral medications against the novel
coronavirus severe acute respiratory syndrome coronavirus 2 (SARS-CoV-2). In order
for a vaccine to be effective, it must be recognized as an antigen by the inoculated
person and elicit an immune response that attacks this antigen. If any vaccine exhibits a weak ability to attack the virus and its effect of eliciting immunity is limited
to a short period of time and moreover if it develops adverse side effects on the
lungs or the antibody-dependent immunopotentiation that enhances infectious diseases may be caused, such vaccine should not be used at all. Many efforts have

been undertaken to accelerate AI research in order to combat this pandemic. Aiming to use AI to predict blueprints for designing universal vaccines against SARS-CoV-2, which contain a sufficiently broad repertoire of T-cell epitopes capable of providing coverage and protection across the global population, Malone et al. [45] have managed to profile the entire SARS-CoV-2 proteome and identify a subset of epitope hotspots that could be harnessed in a vaccine formulation to provide a broad coverage across the global population, by combining the antigen presentation to the infected-host cell surface and immunogenicity predictions of the NEC Immune Profiler with a robust Monte Carlo and digital twin simulation.

The Harvard T.H.Chan School of Public Health and the Human Vaccines Project had announced, on April 14, 2020, a joint effort to use AI models to accelerate vaccine development [46]. A new research consortium jointly managed by UC Berkeley and the University of Illinois at Urbana-Champaign was established and announded, on March 27, 2020, that they are aiming to build AI techniques to assist combat COVID-19 [47]. While a vaccine for SARS-CoV-2 has already entered human trials, the efficacy is unknown. An AI-based approach could help screen compounds to be used as potential adjuvants for the SARS-CoV-2 vaccine, along with screening new compounds based on modeling of potential mutations to the novel coronavirus. This will help us be prepared to develop vaccines as the virus potentially mutates. Furthermore, AI can aid in both screening currently available drugs and accelerating the process of antiviral development to help treat COVID-19. AI algorithms could be trained using the CORD-19 data set and the resulting model could then be used to screen for existing drugs that potentially demonstrate efficacy in the treatment of COVID-19. This screening approach could be used, for example, to identify generic medications that may be effective against COVID-19.

Leveraging the power of AI, trillions of compounds can be screened in a remarkably short amount of time allowing for the rapid identification of drug and vaccine candidates. AI will certainly not be the only factor in the development of a successful antiviral drug or vaccine against SARS-CoV-2. Nevertheless, it is important for the scientific community to realize that we have powerful AI technology which can at the very least speed up the process. In this time of need, it is essential that pharmaceutical companies and basic science laboratories cooperate with industry leaders such as Google and IBM that have powerful AI capabilities. With COVID-19 spreading in our local communities, it is time to utilize our AI capabilities in the race to develop an effective vaccine and antivirals. While AI is not a panacea for all the unprecedented challenges created by this pandemic, it can create innovation in multiple diverse sectors, which may ultimately aid clinicians, patients, and worried citizens alike. Emerging approaches combining integrative medicine with AI could create unique solutions and aid in the fight against this deadly pandemic [48].

A rolling review is one of the regulatory tools that the agency uses to speed up the assessment of a promising medicine or vaccine during a public health emergency. Normally, as has been discussed, all data on a medicine's effectiveness, safety, and quality and all required documents must be submitted at the start of the evaluation in a formal application for marketing authorization. In the case of a rolling review, the CHMP (Committee on Human Medicinal Products) reviews data as they become available from ongoing studies, before deciding that sufficient data are available and that a formal application should be submitted by the company [49]. The EMA (European Medicines Agency) said that large-scale clinical trials involving several thousands of people are ongoing, and results will become available over the coming weeks and months. These results will provide information on how effective the vaccine is in protecting people against coronavirus and will be assessed in later rolling review cycles. All the available data on the safety of the vaccine emerging from these studies, as well as data on its quality (such as its ingredients and the way it is produced), will also be reviewed, the agency stated in an official release. The rolling review will continue until enough evidence is available to support a formal marketing authorization application. EMA will complete its assessment according to its usual standards for quality, safety, and effectiveness. While the overall review timeline cannot be forecast yet, the process should be shorter than a regular evaluation due to the time gained during the rolling review. The rolling review process has been used previously in the assessment of the COVID-19 medicine, Veklury (remdesivir). Once the regulator has enough information to make a decision, the companies will submit a formal application. The agency started the same process last week for another coronavirus vaccine candidate from AstraZeneca Plc and the University of Oxford. The health regulator had started reviewing data on the AstraZeneca COVID-19 vaccine in real time, the first of such moves aimed at speeding up any approval process in the region for a vaccine [49].

10.1.7 Future activities

There seems to be some issues that have been discussed as promising perspectives in an area of AI-powdered drug development.

10.1.7.1 Future developments and concerns for drug discovery

Since the modern ML methods are very fast and can consider several design goals in parallel, the ongoing drug design software was trained to recognize important features and characteristics of known drugs. The obtained models were then used to automatically assemble new molecules with these learned desired properties from scratch [50]. Since drug design algorithms are not able to efficiently search the whole chemical space, the recurrent neural networks (RNNs) are now able to, through learning. Although the RNNs do not really understand chemical structures, they learn rules about how to generate novel character strings that correspond to molecules within the chemical space, so that this approach still requires human input. The 3D image along with speech recognition utilizes ML to screen compounds quickly; however, again it requires a human counterpart [51]. The drug discovery field has always been protective of intellectual property (IP). Given the huge amount of data required for ML to be successful, it is unsurprising that issues with IP are even bigger here [50]. It was noted that IP in AI has two main aspects – subjective and objective. The subjective issues center around to identify the true inventor. The Patent Act, issued in Japan and the United States (which the latter has shifted from the pre-invention principle to the first-to-file principle to identify the true and first inventor since 2013), states that an invention created solely by an AI cannot be protected by the patent act and rather goes to the public domain, although this needs extensive arguments and needs to be debated in detail from the respective field of professionals and expertise [52]. While data privacy and IP are important considerations, AI raises various other ethical considerations. For example, patient data are invaluable to drug discovery research, but patient privacy needs to be maintained [50]. Clearly, AI is already helping drug discovery – it can help identify drug targets, find good molecules from data libraries, suggest chemical modifications, and identify candidates for repurposing before putting the candidate to animal and human clinical trials. It was expected that one of the next big advances will be a much tighter integration with automation that allows us to move from an augmented drug design paradigm where the design chemist takes all the decisions to an autonomous drug design paradigm, where the system can autonomously decide which compound to make next, and fully autonomous laboratories iterate through the design–make–test–analyze cycle of drug discovery without direct human intervention. The result could be the delivery of better starting points for drug discovery faster [50, 51]. The availability of robust data sets and the need for investment to access AI technology could also prove to be hurdles. Mega big data and large amounts of computation needed for AI to work effectively will mean that computation will need to be faster. The age of supercomputers will likely be replaced by a quantum computer or another technology that can do the job in minutes instead of hours [42]. With rapid recent advances in quantum technology, we are close to the threshold of quantum devices whose computational powers can exceed those of classical supercomputers. A quantum computer can be used to elucidate reaction mechanisms in complex chemical systems, using the open problem of biological nitrogen fixation in nitrogenase as an example. It was mentioned that quantum computers will be able to tackle important problems in chemistry without requiring exorbitant resources [53].

10.1.7.2 In silico drug development

In biology and other experimental sciences, an in silico experiment is one performed on computer or via computer simulation. The phrase is pseudo-Latin for "in silicon," referring to silicon in computer chips. In particular, in the healthcare field, it means numerical simulations of biological mechanisms and the efficacy of new drugs. It is also used in medical and dental research area, similar to terms in vivo or in vitro [54].

Drug discovery and development is a very complicated, time-consuming process and there are many factors responsible for the failure of different drugs such as lack of effectiveness, side effects, poor pharmacokinetics, and marketable reasons. ML computational methods for predicting compounds with pharmacological activity, specific pharmacodynamic, and ADMET properties are being increasingly applied in drug discovery and evaluation. Recently, ML techniques such as ANNs, SVMs, and genetic programming have been explored for predicting inhibitors, antagonists, blockers, agonists, activators, and substrates of proteins related to specific therapeutic targets. These methods are particularly useful for screening compound libraries of diverse chemical structures, "noisy" and high-dimensional data to complement QSAR (quality structure–activity relationship) methods, and in cases of unavailable receptor 3D structure to complement structure-based methods. A variety of studies have demonstrated the potential of ML methods for predicting compounds as potential drug candidates [55, 56]. The in silico drug design is a vast field in which the different sides of basic research and practice are combined and inspire each other, modern techniques such as QSAR/QSPR (quality structure–property relationships), structure-based design, combinatorial library design, cheminformatics, bioinformatics, and the increasing number of biological and chemical databases are used in the field [57]. Furthermore, large numbers of the available tools provide a much-developed basis for the design of ligands and inhibitors with preferred specificity. Wadood et al. [58] reviewed the process of in silico drug design. It was concluded that (i) during the process of selection of novel drug candidates many essential steps are taken to eliminate such compounds that have side effects and also show interaction with other drugs; (ii) in silico drug designing software plays an important role to design innovative proteins or drugs in biotechnology or the pharmaceutical field; (iii) the drug designing software and programs are used to examine molecular modeling of gene, gene expression, gene sequence analysis; and 3D structure of proteins and (iv) in silico methods have been of great importance in target identification and in prediction of novel drugs.

Bruno et al. [59] analyzed the early steps of the drug-discovery pipeline, describing the sequence of steps from disease selection to lead optimization and focusing on the most common in silico tools used to assess attrition risks and build a mitigation plan. It was found that (i) a comprehensive list of widely used in silico tools, databases, and public initiatives that can be effectively implemented and used in the drug discovery pipeline has been provided; (ii) a few examples of how these tools can be problem-solving and how they may increase the success rate of a drug discovery and development program have been also provided; and (iii) finally, selected examples where the application of in silico tools had effectively contributed to the development of marketed drugs or clinical candidates will be given. Based on these findings, it was concluded that the in silico toolbox finds great application in every step of early drug discovery: (1) target identification and validation; (2) hit identification; (3) hit-to-lead; (4) lead optimization; and (5) each of these steps has been described in detail, providing a useful overview on the role played by in silico tools in the decision-making process to speed up the discovery of new drugs.

10.1.7.3 Consortium organization

As Wu [64] pointed out that AI-assisted drug discovery is a welcome change to the pharmaceutical industry historically plagued with high cost of R&D that amounts to little payoff, but, making full use of the technology will require collaborative efforts to overcome some inherent challenges in the industry, there is an urgent need to organize a multidisciplinary structure to develop new drugs which are shifting from low-mix high-volume production to high-mix low-volume production partially to facilitate the drugs for precision medicine.

Launched a decade ago, the US FDA's Critical Path Initiative has helped catalyze the formation of many consortia focused on drug development challenges. The term "translational science" (also known as critical path science or regulatory science) is used as a description for the collection of disciplines that will drive the establishment of next-generation development pathways, tools, and regulatory requirements. Translational science requires substantial resource investments, not only in terms of financial commitments, but also in acquiring relevant scientific expertise and large data sets [60]. Activation of industry-academia-government collaboration from drug discovery sites regarding pre-formulation research and its evaluation technology to maximize the effect of new drug candidate compounds, and translational research such as exploratory clinical trial research for quick verification in humans have been advanced.

Consortia are enabling drug discovery in areas that individual organizations are unable to support alone because of the high risk or the need to pool information. Desirable features that can underpin the success of such consortia were discussed [61]. In the past 15 years, large pharma companies have transitioned from an internally focused research model to community-sourcing, collaborative approaches that are now increasingly driving innovation in drug discovery [61, 62]. Such collaborations often begin as one-to-one relationships with a partner organization that performs a high-risk step, such as target validation or lead discovery [63, 64]. In some areas, this type of collaboration has subsequently matured into larger consortia, as increasing numbers of pharma companies have become comfortable with sharing risks, costs, and opportunities with each other, as well as with academic groups, thereby pooling funds and resources and also accessing external funding [60]. Reason and objectives for forming consortia are based on (i) the cost of progressing independently into clinical testing is beyond the resources of an academic or small/medium-sized enterprise, (ii) the likelihood of failure is perceived to be high, (iii) the perceived commercial attractiveness is moderate, and (iv) the consortium's cornerstone of charity and pharma is well known and trusted [60].

It has been a long time since the birth of new drugs, such as diversification of drug discovery fields and open innovation from pharmaceutical companies to academia to meet recent unmet medical needs, and formation of a consortium consisting of many different fields and industries is required. Therefore, every participants of the consortium can share the latest information on pharmacokinetic evaluation, human absorption prediction, formulation, and measurement basic technology in the drug development stage and interact with researchers in the same field. Using AI technology to data analysis or developing platform, the developing new drugs can be accelerated. Organizational format of life intelligence consortia [65] can be either vertical structure based on similar disciplines among industries and academia, horizontal structure composed of mainly pharmaceutical industries, or transdisciplinary structure comprising research agencies, industries, and academia from wider disciplines.

10.2 AI-incorporated drug delivery systems

Employing AI fed by a big data along with IoT technology and nanoscience and nanoengineering, an expanded concept of the DDS (drug delivery system) has been enabled, widening from slow-releasing DDS for lessening administration frequency, self-administrative DDS instead of an oral administration, enhancing targetability with controlled side effect, minor patient–friendly (children or elders) DDS, on-demand-type DDS specifically designed for a patient, DDS that prevents adjustable administration plan according to examine results, DDS with which healthcare providers or health professions can administrate drugs remotely, DDS that can send signals to healthcare providers or health professions whether the patient took medicine or not, and DDS that delivers active ingredients of pharmaceutical products to the right place, in the right amount and in the right time. It is used for various purposes such as enhancing the therapeutic effect, reducing side effects, making it easier to administer, and reducing the number of administrations. This technology is indispensable for maximizing the effects of drugs and reducing adverse side effects. Among them, DDS technology utilizing nanotechnology has been researched and developed for many years. Well known as DDS is the controlled release formulation, which is a technology that controls the rate at which the active ingredient dissolves in the body, and it is effective for a long time, and it can slow down the rise in blood concentration and suppress side effects [66, 67]. There are various DDSs to deliver drugs to a target cells or tissues effectively and efficiently. They include (1) PEG coating, (2) microencapsulation, (3) liposome preparation, (4) ADCs preparation, (5) prodrug, (6) sublingual drug delivery, or (7) transdermal drug delivery system. (1) PEG (polyethylene glycol) coating of PEGylation is a coating the surface of nanoparticles with peg and commonly used approach for improving efficiency of drug and gene delivery to target cells and tissues [66]. (2) Microencapsulation is a process in which a very tiny droplet of particle such as solid, liquid, or even gas can be entrapped, coated, or surrounded with a polymeric particle, and it is a better drug delivery system than conventional drug delivery system accompanied with minimized side effect and targeted action [67]. (3) A liposome preparation is a drug containing an active ingredient encapsulated in a liposome for improving pharmacokinetics, intracellular dynamics, and so on. Liposomes are microvesicles consisting of bilayer membranes of amphipathic lipid molecules with an aqueous fraction inside [68, 69]. (4) Antibody drug conjugates (ADCs) are drugs designed to target specific cancer cells and release a toxic drug into the cancer cell. Antibody drug conjugates work like a "smart bomb" directed against cancer cells. These drugs are composed of three parts: an antibody drug that is specific for the type of cancer being targeted, a cytotoxic chemotherapy drug, and a linker protein to hold the two parts together. The drug is administered, usually through a vein, and the antibody part of the drug targets the specific cancer cell protein that it was made to find [70]. (5) A prodrug is a medication or compound that, after administration, is metabolized (i.e., converted within the body) into a pharmacologically active drug. Instead of administering a drug directly, a corresponding prodrug can be used to improve how the drug is absorbed, distributed, metabolized, and excreted. (6) Sublingual drug delivery is considered to be a promising route for faster and direct absorption of drug into systemic circulation. In buccal cavity, sublingual area is most permeable for drug absorption. The portion of drug absorbed through the sublingual blood vessels bypasses the hepatic first-pass metabolism, which leads to greater bioavailability with better patient compliance [72]. (7) Transdermal drug delivery systems (TDDS) are defined as self-contained, discrete dosage forms that are also known as medicated adhesive patches when patches are applied to the intact skin and deliver the drug through the skin at a controlled rate to the systemic circulation. An advantage of a transdermal drug delivery route over other types of medication delivery such as oral, topical, intravenous, and intramuscular is that the patch provides a controlled release of the medication into the patient, usually through either a porous membrane covering a reservoir of medication or through body heat melting thin layers of medication embedded in the adhesive [73, 74].

It is very difficult to accurately predict the drug release kinetic of the PEG-type drug carrier materials due to environmental variables in the body such as pH and temperature. Boztepe et al. [75] synthesized a pH- and temperature-responsive poly(N-isopropyl acrylamide-co-acrylic acid)/poly(ethylene glycol) (poly(NIPAAm-co-AAc)/PEG) interpenetrating polymer network (IPN) hydrogel by free radical solution polymerization in the presence of poly(NIPAAm-co-AAc) microgels and PEG. It was reported that (i) the synthesized IPN hydrogels showed rapid pH- and temperature-responsive deswelling behavior and (ii) the developed ANN modeling exhibits the best performance in modeling the doxorubicin (DOX) release behavior of poly(NIPAAm-co-AAc)/PEG IPN hydrogels.

The biopharmaceutical classification system (BCS), which was implemented by the US FDA as a scientific framework to predict intestinal drug absorption, is a system to differentiate the drugs substances according to their aqueous solubility and intestinal permeability [76]. When combined with drug product dissolution, the BCS considers three major factors that govern the rate and extent of drug absorption from immediate-release (IR) solid oral dosage forms: dissolution, solubility, and intestinal permeability [77].

In drug delivery, there is often a trade-off between effective killing of the pathogen, and harmful side effects associated with the treatment. Due to the difficulty in testing every dosing scenario experimentally, a computational approach will be helpful to assist with the prediction of effective drug delivery methods. Based on this background, Li et al. [78] have developed a data-driven predictive system, using ML techniques, to determine, in silico, the effectiveness of drug dosing. The system framework is scalable, autonomous, robust, and has the ability to predict the effectiveness of the current drug treatment and the subsequent drug-pathogen dynamics. It was mentioned that (1) the system consisted of a dynamic model incorporating both the drug concentration and pathogen population into distinct states, (2) these states are then analyzed using a temporal model to describe the drug-cell interactions over time, and (3) the dynamic drug–cell interactions are learned in an adaptive fashion and used to make sequential predictions on the effectiveness of the dosing strategy.

Increasing interest has been attracted toward the application of AI technology for analyzing and interpreting the biological or genetic information, accelerated drug discovery, and identification of the selective small-molecule modulators or rare molecules and prediction of their behavior. Application of the automated workflows and databases for rapid analysis of the huge amounts of data and ANNs for development of the novel hypotheses and treatment strategies, prediction of disease progression, and evaluation of the pharmacological profiles of drug candidates may significantly improve treatment outcomes [79]. Target fishing (TF) by rapid prediction or identification of the biological targets might be of great help for linking targets to the novel compounds. AI and TF methods in association with human expertise may indeed revolutionize the current theranostic strategies; meanwhile, validation approaches are necessary to overcome the potential challenges and ensure higher accuracy [79].

10.3 Side effect and drug repositioning

10.3.1 Side effect and its prediction

The drug is approved by the government if it is judged that the balance between the efficacy and the safety is acceptable. Although the notice of side effects has been recommended in the attached document along with lists of active ingredients and nonactive ones, since clinical data collected for the final approval is still limited, there are many side effects reported for the first time after it is commercially available.

Accordingly, the new drug, which has just been approved, is required for a special attention to side effect information and extensive and serous efforts for collecting additional information regarding hidden side effects are conducted by pharmaceutical industries. In recent years, new drugs such as biopharmaceuticals, including antibody drugs of different mechanisms of action from low-molecular-weight drugs as conventional mainstream, have been also approved for practical administration. As a specified part of personalized medical treatment, studies are under way to investigate the ease of therapeutic effects and side effects at the gene level for each patient, and there are several side effects that can be explained, to some extent, pharmacologically. However, in most cases, it would be very hard to predict the risk of side effects for each patient. Due to the side effects of such drugs, even when the drug is used properly, it appears to be impossible to completely prevent the occurrence of side effects, resulting rarely to severe side effects. Hence, pharmaceutical industry should recognize these situations and respond to fulfill explanation obligations on potential adverse chemical reactions of their produced drugs. In most cases, the administrated drug works effectively to target location where it needs, but in other cases, it might affect differently. Such unfavorable effects other than the prim purpose are called the side effects; for example, when you take a cold medicine, chief complaint of runny nose was controlled, but you should feel very sleepiness. Side effects can appear for a variety of reasons: (1) depending on nature of the drug, it may work at a site different from the purpose of treatment, or it may have an effect other than the purpose; (2) depending on how to use the drug, it may occur by mistakes of the time, interval, and amount of drinking, or by drinking with other medicines, foods, and so on; (3) depending on the patient's physical condition and lifestyle, it may occur due to age, gender, weight, preferences, and so on; or (4) depending on the patient's medical condition, you might be susceptible to medicine when you are not feeling well. Hence, side effects are caused by twofold reasons: drug's inherent chemical action and drug's uncontrollable factors such as misuse or improper usage of assigned drugs. Under these circumstances, unfortunately, allergies are unpredictable for when and who will develop allergic reaction and when it might occur [80–83].

Identifying the potential side effects of drugs is crucial in clinical trials in the pharmaceutical industry. The existing side effect prediction methods mainly focus on the chemical and biological properties of drugs. Employing various methods using diverse information such as drug–drug interactions from DrugBank, drug–drug interactions from network, single nucleotide polymorphisms, and side effect anatomical hierarchy as well as chemical structures, indications, and target. Seo et al.'s [84] proposed method is based on the assumption that properties used in drug repositioning studies could be utilized to predict side effects because the phenotypic expression of a side effect is similar to that of the disease. It was reported that (i) the prediction results using the proposed method showed a 3.5% improvement in the area under the curve over that obtained when only chemical, indication, and target features were used and (ii) the random forest model delivered outstanding results for all combinations of feature types. A central issue in drug risk–benefit assessment is identifying frequencies of side effects in humans. Currently, frequencies are experimentally determined in randomized controlled clinical trials. Hence, Galeano et al. [85] presented an ML framework for computationally predicting frequencies of drug side effects with the matrix decomposition algorithm learning latent signatures of drugs and side effects that are both reproducible and biologically interpretable. It was indicated that (i) the usefulness of our approach on 759 structurally and therapeutically diverse drugs and 994 side effects from all human physiological systems were reported; (ii) the current approach can be applied to any drug for which a small number of side effect frequencies have been identified, in order to predict the frequencies of further, yet unidentified, side effects; and (iii) it was shown that the model is informative of the biology underlying drug activity: individual components of the drug signatures are related to the distinct anatomical categories of the drugs and to the specific routes of drug administration.

It was reported that a technology was developed by Fuji Xerox to determine the risk of side effects of medicines using AI, by extracting documents showing the relationship between drugs and side effects. A large amount of paper data is classified using NLP technology, and the part related to side effects is extracted. Pharmaceuticals are marketed through a process of clinical trials and approval review. Side effects have been investigated in the process, but new unknown risks may be discovered in subsequent administration to the patient. Pharmaceutical companies are required to collect information about side effects even after commercial use and usually, a large volume of papers are hand-classified into side effects and related documents. Using NLP technology, it was identified which parts of the paper have descriptions of side effects. With the newly developed technology, the relationship between symptoms and drug names are searched and whether there are side effects is determined. It automatically extracts areas that say that symptoms such as headache, abdominal pain, and numbness in the hands and feet are side effects of the drug. DL is highly accurate, but it is a problem to introduce it because the basis of judgment is not known. On the other hand, Fuji Xerox technology can extract the basis for determining the presence or absence of side effects. Because it can be shown in a way that can be understood by people, it is easy to introduce it as part of the process by which people work [86]. Japan Science and Technology Agency announced that a Top of FormBottom of Form modernity-independent AI system has been successfully developed to predict the side effects, adverse events, and efficacy of the test drug in humans, regardless of the type of drug [87]. The system is characterized by (1) only systemic exhaustive transcriptome of mice administered to the test drug as an index, the structure of the test drug, does not require information such as a target, bias-free system and (2) by stratification of learning data, it is possible to predict the occurrence of side effects by gender and age group. The AI system "hMDB" (humanized Mouse DataBase) is a system that predicts the side effects and efficacy of a substance in humans using systemic transcriptome data from mice treated with a substance. The further advanced system of "hMDB-i" (humanized

Mouse DataBase, indivivized) can predict side effects and adverse events and frequency by gender and age group, and "hMDB/LP method" (hMDB/Link Prediction Method), which can be expected to be used for drug re-positioning [88]. The hMDB predicts the side effects and efficacy of the substance in humans with high accuracy in the following steps: It is a technique that tests only in animals such as mice, and outputs the results of "tests in humans" as a test result. The main procedure is as follows: (1) The computer will learn the systemically comprehensive gene expression patterns that occur in animals such as mice that have administered various drugs currently in circulation, and information on side effects and efficacy in humans (real world data) that have been reported in the world for each drug, and (2) a substance (such as a test drug) that has an unknown action on humans is administered to an animal, and information measuring a systemically comprehensive gene expression pattern is in out to the computer in (1) above; as a result prediction of side effects and efficacy caused by the substance in humans are output. In the prediction of efficacy, it was shown that hMDB can predict not only the known efficacy of the target drug but also the efficacy that has never been reported before, by associating each drug with the adverse events reported in the real world (Link Prediction), suggesting that this method can be used for drug re-positioning [87].

Yao et al. [89] developed an ontology-based model for AI-assisted medicine side-effect prediction, by which three main components, including the drug model, the treatment model, and the AI-assisted prediction model, are presented. To validate the proposed model, an ANN structure is established and trained by 242 traditional Chinese medicine prescriptions. These data are gathered and classified from the most famous ancient traditional Chinese medicine book, and more than one thousand side effect reports, in which two ontology-based attributions, hot and cold, are introduced to evaluate whether the prescription will cause side effect or not. It was reported that (i) the results preliminarily reveal that it is a relationship between the ontology-based attributions and the corresponding predicted indicator that can be learnt by AI for predicting the side effect, which suggests the proposed model has a potential in AI-assisted side effect prediction; (ii) however, it should be noted that the proposed model highly depends on the sufficient clinic data, and hereby, much deeper exploration is important for enhancing the accuracy of the prediction.

Strictly speaking, the term "side effect" contains two distinct meanings: side effects and side reactions. The action of the medicine used for treatment and prevention is called the main action, while the action different from the main action is called a side effect. Side effects in a broad sense include both harmful and nonharmful effects on animals. Therefore, side effects are undesired harmful action, being separate from the action of the drug to cure the disease. There is a medical causality between drugs and side effects, as shown in Figure 10.4 [90]. On the other hand, the side reaction is symptoms such as swelling at the site of vaccination, redness, fever, and rash, which occur after receiving vaccination. Anaphylaxis, which we hear a lot these days, is also one of the side reactions. To this case, there is a causality between vaccination and this type of side reaction. There is also the term "adverse event" along with side effects and side reactions.

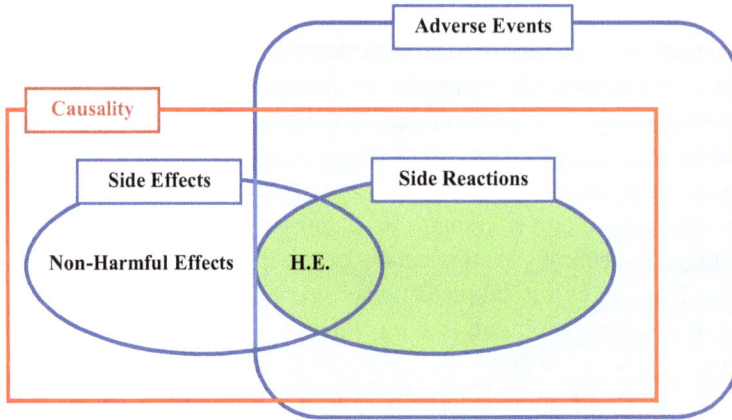

Figure 10.4: Relationship among side effects, side reactions, and adverse events of drug [90].

10.3.2 Drug repositioning

Buvailo [91] introduced an interesting program called the drug repurposing program. Drug repurposing is one of the golden mines for AI-based technologies to drive value since a lot of data is already known about the drug in question. Repurposing previously known drugs or late-stage drug candidates toward new therapeutic areas is a desired strategy for many biopharmaceutical companies as it presents less risk of unexpected toxicity or side effects in human trials, and, likely, less R&D spend [50]. Since, reading, clustering, and interpreting large volumes of textual data is among the most developed use cases for AI-based algorithms, revisiting various types of publications should be inevitable. It comes in handy for the life sciences industry since the number of research publications in the field is growing enormously and it is hard for researchers to sift through vast amounts of data arriving on a daily basis to validate or discard research hypotheses. Development of biomarkers is an important task not only in the context of medical diagnostics but also in the frameworks of drug discovery and development programs. For example, predictive biomarkers are used to identify potential responders to a molecular targeted therapy before the drug is tested in humans [91].

Drug repositioning (or drug repurposing) or redevelopment of existing drug is to find effective efficacy in another disease from therapeutic agents effective for one existing disease. Drugs used for drug re-positioning have already been tested for human safety and pharmacokinetics, so some tests can be skipped, and drug manufacturing methods have been established to reduce development time and R&D costs. However, on the other hand, the possibility of side effects other than the intended efficacy cannot be discarded [92]. Many drugs have been developed and recognized by finding new effects of existing drugs. Characteristic molecular features are often shared among different diseases. Here are just a few to mention. Minoxidil, originally developed as hypertension medicine, is also applied as hair growing agent; bimatoprost as a glaucoma medicine for an eyelash stretch cosmetic; bupropion as an antidepressant for an adjuvant for smoking cessation; rebamipide as a stomach medicine for eye drops for dry eyes; sildenafil, an angina medicine, for an erectile dysfunction and pulmonary hypertension [93]. The drug repositioning utilizes the combined efforts of activity-based or experimental and in silico-based or computational approaches to develop/identify the new uses of drug molecules on a rational basis. It is, therefore, believed to be an emerging strategy where existing medicines, having already been tested safe in humans, are redirected based on a valid target molecule to combat particularly, rare, difficult-to-treat diseases, and neglected diseases [94].

Drug repositioning or repurposing is intended to find alternative uses for a pioneering drug or a drug that is made by another innovator. It mostly involves developing approved or failed compounds. Drug repositioning is expanding in the area of rare and neglected diseases. It is a new way of approaching drug compounds and targets that have been "derisked" during the development stages, which accelerates the process and thus saves money, because the drug could be produced with less effort and marketed with a huge profit margin. Drug repositioning has helped to mitigate failures in drug discovery and has been associated with therapeutic breakthroughs. For example, the thalidomide medicine that had deleterious effects in the past has found a new indication. This is a growth opportunity that brings value to society. However, there are divided interests over the choice of the repurposed drug and the objective of such an endeavor [95]. Drug-repositioning refers to research methods that aim to find new efficacy and put them into practical use, such as existing drugs against which human safety and pharmacokinetics have already been confirmed by actual results. The greatest advantages are "certainty" due to its commercial track record and confirmation of safety and pharmacokinetics in clinical levels, and "low cost" that can use many existing data. In old age diseases, it is necessary to pay particular attention to the side effects of drugs, but drug re-positioning of existing drugs with confirmed safety and side effect information compared to newly developed drugs is considered to be useful for the development of new therapeutic agents for older diseases.

Biotechnological and pharmaceutical companies need to maximize the value of their pipeline, and one of the most cost-effective strategies is to identify new indications for drugs already in the market or in the later stages of clinical development. Even if they are not effective enough for the intended initial indication, these compounds have proven to be safe and, consequently, have to undergo much less regulatory scrutiny. Drug repositioning can no longer rely on serendipity, but has to be a directed, rational and optimized process. Staying close to this philosophy, Anaxomics' TPMS (therapeutic performance mapping system) proprietary technology explores the full potential of systems biology analysis to identify new indications for already existing drugs. By navigating through the pathophysiological maps of diseases, valuable targets and identify compounds can be discovered that modulate TPMS [96]. The systems biology in TPMS allows us to offer two different approaches to drug repurposing, depending on the specific demands of the client [97]: (1) drug-oriented repositioning, in which understanding the full indications potential of a newly characterized target, identification of new uses for individual drugs or combinations from your library, early exploring of the full range of potential indications in candidates undergoing clinical trials and identification of new uses for commercial drugs already in the market are enabled; (2) disease-oriented repositioning, which puts the focus on a pathological condition and attempts to identify existing drugs or combinations of drugs to treat the disease of interest.

Disease-based strategies depend on data related to diseases such as phenotypic traits information, side effects, and indications information as the foundation to predict therapeutic potentials and novel indications for existing drugs. Disease-based strategies are used when there is either insufficient drug-related data available or when the motivation in studying how pharmacological characteristics can contribute to drug repositioning effort concentrated on a particular disease [98]. Although computational drug repositioning can be of enormous benefit to humanity by discovering new indications for approved drugs, speeding up the process of developing new drugs, and giving a second chance to withdrawn and failed drugs [98], there are still several concerned issues, including (1) establishment of new patent strategies, (2) measures for non-adaptive use by later products, (3) drug price countermeasures (low drug price problems based on adaptation of existing products) and (4) the regulatory system for approval affecting the stream of DR significantly [99–102]. Dias et al. [103] employed a network medicine approach to investigate the molecular characteristics of PNDs (psychiatric and neurological disorders) and identify novel drug candidates for repositioning. Using IBM Watson for Drug Discovery, a powerful ML text-mining application, we built knowledge networks containing connections between PNDs and genes or drugs mentioned in the scientific literature published in the past 50 years. This approach revealed several drugs that target key PND-related genes, which have never been used to treat these disorders to date. We validate our framework by detecting drugs that have been undergoing clinical trial for treating some of the PNDs, but have no published results in their support. It was reported that the obtained data provides comprehensive insights into the molecular pathology of PNDs and offers promising drug repositioning candidates for follow-up trials. Liu et al. [104] presented an efficient and easily customized framework for generating and testing multiple candidates for drug repurposing using a retrospective analysis of real-world data. Building upon well-established causal inference and DL methods, the framework emulates randomized clinical trials for drugs present in a large-scale medical claims database. It was demonstrated that the framework on a coronary artery disease cohort of millions of patients, indicating that drugs were successfully identified and drug combinations that substantially improve the coronary artery disease outcomes but have not been indicated for treating coronary artery disease, paving the way for drug repurposing.

10.3.3 Biosimilar and generic drugs

Proteins, which are one of the components of living substances, are closely related to the life phenomena of a wide range of living animals, from viruses to mammals. The variety is enormous, and it is said that there are more than 10 billion species on the Earth's sphere, and human being possesses about 100,000 kinds of proteins. Each protein has functions and specificity essential for life activities, and their defects or deficiencies cause the life homeostasis to collapse. Protein-based drugs and antibody drugs that make use of the inherent functions of proteins in living organisms have been put to practical use in medical field. Simply speaking, drug discovery is the process of first finding a protein that causes a disease, and then finding (designing) a new compound that interacts with the protein to control its function. This basic concept should not be different among branded drugs, biosimilar drugs or generic drugs. Although, in past, a large number of compounds weer prepared to react with proteins one by one to find out if they are effective as drugs, currently, a proper usage of a computer to virtually evaluate binding capability of proteins and compounds and to exmaine further drug action efficacy will enable to narrow down the number of drug candidates [105–109].

In this section, we will be reviewing biosimilar drug and generic drug and differences between them. Generic, biologic, and biosimilar are all terms for non-brand medications (also known as a drug or therapy) determined by the US FDA as equivalent to (the same as) the brand medication. Although biosimilar drugs are often confused with generic drugs, there are a few common factors: (i) both are marketed as cheaper versions of costly branded drugs, (ii) both are available when drug companies' exclusive patents on expensive new drugs expire, as known as off-patent drug, and (iii) both are designed to have the same clinical effect as their pricier counterparts. A biological product, often referred to as a biologic, reviewed, approved and regulated by the FDA and biologics are generally very large and complex molecules that are manufactured by using living organisms, such as microorganisms, plants, or animal cells. Some common types of biologics include proteins, monoclonal antibodies and some vaccines and because biologics are difficult to replicate, a need for biosimilar products was created [105, 106]. Generic drugs have identical active ingredients to the brand product and must demonstrate bioequivalence to the brand product.

Generic drugs are chemically identical to the original branded drug and, as such, cost significantly less because they do not require much testing. In general, generic drugs cost 40–50% less than the brand product; whilst biosimilars are closer to 15–20% cheaper because of the amount the drug manufacturer has to spend on testing. Because biosimilars are made from living organisms, though, and do not contain identical ingredients to their name-brand counterparts, they still require some testing. So, they cost more than generics, but less than the branded biologic [105]. According to the FDA, a generic drug is a medication created to be the same as an already marketed brand-name drug in dosage form, safety, strength, route of administration, quality, performance characteristics, and intended use. These similarities help to demonstrate bioequivalence, which means that a generic medicine works in the same way and provides the same clinical benefit as its brand-name version. In other words, you can take a generic medicine as an equal substitute for its brand-name counterpart. A generic medication can become available when a brand-name medication's patent protections have expired. Generic versions of branded medicines have been around for a long time, as has the process through which generics are introduced to the market: the original brand product has a finite patent life, at the end of which other manufacturers can apply for a license to manufacture and market a generic version [107]. Generic pharmaceuticals require less research and development than their brand-name counterparts. As a result, AI applications for R&D do not seem to be the most prominent solutions for generic drug companies; indicating that despite the lack of precedence, there may be many areas in which AI could help generic drug companies [108].

On the other hand, an FDA-approved biosimilar [105, 106] is a biological product that has been assessed by the FDA to be similar to an existing FDA-approved biologic. Biosimilars are not generic drugs. The FDA refers to the original biologic as the reference product and biosimilars are like the reference product. To be approved, biosimilars must show similar purity, molecular structure, and bioactivity, although there are no clinically meaningful differences between the reference product and the biosimilar based on safety, purity, and potency. The reference product and biosimilars slightly change over time due to the manufacturing process. Branded drugs are either synthetic, meaning they're made from a chemical process, or biological, meaning they're made from living sources. Synthetic branded drugs can be exactly replicated into more affordable generic versions, but because biologics involve large, complex molecules, they cannot. That is where biosimilars come into play. Because there is no assurance that the large protein molecule drugs are chemically identical there was debate about calling large protein molecule drugs generic, which led to the birth of biosimilars. Biosimilars have the same clinical effect as a generic but are only as similar to the original branded drug as validation technologies can confirm [109]. Another key difference is that generics are copies of synthetic drugs, while biosimilars are modeled after drugs that use living organisms as important ingredients. But many experts hope the two will share a critical commonality and that, like generics, biosimilars will dramatically lower the cost of biologic drugs. For a biosimilar drug to receive FDA approval, it must be highly similar to the original biological drug and contain no clinically meaningful differences, although there may be minor differences in clinically inactive ingredients. According to the National Cancer Institute, the biosimilar also must prove to be as safe as, work as well as, and work in the same way as the original drug, and be used in the same way, at the same dose, and for the same condition [110].

Biosimilars are up to 1,000 times the size of small-molecule generic drugs and are far more structurally complex [111]. Additionally, biosimilars are manufactured in living cell lines using processes that cannot be exactly replicated from one manufacturer to the next. Therefore, a biosimilar, unlike a small-molecule generic, is not an exact copy of its reference product. The underlying differences in size, complexity, and manufacturing processes are why biosimilars are fundamentally different from generics. One key reason a biosimilar cannot be identical to its reference biologic is due to post-translational modifications such as glycosylation – the addition of glycans (carbohydrate groups) to a protein within the producing cell. These are unique to each specific cell line and growth conditions and can have a profound impact on the molecule's biological effects, including drug clearance rates and immunogenicity [112, 113]. Therefore, biosimilars are more complicated to develop and the regulatory pathway for approval is more complex than for generic drugs [114].

10.4 Drug information management and pharmacist

10.4.1 Drug information management

Information regarding the marketed drugs (including prescription and over-the-counter drugs) and other marked products includes a record of approval, potential side effects, safety information, and active as well as inactive ingredients [115]. There are several drug information systems as well as database to support the everyday work of pharmacists and to provide a drug information management system that is effective for dosage management and medication guidance [116]. These include (1) DICSDrug Information System, which is a drug information retrieval system that can be flexibly tailored to its installation environment, whether stand-alone or linked to electronic medical records, ordering systems, dispensing systems, and so on. It can be used seamlessly alongside other systems in the modern medical workplace, comfortably providing high reliability and a wealth of information based on a stable environment. (2) DICS-PSPrescription Check System is aimed at the safe use of drugs. Its contents and data maintenance function can be used comprehensively and diversely with hospital information systems such as electronic medical record systems and ordering systems. (3) PICSwebDrug Management & Medication Guidance Support System, offering various kinds of supports for more efficient medication guidance than before, and has dramatically improved operability and visibility. With PICSweb, ward pharmacists in hospitals, who are required to have great expertise in drugs, can offer medication guidance for patients less stress. (4) PICSKsDrug Management & Medication Guidance Support System can be integrated flexibly to various systems. Users can use their input data on PICSks. It reduces the burden of double input tasks, and makes maximum use of the merits of system integration. (5) J-ReporterDrug Check Systems for Drugs Brought by Patients, by which the management of medication brought in is an important task for ward pharmacists in hospitals from the perspective of using drugs appropriately and preventing medical accidents. J-Reporter was created to offer check for drugs brought by patients. (6) Ward MeisterWard Pharmacist Support System has time management function and ward pharmacist work support function. The work time data of ward pharmacists can be output in a CSV file, which is convenient for creating work statistics forms later. Being a web-based system, data input is also possible from any terminal. With ward pharmacist work support function, patients can be monitored by ward, and drug interactions, high risk, and flow rate/dosage can be checked. Ward Meister can be integrated to the medication guidance of Infocom's PICSks. (7) DMEntryDrug Master Management System supports the centralization of the drug master files of systems running in medical institutions, and links to DICS-MASTER, Infocom's medical drug information database for hospitals. This system can improve the efficiency of creating a drug master file. (8) DICS-MASTERDrug Information Database can be used for various purposes: (i) it can be used to check prescriptions on electronic medical record systems, (ii) it can be used as a drug master file for electronic medical records, including various drug codes such as YJ codes, HOT codes etc. and (iii) it can be used for receipt checks and receipt analysis [116].

10.4.2 Prescription errors and management

Prescribing faults and prescription errors are major problems among medication errors. They occur both in general practice and in hospital, and although they are rarely fatal they can affect patients' safety and quality of healthcare [117]. A prescribing fault can arise from the choice of the wrong drug, the wrong dose, the wrong route of administration, and the wrong frequency or duration of treatment, but also from inappropriate or erroneous prescribing in relation to the characteristics of the individual patient or co-existing treatments; it may also depend on inadequate evaluation of potential harm deriving from a given treatment [118]. Errors in dose selection occur most commonly and represent >50% of all prescribing faults [117]. Acquisition of information through error-reporting systems is a prerequisite for preventing prescribing faults and prescription errors, as is the adoption of shared criteria for the appropriateness of procedures. Three major intervention strategies can be adopted: (1) reduction of complexity in the act of prescribing by the introduction of automation; (2) improved prescribers' knowledge by education and the use of online aids, and (3) feedback control systems and monitoring of the effects of interventions

[119]. It was also mentioned that careful evaluation of drug–drug interactions and all types of adverse reactions is necessarily part of a program aimed at improving patient safety and may require monitoring of plasma drug concentrations and evaluation of biomarkers of beneficial or adverse effects [117]. For pursuing these strategies, there could be establishing appropriate computerized systems, which can give significant benefits by guiding the prescription of optimal dosages and should translate into reduced time to therapeutic stabilization, reduced risks of adverse effects, and eventually reduced lengths of hospital stay [119]. Since electronic systems are not yet widely available, are expensive, and require training [120], comprehensive interventions aimed at improving patient safety using a systematic approach are progressing in different institutions, with the use of uniform medication charts, on which all the relevant clinical information is shown along with the prescriptions, so that transcription is abolished. This approach has been validated as a relatively simple alternative to electronic drug prescribing and dispensing systems [121].

Medical errors caused by human error are a serious problem in hospitals and clinics, but research is under way to reduce the occurrence of mistakes with assists from robots equipped with AI [122]. For instance, the medical center at University of California San Francisco (UCSF) has introduced a robotic dispensing medicine system since 2011. The prescription with bar code issued electronically by the doctor is sent to the robot to read the information, pick up the appropriate drug, and dispense the specified amount. The dispensing drug, after the nurse has confirmed that it matches the bar code of the prescription, is given to the patient. There would be two advantages for hospitals: (1) reducing human error in the wrong type and amount of medicine and (2) reducing the labor costs of pharmacists. AI-mounted dispensing robots are able to indicate inconsistencies between symptoms and the action of drugs or to stop medical accidents when a doctor prescribes the wrong medicine accidentally since the AI-driven robot can perform checking drug-to-drug reaction as well as side effects. As many other AI-powered tools, it is highly expected that its accuracy for prescription will be improved as more experience in dispensing. Moreover, Johnson & Johnson's subsidiary developed SEDASYS, an automated anesthesia system that will take the place of anesthesiologists and obtained approval for use by the US FDA in 2014. This AI-powered robot, when using anesthesia in endoscopy of the large intestine (colonoscopy), measures the patient's pulse, blood pressure, and the amount of oxygen are measured and inject automatically required amount of anesthesia. At this moment, since the clinical data of anesthesia by the robot is small, the use authorization of SEDASYS is limited to area which possesses relatively high safety in the endoscopic operation [122].

Research on a project that AI recommends pharmacist continuous guidance for each patient based on the content of the previous medication history and patient profile has been progressing [123]. The research group has begun to develop a system that uses AI to support medication instruction in order to prevent human error such as careless mistakes and ensure patient safety in pharmacist work that is directly related to human life. By linking to an electronic drug history system or the like, AI can learn barious data such as package inserts, drug history, prescriptions, medication notebooks, patient questionnaires, etc, so that the computer can make effective pharmaceutical inferences, based on the patient profile. It is recommended to derive a guidance policy that requires attention such as side effects and concomitant medications, and continue guidance based on previous medication history and guidance. For example, for the prescription content of the patient at his/her first visit, the system can check the complications, concomitant medications, past records on side effects, and so on on the screen. Based on the above information, AI makes a recommendation this patient on taking medication while paying attention to side effects, from the viewpoint of a risk avoidance principle. Recommended guidance sentences focusing on side effects are displayed from a huge combination pattern, and the percentage can be changed at the pharmacist's discretion. At the subsequent visit, the remaining medications, physical condition changes, concomitant medications, and so on were confirmed based on the previous medication instruction. The pharmaceutical reasoning system using AI is expected to contribute not only to avoid patient risk but also to improve the efficiency of pharmacist work such as prevention of omission of description of drug history. It is expected that the system will be put to practical use in the future, but the data input by the pharmacist should be accumulated, and AI will improve the learning ability, so it is expected that the accuracy of the pharmaceutical reasoning system will be further improved. The Pharmacist Support AI Solution system is a service that pharmacists perform at dispensing pharmacies to support operations such as "prescription inspection," "question inquiry," and "medication guidance" and help pharmacists improve the quality of their work and improve the efficiency of their work [124]. For example, pharmacists evaluate the appropriateness of the contents of the prescription and, if necessary, make a question inquiry to the prescribing physician. In addition, they provide medication guidance to explain to the patient the correct use of the drug. Such routine works done by pharmacists' in-person service operations can be extensively supported by utilizing AI. As a method of utilizing AI, a huge amount of dispensing data accumulated from the past, as well as records of question inquiries and medication guidance, are analyzed by pattern learning of AI technology. By using this, it is possible to grasp the necessity of confirmation of prescription and question inquiry [124].

It was reported that Sapporo Medical University, Fujitsu, and Fujitsu Hokuriku Systems recently announced that they will study clinical data of diabetic patients with AI and begin joint research on prescription optimization of oral hypoglycemic drugs to aim to develop a technology for predicting the effect of treatment so that the target value of HbA1c is less than 7.0% and in future this outcome will support the treatment of diabetic patients [125]. First of all, ML will be conducted by removing personal information from medical records, test results, prescription information, and so on of about 5,000 diabetic patients to create a learning model that predicts the effectiveness of treatment. A dataset will be created from prescription data for oral hypoglycemic medicines, as well as test values, for subject patients. Using the dataset created with the previous technology, the organizations will create trained models to predict the effects of treatment, by using ML to train models on the relationship between factors such as test values, types and combinations of medicines, and treatment success of failure results based on patterns of changes in HbA1c levels. Becasue it is essential for the basis of inferences made by AI technology to be easy to understand in computer systems that will form the foundations of society, such as hospital information systems, the organizations will take the selection of algorithms to be used into consideration. The organizations expect that, in the near future, the results obtained from this joint R&D willm enable clinicians to select medicines suited to indivisual patients by displaying information predicted by AI technology, such as the probability that selecting certain oral hypoglycemic medicines will make it easier for the patients to safety and effectively control future blood sugar levels, based on individual characteristics.

10.4.3 Role of pharmacists

In recent years, the rapid progress of AI has become a hot topic, saying that a lot of work will be taken away by AI. The work of pharmacists is also progressing in computerization and mechanization. This is the appropriate timing to re-evaluate works performed by pharmacists and re-position them. With respect to core operations of pharmacists, there are normally three major job descriptions: (1) dispensing business, (2) medicine guidance, and (3) pharmaceutical history management. (1) Dispensing business is the central business of pharmacists, but it is the business of accurately preparing medicines based on a doctor's prescription, and the preparation and package of ointments and water agents are also included in this dispensing business. (2) Guidance on medication represents the provision of information on medicines provided to patients by pharmacists. It is one of the most important works in working as a pharmacist, and it is also a point that must be emphasized most as a family pharmacist. (3) In the pharmaceutical history management, the record of the medication guidance is recorded so that the medicine can be used safely and safely. As explained, the importance of medication guidance is explained, but in order to provide substantial medication guidance, a medical history must be in place. Besides these, there are still other operations such as sales and consultation of over-the-counter top drugs, or home health care [126].

AI has advanced dramatic times in recent years, and how to use AI to streamline work is a challenge in all industries. The situation surrounding pharmacists is not an exception. Before you get worried that your job as a pharmacist will be deprived by AI, let us understand what AI is good at and what we are not good at, and think about how to use it well. What kind of work will pharmacists be required to do in the future when AI becomes popular? [127]. Since AI is already active in the workplace of pharmacists, in the medical field where chronic labor shortages continue, it is urgent to introduce the latest technologies such as AI to improve operational efficiency. In some pharmacies, AI is introduced mainly in the pharmaceutical information business. The characteristic of AI is that it can quickly derive optimal solutions by memories a variety of information. If AI can input a huge amount of data and use the learning function, it will be possible to easily analyze data that has been difficult until now. At medical institutions that have already introduced AI, AI is also moving to create a database of inquiries to DI (drug information) rooms by requiring AI to remember drug attachments and patient drug history. Efforts to improve operational efficiency by utilizing AI are spreading not only to hospitals but also to pharmacies, pharmaceutical companies, and pharmaceutical wholesalers. In the future, AI and pharmacists will coexist, and it will be required to enhance medical services by sharing work. As we have seen, if AI can perform some portion of routine pharmacy works, pharmacists will be able to devote saved time and to focus their resources on home visits and personal work as the family pharmacists that are expected to increase their needs in the future. Therefore, in the future, it is necessary to assign tasks such as the work that AI is good at is left to AI, and the work that only humans can do is done by humans. Verbal communication with patients is an important job that only pharmacists can do. By asking human-like and warm questions, it is possible to confirm the presence or absence of side effects such as worry and forget to drink, and to grasp the patient's condition firmly, which leads to the maintenance of the patient's health [127]. Even if AI does not take the place of pharmacists, it is possible that pharmacists who cannot utilize AI properly will be left behind in the times. In the future, it would be crucial to enhance pharmaceutical knowledge and communication skills while at the same time deepening our understanding of AI should become another important factor [128].

10.5 Placebo effect and nocebo effect

10.5.1 Placebo effect

A placebo is a substance or treatment that is designed to have no therapeutic value. Common placebos include inert tablets (like sugar pills), inert injections (like saline), sham surgery, and other procedures. As mentioned early, among various types of information sources, a clinical report using placebo and double-blind studies is ranked as the most reliable information. Hence, placebo possesses an important position in an entire medicine field. In general, placebos can affect how patients perceive their condition and encourage the body's chemical processes for relieving pain and a few other symptoms but have no impact on the disease itself. Improvements that patients experience after being treated with a placebo can also be due to unrelated factors, such as regression to the mean (a natural recovery from the illness). The use of placebos in clinical medicine raises ethical concerns, especially if they are disguised as an active treatment, as this introduces dishonesty into the doctor–patient relationship and bypasses informed consent. While it was once assumed that this deception was necessary for placebos to have any effect, there is now evidence that placebos can have effects even when the patient is aware that the treatment is a placebo [129].

Determining factors whether a placebo effect is effective or not should include the trust-based relationships between doctors and patients, human factors such as the confidence and prestige of doctors to treat, and situational factors such as hospital atmosphere. Therefore, the flexibility of the patient's mind will give placebo a place to work. Accordingly, placebo is used for psychotherapy to treat mental illness. The cause of the placebo effect is thought to be related, such as the suggestive effect of "prescribed the drug."

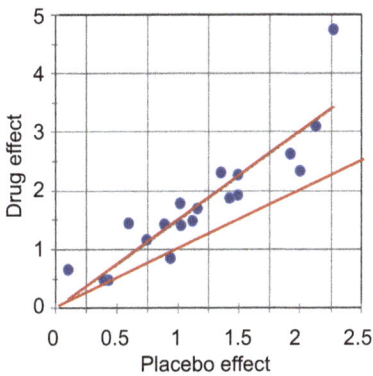

Figure 10.5: The placebo response as a predictor of the drug response [130].

Kirsch et al. [130] studied placebo effects on 2,318 depression patients treated using either an antidepressant medication or a placebo in double-blind clinical trials (19 cases). The calculated correlation between drug response and placebo response was exceptionally high, $r = 0.90$, as seen with red solid line in Figure 10.5, indicating the placebo response was proportionate to the drug response, with remaining variability most likely due to measurement error. The red dotted line indicates one-to-one relationship between these two strategies.

Placebo effects are strongly relied on psychological factors. Differences in the color and size of placed drugs have a subtle effect on the effect. Blue or purple capsules have a stronger sedative or analgesic effect, while red and yellow make mental activity more active. Large capsules are more effective than smaller capsules. If the shape is capsule, it has a stronger effect than tablets. Not only in the fields of medicine and pharmacy, but also in various academic fields such as psychology and sociology, as well as in daily life such as sports, religion, and human relations, the existence of a placebo effect in which expectations and aspirations act positively is recognized. The placebo effect can be found at the label on the wine bottle. If the label looks authentic and liquid is poured from the bottle into a luxurious glass at a top-class restaurant, you will feel like you've had a real Romanée-conti. Since an essence of marketing can be found behind persuading people, the placebo effect has been in marketing area, too [131]. The placebo effect has begun to be involved in the digital health industry. For example, Sasmed, which is developing a smartphone application that is a drug-based insomnia treatment, conducts a comparative and contrasting trial using the placebo effect in clinical trials [132]. Comparative tests using the actually developed drug and the placebo which eliminates the algorithm for performing treatment to test the placebo effect if it would be effective because the prescription drugs were prescribed by a doctor.

Despite efforts to identify characteristics associated with medication-placebo differences in antidepressant trials, few consistent findings have emerged to guide participant selection in drug development settings and differential therapeutics in clinical practice. Limitations in the methodologies used, particularly searching for a single moderator while treating all other variables as noise, may partially explain the failure to generate consistent results. Based on this background, Signal et al. [133] investigated whether interactions between pretreatment patient characteristics, rather than a single-variable solution, may better predict who is most likely to benefit from placebo versus medication. Data were analyzed from 174 patients aged 75 years and older with unipolar depression who were randomly assigned to citalopram or placebo. Model-based recursive partitioning analysis was conducted to identify the most robust significant moderators of placebo versus citalopram response. It was reported that (i) the greatest signal detection between medication and placebo in favor of medication was among patients with fewer years of education (≤12) who suffered from a longer duration of depression since their first episode (>3.47 years) and (ii) compared with medication, placebo had the greatest response for those who were more educated (>12 years), to the point where placebo almost outperformed medication; concluding (iii) ML approaches capable of evaluating the contributions of multiple predictor variables may be a promising methodology for identifying placebo versus medication responders and (iv) duration of depression and education should be considered in the efforts to modulate placebo magnitude in drug development settings and in clinical practice [117]. The biostatistical impact the placebo response has in clinical trials is significant; so significant that it can make a clinical trial fail or skew data to unfavorable outcomes. However, novel statistical approaches to mitigating the placebo response are emerging by creating personalized baselines per patient with AI, ultimately strengthening statistical results. In the interview with Demolle (CEO of Tools4Patient), the effect AI has been discussed in terms of its statistical impact on clinical trial placebo response [133]. With respect to general response of placebo in the clinical trial procedures, it was indicated that (i) the placebo response consists of patients perceiving improvements or experiencing improvements and (ii) some of this symptomatic relief may result from the psychological effects of receiving treatment. As much as 69% of phase II and III clinical trial failures occur due to safety issues or the inability to demonstrate clear superiority of the tested therapy versus a placebo, regularly leading to the premature abandonment of entire development programs [134].

10.5.2 Nocebo effect

It is called the nocebo effect when placebo works negatively, such as shortening life, delaying health recovery, or causing side effects. A phenomenon opposite to the placebo effect has also been observed. When an inactive substance or treatment is administered to a recipient who has an expectation of it having a negative impact, it is known as a nocebo. The nocebo effect occurs when the recipient of an inert substance reports a negative effect or a worsening of symptoms, with the outcome resulting not from the substance itself, but from negative expectations about the treatment. Another negative consequence is that placebos can cause side effects associated with real treatment [129, 135]. Both placebo and nocebo effects are presumably psychogenic, but they can induce measurable changes in the body. One article that reviewed 31 studies on nocebo effects reported a wide range of symptoms that could manifest as nocebo effects, including nausea, stomach pains, itching, bloating, depression, sleep problems, loss of appetite, sexual dysfunction, and severe hypotension [136].

A number of researchers have pointed out that the harm caused by communicating with patients about potential treatment adverse events raises an ethical issue. Informing a patient about what harms a treatment is likely to cause is required to respect autonomy. Yet the way in which potential harms are communicated could cause additional harm, which may violate the ethical principle of non-maleficence. It may be possible that nocebo effects can be reduced while respecting autonomy using different models of informed consent, including the use of a framing effect. In fact, it has been argued that forcing patients to learn about all potential adverse events against their will could violate autonomy. Hence, failure to minimize nocebo side effects in clinical trials and clinical practice raises a number of recently explored ethical issues [137]. Since behavioral, psychoneurobiological, and functional changes occur during nocebo-induced pain processing, Blasini et al. [138] reviewed published human and nonhuman research on algesia and hyperalgesia resulting from negative expectations and nocebo effects. It was reported that (i) negative expectations are formed through verbal suggestions of heightened pain, prior nociceptive and painful experiences, and observation of pain in others, (ii) susceptibility to the nocebo effect can be also influenced by genetic variants, conscious and nonconscious learning processes, personality traits, and psychological factors and (iii) moreover, providers' behaviors, environmental cues and the appearance of medical devices can induce negative expectations that dramatically influence pain perception and processing in a variety of pain modalities and patient populations. Based on these findings, it was concluded that (iv) nocebo studies outline how individual expectations may lead to physiological changes underpinning the central integration and processing of magnified pain signaling, and (v) further research is needed to develop strategies that can identify patients with nocebo-vulnerable pain to optimize the psychosocial and therapeutic context in which the clinical encounter occurs, with the ultimate purpose of improving clinical outcomes [138].

Nocebo effects encompass negative responses to inert interventions in the research setting and negative outcomes with active treatments in the clinical research or practice settings, including new or worsening symptoms and adverse events, stemming from patients' negative expectations and not the pharmacologic action of the treatment itself. Numerous personal, psychosocial, neurobiological, and contextual/environmental factors contribute to the development of nocebo effects, which can impair quality of life and reduce adherence to treatment. Biologics are effective agents widely used in autoimmune disease, but their high cost may limit access for patients. Biosimilar products have gained regulatory approval based on quality, safety, and efficacy comparable to that of originator biologics in rigorous study programs.

Collocal et al. [139] identified gaps in patients' and healthcare professionals' awareness, understanding, and perceptions of biosimilars that may result in negative expectations and nocebo effects, and may diminish their acceptance and clinical benefits and examined features of nocebo effects with biosimilar treatment that inform research and clinical practices. It was mentioned that (i) when biosimilars are introduced to patients as possible treatment options, we recommend adoption of nocebo-reducing strategies to avoid negative expectations, including delivery of balanced information on risk-benefit profiles, framing information to focus on positive attributes, and promoting shared decision-making processes along with patient empowerment and (ii) healthcare professionals confident in their knowledge of biosimilars and aware of bias-inducing factors may help reduce the risk of nocebo effects and improve patients' adherence in proposing biosimilars as treatment for autoimmune diseases such as rheumatoid arthritis and inflammatory bowel disease. Although most studies of nocebo effects have been conducted in the field of pain, these effects have also been demonstrated in other conditions, such as fatigue, gastrointestinal disorders, allergy, and itch [139].

References

[1] Hingorani AD, Kuan V, Finan C, Kruger FA, Gaulton A, Chopade S, Sofat R, MacAllister RJ, Overington JP, Hemingway H, Denaxas S, Prieto D, Casas JP. Improving the odds of drug development success through human genomics: modelling study. Scientific Reports. 2019, 18911, https://doi.org/10.1038/s41598-019-54849-w.

[2] Hughes JP, Rees S, Kalindjuan SB, Philpott KL. Principles of early drug discovery. British Journal of Pharmacology. 2011, 162, 1239–49.

[3] MedicineNet. Drug Approvals – From Invention to Market . . . A 12-Year Trip. https://www.medicinenet.com/script/main/art.asp?articlekey=9877

[4] Thomas DW, Burns J, Audette J, Carroll A, Dow-Hygelund C, Hay M. Clinical Development Success Rates 2006-2015. BIO Industry Analysis. Biotechnology Innovation Organization. 2016. https://www.bio.org/sites/default/files/legacy/bioorg/docs/Clinical%20Development%20Success%20Rates%202006-2015%20-%20BIO,%20Biomedtracker,%20Amplion%202016.pdf.

[5] Wong CH, Siah KW, Lo AW. Estimation of clinical trial success rates and related parameters. Biostatistics. 2019, 20, 273–86.

[6] Parker J, Kohler JC. The success rate of new drug development in clinical trials: Crohn's disease. Journal of Pharmacy and Pharmaceutical Sciences. 2010, 13, 191–7.

[7] Bains W. Failure rates in drug discovery and development: will we ever get any better? 2004. https://www.ddw-online.com/failure-rates-in-drug-discovery-and-development-will-we-ever-get-any-better-1027-200410/.

[8] Wang Y. Extracting. Knowledge from failed development programs. Pharmaceutical Medicine. 2012, 26, 91–6.

[9] Herschel M. Portfolio. Decisions in early development: Don't throw out the baby with the bathwater. Pharmaceutical Medicine. 2012, 26, 77–84.

[10] Schneider P, Walters WP, Plowright AT, Sieroka N, Listgarten J, Goodnow Jr. RA, Fisher J, Jansen JM, Duca JS, Rush TS, Zentgraf M, Hill JE, Krutoholow E, Kohler M, Blaney J, Funatsu K, Luebkemann C, Schneider G. Rethinking drug design in the artificial intelligence era. Nature reviews. Drug discovery. 2020, 19, 353–64.

[11] Brown N, Ertl P, Lewis R, Luksch T, Reker D, Schneider N. Artificial intelligence in chemistry and drug design. Journal of Caided Molecular design. 2020, 34, 709–15.

[12] Oshida Y. Surface Engineering and Technology for Biomedical Implants, Momentum Press, New York NY, 2014.

[13] Zhavoronkov A, Ivanenkov YA, Aliper A, Veselov MS, Aladinskiy VA, Aladinskaya AV, Terentiev VA, Polykovskiy DA, Kuznetsov MD, Asadulaev A, Volkov Y, Zholus A, Shayakhmetov RR, Zhebrak A, Minaeva LI, Zagribelnyy BA, Lee LH, Soll R, Madge D, Xing L, Guo T, Aspuru-Guzik A. Deep learning enables rapid identification of potent DDR1 kinase inhibitors. Nature Biotechnology. 2019, 37, 1038–40.

[14] Stokes JM, Yang K, Swanson K, Jin W, Cubillos-Ruiz A, Donghia NM, MacNair CR, French S, Carfrae LA, Bloom-Ackermann Z, Tran Vm, Chiappino-Pepe A, Badran Ah, Andrews Iw, Chory Ej, Church Gm, Brown Ed, Jaakkola Ts, Barzilay R, Collins Jj. A deep learning approach to antibiotic discovery. Cell. 2020, 180, 688–702.

[15] Yang X, Wang Y, Byrne R, Schneider G, Yang S. Concepts of artificial intelligence for computer-assisted drug discovery. Chemical Reviews. 2019, 119, 10520–94.

[16] Baig MH, Ahmad K, Roy S, Ashraf JM, Mohd Adil, Siddiqui MH, Khan S, Kamal MA, Provazník I, Choi I. Computer aided drug design: success and limitations. Current Pharmaceutical Design. 2016, 22, 572–81.

[17] Yu W, MacKerell AD, Jr. Computer-aided drug design methods. Methods in Molecular Biology (Clifton, N.J.). 2017, 1520, 85–106.

[18] Roberts M, Genway S. How Artificial Intelligence is Transforming Drug Design. In: Drug Discovery World, 2019, https://www.ddw-online.com/how-artificial-intelligence-is-transforming-drug-design-1530-201910/#:~:text=Target%20identification%20and%20validation%20can,associations%2C%20to%20identify%20new%20targets.

[19] Goradia A, Strickling L. Artificial Intelligence and Drug Discovery. 2020; https://www.bblsa.com/industry-insights/2020/1/5/artificial-intelligence-and-drug-discovery-unproven-but-on-target.

[20] Lindsay MA. Target discovery. Nature reviews. Drug discovery. 2003, 2, 831–8.

[21] Sygnature Discovery. Target Validation. https://www.sygnaturediscovery.com/drug-discovery/integrated-drug-discovery/target-validation/.

[22] Zeng X, Zhu S, Lu W, Liu Z, Huang J, Zhou Y, Fang J, Huang Y, Guo H, Li L, Trapp BD, Nussinov R, Eng C, Loscalzoo J, Cheng F. Target identification among known drugs by deep learning from heterogeneous networks. Cell Chemical Biology. 2019, https://pubs.rsc.org/en/content/articlelanding/2020/sc/c9sc04336e#!divAbstract.

[23] Hunter J. The Promise Of Using AI For Target Identification And Working With AstraZeneca. 2019; https://www.benevolent.com/news/the-promise-of-using-ai-for-target-identification-and-working-with-astrazeneca.

[24] Sygnature Discovery. Lead Optimisation. https://www.sygnaturediscovery.com/drug-discovery/integrated-drug-discovery/lead-optimisation/.

[25] Sawyer TK. New screening tools for lead compound identification. Nature Chemical Biology. 2005, 1; https://www.nature.com/articles/nchembio0805-125.

[26] Feng BY, Shelat A, Doman TN, Guy RK, Shoichet BK. High-throughput assays for promiscuous inhibitors. Nature Chemical Biology. 2005, 1, 146–8.

[27] Keserü GM, Gergely M. Makara GM. The influence of lead discovery strategies on the properties of drug candidates. Nature reviews. Drug discovery. 2009, 8, 203–12.

[28] Keinan S, Shipman WJ, Addison E. An effective way to apply ai to the design of new drug lead compounds. Cloud Pharmaceuticals share their experience using AI for novel drug discovery. Pharma iQ, 2019; https://www.pharma-iq.com/pre-clinical-discovery-and-development/articles/an-effective-way-to-apply-ai-to-the-design-of-new-drug-lead-compounds.

[29] Rouse M. Augmented Intelligence. https://whatis.techtarget.com/definition/augmented-intelligence#:~:text=Augmented%20intelligence%20is%20an%20alternative,intelligence%20rather%20than%20replace%20it.

[30] Green CP, Emgkvist O, Pairaudeau G. The convergence of artificial intelligence and chemistry for improved drug discovery. Future Medicinal Chemistry. 2018, 10, https://doi.org/10.4155/fmc-2018-0161.

[31] U.S. Food and Drug Administration. Step 3: Clinical Research. https://www.fda.gov/patients/drug-development-process/step-3-clinical-research.

[32] Akatsuka H. Toward the drug development. https://www.mhlw.go.jp/content/10601000/000490873.pdf.

[33] Gerlovin L, Bonet B, Kambo I, Ricciardi G. The Expanding Role Of Artificial Intelligence In Clinical Research. Cell and Gene. 2020; https://www.clinicalleader.com/doc/the-expanding-role-of-artificial-intelligence-in-clinical-research-0001.

[34] ICON. AI and clinical trials: Artificial Intelligence (AI) technology, combined with big data, hold the potential to solve many key clinical trial challenges. https://www.iconplc.com/insights/digital-disruption/ai-and-clinical-trials/.

[35] Taylor K, Properzi F, Cruz MJ, Ronte H, Haughey J. Intelligent clinical trials: Transforming through AI-enabled engagement. Deloitte Centre for Health Solutions. https://www2.deloitte.com/content/dam/insights/us/articles/22934_intelligent-clinical-trials/DI_Intelligent-clinical-trials.pdf.

[36] Kristensen U. The Many Faces of AI in Clinical Trials. HIT Consultant. 2020; https://hitconsul tant.net/2020/02/28/many-faces-ai-clinical-trials/.

[37] U.S. Food & Drug Administration. 22 Case Studies Where Phase 2 and Phase 3 Trials had Divergent Results. 2017; https://web.media.mit.edu/~pratiks/ai-for-drug-discovery-and-clinical-trials/22-case-studies-where-phase-2-and-phase-3-trials-had-divergent-results.pdf.

[38] Yauney G, Shah P. Artificial Intelligence for Drug Discovery and Clinical Trials. https://web.media.mit.edu/~pratiks/ai-for-drug-discovery-and-clinical-trials/.

[39] Murgia M. AI-designed drug to enter human clinical trial for first time. The Financial Times Limited 2020; https://www.ft.com/content/fe55190e-42bf-11ea-a43a-c4b328d9061c.

[40] Reiss R. Transforming Drug Discovery Through Artificial Intelligence. 2020. https://www.for bes.com/sites/robertreiss/2020/03/03/transforming-drug-discovery-through-artificial-intelligence/#327991316a15.

[41] Marozsan L. The Power of AI in Drug Discovery and Development. 2020. https://www.thermo fisher.com/blog/connectedlab/the-power-of-ai-in-drug-discovery-and-development/.

[42] Wu J. Artificial Intelligence and Drug Discovery: The Pharmaceutical Industry is looking for efficiency and they are getting it with AI. 2020. https://towardsdatascience.com/artificial-intelligence-and-drug-discovery-dbd6b347e695.

[43] Haq I, Muselemu O. Blockchain technology in pharmaceutical industry to prevent counterfeit drugs. International Journal of Computer Applications. 2018, 180, 8–12.

[44] Wallace A. Protection of Personal Data in Blockchain Technology – An investigation on the compatibility of the General Data Protection Regulation and the public blockchain. Stockholm University Department of Law Master's Thesis, 2018; https://www.diva-portal.org/smash/get/diva2:1298747/FULLTEXT01.pdf.

[45] Malone B, Simovski B, Moliné C, Cheng J, Gheorghe M, Fontenelle H, Vardaxis I, Tennøe S, Malmberg J-A, Stratford R, Clancy T. Artificial intelligence predicts the immunogenic landscape of SARS-CoV-2: toward universal blueprints for vaccine designs. bioRXiv. 2020; doi: https://doi.org/10.1101/2020.04.21.052084.

[46] Kent J. New Initiative Uses Artificial Intelligence for Vaccine Development: The Human Immunomics Initiative will develop artificial intelligence tools to speed the development of vaccines for a wide range of diseases. 2020; https://healthitanalytics.com/news/new-initiative-uses-artificial-intelligence-for-vaccine-development.

[47] Kent J. Artificial Intelligence Research Collaborative Targets COVID-19. Health IT Analytics. 2020; https://healthitanalytics.com/news/artificial-intelligence-research-collaborative-targets-covid-19.

[48] Ahuja AS, Reddy VP, Marques O. Artificial intelligence and COVID-19: A multidisciplinary approach. Integrative Medicine Research. 2020, 9; https://doi,org/10.1016/j.imr.2020.100434.

[49] Banerjea A. What is rolling review? Europe starts real-time review of another Covid vaccine. 2020; https://www.livemint.com/news/world/what-is-rolling-review-europe-starts-real-time-review-of-another-covid-vaccine-11601978882290.html.

[50] Lake F. Artificial intelligence in drug discovery: what is new, and what is next? Future Drug Discovery. 2019, 1: https://doi.org/10.4155/fdd-2019-0025.

[51] Kelder R. Commercializing deep neural networks for drug discovery. Eureka Discovery. 2019; https://eureka.criver.com/commercializing-deep-neural-networks-for-drug-discovery/.

[52] Eureka. For AI-enabled drug discoveries, who owns the science? 2019; https://eureka.criver.com/for-ai-enabled-drug-discoveries-who-owns-the-science/.

[53] Reiher M, Wiebe N, Svore KM, Wecker D, Troyer M. Elucidating reaction mechanisms on quantum computers. Proceedings of the National Academy of Sciences of the United States of America. 2017, 114, 7555–60.

[54] In silico. https://en.wikipedia.org/wiki/In_silico.

[55] Dimitar AD, Girinath GP, Mati K. In silico machine learning methods in drug development. Current Topics in Medicinal Chemistry. 2014, 14, 1913–22.

[56] Piñero J, Furlong LI, Sanz F. *In silico* models in drug development: where we are. Current Opinion in Pharmacology. 2018, 42, 111–21.

[57] Bernard D, Coop A, MacKerell AD Jr. Computer-aided drug design: Structure-activity relationships of delta opioid ligands. Drug Des Rev. 2005, 2, 277–91.

[58] Wadood A, Ahmed N, Shah L, Ahmad A, Hassan H, Shams S. In-silico drug design: An approach which revolutionarised the drug discovery process. OA Drug Design & Delivery. 2013, 1, 3, https://pdfs.semanticscholar.org/3a1c/2dd13912b691d370a891b79dbc8df7f0ae69.pdf.

[59] Bruno A, Costantino G, Sartori L, Radi M. The. In silico drug discovery toolbox: Applications in lead discovery and optimization. Current Medicinal Chemistry. 2019, 26, 3838–73.

[60] Woodcock J, Brumfield M, Gill D, Zerhouni E. The driving role of consortia on the critical path to innovative therapies. Nature reviews. Drug discovery. 2014, 13, https://doi.org/10.1038/nrd4462.

[61] Simpson PB, Wilkinson GF. What makes a drug discovery consortium successful?. Nature reviews. Drug discovery. 2020, https://www.nature.com/articles/d41573-020-00079-z.

[62] Reichman M, Simpson PB. Open innovation in early drug discovery: roadmaps and roadblocks. Drug Discovery Today. 2016, 21, 770–88.

[63] Bryans JS, Kettleborough CA, Solari R. Are academic drug discover efforts receiving more recognition with declining industry efficiency?. Expert Opinion on Drug Discovery. 2019, 14, 605–7.

[64] Tralau-Stewart CJ, Wyatt CA, Kleyn DE, Ayad A. Drug discovery: new models for industry–academic partnerships. Drug Discovery Today. 2009, 14, 95–101.

[65] Okuno Y. Overview of life intelligence consortium "LINC". Proceedings for Annual Meeting of The Japanese Pharmacological Society. 2020, 93:2-CS-1; https://www.researchgate.net/publication/339993230_Overview_of_Life_Intelligence_Consortium_LINCraifuinterijensukonsoshiamuLINCnogaiyao.

[66] Suk JS, Xu Q, Kim N, Hanes J, Ensign LM. PEGylation as a strategy for improving nanoparticle-based drug and gene delivery. Advanced Drug Delivery Reviews. 2016, 99, 28–51.

[67] Singh MN, Hemant KSY, Ram M, Shivakumar HG. Microencapsulation: A promising technique for controlled drug delivery. Research in Pharmaceutical Sciences. 2010, 5, 65–77.

[68] Daraee H, Etsemadi A, Kouhi M, Alimirzalu S, Akbarzade A. Application of liposomes in medicine and drug delivery. Artificial Cells, Nanomedicine and Biotechnology. 2016, 44, 381–91.

[69] Zylberberg C, Matosevic S. Pharmaceutical liposomal drug delivery: a review of new delivery systems and a look at the regulatory landscape. Journal of Drug Delivery. 2016, 23, 3319–29.

[70] Walk CM, West HJ. Antibody drug conjugates for cancer treatment. JAMA Oncol. 2019, 10.1001/jamaoncol.2019.3552.

[71] Rautio J, Meanwell NA, Di L, Hageman MJ. The expanding role of prodrugs in contemporary drug design and development. Nature reviews. Drug discovery. 2018, 17, 559–87.

[72] Dodou K. Research and developments in buccal and sublingual drug delivery systems. The Pharmaceutical Journal. The Pharmaceutical Journal. 2012, 288, 446–7.

[73] Prausnitz MR, Robert Langer R. Transdermal drug delivery. Nature Biotechnology. 2008, 26, 1261–8.

[74] Margetts L, Sawyer R. Transdermal drug delivery: principles and opioid therapy. BJA: Continuing Education in Anaesthesia Critical Care & Pain. 2007, 7, 171–6.

[75] Boztepe C, Künkül A, Yüceer M. Application of artificial intelligence in modeling of the doxorubicin release behavior of pH and temperature responsive poly(NIPAAm-co-AAc)-PEG IPN hydrogel. Journal of Drug Delivery Science and Technology. 2020, 57, https://doi.org/10.1016/j.jddst.2020.101603.

[76] Mehta M. Biopharmaceutics Classification System (BCS): Development, Implementation, and Growth, Wiley, 2016.

[77] Demir H, Arica-Yegin B, Oner L. Application of an artificial neural network to predict dissolution data and determine the combined effect of pH and surfactant addition on the solubility and dissolution of the weak acid drug etodolac. Journal of Drug Delivery Science and Technology. 2018, 47, 215–22.

[78] Li YY, Lenaghan SC, Zhang M, Neves NM. A data-driven predictive approach for drug delivery using machine learning techniques. PLos One. 2012, 7, https://doi.org/10.1371/journal.pone.0031724.

[79] Hassanzadeh P, Atyabi F, Dinarvand R. The Significance of Artificial Intelligence in Drug Delivery System Design. In: Advanced Drug Delivery Reviews, 2019, 151–2.

[80] Drug side effects explained; https://www.webmd.com/a-to-z-guides/drug-side-effects-explained.

[81] Drug Side Effects. https://www.drugs.com/sfx/.

[82] DerSarkissian C. Drug Side Effects Explained. 2020. https://www.webmd.com/a-to-z-guides/drug-side-effects-explained.

[83] Side effect. https://en.wikipedia.org/wiki/Side_effect.

[84] Seo S. Lee T, Kim M-h, Yoon Y. Prediction of side effects using comprehensive similarity measures. BioMed International. 2020, https://doi.org/10.1155/2020/1357630.

[85] Galeano D, Li S, Gerstein M, Paccanaro A. Predicting the frequencies of drug side effects. Nature Communications. 2020, 11, https://doi.org/10.1038/s41467-020-18305-y.

[86] AI to judge drug side effects, Fuji Xerox; https://www.nikkei.com/article/DGXMZO57546580R00C20A4X20000/.

[87] Kozawa S, Sagawa F, Endo S, De ALmeida GM, Mistuishi Y, Sato TN. Predicitng Human CLinical Outcomes using Mouse Multi-Organ Transciptome. iScience. 2020, 23; https://doi:10.1016/j.isci.2091.100791.

[88] Kozawa S, Sagawa F, Endo S, De Almeida GM, Mitsuishi Y, Sato TN. Predicting human clinical outcomes using mouse multi-organ transcriptome. iScience. 2020, 23, https://doi.org/10.1016/j.isci.2019.100791.

[89] Yao Y, Wang Z, Li L, Lu K, Liu R, Liu Z, Yan J. An ontology-based artificial intelligence model for medicine side-effect prediction: taking traditional chinese medicine as an example. Computational and Mathematical Methods in Medicine. 2019, https://www.hindawi.com/journals/cmmm/2019/8617503/.

[90] Tamura Y. Differences between side effects and side reactions and adverse events. 2021; https://media.eduone.jp/detail/10322/2/.

[91] Buvailo A. How Big Pharma Adopts AI To Boost Drug Discovery. 2018. https://www.biopharmatrend.com/post/34-biopharmas-hunt-for-artificial-intelligence-who-does-what/.

[92] Pushpakom S, Iorio F, Eyers PA, Escott KJ, Hopper S, Wells A, Doig A, Guilliams T, Latimer J, McNamee C, Norris A, Sanseau P, Cavalla D, Pirmohamed M. Drug repurposing: progress, challenges and recommendations. Nature reviews. Drug discovery. 2019, 18, 41–58.

[93] Yamanishi Y. Data-driven drug discovery and healthcare by AI. https://www.jst.go.jp/kisoken/aip/program/inter/vol2_sympo/slides/part2_1_yamanishi.pdf.

[94] Rudrapal M, Khairnar Sj, Jadhav Ag. Drug Repurposing (DR): An Emerging Approach in Drug Discovery. 2020; https://www.intechopen.com/online-first/drug-repurposing-dr-an-emerging-approach-in-drug-discovery

[95] Osakwe O, Rizvi S. Social Aspects of Drug Discovery, Development and Commercialization. 2016, xvii–xxx.

[96] Drug repositioning; https://www.anaxomics.com/pdf/Anaxomics-DrugRepositioning.pdf

[97] Therapeutic Performance Mapping System (TPMS); https://www.anaxomics.com/tpms.php.

[98] Jarada TN, Rokne JG, Alhajj R. A review of computational drug repositioning: strategies, approaches, opportunities, challenges, and directions. Journal of Chemiinformatics. 2020, 12, https://doi.org/10.1186/s13321-020-00450-7.

[99] Saito S, Yamamoto Y, Hagiwara S. Reopening the door for drug repositioning https://www.jpn-geriat-soc.or.jp/publications/other/pdf/perspective_52_3_200.pdf.

[100] Nishimura Y, Hara H. Drug repositioning: Current advances and future perspectives. Frontiers in Pharmacology. 2018, 20, https://doi.org/10.3389/fphar.2018.01068.

[101] Mucke HAM. Drug repositioning in the mirror of patenting: Surveying and mining uncharted territory. Frontiers in Pharmacology. 2017, https://doi.org/10.3389/fphar.2017.00927.

[102] Nishimura Y, Tagawa M, Ito H, Tsuruma K, Hara H. Overcoming obstacles to drug repositioning in Japan. Frontiers in Pharmacology. 2017, https://doi.org/10.3389/fphar.2017.00729.

[103] Dias TL, Schuch V, Beltrão-Braga Pcb, Daniel Martins-de-souza, Brentani HP, Franco GR, Nakaya HI. Drug repositioning for psychiatric and neurological disorders through a network medicine approach. Translational Psychiatry. 2020, 10, 141, https://doi.org/10.1038/s41398-020-0827-5.

[104] Liu R, Wei L, Zhang P. A deep learning framework for drug repurposing via emulating clinical trials on real-world patient data. Nature Machine Intelligence. 2021, 3, 68–75.

[105] Biosimilars vs. Generics. Quartz. https://quartzbenefits.com/docs/default-source/providers/newsletters/biosimilars-vs-generics.pdf?sfvrsn=82c7e315_0.

[106] FDA. Biosimilars. https://www.fda.gov/drugs/therapeutic-biologics-applications-bla/biosimilars.

[107] Generic and Biosimilar – Medications. https://www.nationalmssociety.org/Treating-MS/Medications/Generic-and-Biosimilar-Medications.

[108] Mejia N. Artificial Intelligence for Generic Drug Companies – Current Applications. 2019, https://emerj.com/ai-sector-overviews/artificial-intelligence-generic-drug-companies/.

[109] Riley S. Generics vs. Biosimilars: Similar but Different Advantages. Drug Discovery & Development. 2017, https://www.drugdiscoverytrends.com/generics-vs-biosimilars-similar-but-different-advantages/.

[110] Biosimilar drug. National Cancer Institute. https://www.cancer.gov/publications/dictionaries/cancer-terms/def/biosimilar-drug.

[111] Lee JF, Litten JB, Grampp G. Comparability and biosimilarity: considerations for the healthcare provider. Current Medical Research and Opinion. 2012, 28, 1053–8.

[112] Camacho LH, Frost CP, Abella E. et al., Biosimilars 101: considerations for U.S. oncologists in clinical practice. Cancer Medicine. 2014, 3, 889–99.

[113] Mellstedt H, Niederwieser D, Ludwig H. The challenge of biosimilars. Annals of Oncology: Official Journal of the European Society for Medical Oncology / ESMO. 2008, 19, 411–9.

[114] Biosimilars Versus Generics. https://www.amgenbiosimilars.com/bioengage/what-are-biosimilars/biologics-vs-generics.

[115] Approval, labeling, side effects, and safety information; https://www.fda.gov/drugs/resources-you-drugs/drug-information-consumers.

[116] Healthcare IT; https://service.infocom.co.jp/healthcare/en/drug/index.html.

[117] Velo GP, Minuz P. Medication errors: prescribing faults and prescription errors. British Journal of Clinical Pharmacology. 2009, 67, 624–8.

[118] Dean B, Schachter M, Vincent C, Barber N. Causes of prescribing errors in hospital inpatients: a prospective study. Lancet. 2002, 359, 1373–8.

[119] Glasziou P, Irwig L, Aronson JK. Evidence-based Medical Monitoring: From Principles to Practice, Wiley-Blackwell, Oxford, 2008.

[120] Durieux P, Trinquart L, Colombet I, Niès J, Walton R, Rajeswaran A, Rège Walther M, Harvey E, Burnand B. Computerized advice on drug dosage to improve prescribing practice. Cochrane Database System Reveiws. 2008, https://doi.org/10.1002/14651858.CD002894.pub2.

[121] Gommans J, McIntosh P, Bee S, Allan W. Improving the quality of written prescriptions in a general hospital: the influence of 10 years of serial audits and targeted interventions. Internal Medicine Journal. 2008, 38, 243–8.

[122] AI robots that increase learning effectiveness and reduce medical errors; https://www.jnews.com/special/health/hea1508.html.

[123] AI recommendations for medication instruction – Risk aversion with pharmacological inference; https://yakuyomi.jp/industry_news/.

[124] AI-assisted pharmacists with prescription examinations and medication guidance; https://it.impress.co.jp/articles/-/18956.

[125] Sapporo Medical University and Fujitsu in Jopint R&D for Diabetes Treatment that Uses AI-based Machine Learning. https://www.fujitsu.com/global/about/resources/news/press-releases/2019/0212-01.html.

[126] With the advent of the AI era, pharmacists will lose their jobs? https://www.38-8931.com/pharma-labo/column/study/future02.php.

[127] The future of pharmacists with the spread of AI; https://www.k-pharmalink.co.jp/contents/beanknowledge/ai-and-pharmacist-from-now-on/.

[128] In an era when pharmacists make use of AI; https://www.yakuji.co.jp/entry73117.html.

[129] Placebo; https://en.wikipedia.org/wiki/Placebo.

[130] Kirsch I, Sapirstein G. Listening to Prozac but Hearing Placebo: A Meta-Analysis of Antidepressant Medication. Prevention and Treatment. 1998.1.1-16; http://psychrights.org/Research/Digest/CriticalThinkRxCites/KirschandSapirstein1998.pdf.

[131] The essence of marketing is the placebo effect; https://blog.goo.ne.jp/shuumei22000/e/e9bae67413e5ebca00a34402b7e727a6.

[132] Insomnia treatment and clinical trials with smartphone app; https://xtech.nikkei.com/dm/atcl/news/16/091304028/?ST=health

[133] Zilcha-Mano S, Roose Sp, Brown Pj, Rutherford Br. A machine learning approach to identifying placebo responders in late-life depression trials. American Journal of Geriatric Psychiatry. 2019, 26, https://www.researchgate.net/publication/322406743_A_Machine_Learning_Approach_to_Identifying_Placebo_Responders_in_Late-Life_Depression_Trials.

[134] Alsumidaie M. AI mitigates clinical trial placebo-response. Applied Clinical Trails. 2019; https://www.appliedclinicaltrialsonline.com/view/ai-mitigates-clinical-trial-placebo-response.

[135] Nocebo; https://en.wikipedia.org/wiki/Nocebo#:~:text=The%20term%20nocebo%20(Latin%20noc%C4%93b%C5%8D,pleasant%2C%20or%20desirable%20effect).

[136] Barber TX. Death by suggestion. A Critical Note. Psychosomatic Medicine. 1961, 23, 153–5.

[137] Howick J. Unethical informed consent caused by overlooking poorly measured nocebo effects. Journal of Medical Ethics. 2020, https://jme.bmj.com/content/early/2020/05/19/medethics-2019-105903.

[138] Blasini M, Corsi N, Klinger R, Colloca L. Nocebo and pain: an overview of the psychoneurobiological mechanisms. Pain Reports. 2017, 2, https://journals.lww.com/painrpts/Fulltext/2017/03000/Nocebo_and_pain__an_overview_of_the.2.aspx.

[139] Collocal L, Panaccione R, Murphy TK. The clinical implications of nocebo effects for biosimilar therapy. Frontiers in Pharmacology. 2019, https://doi.org/10.3389/fphar.2019.01372.

Chapter 11
AI in materials science and development

Materials science is the branch devoted to the discovery and development of new materials. It is an outgrowth of both physics and chemistry, using the periodic table as its grocery store and laws of physics as its cookbook. Unfortunately, because that table is vast and its laws are complicated, materials science has been a historically slow science [1].

Various materials (including metallic materials – alloys and metals, ceramic materials, polymeric materials, composites, and hybrids) have been utilized in both medical and dental fields, if any material clears a certain limitations and strict regulations in biological safety. In addition to these narrowing issues, there should be an engineering/manufacturing challenge; in particular for intraoral applications, devices are needed to be miniaturized to accommodate an occlusal area, leading to development of microtechnology and nanotechnology.

11.1 AI-assisted materials science

Big data has been defined by the three Vs: volume (large amounts of data), variety (data heterogeneity), and velocity (speed of access and analysis) [2, 3]. Analyses of these large datasets have allowed for more powerful approaches toward successful materials science and technology, as well as assessments of healthcare quality and efficiency with the goal of improving patient care. The discovery or development of new materials is one of the driving forces to promote the development of modern society and technology innovation, the traditional materials research mainly depended on the trial-and-error method, which is time-consuming. Recently, AI/ML technologies have made great progress in the research of materials science with the arrival of the big-data era, which gives a deep revolution in human society and advances science greatly. As Juan et al. [4] pointed out, data science handling big data and AI/ML as well as AI/DL interacted well to establish an interdisciplinary field which can be of great assistance for accelerating developing new materials, as shown in upper portion [4] of Figure 11.1. The thus established interconnected area becomes and serves as a core field for further developing materials science and engineering, as seen in lower portion [5] of Figure 11.1, in which classification contains ANN and SVM, statistics contains Bayesian network for posterior probability, clustering has self-organizing map, and k-means clustering, and image processing contains convolutional neural network (CNN) and region-based CNN. There are various projects developed based on big data, such as materials informatics (MI), materials integration, and so on. These projects possess wide approaches. Examples should include, at least, (i) arrangement and prediction of experimental and computational data in complex

https://doi.org/10.1515/9783110717853-012

systems based on data science, (ii) estimation of more accurate coefficients of theoretical and modeling equations that provide feedback of experimental data, (iii) abstract identification of image data, (iv) extraction of specific areas in the image, or (v) determining similarity between data by clustering. These approaches help to objectively organize events based on the complexity of materials engineering on the basis of data science. It can be said that it is the fourth approach of materials engineering after experimentation, theory, and computational engineering. The data science and AI approach does not replace the other three approaches, but plays complementary roles such as complementing them, improving accuracy, and summarizing them [5, 6]. The data science, on which the advanced materials science and engineering is heavily relied, is considered as the fourth paradigm of science, following empirical evidence, scientific theory, and computational science [7].

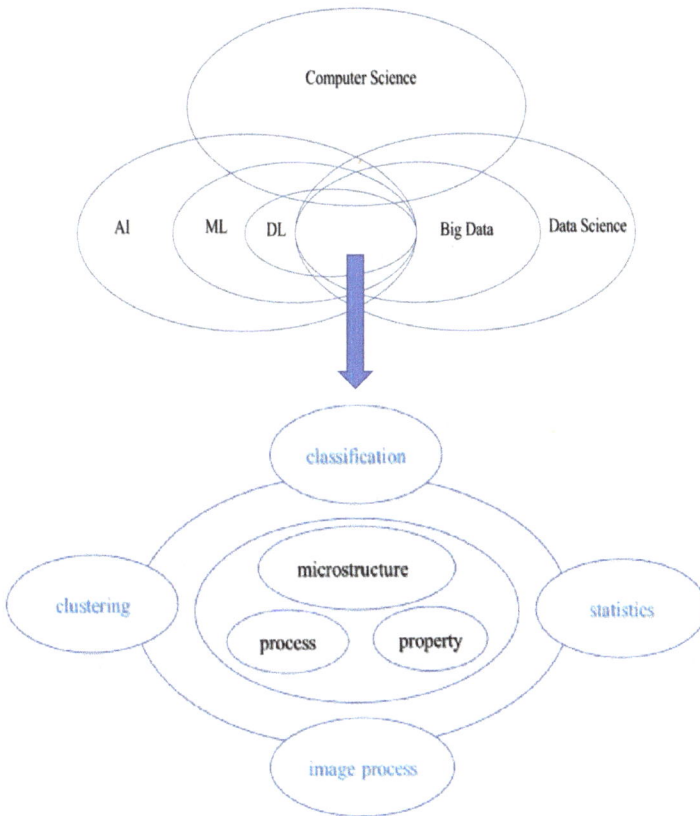

Figure 11.1: Interrelationship among various important disciplines involved in development of materials science and engineering [4, 5].

In recent years, AI has been applied to synthesize new materials and to predict various chemical synthesis [6]. The research process for computational chemistry and materials science has been experiencing generations: (1) the first generation refers to the calculation of "structure–performance interrelationship," which mainly takes advantage of the local optimization algorithm to predict the performance of the materials from the structure; (2) the second is "crystal structure prediction," which mainly adopts global optimization algorithm to predict structure and performance from element composition; and (3) the third generation recognized as "statistically driven design" utilizes ML algorithms to predict the composition, structure, and performance of elements from physical and chemical data [8, 9]. Due to the massive combination spaces of materials, it is difficult to explore all possible combinations in a reasonable time by traditional simulation calculation. However, if a small part of the data is used to train the ML model, and then the model is used to predict the other combinations, the computational complexity will be greatly reduced and the filtering speed will be increased by several orders of magnitude [10]. Despite of the fact that a large number of trial-and-error experiments based on theoretical simulation or chemical scientists' intuition typically leads to dissatisfactory results, the applications of ML models can help a lot by predicting the properties and structures of materials with an acceptable accuracy before synthesis [11, 12]. AI deals with structural data to guide the synthesis route, which has a standard process that allows scientists to design computer programs to deal with synthetic problems [13–15]. Wang et al. [16] realized the inverse exploration of microstructures related to desired target properties (e.g., stress–strain curve, tensile strength, and total elongation) by ML approach. In response to the increasing demand for the highly efficient design of materials, MI has been proposed for using data and computational sciences to extract data features that provide insight into how properties track with microstructure variables. However, the general metrics of microstructural features often ignore the complexities of the microstructure geometry for many properties of interest. Based on this background, Wang et al. [17] introduced an independently developed ML system.

Although AI is expected to make more and more contributions in materials research, there are still outlooks in both software and hardware aspects [6]. Needs for algorithm upgradation was emphasized. Effective training of ML models usually requires abundant data. Such data could come from online databases, published papers, or high-throughput experimental equipment. MI will finally map the relationship between "composition-structure-property-processing-application" through these data.

11.2 Smart materials

In the narrow sense, the shape memory materials such as NiTi alloys [18] are pronouns for smart materials; while in a broad sense, smart materials (which are also called intelligent or responsive materials) are designed materials that have one or more properties that can be significantly changed in a controlled fashion by external stimuli, such as stress, moisture, electric or magnetic fields, light, temperature, pH, or chemical compounds. Since, in a broad definition thereof, smart materials are not limited to metallic materials such as NiTi alloys, but other types of materials such as ceramics, polymers, and ceramic-matrixed and polymer-matrixed composites as well, smart materials exhibit many applications, including sensors and actuators, or artificial muscles, particularly as electroactive polymers, and of course medical and dental applications. Major functions of smart materials are (i) to act simultaneously as actuators and sensors, (ii) to perform controlled mechanical actions without any external mechanism, (iii) to be adaptive with the environmental condition, and (iv) to create the potential for new function development within applications. Depending on the sensing manner, smart materials can be divided into the passive smart materials which can act as sensors but not as actuators or transducers such as fiber optic materials, while the active smart materials which sense a change in the environment and respond to them and possess the capacity to modify their geometric or material properties under the application of electric, thermal or magnetic fields, thereby acquiring an inherent capacity to transduce energy [19, 20].

There are a variety of smart materials available [21–24]. (1) Chromoactive materials are characterized by their ability to change color when faced with an external stimulus. Depending on the stimulus that activates the material, the following types are differentiated: (i) thermochromic materials (the color change occurs after a temperature variation. There is a great variety of colors and temperature ranges, making them one of the most interesting options in the industrial field); (ii) photochromic materials (the color change of the material is generated after exposure to a certain light and generally activated using ultraviolet wavelengths); (iii) hydrochromic materials (the color change occurs when the material comes in contact with water); (iv) electrochromic materials (achieve color changes by passing electrical current through them); (v) halochromic materials (capable of changing color upon changes in pH as pH indicators), and (vi) chemochromic materials (change color in the presence of certain chemical compounds such as hypergols). (2) Self-healing materials are polymers, which, when fractured, are chemically promoted to rebond or to rebound. (3) Magnetic-sensitive materials and magnetorheological fluids are commonly known as "ferrofluids," which perfectly illustrate the shape and presence of magnetic fields. (4) Shape memory alloys are typified by NiTi alloy under certain heat treatment and exhibit various applications from industry [18, 25] to medicine and dentistry [18, 26–29].

11.3 AI-assisted materials design and development

Until recently, development has had major problems, greatly influenced by theory and experimentation, the experience and intuition of researchers, and chance, and is time-consuming and costly. For example, it is said that the filaments* of light bulbs invented by Thomas Edison were not about 6,000 kinds of materials from all over the world, but because the bamboo in the laboratory happened to produce good results, the material was narrowed down to bamboo, and then experimented with 1,200 kinds of bamboo from all over the world, and eventually it was put into production using Japanese bamboo. It is rare to encounter such a coincidence. MI is an effort to solve a situation that relies on the experience, intuition, and chance of researchers through IT.

> *Filament, through the electrical resistance, is heated to incandescence (between 2,200 and 3,000 °C). Carbon filament, carbonized paper, osmium, iridium, and tantalum were used. However, none of those materials have the high temperature resistance, strength, and long life. Tungsten (W) is now the most commonly used filament material. If the wire diameter (0.045mm) is only 1% less than the specified, the life of the lamp may be reduced by as much as 25%. Spacing between coils must also be very accurate to prevent heat concentration at one point and possible shorting.

Materials/alloys design is processed to plan what kind of elemental materials to combine, what kind of synthesis process (temperature, time, or mixing method) to produce the intended material, how to blend the raw materials for making the material, and to determine the operating conditions and to predict the life under each condition. Dialogue robots, voice assistants, or autonomous driving are visible AI applications, while MI is another AI application in the invisible manner. MI is a functional material design technology, combining AI, quantum mechanics computing, and information processing science. More concretely, MI is a field of study that employs the principles of information to materials science and engineering to improve the understanding, use, selection, development, and discovery of materials, and to accelerate materials, products and manufacturing innovations; with association with techniques, tools, and theories drawn from the emerging fields of data science, Internet, computer science and engineering, and digital technologies to the materials science and engineering [30–32]. There are two crucial keys to control the success of development of new materials by MI technologies, (1) select appropriate AI technology according to the amount and type of data available, and (2) appropriately select a "descriptor" that is the dominant parameter in constructing a prediction model [33]. To this end, SVM is one of important factors, which is a supervised ML model that uses classification algorithms for two-group classification problems [34].

Figure 11.2 illustrates a Venn diagram of three major elements for materials innovation ecosystem, which are the core of a multidisciplinary, distributed, collaborative network that couples materials development with manufacturing [35]. To accelerate materials development, the methods should have a high-throughput character, requiring a digital information infrastructure for connectivity.

Increasing the rate of discovery and development of new and improved materials is key to enhancing product development and facilitating mass customization based on emerging technologies such as 3D printing [35] or even 4D printing [36–38]. Acceleration of materials discovery and development has been enabled by advances along

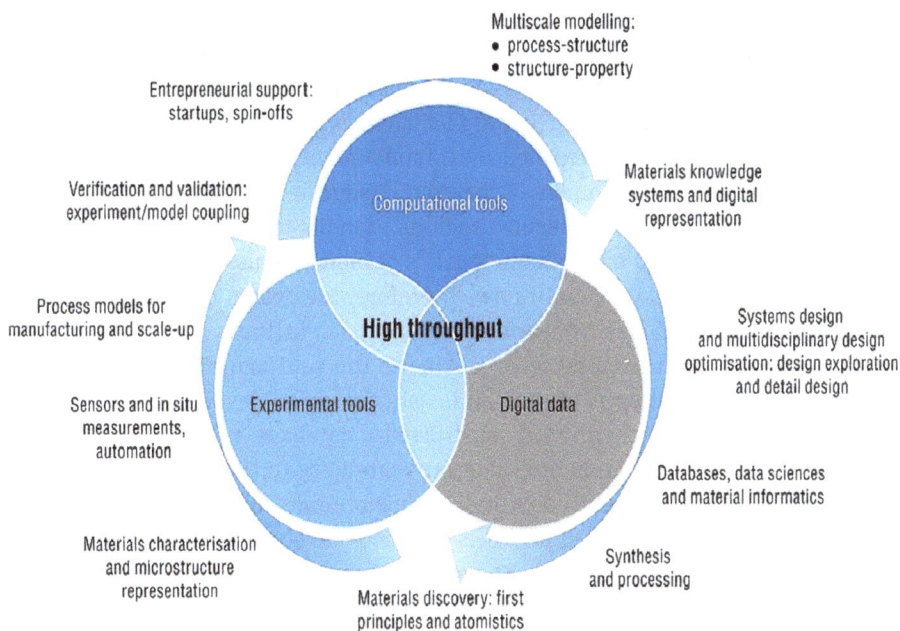

Figure 11.2: Venn diagram of three major elements for materials innovation ecosystem [35].

multiple fronts, including capabilities of scientific instrumentation, high performance computing combined with more predictive computational methods for material structure and properties, and data analytics. The US Materials Genome Initiative has emphasized the need to accelerate the discovery and development of materials to maintain industry competitiveness in new and existing markets. The goal is to build a new kind of coupled experimental, computational, and data sciences infrastructure. The emphasis is on high-throughput methods to accelerate historical sequential processes of serendipitous materials discovery and largely empirical materials development by leveraging computation and modern data sciences and analytics. Historically, it has taken 15–20 years from laboratory discovery of new materials to their deployment in products. Systematic methods for accelerated materials discovery (e.g., penicillin, Teflon, or microwave) and development are still in early stages in the new digital era. The notion of a materials innovation ecosystem is introduced as the framework in which to pursue acceleration of discovery and development of materials consisting of various elements of data sciences, design optimization, manufacturing scale-up and automation, multiscale modeling, and uncertainty quantification with verification and validation. Prospects are bright for realizing a materials innovation ecosystem necessary to integrate new materials with digital manufacturing technologies to achieve new product functionality [35, 39].

There are several unique and interesting examples. Many successful examples of material design utilizing information science have been reported in the field of solid materials such as catalysts and battery materials. In these examples, a database (or big data) of chemical structures and physical properties (material functions) of large-scale materials constructed using computational science plays an important role. On the other hand, the function of biomaterials, such as adhesion of biomolecules and cells, cannot be predicted by computational science, and there are extremely few successful examples of material design in the field of biomaterials. Although adsorption of biomolecules such as proteins to biomaterials and adhesion of cells are the most basic characteristics of biomaterials, quantitative predictions have not been realized until now. Therefore, the design of biomaterials has to rely on the conventional "trial and error" approach (actually making materials, analyzing their properties, and feed backing to the next material design), which was very inefficient in terms of time and cost of material development. Based on the background, Kwaria et al. [40] constructed a database using previously published literature and experimental data on wettability (or surface energy) of organic thin films and protein adsorption to membranes and employed ML with ANN to analyze the correlation between the chemical structure of molecules that make up a membrane and the properties of membranes, as shown in Figure 11.3.

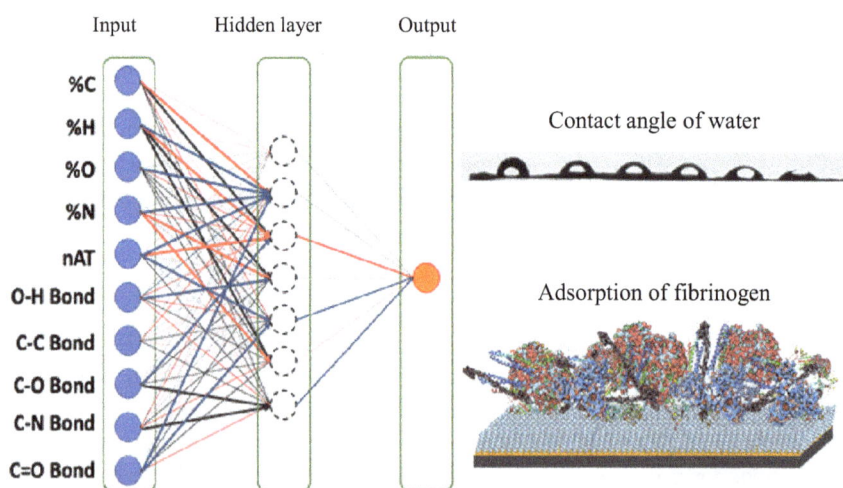

Figure 11.3: Schematic illustration of mechanical learning using ANNs [40].

It was reported that (i) after training the ANN with data of 145 self-assembled monolayers, the ANN became capable of predicting the water contact angle and protein adsorption accurately, (ii) the analysis of the trained ANN quantitatively revealed the importance of each structural parameter for the water contact angle and protein adsorption, providing essential and quantitative information for material design, and (iii) the degree of importance agrees well with our general perception on the physicochemical properties of self-assembled monolayers. As shown in Figure 11.4,

predicted values of both water contact angle and fibrinogen adsorption were linearly correlated very well to the experimentally obtained data with high regression
coefficient, and were described as follows: (1) accurate prediction of material physical properties and interactions with biomolecules from the chemical structure of the
material, (2) construction of databases using previously published data and use
of ML, (3) by screening materials at high speed, the speed of material development
is overwhelmingly accelerated, and (4) a new development method that break away
from the conventional trial-and-error approach to material design [40].

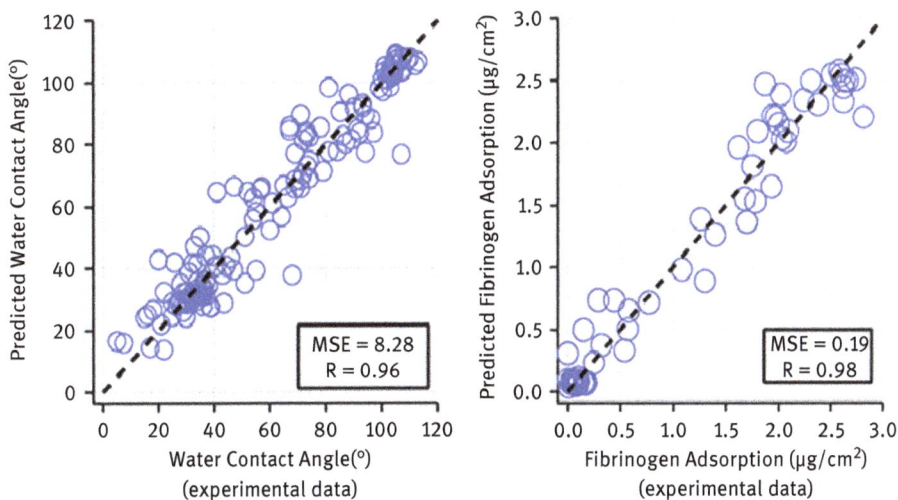

Figure 11.4: Comparison between predicted data and experimentally obtained data on water
contact angle behavior and fibrinogen adsorption phenomenon, indicating a high linear regression
analysis [40].

Researchers use AI to better predict properties of materials, optimize the number of
synthesis, or develop materials faster. Imaging techniques play an essential role in
developing smart materials, better understanding their performance in real applications and optimizing the manufacturing process. From looking at different phases
in alloys to analyzing failures of components and observing different shapes, researchers rely on images to enter the world of unseen microstructures. While image
is a good starting point in any materials analysis, extracting quantitative information is usually the step that takes the innovation of materials to the next level.
Image analysis is obtaining significant information from images (which can be obtained by either optical microscopy, electron microscopy, or microcomputed tomography). In materials research, this can be performed using image segmentation,
where the features of interest are partitioned into multiple regions for easier and
more meaningful analysis [41]. In order to convert images to data in a few minutes,

there are some excellent software available such as ZEISS ZEN Intellesis [42] and MIPAR [43]. ZEISS ZEN Intellesis is a new ML capability that enables researchers to perform advanced analysis of their imaging samples across multiple microscopy methods. The first algorithmic solution introduced by the ZEISS ZEN Intellesis platform makes integrated, easy to use, powerful segmentation for 2D and 3D datasets available to the routine microscopy user. ZEISS ZEN Intellesis allows users to train ML classifications on image datasets, and applies that trained classifier across large, multi-dimensional datasets [42]. MIPAR is a novel image analysis tool that allows users to reliably extract the measurements from their images. Handling challenging applications in everything from materials to life science, MIPAR has facilitated automated solutions to hundreds of practical problems [43]. Figure 11.5 shows that usage of DL on complex analysis of twinned grains resulted in clear augmented image by MIPAR software, while Figure 11.6 demonstrates that MIPAR enhanced textile images by the additive manufacturing feature identification, or nanofiber analysis.

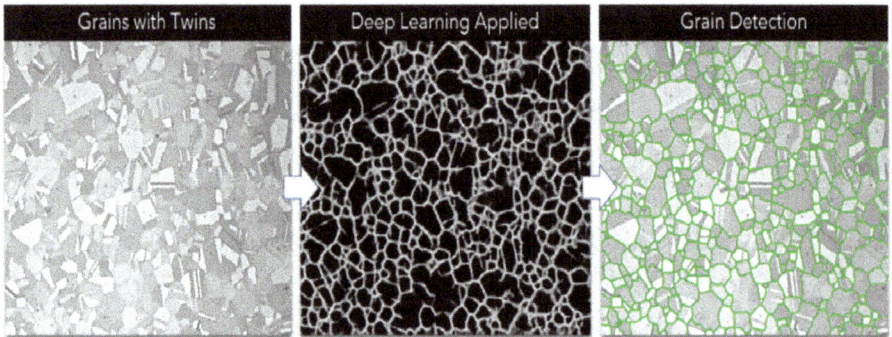

Figure 11.5: Detecting grains while ignoring twins using MIPAR. The model was trained on 25 images in 40 min and applied to the new image in 2 s [41].

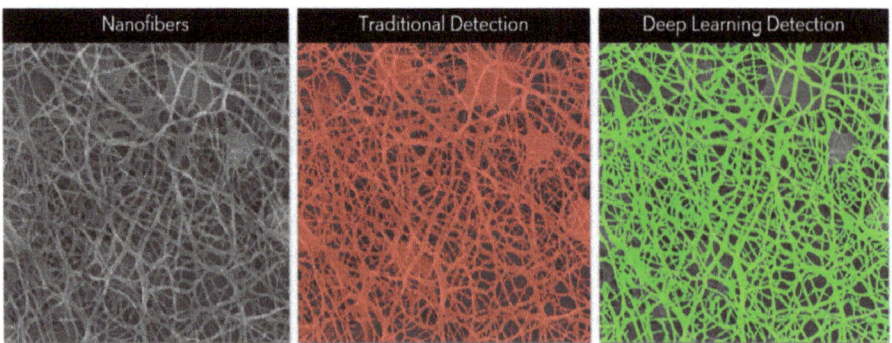

Figure 11.6: Overlapping nanofiber network analysis performed using MIPAR. The model was trained on 36 images in 40 min and applied to the new image in 1.5 s [41].

The rapid increase in global energy demand and the need to replace carbon dioxide (CO_2)-emitting fossil fuels with renewable sources have driven interest in chemical storage of intermittent solar and wind energy [44]. Particularly attractive is the electrochemical reduction of CO_2 to chemical feedstocks, which uses both CO_2 and renewable energy [45, 46]. Although Cu has been the predominant electrocatalyst for this reaction when aiming for more valuable multicarbon products and process improvements have been particularly notable when targeting ethylene, the energy efficiency and productivity (current density) achieved so far still fall below the values required to produce ethylene at cost-competitive prices. Zhong et al. [44] described that Cu–Al electrocatalysts, identified using density functional theory calculations in combination with active ML, that efficiently reduce CO_2 to ethylene with the highest Faradaic efficiency reported so far. This Faradaic efficiency of over 80% (compared to about 66% for pure Cu) is achieved at a current density of 400 mA/cm^2 (at 1.5 V vs. a reversible hydrogen electrode) and a cathodic-side (half-cell) ethylene power conversion efficiency of 55 ± 2% at 150 mA/cm^2. Based on these background, computational studies were conducted and it was suggested that (i) the Cu–Al alloys provide multiple sites and surface orientations with near-optimal CO binding for both efficient and selective CO_2 reduction; (ii) furthermore, in situ X-ray absorption measurements reveal that Cu and Al enable a favorable Cu coordination environment that enhances C–C dimerization, indicating that these findings illustrate the value of computation and ML in guiding the experimental exploration of multimetallic systems that go beyond the limitations of conventional single-metal electrocatalysts [44].

The above process can be applied to develop new drugs, although there should be additional checkpoints such as clinical evaluations, new drug application, and approval, as discussed in Chapter 10.

References

[1] Diamandis PH, Kotler S. The Future Is Faster Than You Think. Simon & Schuster, New York NY, 2020.

[2] Mooney SJ, Westreich DJ, El-Sayed AM. Commentary. Epidemiology. 2015, 26, 390–4.

[3] Price WN, Nicholson Price W, Glenn Cohen I. Privacy in the age of medical big data. Nature medicine. 2019, 25, 37–43.

[4] Juan Y, Dai Y, Yang Y, Zhang J. Accelerating materials discovery using machine learning. Journal of Materials Science & Technology. 2021, 79, 178–90.

[5] Adachi Y Utilization of Artificial Intelligence in Materials Engineering. https://www.research gate.net/publication/307902051_AI-Materials_Science_and_Engineering_-Microstructure _Recognition_by_Deep_Learning-.

[6] Sha W, Guo Y, Yuan Q, Tang S, Zhang X, Lu S, Guo X, Cao Y-C, Cheng S Artificial Intelligence to Power the Future of Materials Science and Engineering. Advanced Intelligent Systems. 2020; https://doi.org/10.1002/aisy.201900143.

[7]   Hey T, Tansley S, Tolle K. The fourth paradigm: Data-intensive scientific discovery. Microsoft Research. 2009; https://www.microsoft.com/en-us/research/publication/fourth-paradigm-data-intensive-scientific-discovery/.

[8]   Schmidt M, Lipson H. Distilling free-form natural laws from experimental data. Science. 2009, 324, 81–5.

[9]   Salmenjoki H, Alava MJ, Laurson L. Machine learning plastic deformation of crystals. Nature Communications. 2018, 9, 5307, https://doi.org/10.1038/s41467-018-07737-2.

[10]   Panapitiya G, Avendano-Franco G, Ren P, Wen X, Li Y, Lewis JP. Machine-learning prediction of CO adsorption in Thiolated, Ag-Alloyed Au Nanoclusters. Journal of the American Chemical Society. 2018, 140, 17508–14.

[11]   Sendek AD, Cubuk ED, Antoniuk ER, Cheon G, Cui Y, Reed EJ. Machine learning-assisted discovery of solid li-ion conducting materials. Chemistry of Materials. 2018, 31, 342–52.

[12]   Sun M, Wu T, Xue Y, Dougherty AW, Huang B, Li Y, Yan C-H. Mapping of atomic catalyst on graphdiyne. Nano Energy. 2019, 62, 754–63.

[13]   Sanchez-Lengeling B, Aspuru-Guzik A. Inverse molecular design using machine learning: Generative models for matter engineering. Science. 2018, 361, 360–5.

[14]   Xie T, France-Lanord A, Wang Y, Shao-Horn Y, Grossman JC. Graph dynamical networks for unsupervised learning of atomic scale dynamics in materials. Nature Communications. 2019, 10, 2667, https://doi.org/10.1038/s41467-019-10663-6.

[15]   De Almeida AF, Moreira R, Rodrigues T. Synthetic organic chemistry driven by artificial intelligence. Nature Reviews Chemistry. 2019, 3, 589–604.

[16]   Wang Z-L, Adachi Y. Property prediction and properties-to-microstructure inverse analysis of steels by a machine-learning approach. Material Science and Engineering A. 2019, 744, 661–7.

[17]   Wang Z-L, Ogawa T, Adachi Y. A machine learning tool for materials informatics. Journal of Advanced Theory and Simulations. 2019, 3, 1900177, https://doi.org/10.1002/adts.201900177.

[18]   Oshida Y, Tominaga T. Nickel-Titanium Materials – Biomedical Applications. De Gruyter, Berlin, 2020.

[19]   Fairweather JA. Designing with Active Materials: An Impedance Based Approach. 1st UMI, New York, 1998.

[20]   Kamila S. Introduction, classification and applications of smart materials: An overview. American Journal of Applied Sciences. 2013, 10, 876–80.

[21]   Smart Material. https://www.smart-material.com/index.html.

[22]   Smart materials, discover the materials with which we will shape the future. https://www.iberdrola.com/innovation/smart-materials-applications-examples.

[23]   Types and Applications of Smart Materials. https://www.bareconductive.com/news/types-and-applications-of-smart-materials/.

[24]   Types of Smart Materials. https://www.mide.com/smart-materials.

[25]   https://www.sculpteo.com/en/3d-learning-hub/best-articles-about-3d-printing/smart-materials/. Smart materials: What is it and how to use them?

[26]   Petrini L, Migliavacca F. Biomedical applications of shape memory alloys. Journal of Metallurgy. 2011, http://dx.doi.org/10.1155/2011/501483.

[27]   Aihara H, Zider J, Fanton G, Duerig T. Combustion synthesis porous nitinol for biomedical applications. International Journal of Biomaterials. 2019, https://doi.org/10.1155/2019/4307461.

[28]   Bahami V, Ahuja B. Biosmart materials: Breaking new ground in dentistry. The Scientific World Journal. 2014, http://dx.doi.org/10.1155/2014/986912.

[29] Jain P, Kaul R, Saha S, Sarkar S. Smart materials-making pediatric dentistry bio-smart. International Journal of Pedodontic Rehabilitation. 2017, 2, 55–9.

[30] Ramakrishna S. 9 co-authors, Materials informatics. Journal of Intelligent Manufacturing. 2019, 30, 2307–26.

[31] Ball P. Using Artificial Intelligence to Accelerate Materials Development. Cambridge University Press, 2019, https://doi.org/10.1557/mrs.2019.113.

[32] Leveraging Artificial Intelligence for Materials Design and Production, 2019 Markets – Using Advanced Algorithms to Accelerate Material Design and Discovery Procedures. 2019. https://www.prnewswire.com/news-releases/leveraging-artificial-intelligence-for-materials-design-and-production-2019-marketsusing-advanced-algorithms-to-accelerate-material-design-and-discovery-procedures-300939667.html.

[33] Sakai A Material development using data and AI. 2020. https://xtech.nikkei.com/atcl/nxt/mag/ne/18/00030/00003/.

[34] Stecanella B. An Introduction to Support Vector Machines (SVM). 2017; https://monkeylearn.com/blog/introduction-to-support-vector-machines-svm/.

[35] McDowell DL, Kalidindi SR. The Materials Innovation Ecosystem: A Key Enabler for the Materials Genome Initiative. Cambridge University Press, 2016, https://www.cambridge.org/core/journals/mrs-bulletin/article/abs/materials-innovation-ecosystem-a-key-enabler-for-the-materials-genome-initiative/179CF25E8B3101FBD6E81380F7364305.

[36] 4D Printing: A technology coming from the future. https://www.sculpteo.com/en/3d-learning-hub/best-articles-about-3d-printing/4d-printing-technology/.

[37] Momeni F, Liu X, Ni J. A review of 4D printing. Materials & Design. 2017, 122, 42–79.

[38] Sossou G, Demoly F, Montavon G, Gomes G. Design for 4D printing: rapidly exploring the design space around smart materials. Procedia CIRP. 2018, 70, 120–5.

[39] McDowell DL, Paden CN Chapter 6. Revolutionising product design and performance with materials innovation. https://www.oecd-ilibrary.org/sites/9789264271036-10-en/index.html?itemId=/content/component/9789264271036-10-en.

[40] Kwaria RJ, Mondarte EAQ, Tahara H, Chang R, Hatashi T. Data-driven prediction of protein adsorption on self-assembled monolayers toward material screening and design. ACS Biomaterials Science & Engineering. 2020, 14, 4949–56.

[41] Stratulat A. Artificial Intelligence for Smart Materials – an Image Analysis Approach. 2020; https://matmatch.com/blog/artificial-intelligence-for-smart-materials-an-image-analysis-approach/.

[42] Barnett R, Stratulat A, Andrew M Advanced Segmentation for Industrial Materials using Machine Learning. 2020. https://www.zeiss.com/intellesis.

[43] Sosa J. Using MIPAR software to save time and costs in image analysis. MIPAR Image Analysis. 2020, https://www.azom.com/article.aspx?ArticleID=19315.

[44] Zhong M. 21 co-authors, Accelerated discovery of CO2 electrocatalysts using active machine learning. Nature. 2020, 581, 178–83.

[45] Jeanty P, Scherer C, Magori E, Wiesner-Fleischer K, Hinrichsen O, Fleischer M. Upscaling and continuous operation of electrochemical CO2 to CO conversion in aqueous solutions on silver gas diffusion electrodes. Journal of CO2 Utilization. 2018, 24, 454–62.

[46] Dinh C. 15 co-authors, CO2 electroreduction to ethylene via hydroxide-mediated copper catalysis at an abrupt interface. Science. 2018, 360, 783–7.

Chapter 12
AI in COVID-19 era, infodemic, and post-COVID-19 era

12.1 With COVID-19

Healthcare delivery requires the support of new technologies like AI, IoT, big data, and AI/ML to fight and look ahead against the new diseases, and Vaishya et al. [1] reviewed the role of AI as a decisive technology to analyze, prepare us for prevention, and fight with COVID-19 and other pandemics. The rapid review of the literature is done on the database of Pubmed, Scopus, and Google Scholar using the keyword of COVID-19 or coronavirus (CoV) and AI. They collected the latest information regarding AI for COVID-19, then analyzed the same to identify its possible application for this disease. Seven important areas are listed to which AI technologies have been applied in COVID-19 pandemic [1]. (1) For early detection and diagnosis of the infection, AI can quickly analyze irregular symptom and other "red flags" and thus alarm the patients and the healthcare authorities [2]. It helps to provide faster decision-making, which is cost-effective. It helps to develop a new diagnosis and management system for the COVID-19 cases, through useful algorithms. AI is helpful in the diagnosis of the infected cases with the help of medical imaging technologies like CT and MRI scan of human body parts. (2) For monitoring the treatment, AI can build an intelligent platform for automatic monitoring and prediction of the spread of this virus. A neural network can also be developed to extract the visual features of this disease, and this would help in proper monitoring and treatment of the affected individuals [3, 4]. It has the capability of providing day-to-day updates of the patients and also to provide solutions to be followed in COVID-19 pandemic. (3) For contact tracing of the individuals, AI can help analyze the level of infection by this virus by identifying the clusters and "hot spots" and can successfully do the contact tracing of the individuals and also to monitor them. It can predict the future course of this disease and likely reappearance. (4) For projection of cases and mortality, AI technology can track and forecast the nature of the virus from the available data, social media, and media platforms, about the risks of the infection and its likely spread. Further, it can predict the number of positive cases and death in any region. AI can help identify the most vulnerable regions, people, and countries and take measures accordingly. (5) For development of drugs and vaccines, AI is used for drug research by analyzing the available data on COVID-19. It is useful for drug delivery design and development. This technology is used in speeding up drug testing in real-time, where standard testing takes plenty of time and hence helps to accelerate this process significantly, which may not be possible by a human [2]. It can help to identify useful drugs for the treatment of COVID-19

https://doi.org/10.1515/9783110717853-013

patients. It has become a powerful tool for diagnostic test designs and vaccination development [5–7]. AI helps in developing vaccines and treatments at much of faster rate than usual and is also helpful for clinical trials during the development of the vaccine. (6) For reducing the workload of healthcare workers due to a sudden and massive increase in the numbers of patients during COVID-19 pandemic; healthcare professionals have a very high workload. Here, AI is used to reduce the workload of healthcare workers [8, 9]. It helps in early diagnosis and providing treatment at an early stage using digital approaches and decision science, offers the best training to students and doctors regarding this new disease [10]. AI can impact future patient care and address more potential challenges which reduce the workload of the doctors. Finally, (7) for prevention of the disease with the help of real-time data analysis, AI can provide updated information which is helpful in the prevention of this disease. It can be used to predict the probable sites of infection, the influx of the virus, need for beds and healthcare professionals during this crisis [1]. AI is helpful for the future virus and diseases prevention, with the help of previous mentored data over data prevalent at different time. It identifies traits, causes, and reasons for the spread of infection. In future, this will become an important technology to fight against other epidemics and pandemics. It can provide a preventive measure and fight against many other diseases. In future, AI will play a vital role in providing more predictive and preventive healthcare. The features of infected lungs and hearts seen on medical images can help assess disease severity, predict response to treatment, and improve patient outcomes. However, a major challenge is to rapidly and accurately identify these signatures and to evaluate this information in combination with many other clinical symptoms and tests. The Medical Imaging and Data Resource Center (MIDRC) goals are to lead the development and implementation of new diagnostics, including AI/ML algorithms, that will allow rapid and accurate assessment of disease status and help physicians optimize patient treatment, with gathering a large repository of COVID-19 chest images nationally and internationally. The MIDRC will facilitate rapid and flexible collection, analysis, and dissemination of imaging and associated clinical data [11].

The organized World Economic Forum, in the fight against COVID-19, has been quick to apply their ML expertise in several areas: scaling customer communications, understanding how COVID-19 spreads, and speeding up research and treatment. The organizations took two important actions: (1) understanding how COVID-19 spreads and (2) speeding up research and treatment. ML is helping researchers and practitioners analyze large volumes of data to forecast the spread of COVID-19, in order to act as an early warning system for future pandemics and to identify vulnerable populations, including Chan Zuckerberg Biohub, AWS Diagnostic Development Initiative, BlueDot, or Closedloop. In parallel to these activities, healthcare providers and researchers are faced with an exponentially increasing volume of information about COVID-19, which makes it difficult to derive insights that can inform treatment, including COVID-19 Search, COVID-19 open research dataset, UC San Diego Health, or

BenevolentAI [12]. This is in part due to advancements in rapid genome sequencing technologies. AI can then use this information to predict which compounds are most likely to be effective against a specific target. But for this to work, the quality of the data is essential. Exscientia, based in Oxford, has a two-pronged approach to drug discovery: an in-house lab for data generation and experimentation, and an ML platform for designing experiments. AI methods have already yielded results for two companies. British drug discovery company BenevolentAI and Germany-based Innoplexus identified potential therapeutic options for COVID-19 patients earlier this year. BenevolentAI's approach integrates many types of data to generate predictive models for a wide range of diseases, which are in turn improved with feedback from experimental data. Innoplexus uses a similar approach but with a broader data input consisting of published papers, online data from presentations, symposiums and conferences, clinical trial data, and publicly available hospital data as well as unpublished datasets. In March, the company released an analysis of potential COVID-19 therapeutics as well as drug combinations [13].

AI could play a decisive role in stopping the COVID-19 pandemic. To give the technology a push, the MIT-IBM Watson AI Lab is funding 10 projects at MIT aimed at advancing AI's transformative potential for society. The research will target the immediate public health and economic challenges of this moment. But it could have a lasting impact on how we evaluate and respond to risk long after the crisis has passed. The 10 research projects include (1) early detection of sepsis in COVID-19 patients, (2) designing proteins to block SARS-CoV-2, (3) saving lives while restarting the US economy, (4) materials to make the best face masks (5) treating COVID-19 with repurposed drugs, (6) a privacy-first approach to automated contact tracing, (7) overcoming manufacturing and supply hurdles to provide global access to a CoV vaccine, (8) leveraging electronic medical records to find a treatment for COVID-19, (9) finding better ways to treat COVID-19 patients on ventilators, and (10) returning to normal via targeted lockdowns, personalized treatments, and mass testing. An inexpensive, spectroscopy-based test will also be developed for COVID-19 that can deliver results in minutes and pave the way for mass testing. The project will draw on clinical data from four hospitals in the US and Europe, including Codogno Hospital, which reported Italy's first infection [14].

Current diagnostic tests for COVID-19 detect viral ribonucleic acid to determine whether someone does or does not have the virus – but they do not provide clues as to how sick a COVID-positive patient may become. Identifying and monitoring those at risk for severe cases could help hospitals prioritize care and allocate resources like intensive care unit beds and ventilators. Likewise, knowing who is at low risk for complications could help reduce hospital admissions while these patients are safely managed at home. Based on this background, McRae et al. [15] used AI application to assess risk factors and key biomarkers from blood tests, producing a COVID-19 severity score. Using data from 160 hospitalized COVID-19 patients in Wuhan, China, the researchers identified four biomarkers measured in blood tests that were significantly elevated in patients who died versus those who recovered: C-reactive protein, myoglobin, procalcitonin, and cardiac troponin I (cTnI). These biomarkers can signal complications that are relevant to COVID-19, including acute inflammation, lower respiratory tract infection, and poor cardiovascular health. The researchers then built a model using the biomarkers as well as age and sex, two established risk factors. They trained the model using a ML algorithm, a type of AI, to define the patterns of COVID-19 disease and predict its severity. When a patient's biomarkers and risk factors are entered into the model, it produces a numerical COVID-19 severity score ranging from 0 (mild or moderate) to 100 (critical). The model was validated using data from 12 hospitalized COVID-19 patients from Shenzhen, China, which confirmed that the model's severity scores were significantly higher for the patients that died versus those who were discharged. It was found that (i) the median COVID-19 severity score was significantly lower for the group that recovered versus the group that died from COVID-19 complications (9 vs. 59, respectively), (ii) the area under the curve (AUC) value for the COVID-19 severity score was 0.94, demonstrating strong potential for its utility in identifying patients with increased risk of mortality, (iii) in the analysis of patients with hypertension, as expected, cardiac biomarkers had a large effect on the COVID-19 severity score. The development and distribution of a portable, affordable, widely distributed smart sensor technology with anticipated availability/readiness within months promises to be an important solution for the management of the current CoV crisis, as well as an adaptable tool to combat future viral or biological threats. Likewise, in addition to this COVID-19 severity score, a sustaining contribution of this work may be in the development of an acute respiratory distress syndrome clinical decision support tool for other infectious viral agents, bacteria, and parasites [15].

Chest imaging is not currently a standard method for diagnosing COVID-19, but the technology has helped providers exclude other possible causes for COVID-19-like symptoms, confirming a diagnosis made by another means, or providing critical data for monitoring a patient's progress. AI has the potential to expand the role of chest imaging in COVID-19 beyond diagnosis to enable risk stratification, treatment monitoring, and discovery of novel therapeutic targets. AI's power to generate models from large volumes of information – fusing molecular, clinical, epidemiological, and imaging data – may accelerate solutions to detect, contain, and treat COVID-19. In the future, AI could help researchers discover disease progression across different populations, such as in patients with chronic lung conditions and long-term smokers. AI has successfully harnessed radiomics to characterize biological underpinnings of various disease processes, such as cancer, to predict response to treatment [16]. Of all the various technologies that make up what is collectively referred to in the singular as AI, perhaps the one with the greatest potential benefit in the fight against COVID-19 is ML, combining large amounts of data with fast, iterative processing and intelligent algorithms that allow the software to learn automatically from patterns or features in the data. Benefits of AI fighting against the COVID-19 should include, as many articles already mentioned, early detection, and epidemic analysis. In addition to this, Weldon [17] added triage and diagnosis, which must be is one of the strongest use cases for ML for COVID-19. Temperature, heart rate (HR), respiration, and blood oxygen are vital signs that can be taken with a high degree of accuracy through non-invasive, remote sensors. Other health signs, including voice, movement, weight, and toileting, can also be effectively monitored with remote sensors. Combined, ML models are being created to monitor and predict the onset of adverse events and the progression of disease. These remote sensors and models will ensure that we are taking care of patients in the most appropriate facility, and that the most vulnerable patients are identified quickly and monitored effectively [17]. As the COVID-19 pandemic continues to infect people across the world, a technological application already familiar to many in the biotech field is lending a key supporting role in the fight to treat and stop it: AI. AI is currently being used by many companies to identify and screen existing drugs that could be repurposed to treat COVID-19, aid clinical trials, sift through trial data, and scour through patient EMRs. The power of AI in COVID-19 is that it has been used to generate actionable information – some of which would be impossible without AI – much more quickly than before. One promising AI application is repurposing existing drugs that could be effective in treating COVID-19, in particular probing molecular structures, by the three-dimensional pocket of critical proteins in the viral infection pathway. AI can discover new drug candidate against COVID-19 [18].

With the limited availability of testing for the presence of the SARS-CoV-2 virus and concerns surrounding the accuracy of existing methods, other means of identifying patients are urgently needed. Previous studies showing a correlation between certain laboratory tests and diagnosis suggest an alternative method based on an ensemble of tests. Accordingly, Bayat et al. [19] trained an ML model to analyze the correlation between SARS-CoV-2 test results and 20 routine laboratory tests collected within a 2-day period around the SARS-CoV-2 test date. The model to compare SARS-CoV-2 positive and negative patients was employed. It was obtained that, in a cohort of 75,991 veteran inpatients and outpatients who tested for SARS-CoV-2 in the months of March through July 2020, 7,335 of whom were positive by reverse transcription polymerase chain reaction (RT-PCR) or antigen testing, and who had at least 15 of 20 lab results within the window period, the present ML model predicted the results of the SARS-CoV-2 test with a specificity of 86.8%, a sensitivity of 82.4%, and an overall accuracy of 86.4% (with a 95% confidence interval of (86.0%, 86.9%)). It was therefore concluded that (i) although molecular-based and antibody tests remain the reference standard method for confirming a SARS-CoV-2 diagnosis, their clinical sensitivity is not well known, and (ii) the model described herein may provide a complementary method of determining SARS-CoV-2 infection status, based on a fully independent set of indicators, that can help confirm results from other tests as well as identify positive cases missed by molecular testing like PCR test [19].

Levitan [20] mentioned that (i) a significant number of hospital admissions not due to COVID-19 were actually found to be suffering from pneumonia due to COVID-19 upon having their chest X-rays taken, most with severely low oxygen saturation levels of only 50%, (ii) unlike normal pneumonia, in which patients will feel chest pain and significant breathing difficulties, initially COVID-19 pneumonia causes oxygen deprivation that is difficult to detect since the patients do not experience any noticeable breathing difficulties, hence causing a condition which was termed as "silent" hypoxia, and (iii) by the time COVID-19 patients realize they are short of breath, their conditions have already significantly deteriorated into moderate-to-severe levels of pneumonia. Analysis of COVID-19 pneumonia patients revealed that the virus initially attacks the lungs in a different way. The air sacs in COVID-19 patients' lungs do not fill with fluid or pus as in normal pneumonia infections but rather the virus only causes the air sacs to collapse, thereby reducing the oxygen levels that lead to hypoxia in these patients but still maintains the lungs' normal ability to expel carbon dioxide. Consequently, the still-efficient removal of carbon dioxide is the reason why COVID-19 patients do not feel shortness of breath in the initial stages of COVID-19 pneumonia [20]. It is possible to perform this early detection of silent hypoxia in COVID-19 patients using smartphones [21]. Tayfur et al. [22] reported that oxygen saturation (SaO2) readings measured using smartphones correlate highly with readings obtained using medical-grade pulse oximeters. Using a smartphone, pulse oximetry readings were compared to two medical-grade devices commonly found in emergency rooms of 101 patients and reported that (i) there was a high correlation between HR measured by smartphone and HR measured by the vital signs monitor with correlation of 95%, and (ii) SaO2 values obtained by smartphone were highly correlated with those by the arterial blood gas with correlation of 95%. It was then concluded that (i) the HR and SaO2 values obtained by smartphone were found to be consistent with the measurements of the reference devices, and (ii) with the growing use of smartphone technology in the health field, we foresee that patients will be able to make their own triage assessment before presenting to the hospital [22].

Patients with influenza and SARS-CoV-2/COVID-19 infections have different clinical course and outcomes. Yanamala et al. [23] developed and validated a supervised ML pipeline to distinguish the two viral infections using the available vital signs and demographic dataset from the first hospital/emergency room encounters of 3,883 patients who had confirmed diagnoses of influenza A/B, COVID-19, or negative laboratory test results. It was reported that the models were able to achieve an area under the receiver operating characteristic curve (AUC-ROC) of at least 97% using our multiclass classifier. The predictive models were externally validated on 15,697 encounters in 3,125 patients available on TrinetX database that contains patient-level data from different healthcare organizations. It was found that the influenza versus COVID-19-positive model had an AUC of 98%, and 92% on the internal and external test sets, respectively, indicating the potentials of ML models for accurately distinguishing the two viral infections.

To prevent shortages, researchers have come up with a new way to design synthetic routes to drugs now being tested in some COVID-19 clinical trials, using AI software [24]. As scientists uncover drugs that can treat CoV infections, demand will almost certainly outstrip supplies – as is already happening with the antiviral Remdesivir* [25]. The AI-planned new recipes (for 11 medicines so far) could help manufacturers produce medications whose syntheses are tightly held trade secrets. And because the new methods use cheap, readily available starting materials, licensed drug suppliers could quickly ramp up production of any promising therapies [24].

*Remdesivir is an intravenous nucleotide prodrug of an adenosine analog. It has been demonstrated in-vitro activity against SARS-CoV-2. Remdesivir is approved by the FDA for the treatment of COVID-19 in hospitalized adult and pediatric patients (aged ≥12 years and weighing ≥40 kg). This drug should be administered in a hospital or a healthcare setting that can provide a similar level of care to an inpatient hospital[25].

12.2 Some concerns and infodemic

12.2.1 Concerns

Although there can be found many positive contributions of AI technologies against COVD-19 pandemic, as reviewed in the above, there are still some skeptical opinions and concerns. While ML might hold promise as a powerful medical tool, statisticians warn that current models are severely flawed. For years, many AI enthusiasts and researchers have promised that ML will change modern medicine. The COVID-19 literature has grown in much the same way as the disease's transmission: exponentially. The NIH's COVID-19 Portfolio, a website that tracks papers related to the SARS-CoV-2 and the disease it causes, lists more than 28,000 articles, indicating that these volumes of publications are far too many for any researcher to read, as shown in Figure 12.1 [26].

Whitten [27] mentioned that (i) thousands of algorithms have been developed to diagnose conditions like cancer, heart disease, and psychiatric disorders and now, algorithms are being trained to detect COVID-19 by recognizing patterns in CT scans and X-ray images of the lungs, (ii) many of these models aim to predict which patients will have the most severe outcomes and who will need a ventilator, and (iii) the excitement is palpable; if these models are accurate, they could offer doctors a huge leg up in testing and treating patients with the CoV; however the allure of AI-aided medicine for the treatment of real COVID-19 patients appears far off, and (iv) a group of statisticians around the world are concerned about the quality of the vast majority of ML models and the harm they may cause if hospitals adopt them any time soon. Wynants et al. [28] reviewed and evaluated the validity and usefulness of published and preprint reports of prediction models for diagnosing COVID-19 in patients with suspected infection, for prognosis of patients with COVID-19, and for

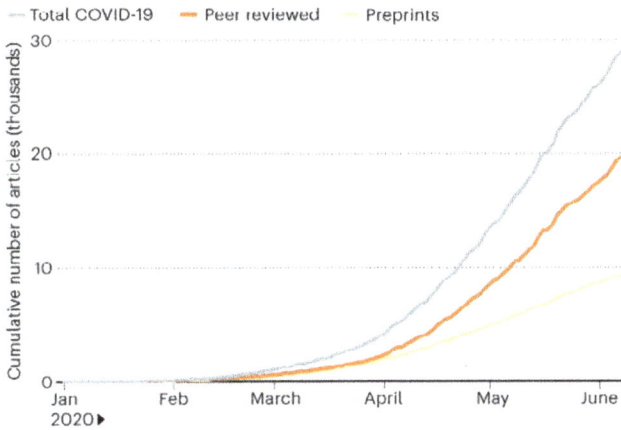

Figure 12.1: Explosive growth of publications on COVID-19 [26].

detecting people in the general population at increased risk of COVID-19 infection or being admitted to hospital with the disease. Data sources were PubMed and Embase through Ovid, up to 1 July 2020, supplemented with arXiv, medRxiv, and bioRxiv up to 5 May 2020. It was concluded that (i) almost all published prediction models are poorly reported, due to a serious lack of data and necessary expertise from a wide array of research fields, and (ii) prediction model authors should adhere to the transparent reporting of a multivariable prediction model for individual prognosis or diagnosis reporting guideline. Based on a fact that DL offers considerable promise for medical diagnostics, Liu et al. [29] evaluated the diagnostic accuracy of DL algorithms versus healthcare professionals in classifying diseases using medical imaging. It was reported that, (i) although the diagnostic performance of DL models were found to be equivalent to that of healthcare professionals, a major finding of the review is that few studies presented externally validated results or compared the performance of DL models and healthcare professionals using the same sample, (ii) poor reporting is prevalent in DL studies, which limits reliable interpretation of the reported diagnostic accuracy, and (iii) new reporting standards that address specific challenges of DL could improve future studies, enabling greater confidence in the results of future evaluations of this promising technology. Additionally, a lack of transparency is a crucial issue found in most of published articles [27].

The next big area where AI was supposed to help is modeling. For a time, the entire technology news cycle was dominated by headlines declaring that AI had first discovered the COVID-19 threat and ML would determine exactly how the virus would spread. Greene [30] described that (i) unfortunately modeling a pandemic is not an exact science, because models started with guesses and were subsequently trained on up-to-date data from the unfolding pandemic and (ii) AI models have not proven much better than our best guesses, and they can only show us a tiny part of the overall picture because we are only working with the data we can actually see, (iii) up to 80% of COVID-19 carriers are asymptomatic and a mere fraction of all possible carriers have been tested. Private and public entities around the world, particularly in the healthcare and governance sectors, are developing and deploying a range of AI systems in emergency response to COVID-19. Some of these systems work to track and predict its spread; others support medical response or help maintain social control. Indeed, AI systems can reduce strain on overwhelmed healthcare systems; help save lives by quickly diagnosing patients, and assessing health declines or progress; and limit the virus's spread. But there is a problem. Smith et al. [31] pointed out that the algorithms driving these systems are human creations, and as such, they are subject to biases that can deepen societal inequities and pose risks to businesses and society more broadly and suggested a number of ways nonprofit and business leaders can help ensure that they develop, manage, and use transformative AI equitably and responsibly. Various AI systems are proving incredibly valuable to tackling the pandemic, and others hold immense promise. But leaders must take care to develop, manage, and use this technology responsibly and equitably; the risks of discrimination and deepening inequality are simply unacceptable. Five actions are listed to take now: (1) demand transparency and explanation of the AI system, (2) join and promote multidisciplinary ethics working groups or councils to inform response to COVID-19, (3) build partnerships to fill health-data gaps in ways that protect and empower local communities, (4) advance research and innovation while emphasizing diversity and inclusion, and (5) resist the urge to prioritize efficiency at the cost of justice and equity [31].

12.2.2 Infodemic

Infodemic is a portmanteau of "information" and "epidemic" that typically refers to a rapid and far-reaching spread of both accurate and inaccurate information such as fake news on the web (especially social media) about something, such as a disease. As facts, rumors and fears mix and disperse causing people in panic and social chaotic behavior and it becomes difficult to learn essential information about an issue [32–34]. Here is an evidence. According to a recent analysis, an average of 367 COVID-19 papers are being published every week, with a median time from submission to acceptance of just 6 days (compared with 84 days for non-COVID-19 content) [35]. These unprecedented peer-review turnaround times – and in some cases relaxed editorial standards – are justifiable in a context where new information may accelerate knowledge and solutions to the emerging global medico-socio-economic disaster, but they also risk the release of preliminary or flawed publications that can mislead research and development efforts, compromise clinical practice, and misinform policy makers. Hence, Levin et al. raised a question whether traditional computational analysis and ML can help compensate for inadequate peer review of drug-repurposing papers in the context of an infodemic [36]. It was mentioned that (i) AI/ML has a critical role in bolstering data to supplement peer-reviewed papers and (ii) it was called for the rapid development and prospective validation of comprehensive, explainable AI/ML systems that use preclinical and clinical data, capable of not only rapidly predicting clinical trial outcomes, but also highlighting possible flaws in published work and the features contributing to the increased probability of failure.

The latest technology to fight the COVID-19 with medical AI is updated every day, bringing together the wisdom of humanity. On the other hand, there are reports that raise the alarm about the ethical issues of a rapid AI approach. Cave et al. [37] stressed the importance of adopting a consistent ethical approach to AI utilization from the very beginning, as there is a risk that the benefits of technology will be offset by adverse events in rapid and large-scale AI deployments and that society as a whole will lose confidence in AI. They listed up several key points: (i) AI-based technologies promise benefits for tackling a pandemic like COVID-19, but also raise ethical challenges for developers and decision makers, (ii) if an ethical approach is not taken, the risks increase of unintended harmful consequences and a loss of stakeholder trust, (iii) ethical challenges from the use of AI systems arise because they often require large amounts of personal data; automate decisions previously made by humans; and can be highly complex and opaque, (iv) the four pillars of biomedical ethics – beneficence, non-maleficence, autonomy, and justice – are a helpful way of seeing how these challenges can arise in public health, and (v) open and transparent communication with diverse stakeholder groups during development of AI systems is the best way of tackling these challenges [37]. Leslie et al. [38] stated that AI design itself argues for risks susceptible to biases that encourage existing inequalities and concluded that AI could make a valuable contribution to clinical, research, and public health tools in the fight against COVID-19. The widespread sense of urgency to innovate, however, should be tempered by the need to consider existing health inequalities, disproportionate pandemic vulnerability, sociotechnical determinants of algorithmic discrimination, and the serious consequences of clinical and epidemiological AI applications. Without this consideration, patterns of systemic health inequity and bias will enter AI systems dedicated to tackling the pandemic, amplifying inequality, and subjecting disadvantaged communities to increasingly disproportionate harm. It also mentioned the following key messages: (i) the impact of COVID-19 has fallen disproportionately on disadvantaged and vulnerable communities, and the use of AI technologies to combat the pandemic risks compounding these inequities, (ii) AI systems can introduce or reflect bias and discrimination in three ways: in patterns of health discrimination that become entrenched in datasets, in data representativeness, and in human choices made during the design, development, and deployment of these systems, (iii) the use of AI threatens to exacerbate the disparate effect of COVID-19 on marginalized, underrepresented, and vulnerable groups, particularly black, Asian, and other minoritized ethnic people, older populations, and those of lower socioeconomic status, and (iv) to mitigate the compounding effects of AI on inequalities associated with covid-19, decision makers, technology developers, and health officials must account for the potential biases and inequities at all stages of the AI process [38].

12.3 Post COVID-19

Anonymous bacteriologist mentions that predictions from corona mutant amplification are sufficiently likely to cause pandemics such as COVID-23 or COVID-36, as we have not yet seen the end of COVID-19. COVID-19 pandemic has changed the world, the country, the society, the home, and the individual to different extent. There are certain situations that cannot be back to the original condition any longer, and there are a lot of organizations which cannot help receiving the transition of the "New Normal" without resistance. This is what we all need to face. AI technology is definitely expected to contribute its power in different sectors. In this section, we will discuss areas in which AI technology can be devoted for better new normal lives of people.

As a global trend, there are two major changes to be notified; (1) the flow of digital shift due to coronal disaster is irreversible and accelerated, and (2) globalization is not reversed but going slow down. Much has been written about how the COVID-19 pandemic has accelerated the pace of digital transformation, including rapid migration to the cloud and the move to work from home [39]. A recent survey by IBM confirms this by revealing that the majority of global businesses have accelerated their DX plans due to the COVID-19 pandemic [40]. About the globalization, inside the border between countries, people awaken to solidarity, and strong centripetal force works; while there will be a strong centrifugal force outside the borderline, and an international divide will be spreading slowly and steadily. Will globalization be reversed? Probably not, it is too important an economic development for that to happen, but it could well be slowed down [41]. In conclusion, COVID-19 looks like a "bend but won't break crisis" for globalization. International flows are plummeting, but globalization – and opposition to globalization – will continue to present business opportunities and challenges. Careful attention to the drivers of globalization's future can help companies navigate through and even profit from globalization's turbulence. A volatile world of partially connected national economies expands possibilities for global strategy even as it complicates the management of multinational firms [42]. Elements constructing a new normal can include (1) advent of DX society, (2) transformation of organization and employment systems, (3) shift to sustainable society, (4) increase in personal discretion and responsibility, and (5) changing balance between individuals and society. Dawson [43] mentioned that, while it might still be too early to make predictions, there are dozens of articles on the web predicting how the COVID-19 pandemic might change our long-term behavior. Here are some of the more interesting predictions: (i) an outcry for better home broadband, (ii) digital meetings are here to stay, (iii) demand for faster upload speeds, (iv) telemedicine has arrived, (v) expect contactless payments, (vi) more telecommuting, and (vii) a reboot for corporate security.

It was reported that business and technology leaders have pointed to a range of opportunities provided by AI to support the economy's revitalization, including new efficiencies and new business models [44]. But a lack of understanding in the boardroom, legacy technologies, and an under-skilled workforce all represent major challenges. A new report entitled AI in a post-COVID-19 world, has revealed that 72% of leaders feel positive about the role that AI will play in the future, with the number one expectation being that it will make business processes more efficient (74%). Fifty-five percent of leaders have suggested that AI will help to create new business models, and 54% expect it to enable the creation of new products and services. But respondents have pointed to a range of barriers to AI achieving its potential, led by a lack of understanding or commitment towards investing at board level – feared by 59% of respondents. The legacy processes and technologies within businesses that do not support AI (50%), and the lack of relevant skills within the workforce (48%), were also big concerns. Sixty percent of respondents reported that their organization currently uses AI; a further 52% are currently planning an implementation. ML is the key technology of the moment – it is already being used in many companies (70%), and 63% are planning further integrations. The other key technologies pointed to by respondents were data science, predictive analytics, and chatbots [44].

COVID-19 pandemic is the ultimate catalyst for DX and will greatly accelerate several major trends that were already well underway before the pandemic. The COVID-19 pandemic will have a lasting effect not only on our economy, but also on how we go about our daily lives, and things are not likely to return to pre-pandemic norms. While this pandemic has forced many businesses to reduce or suspend operations, affecting their bottom line, it has helped to accelerate the development of several emerging technologies. This is especially true for innovations that reduce human-to-human contact, automate processes, and increase productivity amid social distancing [45]. Post-corona AI will be interconnected and accelerated with cloud computing, voice user interface, virtual reality/augmented reality, 5G networks, IoT, cybersecurity, and blockchain platforms [45, 46], as illustrated in Figure 12.2 [45]. If AI is rephrased as digital technology ecosystem, the similar infrastructure configuration as Figure 12.2 can be seen [47].

There are several profound insights on healthcare in the post-corona era. The COVID-19 pandemic has provided countless lessons and clues on how we could prepare better for the future. It has unearthed or reinforced consumer behaviors and actions which may fundamentally alter healthcare as we know it. Shrikhande et al. [48] mentioned that the post-corona healthcare is mainly supported by accessibility, affordability, and quality and these are basically relied on detailed technologies such as (1) telemedicine, (2) point-of-care diagnostics, (3) patient education and awareness platforms, (4) self-health management enablers, (5) affordable medical devices, and (6)

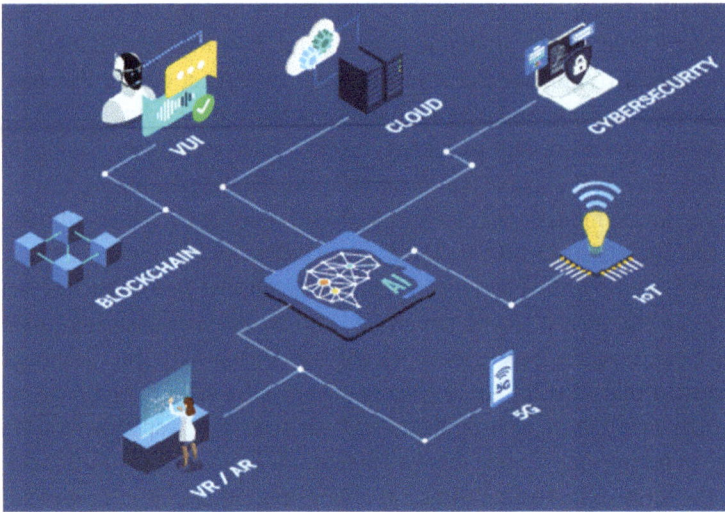

Figure 12.2: Infrastructural technologies supporting the post-corona AI era [45].

EHRs and analytics, as shown in Figure 12.3. It was summarized that (i) these technologies will make healthcare more personalized, accessible, and affordable; however, governments also need to play a role in unlocking the potential of these technologies, (ii) there is no doubt that COVID-19 will leave a significant imprint on how healthcare is funded and financed, provided and consumed; but, with great challenges, come great opportunities, (iii) evidence suggests that the most resilient businesses emerge in times of recession, (iv) hopefully, we will be better prepared

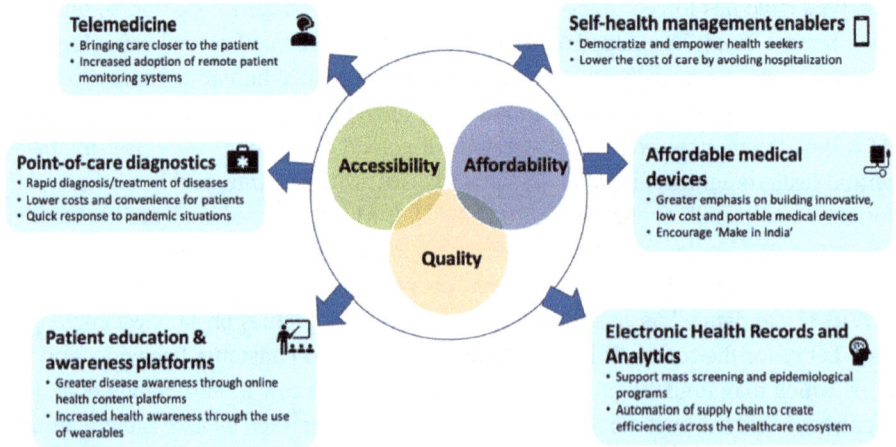

Figure 12.3: The future of healthcare in a post-COVID-19 world [48].

when (we hope it is an "if") a pandemic occurs again, and (v) hopefully, we do not forget the lessons we have learnt over the past few weeks [48].

Hughes-Cromwick et al. [49] mentioned, as what we have been learning through the current corona crisis for the future, that there are three challenges: (1) closer integration of public health infrastructure and medical care, particularly to ensure adequate public health supports for nursing homes, safety-net providers, and others who serve high percentages of Medicaid patients; (2) enhanced federal supports for state and local health departments, with a focus on public health preparedness and prevention; and (3) improved public health supports for undocumented immigrants and other vulnerable populations that face elevated COVID risks.

The current COVID-19 virus is the third CoV outbreak of international concern in 20 years, after the severe acute respiratory syndrome and the Middle-East respiratory syndrome in addition to other viral outbreaks such as Zika virus and Ebola virus over the last decade. The effects of the COVID-19 pandemic globally are striking as it impacts greatly the social, political, economic, and healthcare aspects of many countries. The toll of this pandemic quantified with human lives and suffering, the psychosocial impact, and the economic slowdown constitute strong reasons to translate experiences into actionable lessons, not simply to prevent similar future crises, but rather to improve the whole spectrum of population health and healthcare delivery [50]. Jazieh et al. [50] concluded the article that (i) the COVID-19 outbreak serves as a reminder that proactive planning for healthcare emergencies as well as an intensified commitment to global public health preparedness remains necessary, (ii) the lessons learned on the limitations of extant healthcare systems and their capacity to respond to infectious disease epidemics in the twenty-first century should be considered, enabling the transformation of future healthcare, (iii) in addition, the realization that technologically empowered solutions can be implemented and work well-should constitute the benchmark for the greater integration of such technologies as part of routine healthcare design and provision, (iv) optimal outcomes can be attained where both patients and healthcare providers become active participants in this process; however, for that to be achieved, ethical, regulatory, and legal concerns that emerged during this pandemic need to be addressed, and (v) the current global experiences lay the foundation for a significant post-COVID-19 healthcare transformation, so that systems can better prepare to address the next global threat(s) of the twenty-first century [50].

The pandemic accelerated one of the largest trends in healthcare today: the movement of high-quality healthcare from the hospital and into homes and the community. Rkowski [51] concluded that (i) the COVID-19 experience unmasked major deficiencies and inflexibility in our facility-based healthcare system, (ii) the crisis created a unique opportunity to fully reimagine and transform our healthcare delivery system that will be co-created with hospitals and rely on a new, more decentralized paradigm, supported by technology and advanced logistics, and (iii) this transformation has already begun and will accelerate and reach a tipping point, as the tailwinds of increased costs, consumerism, and technology push us to a new future.

COVID-19 pandemic accelerates the further development of the bioinformatics, which will be reviewed in Chapter 13.

References

[1] Vaishya R, Javaid M, Khan IH, Haleem A. Artificial Intelligence (AI) applications for COVID-19 pandemic. Diabetes & Metabolic Syndrome: Clinical Research & Reviews. 2020, 14, 337–9.

[2] Ai T, Yang Z, Hou H, Zhan C, Chen C, Lv W, Tao Q, Sun Z, Xia L. Correlation of chest CT and RT-PCR testing in coronavirus disease 2019 (COVID-19) in China: a report of 1014 cases. Radiology. 2020, 296. doi: https://doi.org/10.1148/radiol.2020200642.

[3] Haleem A, Vaishya R, Javaid M, Khan IH. Artificial Intelligence (AI) applications in orthopaedics: an innovative technology to embrace. Journal of Clinical Orthopaedics and Trauma. 2019. doi: https://doi.org/10.1016/j.jcot.2019.06.012.

[4] Stebbing J, Phelan A, Griffin I, Tucker C, Oechsle O, Smith D, Richardson P. COVID-19: combining antiviral and anti-inflammatory treatments. The Lancet Infectious Diseases. 2020, 20, 400–2.

[5] Sohrabi C, Alsafi Z, O'Neill N, Khan M, Kerwan A, Al-Jabir A, Iosifidis C, Agha R. World Health Organization declares global emergency: a review of the 2019 novel coronavirus (COVID-19). International Journal of Surgery. 2020, 76, 71–6.

[6] Chen S, Yang J, Yang W, Wang C. Bärnighausen T. COVID-19 control in China during mass population movements at New Year. Lancet. 2020. doi: https://doi.org/10.1016/S0140-6736(20)30421-9.

[7] Bobdey S, Ray S. Going viral–COVID-19 impact assessment: a perspective beyond clinical practice. Journal of Marine Medical Society. 2020, 22, 9–12.

[8] Pirouz B, Haghshenas Sina S, Haghshenas Sami S, Piro P. Investigating a serious challenge in the sustainable development process: analysis of confirmed cases of COVID-19 (new type of coronavirus) through a binary classification using artificial intelligence and regression analysis. Sustainability. 2020, 12, 2427. doi: https://doi.org/10.3390/su12062427.

[9] Li L and 17 co-authors. Artificial intelligence distinguishes COVID-19 from community-acquired pneumonia on chest CT. Radiology. 2020, 2009, 05. doi: 10.1148/radiol.2020200905.

[10] Gupta R, Misra A. Contentious issues and evolving concepts in the clinical presentation and management of patients with COVID-19 infection with reference to use of therapeutic and other drugs used in Co-morbid diseases (Hypertension, diabetes etc). Diabetes & Metabolic Syndrome: Clinical Research. 2020, 14, 251–4.

[11] NIH harnesses AI for COVID-19 diagnosis, treatment, and monitoring. 2020. https://www.nih.gov/news-events/news-releases/nih-harnesses-ai-covid-19-diagnosis-treatment-monitoring.

[12] Sivasubramanian S. How AI and machine learning are helping to fight COVID-19. 2020. https://www.weforum.org/agenda/2020/05/how-ai-and-machine-learning-are-helping-to-fight-covid–19/.

[13] Kirk D. Artificial Intelligence: A Superpower in the Fight Against Covid-19. 2020, https://www.labiotech.eu/ai/artificial-intelligence-covid–19/.

[14] Martineau K. Marshaling artificial intelligence in the fight against Covid-19. 2020, http://news.mit.edu/2020/mit-marshaling-artificial-intelligence-fight-against-covid–19-0519.

[15] McRae MP. and 14 co-authors. Clinical decision support tool and rapid point-of-care platform for determining disease severity in patients with COVID-19. Lab on a Chip. 2020. https://pubs.rsc.org/en/content/articlelanding/2020/lc/d0lc00373e#!divAbstract.

[16] Kundu S, Elhalawani H, Gichoya JW, Kahn CE Jr. How Might AI and Chest Imaging Help Unravel COVID-19's Mysteries?. Radiology: Artificial Intelligence. 2020, 2. doi: https://doi.org/10.1148/ryai.2020200053.

[17] Weldon D. How AI can be a COVID-19 game-changer. https://techbeacon.com/enterprise-it/how-ai-can-be-covid-19-game-changer.

[18] Block J. COVID-19 Puts Spotlight on Artificial Intelligence. 2020. https://www.genengnews.com/gen-edge/covid-19-puts-spotlight-on-artificial-intelligence/.

[19] Bayat V and 9 co-authors. A severe acute respiratory syndrome Coronavirus 2 (SARS-CoV-2) prediction model from standard laboratory tests. Clinical Infectious Diseases.2020, ciaa1175; https://doi.org/10.1093/cid/ciaa1175.

[20] Levitan R. The Infection That's Silently Killing Coronavirus Patients. 2020, https://www.nytimes.com/2020/04/20/opinion/sunday/coronavirus-testing-pneumonia.html.

[21] Teo J. Early detection of silent hypoxia in Covid-19 Pneumonia using smartphone pulse oximetry. Journal of Medical Systems. 2020, 44, 134. doi: 10.1007/s10916-020-01587-6.

[22] Tayfur İ, Afacan MA. Reliability of smartphone measurements of vital parameters: A prospective study using a reference method. American Journal of Emergency Medicine. 2019, 37, 1527–30.

[23] Yanamala N. and 9 co-authors. A Vital Sign-based Prediction Algorithm for Differentiating COVID-19 Versus Seasonal Influenza in Hospitalized Patients. medRxiv. 2021, doi: https://doi.org/10.1101/2021.01.13.21249540.

[24] Servie RF. AI invents new 'recipes' for potential COVID-19 drugs. 2020; https://www.sciencemag.org/news/2020/08/ai-invents-new-recipes-potential-covid-19-drugs.

[25] Remdesivir. 2020, NIH: COVID-19 Treatment Guidelines. https://www.covid19treatmentguidelines.nih.gov/antiviral-therapy/remdesivir/.

[26] Hutson M. Artificial-intelligence tools aim to tame the coronavirus literature. Technology Future. 2020, https://www.nature.com/articles/d41586-020-01733-7.

[27] Whitten A. Is AI Ready to Help Diagnose COVID-19? 2020, https://www.discovermagazine.com/technology/is-ai-ready-to-help-diagnose-covid–19.

[28] Wynants L. and 46 co-authors. Prediction models for diagnosis and prognosis of covid-19: systematic review and critical appraisal. BMJ. 2020, 369. doi: https://doi.org/10.1136/bmj.m1328.

[29] Liu X. and 16 co-authors. A comparison of deep learning performance against health-care professionals in detecting diseases from medical imaging: a systematic review and meta-analysis. The Lancet Digital Health. 2019, 1, e271–97.

[30] Greene T. Here's why AI didn't save us from COVID-19. 2020, https://thenextweb.com/neural/2020/07/24/heres-why-ai-didnt-save-us-from-covid–19/.

[31] Smith G, Rustagi I. The Problem With COVID-19 Artificial Intelligence Solutions and How to Fix Them. Stanford Social Innovation Review. 2020, https://ssir.org/articles/entry/the_problem_with_covid_19_artificial_intelligence_solutions_and_how_to_fix_them.

[32] https://en.wikipedia.org/wiki/Infodemic.

[33] Zimmer B. 'Infodemic': When Unreliable Information Spreads Far and Wide. 2020. https://www.wsj.com/articles/infodemic-when-unreliable-information-spreads-far-and-wide–11583430244.

[34] Cinelli M, Quattrociocchi W, Galeazzi A, Valensise CM, Brugnoli E, Schmidt AL, Zola P, Zollo F, Scala A. The COVID-19 social media infodemic. Scientific Reports. 2020, 10, 16598. doi: https://doi.org/10.1038/s41598-020-73510-5.

[35] Palayew A, Norgaard O, Safreed-Harmon K, Andersen TH, Rasmussen LN, Lazarus JV. Pandemic publishing poses a new COVID-19 challenge. Nature Human Behaviour. 2020, 4, 666–9.

[36] Levin JM, Oprea TI, Davidovich S, Clozel T, Overington JP, Vanhaelen Q, Cantor CR, Bischof E, Zhavoronkov A. Artificial intelligence, drug repurposing and peer review. Nature Biotechnology. 2020, 38, 1127–31.

[37] Cave S, Whittlestone J, Nyrup R, hEigeartaigh SO, Calvo RA. Using AI ethically to tackle covid-19. British Medical Journal. 2021, 372. doi: https://doi.org/10.1136/bmj.n364.

[38] Leslie D, Mazumder A, Peppin A, Wolters MK, Hagerty A. Does "AI" stand for augmenting inequality in the era of covid-19 healthcare?. British Medical Journal. 2021, 372. doi: https://doi.org/10.1136/bmj.n304.

[39] Kovar J. COVID-19 Accelerating Digital Transformation: McKinsey. 2020. https://www.crn.com/slide-shows/channel-programs/covid-19-accelerating-digital-transformation-mckinsey.

[40] McClean T. Covid-19 Has Accelerated Digital Transformation – With AI Playing A Key Role. 2020. https://www.forbes.com/sites/forbestechcouncil/2020/11/04/covid-19-has-acceler ated-digital-transformation—with-ai-playing-a-key-role/?sh=3129c7745a58.

[41] Bloom J. Will coronavirus reverse globalisation? 2020. https://www.bbc.com/news/busi ness-52104978.

[42] Altman SA. Will Covid-19 Have a Lasting Impact on Globalization? 2020. https://hbr.org/2020/05/will-covid-19-have-a-lasting-impact-on-globalization.

[43] Dawson D. Predictions for a Post-COVID-19 World. 2020. http://www.circleid.com/posts/20200420-predictions-for-a-post-covid-19-world/.

[44] AI in a post-COVID-19 world. The AI Journal. https://aijourn.com/report/ai-in-a-post-covid-19-world/.

[45] Banafa A. 8 Key Tech Trends in a Post-COVID-19 World. Technology Digital World. 2020. https://www.bbvaopenmind.com/en/technology/digital-world/8-key-tech-trends-in-a-post-covid-19-world/.

[46] Smith E. How will AI Accelerate After COVID-19? 2020. https://www.iotforall.com/how-will-ai-accelerate-after-covid-19.

[47] Lundin N. Preparing for China's "Post-Covid-19 Era" – Digital transformation for growth and for good. 2020. https://sweden-science-innovation.blog/beijing/preparing-for-chinas-post-covid-19-era-digital-transformation-for-growth-and-for-good/.

[48] Shrikhande P, Thiruppathy P. Healthcare in a post-COVID-19 world: investable models. 2020, https://koisinvest.com/post/healthcare-in-a-post-covid-19-world-investable-models/.

[49] Hughes-Cromwick P, Miller G, Beaudin-Seiler B. Higher Health Care Value Post COVID-19. 2020. https://www.healthaffairs.org/do/10.1377/hblog20200930.623602/full/.

[50] Jazieh R, Kozlakidis Z. Healthcare transformation in the post-coronavirus pandemic era. Front Med (Lausanne). 2020, 7, 429. doi: https://doi.org/10.3389/fmed.2020.00429.

[51] Rakowski R. A Vision of Healthcare In a Post-COVID-19 World. Scientific American. 2020. https://www.scientificamerican.com/custom-media/a-vision-of-healthcare-in-a-post-covid-19-world/.

Chapter 13
AI in future

Seeds of modern AI have been fostered by classical philosophers. Greek poet Homer described in his epic poem Iiad (in italic font) to creat a living thing in the form of a human being. French philosopher Pascal, with a purpose of completion of a mechanical computer, tried to explain the human thought process as a mechanical manipulation of symbols growing up. Along this extended line, in the 1940s, a machine based on the abstract essence of mathematical reasoning, a programmable digital computer, was invented. With the development of computers, several attempts have been made to create machines that perform human intellectual activities, which have been discussed in fields such as philosophy, mathematics, logic, and psychology. Inspired by the device and the thinking behind it, a handful of scientists began to seriously discuss the possibility of building an electronic brain [1]. The word "artificial intelligence" was defined as "a computer program with functions close to that of the human brain," and here it became the birth of the modern AI era. It was already more than half a century ago.

Any technology can be characterized by either seeds-oriented technology or needs-oriented technology. "Needs-oriented" is a style in which products are created in pursuit of what consumers are looking for and the needs that have already become apparent, and the superiority of the products and services is appealed and marketed. Needs-oriented products and services are all in high demand, and consumer demand is uninterrupted. In addition, even if it is an already available product, it can meet the needs of consumers if it is a product that is more functional and cheaper than other companies. Needs-oriented examples include smartphones, cars, and clothing. "Seeds-oriented" is, on the other hand, a style in which companies create new products and services based on their technologies and ideas to create new markets. Examples of seeds-oriented products include iPhones, Google Glass, and autonomous cars. These products were marketed as innovative products with no analogues and were a huge hit by winning the needs of consumers. In order for seeds-oriented products and services to sell, it is important to meet the potential needs of consumers [2–4]. These two distinct technologies are ever perpetuating in a circle with consumers and producers as their vehicles, as illustrated in Figure 13.1 [2].

Briefly, when we look at the half-century history of modern AI era, there is the first boom when symbol processing was accomplished, followed by AI's winter when lack of common sense was recognized. Then the second boom came when knowledge processing was achieved, followed by the second winter when the tacit knowledge cannot be processed. Finally, the third boom came when the tacit knowledge was enabled to be processed by deep learning (DL). With these colorful backgrounds, AI technology

https://doi.org/10.1515/9783110717853-014

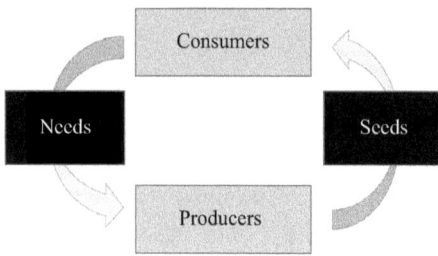

Figure 13.1: Perpetuating circulation between seeds and needs among consumers and producers [2].

appears to present itself in an amphoteric nature, carrying both seeds-oriented and needs-oriented technologies.

This book has been witnessing a real-time ever-changing society being affected by the unprecedented COVID-19 pandemic crisis, and at the same time, wise human beings are challenging this crisis and reestablishing (or resetting) society in better and sustainable ways. While looking for a future no matter what might be the object, there should always be optimistic or pessimistic sides of opinions. In the following sections, we will be reviewing these opinions along with several new concepts accompanying new generation of AI technology.

13.1 Present and future of AI technology

From the definition, it is known that AI is a computer that thinks for itself, but none has yet invented the thus defined AI. The current AI is a byproduct of the process of creating a computer that thinks like a human being. For example, speech recognition, character recognition, NLP, and search engines are active as useful things in the world from AI research and technology. In the future, attention is focused on research and developments on DL toward the originally defined AI technology.

13.1.1 Reinforcement learning (RL)

Reinforcement learning (RL) is an algorithm or programming that uses a system of cumulative rewards and punishments to train algorithms and one of the hottest research topics, currently. RL is one of three basic ML paradigms (alongside supervised leaning and unsupervised learning) in which a computer learns to perform a task through repeated trial-and-error interactions in a dynamic environment. This learning approach enables the computer to make a series of decisions that maximize a reward metric for the task without human intervention and without being explicitly programmed to achieve the task [5–7]. Humans and most animals learn from past experiences. RL occurs when the machine uses previous data to evolve and learn. Let us take an example of training a dog (*agent*) to compete task within an *environment*, which includes the surroundings of the dog as well as the trainer. First, the trainer issues a command or cue, which the dog observes. The dog then responds by taking an *action*. If the action is close to the desired behavior, the trainer will likely provide a *reward*, such as a food treat or a toy; otherwise, no reward or a negative reward will be provided. At the beginning of training, the dog will likely do more random actions like rolling over when the command given is "sit," as it is trying to associate specific observations with actions and rewards. This association, or mapping, between observations and actions is called *policy* [6, 8]. Figure 13.2 illustrates an interaction between agent and environment, where environment represents physical world in which the agent operates, state represents current situation of the agent, reward represents a feedback from the environment, policy is a method to map the agent's state to actions, and *value* describes future reward that an agent would receive by taking an action in a particular state [7].

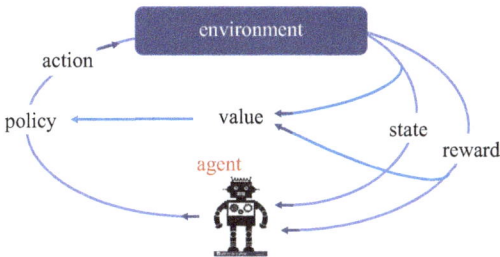

Figure 13.2: Agent–environment interaction [7].

There are three approaches to implement a RL algorithm: (1) value-based, (2) policy-based, and (3) model-based [9]. Model-based algorithm is further divided into (i) model-based RL and (ii) model-free RL [8]. Two kinds of RL methods are [9, 10]: positive RL, which is defined as an event that occurs because of specific behavior. It increases the strength and the frequency of the behavior and impacts positively on the action taken by the agent. The other type is negative RL, which is defined as strengthening of behavior that occurs because of a negative condition which should have stopped or avoided. It helps you to define the minimum stand of performance. However, the drawback of this method is that it provides enough to meet up the minimum behavior.

The benefit of a medical imaging examination in terms of its ability to yield an accurate diagnosis depends on the quality of both the image acquisition and the image interpretation. During the past century, radiology has grown tremendously due to advances in image detector systems and computer technology. On the other side, the image interpretation is an error-prone task, leading to the development of novel devices and techniques to assist the specialist in the diagnosis achievement.

In RL, the learner is not told which action to take, but instead must find which actions yield a better reward after trying them. The most distinguishing features of RL are trial-and-error search and delayed reward. In an ML community, RL has been used to solve many complex tasks normally thought of as quite cognitive. Netto et al. [11] investigated the adequacy of the RL technique to classify lesions based on medical image, by showing the application of this technique with the goal of lung nodules classification between malignant and benign and using a set of 3D geometric measures extracted from the lung lesion's CT images. It was reported that the application of RL to solve the problem of lung nodules classification used the 3D geometric nodules characteristics to guide the classification and the obtained results are very encouraging, demonstrating that the RL classifier using characteristics of the nodules' geometry can effectively classify benign from malignant lung nodules based on CT images [11].

In spite of remarkable applications of RL [12–14], there are still fundamental issues and challenges emphasized when it comes to medical field [15]. First of all, learning and evaluating on primary observational data should be considered. Most importantly, it would be unethical to utilize patients for the purposes of training the RL algorithm. In addition, it would be costly and would likely take years to complete. Therefore, it is necessary to learn from observational historical data. In RL literature, this is referred to as "off-policy evaluation." Many RL algorithms such as Q-learning (or a model-free RL) can, "in theory," learn the optimal policy effectively in the off-policy context. As it was mentioned [16], it is tricky to conduct accurate evaluation on these learned policies. In the absence of data on patients with high acuity who received no treatment, the RL algorithm concludes that trying something rarely or never performed may be better than treatments known to have poor outcomes [16]. This goes back to the classic problem of correlation not being equal to causation. While it is rather obvious here that the model has problems in other settings, it may be much more subtle and not easily detectable without proper evaluation [15]. The next issue is a partial observability. In medicine, it is almost impossible to observe everything going on in the body. We can take blood pressure, temperature, SO_2, and simple measures at almost every interval but these are all signals and not the ground truth about the patient. Additionally, this information is just data at one point in a time frame. It is a difficult problem in healthcare where there is a lot unknown about the patient at every time step. This issue contains (1) reward function, (2) more data required, and (3) non-stationary data. Regarding the reward function, finding a good one is challenging in many real-world problems. Healthcare is no exception to this as it is often hard to find a reward function that balances short-term improvement with overall long-term success. For instance, periodic improvements in blood pressure may not cause improvements in outcome in the case of sepsis. In contrast, having just one reward given at the end (survival or death) means a very long sequence without any intermediate feedback for the agent. It is often hard to determine which actions resulted in the reward or penalty. For need of more data, almost all the major breakthroughs in deep RL have been trained on years' worth of simulated data. Obviously, this is less of a problem when you can generate data easily through simulators. Since data for specific treatments is often scarce to begin with, the data that is there takes tremendous effort to annotate, and due to HIPPA compliance and protected/personal health information (PHI), hospitals/clinics are very wary of sharing their data at all. This creates problems for applying deep RL to healthcare. As for the non-stationary data, healthcare data by nature is non-stationary and dynamic. For instance, patients will likely have symptoms recorded at inconsistent intervals and some patients will have their vitals recorded more than others. Treatment objectives may change overtime as well [15, 16]. Godfried [15] concluded that the hardest and most prominent is the problem of evaluating RL algorithms effectively in healthcare scenarios. Other challenges relate to the amount of data needed for training, to the non-stationary nature of the data, and to the fact that it is only partially observable.

Precision medicine requires individualized treatment for subjects with different clinical characteristics. ML methods have witnessed rapid progress in recent years, which can be employed to make individualized treatment in clinical practice. In clinical medicine, the RL method can be used to assign optimal regime to patients with distinct characteristics. In the field of statistics, RL has been widely investigated, aiming to identify an optimal dynamic treatment regime. Q-learning is among the earliest methods to identify optimal dynamic treatment regime, which fits linear outcome models in a recursive manner. The advantage is its easy interpretation, and it can be performed in most statistical softwares. However, it suffers from the risk of misspecification of the linear model. More recently, some other methods, which do not heavily depend on model specification, have been developed such as inverse-probability weighted estimator (IPWE) and augmented IPWE (AIPWE). Based on this background in current medical fields, Zhang [17] discussed the basic ideas of these methods and shows how to perform the learning algorithm within RL environment. It was concluded that (i) exploring dynamic treatment borrows ideas from the RL algorithm from computer science; (ii) the Q-learning algorithm is easy to interpret for domain experts, however, it is limited by the risk of misspecification of the linear outcome model; (iii) the IPWE overcomes this problem by non-parametric modeling that the mean outcome is estimated by weighting the observed outcome; (iv) the AIPWE borrows information from both the propensity model and mean outcome model and enjoys the property of double robustness [18]; however, this statistical method does not allow the action space to have multiple levels; and (v) in real clinical practice, the interest may focus on the combinations of treatment regimes, giving rise to a multidimensional action space [17].

RL, as a subfield of ML, aims at empowering one's capabilities in behavioral decision-making by using interactive experience with the world and an evaluative feedback. Such distinctive features make RL technique a suitable candidate for developing powerful solutions in a variety of healthcare domains, where diagnosing decisions or treatment regimes are usually characterized by a prolonged and sequential procedure.

Yu [19] discussed the broad applications of RL techniques in healthcare domains in order to provide the research community with systematic understanding of theoretical foundations, enabling methods and techniques, existing challenges, and new insights of this emerging paradigm. The challenges and open issues in the current research have been discussed in detail from the perspectives of basic components constituting an RL process (i.e., states, actions, rewards, policies, and models), and fundamental issues in RL research (i.e., the exploration–exploitation dilemma and credit assignment). It should be emphasized that [19], although each of these challenging issues has been investigated extensively in the RL community for a long time, achieving remarkably successful solutions, it might be problematic to directly apply these solutions in the healthcare setting due to the inherent complexity in processes of medical data processing and policy learning. In fact, the unique features embodied in the clinical or medical decision-making process urgently call for development of more advanced RL methods that are really suitable for real-life healthcare problems. Interpretable learning, transfer learning, as well as small-data learning are the three theoretical directions that require more effort in order to make substantial progress. The application of the RL in healthcare is at the intersection of computer science and medicine. Such transdisciplinary research requires a concerted effort from ML researchers and clinicians who are directly involved in patient care and medical decision-makings. Driven by both substantial progress in theories and techniques in the RL research as well as practical demands from healthcare practitioners and managers, this situation is now changing rapidly, and recent years have witnessed a surge of interest in the paradigm of applying RL in healthcare, which can be supported by the dramatic increase in the number of publications on this topic in the past few years [19].

13.1.2 Generative adversarial networks (GAN)

GAN can be considered as an extension of RL. GAN is a network wherein two NNs compete with one another and have the ability to capture, analyze, and copy the trends and variations within a dataset [6]. This technique can prove useful in criminal identification, in which eyewitnesses or policemen can create an avatar of the criminal by choosing from a set of options. This is easier as compared to the regular sketches, which can be quite stressful as drawing an image on the basis of plain oral descriptions might not be fully accurate. This technique also allows the cartoon avatar to be translated into a photo again, which looks closely like a face. This is important, and human memory fades sharply when spoken, so montage face paintings have witnesses write caricatures before they speak. Language reporting of observed faces has a "language concealment effect" that works obstructively in later reacknowledged memories, so that, in criminal investigations, witnesses are asked to testify to the characteristics of the perpetrator's face by drawing a montage.

In medicine, GANs are associated with the creation of synthetic healthcare data for AI projects to overcome related data deficiency or inaccessibility challenges [20]. Bresnick [21] reported that Nvidia had partnered with the Mayo Clinic and the MGH & BWH Center for Clinical Data Science to construct abnormal MRI scans with GANs that can be used to train its DL model. Researchers involved in the project believe that the use of GANs would not only guarantee the production of low-cost and diverse data that has no privacy concern and can be freely shared across institutions. It also allows them to make changes to the size of a tumor or its location, to come up with millions of different combinations which would otherwise be hard to achieve in organic images. It was also reported [22] that Insilico Medicine, an American biotechnology company which specializes in drug discovery and aging research, had combined GANs with RL to build an adversarial threshold neural computer (ATNC) model for the design of de novo organic molecules that come with the required pharmacological properties to facilitate the drug discovery process. It was followed by demonstration that AI trained using GAN-generated synthetic data for tissue recognition can reach accuracy level (i.e., 98.83%) equal, if not superior, to human experts.

Due to its unique technological features associated with GANs, there are research works for augmentation of images [23–27]. To achieve generalizable DL models, large amounts of data are needed. Standard data augmentation is a method to increase generalizability and is routinely performed. GANs offer a novel method of data augmentation. Sanfort et al. [24] evaluated the use of CycleGAN for data augmentation in CT segmentation tasks. Using a large image database, a CycleGAN was trained to transform contrast CT images into non-contrast images, then the trained CycleGAN was employed to augment training using these synthetic non-contrast images. The segmentation performance of a U-Net trained on the original dataset was compared to a U-Net trained on the combined dataset of original data and synthetic non-contrast images. It was reported that in several CT segmentation tasks performance is improved significantly, especially in out-of-distribution (non-contrast CT) data. For example, when training the model with standard augmentation techniques, performance of segmentation of the kidneys on out-of-distribution non-contrast images was dramatically lower than for in-distribution data (Dice score of 0.09 vs. 0.94 for out-of-distribution vs. in-distribution data, respectively). When the kidney model was trained with CycleGAN augmentation techniques, the out-of-distribution (non-contrast) performance increased dramatically (from a Dice score of 0.09 to 0.66). Improvements for the liver and spleen were smaller, from 0.86 to 0.89 and 0.65 to 0.69, respectively. Based on these findings, it was suggested that this method will be valuable to medical imaging researchers to reduce manual segmentation effort and cost in CT imaging [24].

The most prevalent applications of GANs in healthcare involve medical imaging [25]. It was mentioned that (i) given that GANs have shown promise in the space of image segmentation, image synthesis, and discrete patient data synthesis, they have the potential to revolutionize healthcare analytics; (ii) the method of image segmentation in medicine can be extended to foreign object identification in medical images, detection of additional types of tumor growths, and precise identification of organ structure; (iii) toward the latter, image segmentation using GANs can be used to give precise structural details of the brain, liver, chest, and abdomen in MRIs. Further, GANs show strong promise in the space of medical image synthesis; (iv) in many instances, medical imaging analysis is limited by lack of data and/ or high cost of authentic data; (v) GANs can circumvent these issues by enabling researchers and doctors to work with high-quality, realistic, synthetic images, and this can significantly improve disease diagnosis, prognosis, and analysis; (vi) GANs show significant potential in the space of patient data privacy, as it provides a more reliable way to implicitly map real patient data to synthetic data; and (vii) this implicit map improves the data privacy of patients since it is not a typical one-to-one map and is thereby difficult to recover explicitly. Improvements in medical image segmentation, image synthesis, and data anonymization are all stepping-stones toward improving the efficiency and reliability of healthcare informatics [25].

DL algorithms produces state-of-the-art results for different ML and computer vision tasks. To perform well on a given task, these algorithms require large dataset for training; however, DL algorithms lack generalization and suffer from over-fitting whenever trained on small datasets, especially when one is dealing with medical images [26]. For supervised image analysis in medical imaging, having image data along with their corresponding annotated ground-truths is costly as well as time-consuming since annotations of the data is done by medical experts manually. Accordingly, Iqbal et al. [26] proposed a new GAN for medical imaging (MI-GAN). The MI-GAN generates synthetic medical images and their segmented masks, which can then be used for the application of supervised analysis of medical images. MI-GAN for synthesis of retinal images was presented. It was reported that (i) the proposed method generates precise segmented images better than the existing techniques, and (ii) the proposed model achieves a Dice coefficient of 0.837 on STARE dataset and 0.832 on DRIVE dataset with state-of-the-art performances on both the datasets.

Image-to-image translation is considered as a new frontier in the field of medical image analysis, with numerous potential applications. However, a large portion of recent approaches offers individualized solutions based on specialized task-specific architectures or require refinement through non-end-to-end training. Armanious et al. [27] proposed a new framework, named MedGAN, for medical image-to-image translation which operates on the image level in an end-to-end manner. MedGAN builds upon recent advances in the field of GANs by merging the adversarial framework with a new combination of non-adversarial losses. A discriminator network was utilized as a trainable feature extractor, which penalizes the discrepancy between the translated medical images and the desired modalities. Moreover, style-transfer losses are utilized to match the textures and fine structures of the desired target images to the translated images. A new generator architecture (titled CasNet) was used, which enhances the sharpness of the translated medical outputs through progressive refinement via encoder–decoder pairs. Without any application-specific modifications, we apply MedGAN on three different tasks: PET-CT translation, correction of MR motion artefacts, and PET image denoising. Perceptual analysis by radiologists and quantitative evaluations illustrate that the MedGAN outperforms other existing translation approaches [27].

13.1.3 Drastic change in employment sector and society

Nowadays many companies are using robotic arms in the routine operational aspects of manufacturing (assembly line operations), where employees can put more focus on the critical aspects of their jobs. A recent study shows robots will take over more than 20 million jobs (or 33%, or more than 700 different types of jobs) by 2030, thus creating mass unemployment. One advantage of this could be that robots could take up jobs that pose danger to human life, such as welding. The process of welding emits toxic substances and an extremely loud noise, which is harmful to the health of a human doing that job.

The era of one home robot in the family will be soon arrived. Robots equipped with advanced AI that can perform housework are popularized in homes, and people are freed from housework. The industrial robots mentioned earlier have already become a must at various sites and are expected to continue to grow at a high level in the future due to the use of AI and sensor technology. An era of cashless world will also soon arrive, so that people can live anywhere in the world without carrying wallet. Internationally standardized electronic money and identity management technologies have become popular, and highly secure and convenient services can be used all over the world. In parallel to ever-reforming working styles, there are various changes in many social and lifestyle sectors. People can communicate to others from all over the world with headphones equipped with automatic translation function using speech recognition technology. Technology that can reproduce stereoscopic images, sounds, scents, and even touch is realized by headgear, bringing the real world and the cyber world very close, by using VR technologies fully. A curved liquid crystal display (LCD) monitor using hard coat film has already been developed, and it is said that it is close to being mounted on a smartphone. Hence, it is predicted that a display that can be rounded like paper will be developed, and news and video can be seen just by carrying it around [28]. By checking the distribution history/record and allergy information from production sides when shopping, food safety is ensured by electronic tags affixed to food. Grocery records (on food material, amount per grocery, frequency of shopping, etc.) will be automatically analyzed in terms of nutritional balance, and additional suggestions on what type of food material(s) will be missing will be informed to customer(s) for next shopping.

Iluchi, in 1973 (which was between the first AI boom era and the second AI boom era; in other words, in the first AI winter era), published a book entitled *Tools for Conviviality*, criticizing "useful tools" that lose human instincts and skills, and calling for the need for "tools to make the most of human beings" [29]. Iluchi introduced the concept of a multidimensional balance of human life, which can serve as a framework for evaluating man's relation to his tools. In each of several dimensions of this balance, it is possible to identify a natural scale. When an enterprise grows beyond a certain point on this scale, it first frustrates the end for which it was originally designed and then rapidly becomes a threat to society itself. These scales must be identified, and the parameters of human endeavors within which human life remains viable must be explored. Society can be destroyed when further growth of mass production renders the milieu hostile, when it extinguishes the free use of the natural abilities of society's members, when it isolates people from each other and locks them into a man-made shell, when it undermines the texture of community by promoting extreme social polarization and splintering specialization, or when cancerous acceleration enforces social change at a rate that rules out legal, cultural, and political precedents as formal guidelines to present behavior. Corporate endeavors which thus threaten society cannot be tolerated. At this point, it becomes irrelevant whether an enterprise is nominally owned by individuals, corporations, or the slate, because no form of management can make such fundamental destruction serve a social purpose [29].

13.1.4 Automated transportation

The transportation domain (including all technologies for moving people and cargo) is now applying AI in mission-critical tasks (e.g., self-driving vehicles carrying passengers) where the reliability and safety of an AI system will be under question from the general public. Major challenges in the transportation industry like capacity problems, safety, reliability, environmental pollution, and wasted energy are providing ample opportunity (and potential for high return on investment ROI) for AI innovation [30].

AI is one of the best technologies supporting the transportation industry in many ways. It was a proven technology that transformed the transportation sector tremendously. In addition to making our lives easier, AI technology helps to make all modes of human transportation system safer and more efficient. With the introduction of AI in transportation, the industry has progressed in such a level that vehicles are moving without any assistance from humans. Very soon, AI technology in transport will bring huge revolutions, not only to the vehicles but also to the complete ecosystem. Let us take some examples of AI use cases in transportation [31]. Top six AI use cases in transportation sector are listed as follows: (1) AI for autonomous or self-driving vehicles, (2) AI application in traffic management, (3) reduce passenger wait times, (4) drone taxis, (5) remote-controlled cargo ships, and (6) AI in trucking [31]. The technology improved the transportation sector in multiple ways.

Implementation of AI technology in transportation will give unbelievable results than expected. These are predictions of AI-assisted transportation technologies: (1) Enhanced logistics utilizing real-time and previous data: By seeing real-time and previous data regarding traffic, weather, waiting time, stop maintenance, and so on, the system can plan the route smartly. (2) Forecasting vehicle maintenance: The systems can forecast vehicle maintenance with the help of smart dashboards and sensors. You can use telematic solutions to take care of the vehicles. It will send you alerts and conditions of the track and vehicles so that you can save yourself from probable hazards. (3) Supply chain optimization through data-driven software: With a small effort, you can maintain your workflow efficiently. You can strengthen your supply chain experts by utilizing smart tools experienced on big data to balance the demand and the supply, manage timings of the staff, and automate the customer service with the help of chatbots. (4) Track planner with traffic expectation and real-time updates: While traffic congestions are inescapable, you can predict the traffic jams and adjust your path accordingly. (5) Intelligent staff management: To advance the productivity and safety of drivers, intelligent driver behavior monitoring applications will be useful. Using data captured by the IoT sensors, you can manage your fleet through data-controlled dashboards for smart trucks, ships, and train connectivity [31].

Automated transportation has six levels, which basically represent the extent of autonomy achieved [6]: (1) At level 0, the driver is responsible for performing all tasks to drive the car – from applying brakes to changing gear to control the steering. (2) Level 1 is driver assistance, where the driver assistance systems support the driver but do not take full control. One such feature is the park assist feature. Here, the driver only takes care of the car's speed, while the car controls the steering. (3) Level 2 is when the car can drive alone, but the driver has to be present in case the system fails. Both Tesla's Autopilot and Nissan's ProPilot provide the steering, acceleration, and braking systems, but the driver has to be able to intervene in case of a failure. Here, the driver still needs to be alert and keep an eye on the road. (4) At level 3, the driver can entirely disengage from driving; however, the driver has to be present to handle any unforeseen failures. Audi's A8L can take up full driving responsibility in slow-moving traffic. This was the first car to claim level 3 autonomy. (5) We can activate full self-driving mode at level 4 in certain conditions only, like cities and states. They can drive independently but do require a driver. Google's Waymo project is one such car, which has been operating in the US driver-free for some time now. (6) Level 5 is the ultimate level of autonomous transportation, which requires zero human interaction to maneuver. One example of such cars can be a robotic taxi [5].

AI has been very effective in accelerating efficiency, accessibility, and safety of transportation means. At the same time, it would be worthy to listen to what Wilde had proposed, a theory under the name "risk homeostasis" [32]. The term "homeostasis" is originally used in physiology, to define the environment in the body, such as body temperature and blood concentration, which is kept within a certain range regardless of changes in the external environment such as temperature and humidity. No strategy for countermeasure design or future directions of research in the areas of human behavior which leads to traffic accidents or lifestyle-related diseases can be rationally developed without an acceptable working theory of human behavior in these domains. To this end, Wilde made an attempt to conceptually integrate the available evidence with respect to the role of human behavior in the causation of road accidents. From this integrative effort, it would seem that the accident rate is ultimately dependent on one factor only, the target level of risk in the population concerned, which acts as the reference variable in a homeostatic process, relating accident rate to human motivation. Various policy tactics for the purpose of modifying this target level of risk have been pointed out and the theory of risk homeostasis has been speculatively extended to the areas of lifestyle-dependent morbidity and mortality. In a nutshell, the theory is that even if the safety of the car is increased, the driver will drive with a high risk in search of profit as much as it becomes safe, so the probability of an accident as a result will be kept within a certain range [32].

13.2 Ecosystem

About 30 years ago, Rykiel [33] analyzed the AI ecology and stated that, as it is currently taught and practiced, ecology reflects in large measure the collection of ever more data. Ecologists have a variety of tools for collecting and analyzing data, but relatively few tools that facilitate ecological reasoning. Up to this time, simulation models have been the basic means of organizing ecological knowledge in a way that can be rapidly processed by computer. The areas of ecological science, in which this technology is likely to prove important, include: modeling and simulation, integration of qualitative and quantitative knowledge, theoretical development, and natural resource management. Researchers and managers in both basic and applied ecology will be affected by these developments as AI-derived technologies are added to the ecological toolkit. It was concluded that AI techniques are now being investigated for application to ecological science. At this very early stage of development, research is focused on how to use AI/ES technologies to further ecological research. The role that this technology can play in ecology will be decided over the course of the next decade. Many current expectations will fall by the wayside, but some will be realized and lead to useful tools for ecological research and for the application of ecological knowledge to management [33]. As time progresses, "ecological AI," or the realization of intelligence in the environment, has emerged from cognitive science and biology. It is oriented to realize AI from the body and ecology of organisms in the environment. AI with the body does not just make decisions in information processing. AI is expected to adapt its own physical movements to the environment. Such intelligence works in unconsciousness in the deeper layers of mind [34].

AI ecosystems including concepts such as ML, DL, and artificial narrow intelligence are comprised of various components that can be very confusing. The age of AI has become a reality. AI is essentially a simulation of human intelligence processes demonstrated by machines and systems. Unlike its fictional counterparts, AI today is created to assist us with our daily tasks and has significantly developed in recent years. As this technology continues to develop and grow, so does our standard of living. Over the past few years, this new and emerging technology has begun to connect with each other, building a far more advanced and powerful system known as an open AI ecosystem. This ecosystem is able to connect all of our technologies together whether it may be our mobile devices, computers, coffee machines, and household lights [35, 36]. It is poised to transform fields from earthquake prediction to cancer detection to self-driving cars, and now scientists are unleashing the power of DL on a new field – ecology. This ecosystem includes terms such as general AI, artificial narrow intelligence, ML, DL, and so many others [37, 38]. In order to develop AI ecosystem, assistance from a stakeholder is needed [39]. A stakeholder is a party that has an interest in a company and can either affect or be affected by the business. The primary stakeholders in a typical corporation are its investors, employees, customers, and suppliers. However, with the increasing attention on corporate social responsibility, the concept has been extended to include communities, governments, and trade associations [40]. Such stakeholder can include AI consumer, AI provider, AI auditor, AI adaptor, AI carrier, AI community, AI integrator, and AI enabler [39].

Recently, a lot of hype has been generated around DL, a group of AI approaches able to break accuracy records in pattern recognition. Over the course of just a few years, DL revolutionized several research fields such as bioinformatics or medicine. Yet such a surge of tools and knowledge is still in its infancy in ecology despite the ever-growing size and the complexity of ecological datasets [41].

13.3 Bioinformatics and ALife

13.3.1 Bioinformatics

Bioinformatics can be also defined as an umbrella concept dealing with all the applications of information technology to the field of molecular biology [42, 43]. Bioinformatics, as a new interdisciplinary approach, combines mathematics, information science, and biology and helps answer biological questions [44, 45], as illustrated in Figure 13.3 [44]. The main components of bioinformatics are (1) the development of software tools and algorithms, and (2) the analysis and interpretation of biological data by using a variety of software tools and particular algorithms. An nprecedented wealth of biological data has been generated by the human genome project and sequencing projects in other organisms. The huge demand for analysis and interpretation of these data is being managed by the evolving science of bioinformatics [44]. Bioinformatics is defined as the application of tools of computation and analysis to the capture and interpretation of biological data. As shown in Figure 13.3, it is an interdisciplinary field, which harnesses computer science, mathematics, physics, and biology; bioinformatics is essential for management of data in modern biology and medicine [44].

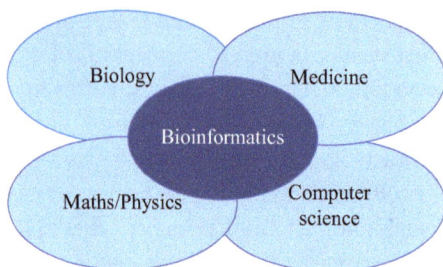

Figure 13.3: Interaction of disciplines that have contributed to the formation of bioinformatics [44].

Bioinformatics is a subdiscipline of biology and computer science, concerned with the acquisition, storage, analysis, and dissemination of biological data, most often DNA and amino acid sequences. Bioinformatics uses computer programs for a variety of applications, including determining gene and protein functions, establishing evolutionary relationships, and predicting the three-dimensional shapes of proteins [46]. GAN has been applied in medical informatics and bioinformatics [47].

Three new disciplines (i.e., genomics, proteomics, and bioinformatics) have emerged and rapidly developed over the past couple of decades to decipher life on a larger scale, at a greater speed, and for a deeper insight. Genomic and proteomic studies generate information on genomic sequence and protein properties, respectively, and bioinformatic studies provide methods and tools for analyses of genomic and proteomic data. Essentially, bioinformatics is the application of computer science and information technology to the fields of biology [48]. With the appearance of next-generation sequencing (NGS) technologies [49], bioinformatics plays increasingly more important roles in handling enormous amounts of biological data. The third and newer generation sequencing technologies are now gradually in place to bring about another round of revolution in biological and medical research by producing unprecedented quality and quantity of genomic and proteomic data; bioinformatics thus faces huge challenges in handling these data with sufficient capacity in both speed and accuracy, and great opportunities to make new discoveries [49].

There are several unique applications of bioinformatic technologies. For a protein to be a suitable target, its appropriate role in the disease process is only one factor [50]. The protein has to be suitable for drug design. This implies an appropriate shape of the binding pocket to be able to bind to small-molecule drugs. Also, the protein must allow for the development of specific drugs, that is, there should be no other proteins with similar binding pockets to which the drug may bind and cause side effects [50]. Bioinformatics is basically the study of informatic processes in biotic systems [51]. Actually what constitutes bioinformatics is not entirely clear and arguably varies depending on who tries to define it. This chapter discusses the considerable progress in infectious diseases research that has been made in recent years using various "omics" case studies. Bioinformatics is tasked with making sense of it, mining it, storing it, disseminating it, and ensuring valid biological conclusions can be drawn from it. This chapter discusses the current state of play of bioinformatics related to genomics and transcriptomics, and briefs metagenomics that finds use in infectious disease research as well as the random sequencing of genomes from a variety of organisms. This chapter explains the various possibilities of pan-genome, transcriptional reshaping, and also enormous progress of proteomics study. Bioinformatic algorithms and tools are crucial tools in analyzing the data. The chapter also attempts to provide some details on the various problems and solutions in bioinformatics that current-day scientists face while concentrating on second-generation sequencing strategies [51].

Apart from analysis of genome sequence data, bioinformatics is now being used for a vast array of other important tasks. They should include analysis of gene bvariation and expression, analysis and prediction of gene and protein structure and function, and detection of gene regulation networks. Fuethermore the following tasks should be included such as simulation environments for whole cell modeling, complex modeling of gene regulatory dynamics and networks, and presentation and analysis of molecular pathways in order to undersatnd gene-disease interactions [44, 52]. Although on a smaller scale, simpler bioinformatic tasks valuable to the clinical researcher can vary from designing primers (short oligonucleotide sequences needed for DNA amplification in polymerase chain reaction experiments) to predicting the function of gene products [44]. The practice of studying genetic disorders is changing from investigation of single genes in isolation to discovering cellular networks of genes, understanding their complex interactions, and identifying their role in disease [44, 53]. As a result of this, a whole new age of individually tailored medicine will emerge. Bioinformatics will guide and help molecular biologists and clinical researchers to capitalize on the advantages brought by computational biology [54]. The clinical research teams that will be most successful in the coming decades will be those that can switch effortlessly between the laboratory bench, clinical practice, and the use of these sophisticated computational tools [44]. Surprisingly, discussions of ethical tensions and public policy choices in this field have not yet fully emerged. Indeed, all the ethical discussions in this field are historically confined to intellectual property and patents [54].

13.3.2 ALife

Although learning is a method that depends on past data, it cannot exceed the past (although interpolation maybe possible, but extrapolation is very difficult), hence it is not universal and almighty. Intelligence is a complex system, and its object is also a complex system. In complex systems, complete information cannot be obtained, and algorithms (mechanical procedures to guarantee the correctness of calculation results) cannot be constructed. Function of the intelligence is thought to treat such a system well. Man is good at that process, and such process is required to the AI technology [55]. There is a field of study wherein researchers examine systems related to natural life, its processes, and its evolution, through the use of simulations with computer models, robotics, and biochemistry. It studies the fundamental processes of living systems in artificial environments in order to gain a deeper understanding of the complex information processing that defines such systems. It is called "artificial life: ALife" [56], which is the study of synthetic systems that exhibit behaviors characteristic of natural living systems, and it complements biology in attempting to "reproduce" biological phenomena. These topics are broad, but often include evolutionary dynamics, emergent properties of collective systems, biomimicry, as well as related issues about the philosophy of the nature of life and the use of lifelike properties in artistic works. It is clearly differentiated between AI and ALife although both appear to look at the same subject: AI is mainly pursuing "automation" while ALife pursues "autonomy" – a system that determines intentions and actions on its own, like living things in nature. Accordingly, research approach appears to show opposite nature to each other between AI and ALife. AI is a top-down nature that gives commands from top and AI researchers ponder the nature of intelligence by trying to build intelligent systems from scratch [55].

ALife attempts to understand the essential general properties of living systems by synthesizing lifelike behavior in software, hardware, and biochemicals. As many of the essential abstract properties of living systems (e.g., autonomous adaptive and intelligent behavior) are also studied by cognitive science, ALife and cognitive science have an essential overlap. This review highlights the state-of-the-art AL with respect to dynamical hierarchies, molecular self-organization, evolutionary robotics, the evolution of complexity and language, and other practical applications. It also speculates about future connections between ALife and cognitive science [57]. Namely, if you look only at the parts called "people," you do not know anything about the relationship between people. The relation between people is understood only after a group of people exists. In the same way, ALife is often not understood even if only individual elements are observed, and it is an approach to create one phenomenon by making full use of software (computer simulation), hardware (robot), wetware (biochemical reaction), and observing them. There are three main approaches: (1) hard ALife, concerned with hardware, covering robotics and new computing architectures, (2) soft ALife, concerned with software, covering computer simulations (including AI), and (3) wet ALife, concerned with wetware, covering chemistry and biology [55]. In addition to these three ALife, there is an Art ALife [58].

13.4 Medicine

As AI is growing exponentially, healthcare system and practice, and drug discovery will be well advanced and accelerated. Scientists are working on robots that can provide medical care to senior citizens. As robotics is now moving toward a level that is higher than that of human efficiency, there will come a day when we find robots reminding our parents/grandparents to take medicines or assist them in carrying out tasks involving motor functions. By combining biotechnology and nanotechnology, drug delivery technology will be further advanced. At the same time, an overnight health monitoring system can be developed. By taking one small capsule before the bedtime, all necessary vital data will be analyzed and sent to his/her physician to manage health condition of, particularly, an elder parson.

13.4.1 Volumetric display technology

Now moving from extensive existing technologies to explosive applications of new technoligies, there are volumetric displays to create visual representations of objects in three dimensions, with an almost 360-degree spherical viewing angle in which the image changes as the viewer moves around [59]. The most expected practical application of 3D display technology is the medical field. However, in medical care, there is a high demand for stereoscopic vision because it is necessary to accurately perform diagnosis and treatment. Today, diagnosis by X-ray stereoscopic television is the most popular, and stereoscopic vision of so-called X-ray fluoroscopy, which observes X-ray images as videos in real-time, has been studied from an early stage. Recently, attempts to apply high-quality, faithful color reproducibility, and realistic image systems, which are characteristic of stereoscopic high definition, to medicine have attracted attention. Current technologies of X-ray CT scan and MRI enable to reveal internal conditions of each part of internal organs. However, the image displayed on the monitor only displays a specific cross section, and the doctor is currently diagnosing it while imagining a 3D image by looking at these 2D images. For this reason, it is required to develop a system that enables stereoscopic display rather than displaying cross-sectional images obtained by X-ray CT [60]. Volumetric display technology can be divided into two categories: swept volume displays and static volume displays [59, 61, 62]. Swept-volume displays use the persistence of human vision to recreate volumetric images from rapidly projected 2D "slices."

Displays based on swept-volume surfaces, holography, optophoretics, plasmonics, or lenticular lenslets can create 3D visual content without the need for glasses or additional instrumentation, although they are slow, have limited persistence-of-vision capabilities, and most importantly, rely on operating principles that cannot produce tactile and auditive content as well. Static volume displays use no major moving parts to display images but rather rely on a 3D volume of active elements (volumetric picture elements or voxels) changing color (or transparency) to display a solid option.

In national defense program, there are several applications of 3D display technologies such as military simulation and training, automotive/aerospace decision-making, as well as homeland security [63]. It is mentioned that this technology is There should be urgent need for combat medicine on the battle field, where immediate ifesaving measures are required and more extensive and longer-term care should be available. Each of these requires systems for the display of medical imaging data or the display of data for education and training. While many of the systems used are commercially available, others are developed to meet the unique needs of the military in terms of mobility and durability. Traditionally, the level of care in the battlefield is limited to a "buddy system" when medical personnel are not present. This practice is rapidly changing as advances in technology are providing greater capability in medical care closer to the battlefield. For example, advanced imaging systems have become more portable, making more detailed medical data of an injured soldier available to be transmitted rearward and allow clinicians to assist in medical decision-making. These advances will continue to expand the need for glasses-free 3D volumetric displays that can represent the data available more accurately [64].

Stereoscopic displays can potentially improve many aspects of medicine. However, weighing the advantages and disadvantages of such displays remains difficult, and more insight is needed to evaluate whether stereoscopic displays are worth adopting. In this article, we begin with a review of monocular and binocular depth cues. This knowledge has then been applied to examine how stereoscopic displays can potentially benefit diagnostic imaging, medical training, and surgery. It is apparent that the binocular depth information afforded by stereo displays (1) aid the detection of diagnostically relevant shapes, orientations, and positions of anatomical features, especially when monocular cues are absent or unreliable, (2) help novice surgeons orient themselves in the surgical landscape and perform complicated tasks, and (3) improve the three-dimensional anatomical understanding of students with low visual-spatial skills. The drawbacks of stereo displays are also discussed, including extra eyewear, potential three-dimensional misperceptions, and the hurdle of overcoming familiarity with existing techniques [64]. Empirical studies concerning the effectiveness of stereoscopic displays in medicine were reviewed [65]. The domains covered in this review should include diagnosis, preoperative planning, minimally invasive surgery (MIS), and training/teaching. For diagnosis, stereoscopic viewing of medical data has been shown to improve the sensitivity of tumor detection in breast imaging and to improve the visualization of internal structures in 3D ultrasound. For MRI and CT data, where images are frequently rendered in 3D perspective, the added value of binocular depth has not yet been convincingly demonstrated. For MIS, stereoscopic displays decrease surgery time and increase accuracy of surgical procedures when the resolution of the stereoscopic displays is comparable to that of 2D displays. Training and surgical planning already use computer simulations; more research however is needed to assess the potential benefits of stereoscopic displays in those applications. Overall, there is a clear need for more empirical evidence that quantifies the added value of stereoscopic displays in medical domains [65]. The conventional 2D and glasses-assisted 3D display systems can no longer meet the clinical requirements with the development of minimally invasive video-assisted thoracoscopic surgery (VATS). The glasses-free 3D display technology adopts both lenticular lens technology and face-tracking and face-positioning systems and offers high brightness, large viewing area, and strong anti-interference capability, which significantly improves the operator's experience. When applied in VATS, it has many advantages including good display depth, convenience for performing complex and fine operations, and short learning curve. This novel display technology will greatly promote the development of MIS procedure [66].

Medical imaging (X-ray, MRI, ultrasounds, etc.) is used to create visual representations of the interior of the body for clinical analysis and medical intervention of complex diseases in a short period of time. Traditional medical imaging systems provide 2D visual representations of human organs, while more advanced digital medical imaging systems (e.g., X-ray CT) can create both 2D and in many cases 3D images of human organs. Systems capable of 3D digital medical imaging are currently only a small part of the overall medical imaging market, but it has effectively doubled in size over the last two years and is already rapidly expanding into practice areas such as oncology, orthopedics, obstetrics/gynecology, cardiology, and dentistry [67]. Displays are an integral part of these digital medical imaging systems. Current systems are being provided with displays that can only visually represent the imaging data collected in 2D or at best in simulated 3D on 2D. Freitag [67] made conclusive comments that medicine is one of the most rapidly changing industries in the world as innovation and technological advancements come to light every day, and 3D medical imaging is no different. The technology is poised to change how we diagnose and treat a plethora of medical scenarios – where a picture is worth a thousand words, a 3D image could save a life. Sun [68] also mentioned that 3D-printed models have been used in many medical areas ranging from accurate replication of anatomy and pathology to assisting in presurgical planning and simulation of complex surgical or interventional procedures; they serve as a useful tool for education of medical students and patients, and improve doctor–patient communication. Furthermore, patient-specific 3D-printed models can be used as a cost-effective tool to develop optimal CT scanning protocols. Current research should go beyond investigation of the model accuracy to focus more on clinical trials regarding the impact of 3D printing on clinical decision-making and patient's outcomes. Prospective studies with inclusion of more cases at multicenter sites are desirable to validate these findings. With further technical improvements in 3D printing techniques and reductions in printing cost and image post-processing time, 3D printing will be incorporated into routine clinical diagnosis in the near future.

13.4.2 Parity mirror imaging

This parity innovation was born due to a successful collaboration of 3D display technology and holography to create a floating image in air [69, 70]. Referring to Figure 13.4 [71], parity mirror is composed of imaging optical elements of many quadrature mirrors like prism. When a light emitter (light source) is placed under the parity mirror, it reflects twice inside the parity mirror, and the aerial image is displayed in a position symmetrical to the light source. The light source can be a LCD or anything that shines LED light on the photo.

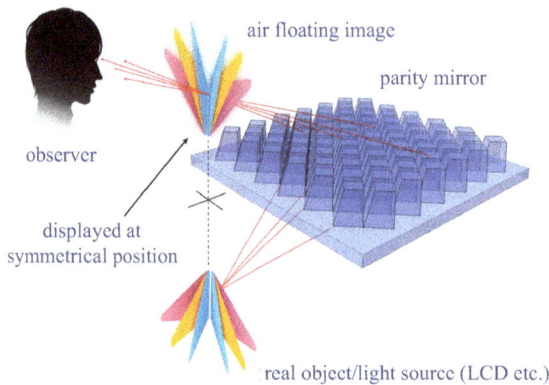

Figure 13.4: Principle of parity mirror to create a floating image in air [71].

This technology is characterized by (i) no need for special glasses, and (ii) aerial image is distortion-free, so that applicability should be versatile. As demonstrated in Figure 13.5, non-touch type numpad (numeric keypad) is installed in automated teller machine (ATM) registration keypad board (a) or temperature and water-amount controller for hand-washing device (b) [72].

These ideas can be extended to a variety of applications, in particular, any devices which prefer non-contact operation due to COVID-19 virus. They can inlcude a switch/control panel for numerous unknown people to touch such as facilities in rest rooms or elevator push buttons, a control panel for machineries with oily or dirty hands, or cooking appliances with oily hands. Someday, we can see parity mirror

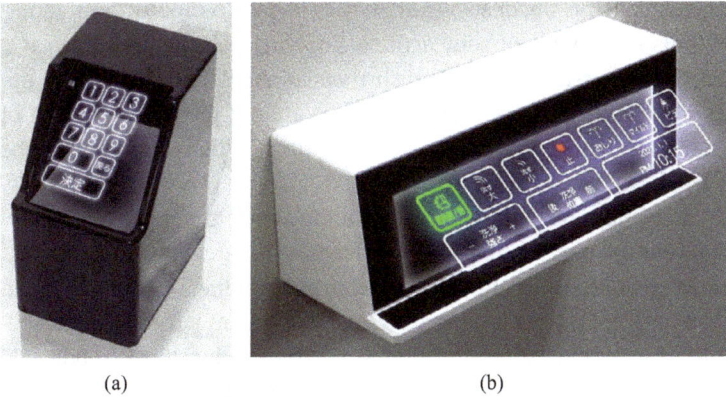

(a)   (b)

Figure 13.5: Parity mirror installed non-touch type keyboard controller [72].

installed non-touch type registering panel at clinics or hospital reception desks. Or more practically, we may see this device in surgical operation rooms.

13.5 Brain–computer interface (BCI)

13.5.1 Consciousness of robot

Ideas, such as what if robots can be given consciousness (in other words, AGI) and minds to AI or the coupling of the BCI, seem to be human egos and have been the subject of great debate. robots can be given consciousness (in other words, AGI) and minds to AI or the coupling of the BCI have been the subject of great debate. At the same time, there is a local development in terms of research, and there are also some births of a project. These are the contents which need to be thought before moving on to the examination of singularity which is the next topic. Artificial consciousness (AC) is a research area related to AI and cognitive robotics, with the aim of raising awareness of artifacts created by technology and is also called as machine consciousness (MC) or synthetic consciousness (SC). In science fiction, robots and AI have been portrayed as AC beings. AC is also an interesting philosophical problem. As research on genetics, brain science, and information processing progressed, the possibility of having an artificial existence with consciousness emerged [73]. Neuroscience hypothesizes that consciousness is generated by the interoperation of various parts of the brain, called the neural correlates of consciousness (NCC), though there are challenges to that perspective. Proponents of AC believe it is possible to construct systems (e.g., computer systems) that can emulate this NCC interoperation [74]. AC concepts are also pondered in the philosophy of AI through questions about mind, consciousness, and mental states [75, 76]. There are two types of actions, namely, unconscious activity and conscious activity. Unconscious activity is linked to input from the out-side world and outputs in the form of body actions, implying that the unconscious is directly connected to the outside world. English translation: When recognizing ap-ples, you can also think of them in mind, even if they are not presented in front of you, indicating that consciousness is separated from the outside world and can be active only in the head.

There is a self-modeling concept. Animals sustain the ability to operate after injury by creating qualitatively different compensatory behaviors. Although such robustness would be desirable in engineered systems, most machines fail in the face of unexpected damage. A robot can recover from such change autonomously, through continuous self-modeling. A four-legged machine uses actuation–sensation relationships to indirectly infer its own structure, and it then uses this self-model to generate forward locomotion. When a leg part is removed, it adapts to the self-model, leading to the generation of alter-native gaits. This concept may help develop more robust machines and shed light on self-modeling in animals [77]. Humans are masters of a self-reflection, and we use our past memories of life experiences to create an imaginary version of ourselves which act out these actions before we do. We refer to this imaginary version of ourselves as a "self-model." While the creation of this self-model is second nature in humans, it is something that robots have been unable to learn themselves, instead relying on labor-intensive hand-crafted simulators which inevitably become outdated as the robot develops. Once the robot has formed a reliable self-model, it has the ability to use that self-model in the same way modern robots use simulators. The robot can plan the self-model and then use the plan formulated in the self-model to perform tasks successfully in real life. It was also indicated that these self-models are informative enough to be able to perform a pick and place task on an articulated robotic arm successfully in closed or open loop. In closed loop, our self-model had a divergence from reality less than that of a hand-coded simulator [78].

In psychology, mental time travel (also called "chronosthesia") is the capacity to mentally reconstruct personal events from the past (episodic memory) as well as to imagine possible scenarios in the future (episodic foresight / episodic future thinking). The term was coined by Endel Tulving in 1985, as was the largely synony-mous term "chronosthesia." Mental time travel has been studied by psychologists, cognitive neuroscientists, and philosophers in a variety of other academic disci-plines. Major areas of interest include the nature of the relationship between mem-ory and foresight, the evolution of the ability (including whether it is uniquely human or shared with other animals), its development in young children, its under-lying brain mechanisms, as well as its potential links to consciousness, the self, and free will [79]. This amounts to a philosophical concept called "mental time travel," the ability to imagine that the system itself is in a new situation that has never been experienced before, or it seems to reexperience the experience of the past now [80]. There are dynamic changes and various time delays in body movement, and there are a great variety of elements intertwined. Now that we can simulate the movement of the body, we will ask ourselves, "Can robots simulate their own mental activities?" The problem comes out. This is a higher level simulation, and the question is whether we can create a system that simulates thought processes rather than just motion. In addition to determining what robots and AI are looking at and what the world is like, it is also necessary to understand what agents other than yourself can and cannot do, how other agents think, and what simulations they are doing.

Biologically, it is also said that it may be possible to artificially create life by incorporating an artificial genome with genetic information necessary for the human brain into the cells of a suitable host, and such ALife is likely to be conscious. However, what attributes in the life entity create consciousness? Can we not make something similar from nonbiological parts? Can the technology for designing computers create such a conscious body? Would such an act not be ethically a problem? Various problems are conceived [81]. The pattern (texture) is the same, but other unique feeling of the car brand is Qualia. These sensations are difficult to express in words, using very small visual differences as clues. It is hard to express the taste of chocolate in words, and so is the taste of red wine. These qualitative sensations are based on your past experiences when they arise. However, it is difficult to convey this to others. However, the fact that AI has a subjective sense creates a completely different ethical challenge. If AI becomes self-conscious and even subjective, we need to consider the pain and suffering that AI feels [81].

AI is at a turning point, with a substantial increase in projects aiming to implement sophisticated forms of human intelligence in machines. Haladjian et al. [82] conducted research to model specific forms of intelligence through brute-force search heuristics and also reproduce features of human perception and cognition, including emotions. Such goals have implications for AC, with some arguing that it will be achievable once we overcome short-term engineering challenges. We believe, however, that phenomenal consciousness cannot be implemented in machines. This becomes clear when considering emotions and examining the dissociation between consciousness and attention in humans. While we may be able to program ethical behavior based on rules and ML, we will never be able to reproduce emotions or empathy by programming such control systems – these will be merely simulations. Arguments in favor of this claim include considerations about evolution, the neuropsychological aspects of emotions, and the dissociation between attention and consciousness found in humans. Ultimately, we are far from achieving AC [82].

Consciousness plays an important role in debates around the mind–body problem, the controversy over strong vs. weak AI, and bioethics. Strikingly, however, it is not prominent in current debates on ethical aspects of AI and robotics. This text explores this lack and makes two claims: we need to talk more about AC and we need to talk more about the lack of consciousness in current robots and AI [83]. This may change in the future, however. Then it may be plausible to think about a concept of "robothood" and ascribe moral status to these future robots, based on their capabilities. There is already an interesting and controversial discussion going on about ascribing legal personhood to robots [84]. For the debate on the moral and legal status of robots, but also for the broader question of how to respond to and interact with machines, a better understanding of AC, artificial rationality, artificial sentience, and similar concepts is needed. Hildt [83] mentioned that we need to talk more about AC and the lack of consciousness in current AI and robots. In this, focusing on third-person definitions of AC and access consciousness will prove particularly helpful [83].

We should not be creating conscious, humanoid agents but an entirely new sort of entity, rather like oracles, with no conscience, no fear of death, no distracting loves and hates [85]. We do not need AC agents, rather we need intelligent tools. Tools do not have rights and should not have feelings that could be hurt or be able to respond with resentment to "abuses" rained on them by inept users. Dennet [85] mentioned that it will be hard enough learning to live with them without distracting ourselves with fantasies about the singularity in which these AIs will enslave us, literally. The human use of human beings will soon be changed – once again – forever, but we can take the tiller and steer between some of the hazards if we take responsibility for our trajectory.

13.5.2 Human–robot interaction

A BCI, sometimes called as a neural control interface (NCI), mind–machine interface (MMI), direct neural interface (DNI), or brain–machine interface (BMI), is a direct communication pathway between an enhanced or wired brain and an external device. BCIs are often directed at researching, mapping, assisting, augmenting, or repairing human cognitive or sensory-motor functions [86]. Due to the cortical plasticity of the brain, signals from implanted prostheses can, after adaptation, be handled by the brain like natural sensor or effector channels. Cross-correlation between a trigger-averaged event-related potential (ERP) template and continuous electrocorticogram was used to detect movement-related ERPs. The accuracy of ERP detection for the five best subjects (of 17 studied) had hit percentages >90% and false positive percentages <10%. These cases were considered appropriate for operation of a direct brain interface [86]. Recently, studies in human–computer interaction has shown high levels of success in classifying mental states (relaxed, neutral, concentrating), mental emotional states (negative, neutral, positive), and thalamocortical dysrhythmia (TCD) [87]. BCIs have shown great prospects as real-time bidirectional links between living brains and actuators. AI, which can advance the analysis and decoding of neural activity, has turbocharged the field of BCIs. Over the past decade, a wide range of BCI applications with AI assistance have emerged. These "smart" BCIs including motor and sensory BCIs have shown notable clinical success, improved the quality of paralyzed patients' lives, expanded the athletic ability of common people, and accelerated the evolution of robots and neurophysiological discoveries. However, despite technological improvements, challenges remain with regard to the long training periods, real-time feedback, and monitoring of BCIs. In this article, the authors review the current state of AI as applied to BCIs and describe advances in BCI applications, their challenges, and where they could be headed in the future [88].

Progress in neurotechnology is critical to improve our understanding of the human brain and improve the delivery of neurorehabilitation and mental health services at the global level. Meanwhile, advances at the interplay between neuroscience and AI are rapidly augmenting the computational resources of neurodevices. As neurotechnology becomes more socially pervasive and computationally powerful, several experts have called for preparing the ethical terrain and charting a route ahead for science and policy. In addition, calibrated policy responses should take into account issues of fairness and equality. BMI should be fairly distributed and should not exacerbate preexisting socioeconomic inequalities. While access to BMI-mediated health solutions should be as widespread as possible, open-development initiatives including hackathons, open-source platforms (e.g., Open BCI), and citizen-led data-sharing initiatives should be incentivized. In parallel, the growing involvement of for-profit corporations in BMI development urges us to assess the democratic accountability of company-driven technology development. In a not-too-distant future where BMI will likely be widespread, there will be an increasing need to maintain trust in data donation among individual citizens. This could be obtained through clear rules for data collection and secondary use, enhanced data protection infrastructures, public engagement, and neurorights enforcement [89, 90].

BCI is just what the name implies – it is directly connecting the brain and the computer. There is no need for eyes, ears, keyboards, or voice instructions. BCI perceives your thoughts and responds with actions that can be directly inputted to the brain. Most early prototypes have been developed as brain-implanted sensors for those with physical impairments. However, more advanced technologies such as Neuralink – originally founded by Elon Musk – take a different approach. These approaches use external or implanted sensors to interpret thoughts in language form, not articulated vocally, and direct them into a computer running an AI algorithm. Like a lot of BCIs, Neuralink's was framed initially as a way to help people with neurological disorders, but Musk is looking further out, claiming that Neuralink could be used to allow humans a direct interface with AI, so that humans are not eventually outpaced by AI. It might be that the only way to stop ourselves becoming outclassed by machines is to link up with them – if we cannot beat them, Musk's thinking goes, we may have to join them [91].

With approximately 100 billion neural connections, our brains are capable of processing billions of bits of information per second. The latest trend in unlocking the mysteries of the mind is BCI. In the public sector, initiatives such as the Human Brain Project have sought to accelerate research that can help us learn more about our own brains, in order to be able to better treat diseases and improve cognitive functioning. In the private sector, a number of companies are working to develop effective BMI for a wide range of uses. Read on to learn more about seven companies focused on direct communication from human brains to machines and the technologies and approaches they are using to make the future arrive faster. Up until now there are seven noticeable systems for BCI, including Neuralink, Neurable, Emotiv, Kernel, NextMind, MELTIN MMI, and BitBrain [92]. BCIs are slowly moving into the mass market. In the next few years, we might be able to control our PowerPoint presentation or Excel files using only our brains. And, companies may want to use BCI technology to monitor the attention levels and mental states of their employees. Obviously, there are myriad ethical questions and concerns surrounding the use of BCI technology in the workplace. The technology is well ahead of the policies and regulations that would need to be put in place. But, it is time for business leaders to start building a BCI strategy as soon as possible to address the potential risks and benefits [93]. BCIs refer to devices that allow users to interact with computers, measuring brain activity through electroencephalogram (EEG), which recognizes the energy and frequency patterns of the brain. There are currently two types of BCIs: invasive and non-invasive. By combining knowledge from AI, and specifically ML, BCIs have become a vital tool in aiding the accuracy and reliability of usability testing and user experience research, allowing us to talk about a new era of human factor design. BCI combined with AI-powered algorithms and neuroscience analysis is a relatively new method that attempts to revolutionize and replace the traditional usability testing of medical devices which often includes questionnaires, interviews, surveys, heuristic evaluations, and think-aloud protocols amongst other widely used evaluations [94]. Over the past years, there has been an influx in the complexity of new medical devices being introduced to the market, and thus a greater need to acquire bias-free, objective, and fast usability testing. Adding to these, over the last 10 years, there have been 2 million injuries and 80,000 deaths related to faulty medical devices, out of which 36% are attributed to poor usability testing. Objective and reliable usability testing is thus a matter of life and death. By using a BCI application such as EEG signal processing, the cognitive load and certain emotions such as frustration, joy, and relaxation of patients or healthcare providers can directly be extracted, in real-time, leading to better decisions in the manufacturing and design of a medical device. The device can be optimized early during the usability testing process, avoiding extra costs and long waiting time for approval. This new method attempts to increase the safety of medical devices that are being introduced to the market [94].

Overall, researchers broadly agree that current machines and robots are not conscious – in spite of a huge amount of science fiction depictions that seem to suggest otherwise. Consciousness-related questions may be expected to arise most easily with social robots and human–robot social interaction. Social robots (defined as a physically embodied, autonomous agent that communicates and interacts with humans on a social level [95]) have several characteristics that make them special for humans: They are capable of limited decision-making and learning, can exhibit behavior, and interact with people. In addition, capabilities like nonverbal immediacy of robot social behavior, speech recognition and verbal communication, facial expressions, and a perceived "personality" of robots play important roles in how humans respond to robots [83].

13.6 Prediction and singularity

13.6.1 Prediction before singularity

There is so much that AI is being used for and so much more potential that it is hard to picture our future without it, especially when it comes to business. From workflow management tools to trend predictions and even the way brands purchase ads, ML technologies are driving increases in productivity like never before. AI can collect and organize large amounts of information to make insights and guesses that are beyond the human capabilities of manual processing. It also increases organizational efficiencies yet reduces the likelihood of a mistake and detects irregular patterns, like spam and fraud, to warn business in real-time about a suspicious activity, among many other things. AI is said to reduce costs in many ways; for example, by "training" machines to handle incoming customer support calls and replacing many jobs in that way. It is also known that if your business does not use AI, it is probably falling behind competitively. AI has become so important and advanced that a Japanese venture capital firm made history by being the first company to nominate an AI board member for its capabilities to predict market trends faster than a human. AI is and will be a commonplace in every aspect of life, like the future of self-driving cars, more accurate weather predictions, or earlier health diagnosis, just to name a few [96].

There are several predictions on AI technology for near future. While creating a self-teaching and conscious superintelligence might be the ultimate goal for some researchers, it is not the whole truth. Ruuse [97] pointed out that the scope of AI is much broader, including technologies like VAs, NLP, ML platforms, and many other, as seen in Figures 1–4. With these technologies, it was predicted that (1) computers will solve all problems known to the human race (we will create systems and robots, which are smarter than us. According to Ray Kurzweil, one of the most-known futurists, computers will have the same level of intelligence as humans by 2045, and this point of evolution is called by some scientists as the singularity); (2) machines will become our best friends, advisors, and caretakers (as machines get more intelligent and can better adapt to its users, people may end up preferring to interact with machines over people. Moreover, robots and AI could provide excellent care and support for the elderly, the ill, and disabled); (3) we could accurately predict the future, based on data and high-level analytics (learning – including ML – is using the past to make predictions about the future. We already have enough data and analytic systems to make fairly accurate predictions); (4) we will all become cyborgs, enhanced by technology (our minds and bodies will be enhanced by prosthetics and implants, giving us infallible functions and motorics. Technology could cure us of deadly injuries, replace our limbs and organs, or give us sensory abilities far beyond existing vision, hearing, and manipulation. These future systems will be enabled via robotics, augmented reality, neuroscience, 3D printing, programming, material design, etc.); (5) we will become superhumans, living in symbiosis with AI (instead of being physically enhanced by technology, we will be living in symbiosis with superintelligent systems. We will connect our brains directly to them, thus expanding who we are. Imagine being able to access a computer for thousands or even millions of times more powerful than your own brain. Humans and AI have potential to create combined systems that are smarter than either alone); (6) we will be able to upload our minds to the cloud and reach immortality (instead of only connecting yourself to the cloud and using its resources, you will be able to upload your whole consciousness into the cloud or virtual reality. You could live out your wildest dreams and become immortal. The idea is that consciousness is the product of an individual's neural activity, and if all of the "data" in a brain (memories, thought patterns, etc.) could be "copied" into a digital realm, life would prolong for infinity); (7) humans do not have to work anymore (instead of fearing that "the robots will take our jobs," there may be a different perspective. Of course, in our current society, if someone would take our job – it would be dreadful. We fear what we do not know and cannot imagine); and (8) we will have to rethink the value and purpose of the human mind and body (there are two kinds of wealth in the world – labor and capital. Via labor we create capital. What if machines will be able to do all the work for us? This would drastically undercut human value. We as a society would have to find a new way of valuing people, not just for the work they do. Of course, we will adapt. But how? What would be the point of human life if there would be no need to think, no need for physical upkeep, no need to work, and even no need to socialize with other people? It all might seem unrealistic and far-fetched. But did people 30 years ago imagine that you would be in constant connection with the world via the Internet and mobile devices? Futurists believe that reaching singularity (creation of artificial superintelligence) is predictable, but we cannot know for sure what happens next. Will it mean extinction or transcendence, or something in between?) [97].

Press [98] collected opinions on AI predictions from 120 senior executives. Given the current state of AI technology, what they foresee is so widely varied. Before introducing summarized opinions, let us listen what Wiebe (Industrial Intelligence Consulting,

Teradata) stated that what the world is calling AI today will split into several areas in 2020, which someone in marketing will inevitably create pithier names for, including RPA, automated feature engineering and selection, perception AI which is the automation and refinement of physical perception, and resource allocation AI, the marriage of optimization technologies to sense and respond to demands in real time. The greatest potential for AI is not AI, but augmented intelligence. Hence, AI will be well balanced with artificial humanity (AH), which accounts for the emotional (and sometimes irrational) drivers behind human decision-making and the concept of empathy for which voice assistant adoption will be massively improved. AI chips will be improving vehicles' abilities to process visual data more efficiently, paving the way for autonomous vehicles of the future. Explainable AI and data maturity are still improving. AI at edge and IoT processing in conjunction with cloud connectivity and self-learning models, and 5 G technological environment will all be integrated to form contextual intelligent systems. AI will dramatically improve the employee experience by increasing the amount of automation in the workplace, only a small percentage of jobs will be completely replaced; doing work accurately enough to replace tasks have been previously completed only by humans. Healthcare organizations will leverage AI to assist medical staff in providing patients with possible prognoses for their symptoms. Higher education will continue to see the value of implementing AI-based solutions, with assists from critical thinking, collaboration, communication, and creativity. Businesses that embrace conversational interactions with customers will see increased team efficiency, stronger customer relationships, faster growth, and many promising applications of AI and ML technologies and tools in more personalized services [98].

Marr [99] listed five things that we can expect to happen: (1) AI increasingly becomes a matter of international politics; (2) a move toward "transparent AI"; (3) AI and automation drilling deeper into every business; (4) more jobs will be created by AI than will be lost to it; and (5) AI assistants will become truly useful. In 2018, AI and ML were making headlines, and there are likely to be more sensationalized claims about robots wanting to take our jobs or even destroy us. However, stories about real innovation and progress should start to receive more prominence as the promise of the smart, learning machines increasingly begins to bear fruit. Here are my predictions of what was foreseen in 2018: (1) There will be less hype and hot air about AI – but a lot more action; (2) more money will pour into AI enterprise projects than ever before; (3) a lot of AI projects will fail, in a costly manner; (4) the way we interact with machines will continue to shift toward voice; and (5) robots will become more closely involved in looking after our health and well-being [100].

AI industries (including top AI companies like GAFA) estimated their investment for AI R&D to reach 118 billion dollars in 2025; however, there is a flip side to this AI coin. As stated by Stephen Hawking, "AI likely the best or worst thing to happen to humanity." Even under such complicated situation, top five AI predictions in 2020 were listed [101]: (1) human augmentation, which can be purely in the physical by increasing their capabilities using various gadgets; (2) multi-experience, which can be done using technologies like VR, AR, and MR, can provide multidimensional experiences that were never seen before! This means that people can experience interactions with technology that are multisensory in nature; (3) explainability, humanity and AI ethics which is directly and indirectly related to data bias against other rekigions, genders, nayionaoities and so on, as humans do. (4) hyperautomation, which is the capacity of machines to use ML along with various automation tools to become even more independent and makes use of a partnership between humans and machines to train the automation tools so that they can make independent decisions using AI and do tasks that only humans could do before. (5) autonomous things, which we will keep learning from each other and soon you may see autonomous robots a lot in public [101].

Smarter technologies in our factories and workplaces and connected machines that will interact, visualize the entire production chain, and make decisions autonomously is just a couple of the ways in which the Industrial Revolution will cause advancements in business. One of the greatest promises that the fourth Industrial Revolution brings is the potential to improve the quality of life for the world's population and raise income levels. Our workplaces and organizations are becoming "smarter" and more efficient as machines; humans are starting to work together, and we use connected devices to enhance our supply chains and warehouses [95]. Talwar et al. [102] listed seven possible stages through which AI might develop, highlighting the types of applications we might see in the next 15–20 years: (1) rule-based systems – domestic applications and RPA software that surrounds us everywhere, every day; (2) context awareness and retention – algorithms that build up a body of information that is used and updated by machines, for example, chatbots and robo-advisors; (3) domain-specific expertise – machines that can develop expertise in a specific field that extends beyond the capability of humans because of all the informational access they can quickly get to, to reach a decision; (4) reasoning machines – these algorithms have a "theory of mind," some ability to attribute mental states to themselves and others. They have a sense of belief, intention, knowledge, and are aware of how their own logic works. Hence, they have the capacity to reason, negotiate, and interact with humano and other machines; (5) self-aware systems – the goal for those working in the AI field is to create and develop systems with human-like intelligence. There is no such evidence of that today but some say that there will be in as little as five years while others believe we may never achieve that level of intelligence; (6) artificial superintelligence – developing AI algorithms that are capable of outperforming the smartest of humans in every single domain; (7) singularity and transcendence – a development path enabled by ASI that could lead to a massive expansion of human capability, where one day we might be sufficiently augmented and enhanced such that humans could connect their brains to each other and to a future successor of the current internet [102].

Furthermore, AI in the next 20 years and beyond was envisioned. In a period of 2020–2025, (i) between 70% and 90% of all initial customer interactions are likely to be conducted or managed by AI; (ii) product development in a range of sectors from fashion items and consumer goods to manufacturing equipment could increasingly be undertaken and tested by AI; (iii) individuals will be able to define and design the personalized products and services they require in sectors ranging from travel to banking, savings, and insurance; (iv) the technology is likely to be deployed across all government agencies and legal systems – with only the most complex cases requiring a human judge and full court proceedings; (v) autonomous vehicles will start appearing in many cities across the world, and (vi) our intelligent assistants could now be managing large parts of our lives from travel planning to compiling the information we need prior to a meeting. In a period of 2026–2035, (i) globally approved, smart crypto tokens may be accepted alongside fiat currencies as we edge toward a single global medium of exchange; (ii) AI is likely to have penetrated every commercial sector; (iii) the evolution of AI could see the emergence of a wide range of fully automated decentralized autonomous organization (DAO) businesses including banks, travel agents, and insurance companies; (iv) scientific breakthroughs could enable us to develop artificial animal and ecosystem intelligence; (v) the emergence of self-aware and self-replicating software systems and robots; (vi) there is a reasonable possibility of achieving AGI, (vii) there is a small chance of creating ASI; and (viii) the singularity remains an unlikely possibility in this timeframe [102].

Kurzweil [103] predicts that by 2045, we will be able to multiply our intelligence a billionfold by linking wirelessly from our neocortex to a synthetic neocortex in the cloud. This will essentially cause a melding of humans and machines. Not only will we be able to connect with machines via the cloud, we will be able to connect to another person's neocortex. This could enhance the overall human experience and allow us to discover various unexplored aspects of humanity. Though we are years away from ASI, researchers predict that the leap from AGI to ASI will be a short one. No one really knows when the first sentient computer life form is going to arrive. But as narrow AI gets increasingly sophisticated and capable, we can begin to envision a future that is driven by both machines and humans; one in which we are much more intelligent, conscious, and self-aware [104].

In the future, if AI research develops more, robots will be able to talk freely to each other. However, the conversation made between and among robots would be probably done only for information transmission. It is available to convey information to the other party, or to compare the opinions of robot A and robot B, and to draw a conclusion on which to be chosen, but free talk will be difficult.

13.6.2 Singularity and beyond

The singularity (or technological singularity) is a hypothetical point in time at which technological growth becomes uncontrollable and irreversible, resulting in unforeseeable changes to human civilization. According to the most popular version of the singularity hypothesis, called intelligence explosion, an upgradable intelligent agent will eventually enter a runaway reaction of self-improvement cycles, each new and more intelligent generation appearing more and more rapidly, causing an explosion in intelligence and resulting in a powerful superintelligence that qualitatively far surpasses all human intelligence [105]. This is a transition from supercomputers to a superintelligence. Provided that the IQ is a reasonable scale to measure the intelligence, it is said that humans are quite smart when they have an IQ of 100; Einstein and Leonardo Da Vinci are said to have an IQ of 200. The IQ of AI today is still far from human. But how much IQ do you think AI will go to 10 years from now? It is said to reach 10,000, far beyond Einstein. Ten years later, the IQ will exceed 100 million, and by year 2045, it will exceed the intelligence of all mankind [106, 107]. This is the "2045 problem" that human beings developed. None knows if this super intelligence can be controlled, similar to when human developed nuclear weapons for the first time. There is a real concern that humans may give robots infinite intelligence, and humanity may be dominated by robots.

And if singularity has a strong "idea that AI and humans are united," the concept of "extended intelligence (EI or XI)" as a further evolutionary system has also been proposed [107]. This is different from self-judging AI, rather it is an intelligence that programs everything so that humans make ethical decisions to machines. So to speak, it is an integration of human beings and AI. In this way, AI algorithms are never uniform, and multi-level possibilities are being discussed and developed. Eventually, Ito [108] proposed a kind of EI, understanding intelligence as a fundamentally distributed phenomenon. EI research currently includes: (1) connecting electronics to human neurons to augment the brain and our nervous system, (2) using ML to understand how our brains understand music, and to leverage that knowledge to enhance individual expression and establish new models of massive collaboration, (3) if the best human or computer chess players can be dominated by human–computer teams including amateurs working with laptops, how can we begin to understand the interface and interaction for those teams? How can we get machines to raise analysis for human evaluation, rather than supplanting it? (4) ML is mostly conducted by an engineer tweaking data and learning algorithms, later testing this in the real world. This augments human decision-making and makes the ML training more effective, with greater context, (5) building networked intelligence, studying how networks think, and how they are smarter than individuals, (6) developing humans and machine interfaces through sociable robots and learning technologies for children, (7) developing "society in the loop," pulling ethics and social norms from communities to train machines, testing the machines with society, in a kind of ethical Turing test, (8) developing wearable interfaces that can influence human behavior through consciously perceivable and subliminal I/O signals, (9) extending human perception and intent through pervasively networked sensors and actuators, using distributed intelligence to extend the concept of "presence," (10) incorporating human-centered emotional intelligence into design tools so that the "conversation" the designer has with the tool is more like a conversation with another designer than interactions around geometric primitives (e.g., "Can we make this more comforting?"), (11) developing a personal autonomous vehicle that can understand, predict, and respond to the actions of pedestrians; communicate its intentions to humans in a natural and non-threatening way; and augment the senses of the rider to help increase safety, (12) providing emotional intelligence in human–computer systems, especially to support social-emotional states such as motivation, positive affect, interest, and engagement. For example, a wearable system designed to help a person forecast mental health (mood) or physical health changes will need to sustain a long-term non-annoying interaction with the person in order to get the months and years of data needed for successful prediction, (13) using AI and crowdsourcing for understanding and improving the health and well-being of individuals, (14) collaborating with the Camera Culture Group on AI and crowdsourcing for understanding and improving our cities, (15) macroconnections have also developed Data Viz Engines such as the OEC, Dataviva, Pantheon, and Immersion, which served nearly 5 million people last year. These tools augment networked intelligence by helping people access the data that large groups of individuals generate, and that are needed to have a panoptic view of large social and economic systems. And (16) collaborating with Canan Dagdeviren to explore novel materials, mechanics, device designs, and fabrication strategies to bridge the boundaries between brain and electronics. Further, developing devices can be twisted, folded, stretched/flexed, wrapped onto curvilinear brain tissue, and implanted without damage or significant alteration in the device's performance. There are still researches toward a vision of brain probes that can communicate with external and internal electronic components [108].

This aspect consisting of a heterogeneous group of projects is a feature of the media lab. However, it is the embodiment of the field of EI respectively. That is, intelligence, ideas, analysis, and behavior are the whole that does not fit into the system of one neuron or code. All these projects set out this core philosophy as a value together with research subjects and believe that this method is the true approach of intelligence [109].

13.7 Quantum computers

All computing systems rely on a fundamental ability to store and manipulate information. Current computers manipulate individual bits, which store information as binary 0 and 1 states. Quantum computers leverage quantum mechanical phenomena to manipulate information. To do this, they rely on qubits (or quantum bits).

The quantum is a digitized minimal unit and possesses dual characters of particle and wave. Besides, quantum waves are not the three-dimensional space in which we live, but the inhabitants of the abstract space of the Hilbert space of the infinite dimension. Quantum computers make full use of three unique superpowers of qubits: superposition, interference, and entanglement. The result is that a series of qubits can represent different things simultaneously. There are two types of quantum computers: quantum gate type (or quantum circuit type) and quantum annealing type. With the former type, multiple calculations can be achieved at the same time, while the latter type is suitable for optimization [110–112]. Three equally important elements – bits (which are information infrastructure for today's computer and supercomputer), qubits (which are new information infrastructure), and neurons (which constructs neuron architecture as an AI infrastructure) – are nicely conversed to an integrated information infrastructure, which is also externally orchestrated with hybrid cloud and AI-assisted programming, as demonstrated in Figure 13.6 [113].

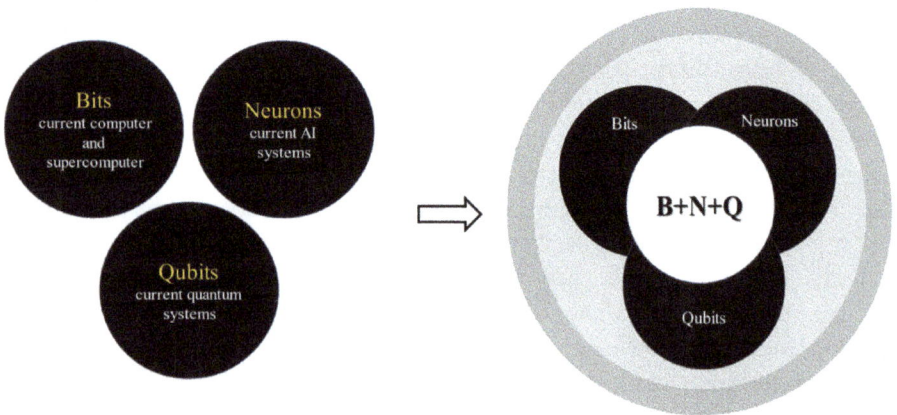

Figure 13.6: Conversion of bits, qubits, and neurons to form integrated information infrastructure [113].

There are mainly three obstacles before the true realization of the quantum computer technologies: (1) error immunity problem to eliminate effect of noise affecting the calculation process from the calculation results, (2) realization of universal computing, and (3) establishment of scalability, so that even if the number of qubits increase, a technology that does not increase experimental difficulties exponentially should be established [114, 115]. Quantum ML is the biggest impact of quantum computers on AI. Fueled by increasing computer power and algorithmic advances, ML techniques have become powerful tools for finding patterns in data. Quantum systems produce atypical patterns that classical systems are thought not to produce efficiently, so it is reasonable to postulate that quantum computers may outperform classical computers on ML tasks. The field of quantum ML explores how to devise and implement quantum software that could enable ML that is faster than that of classical computers. Recent work has produced quantum algorithms that could act as the building blocks of ML programs, but the hardware and software challenges are still considerable [116].

As mentioned previously, noise and decoherence are major problems for quantum computers. Even if we leverage the power of quantum firmware, unavoibale error will limit the scale of machine that we can build. But the good news is that clever algorithm designers have started to discover new problems which can potentially be solved – problems that could have huge impacts – with noisy intermediate scale quantum computers (NISQ) machines [117–119]. Suchara et al. [118] indicated using a hybrid quantum-classical architecture where larger quantum circuits are broken into smaller sub-circuits that are evaluated separately, either using a quantum processor or a quantum simulator running on a classical supercomputer. Circuit compilation techniques that determine which qubits are simulated classically will greatly impact the system performance as well as provide a trade-off between circuit reliability and runtime.

Considering recent advancements and successes in the development of efficient quantum algorithms for electronic structure calculations, alongside impressive results using ML techniques for computation, hybridizing quantum computing with ML for the intent of performing electronic structure calculations is a natural progression. Xia et al. [120] reported a hybrid quantum algorithm employing a restricted Boltzmann machine to obtain accurate molecular potential energy surfaces, and mentioned that (i) by exploiting a quantum algorithm to help optimize the underlying objective function, an efficient procedure for the calculation of the electronic ground state energy for a small molecule system was obtained; and (ii) high accuracy for the ground state energy for H_2, LiH, and H_2O at a specific location on its potential energy surface with a finite basis set was achieved. With the future availability of larger-scale quantum computers, quantum ML techniques are set to become powerful tools to obtain accurate values for electronic structures. These algorithms would be useful in predicting multiple particle interactions in chemical reactions, including a new drug development and new materials [121].

IBM CEO – Arvind Krishna – mentioned, during the "World Digital Summit 2021" (June 8, 2021), that quantum computers will be realized within 3 years.

References

[1] History of artificial intelligence; https://en.wikipedia.org/wiki/History_of_artificial_intelligence.

[2] Kurosu M. Needs oriented and seeds oriented. 2002. https://u-site.jp/lecture/20020115.

[3] Yamauchi T, Kobayashi H, Kobayashi T. Developing Needs-Oriented Technologies with Bio-TRIZ: An Example of a New Lifestyle in Kitakami Achieved through Japanese Umbrellas. 2019. https://www.taylorfrancis.com/chapters/developing-needs-oriented-technologies-bio-triz-example-new-lifestyle-kitakami-achieved-japanese-umbrellas-takeshi-yamauchi-hidetoshi-kobayashi-toru-kobayashi/e/10.1201/b22418-10.

[4] Matsumoto Y, Nishida Y, Motomura Y, Okawa Y. A concept of needs-oriented design and evaluation of assistive robots based on ICF. IEEE International Conference on Rehabilitation Robotics. 2011, 5975437. doi: 10.1109/ICORR.2011.5975437.

[5] Three Things to Know About Reinforcement Learning. https://www.kdnuggets.com/2019/10/mathworks-reinforcement-learning.html.

[6] Future of AI – 'Coz it's time to get high on AI. https://data-flair.training/blogs/future-of-ai/.

[7] Bhatt S. 5 Things You Need to Know about Reinforcement Learning. 2018. https://www.kdnuggets.com/2018/03/5-things-reinforcement-learning.html.

[8] Moni R. Reinforcement Learning algorithms – an intuitive overview. 2019. https://medium.com/@SmartLabAI/reinforcement-learning-algorithms-an-intuitive-overview-904e2dff5bbc.

[9] Reinforcement Learning: What is, Algorithms, Applications, Example. https://www.guru99.com/reinforcement-learning-tutorial.html

[10] Nicholson C. A Beginner's Guide to Deep Reinforcement Learning. https://wiki.pathmind.com/deep-reinforcement-learning.

[11] Netto SMB, Leite VRC, Silva AC, De Paiva AC, De Neto A A. Application on reinforcement learning for diagnosis based on medical image, reinforcement learning, cornelius weber, mark elshaw and norbert michael mayer. IntechOpen. 2008. doi: 10.5772/5291.

[12] Jonsson A. Deep reinforcement learning in medicine. Kidney Diseases. 2019, 5, 18–22. doi: https://doi.org/10.1159/000492670.

[13] Gottesman O, Johansson F, Komorowski M, Faisal A, Sontag D, Doshi-Velez F, Celi LA. Guidelines for reinforcement learning in healthcare. Nature Medicine. 2019, 25, 16–18.

[14] Coronato A, Naeem M, De Pietro G, Paragliola G. Reinforcement learning for intelligent healthcare applications: A survey. Artificial Intelligence in Medicine. 2020, 109, 101964. doi: https://doi.org/10.1016/j.artmed.2020.101964.

[15] Godfried I. A review of recent reinforcement learning applications to healthcare. 2018. https://towardsdatascience.com/a-review-of-recent-reinforcment-learning-applications-to-healthcare-1f8357600407.

[16] Gottesman O. and 19 co-authors. Evaluating Reinforcement Learning Algorithms in Observational Health Settings. https://arxiv.org/pdf/1805.12298.pdf.

[17] Zhang Z. Reinforcement learning in clinical medicine: a method to optimize dynamic treatment regime over time. Annals of Translational Medicine. 2019, 7. doi: 10.21037/atm.2019.06.75.

[18] Zhang D, Tsiatis AA, Laber EB, Davidian M. A robust method for estimating optimal treatment regimes. Biometrics. 2012, 68, 1010–18.

[19] Yu C, Liu J, Nemati S. Reinforcement Learning in Healthcare: A Survey. 2019; https://arxiv.org/pdf/1908.08796.pdf.

[20] Tang H. GANs in medicine. Artificial intelligence in Medicine. 2020. https://ai-med.io/ai-in-medicine/gans-in-medicine/.

[21] Bresnick J. MRI Images Created by AI Could Help Train Deep Learning Models. 2018, https://healthitanalytics.com/news/mri-images-created-by-ai-could-help-train-deep-learning-models.

[22] EurekAlert. Combining GANs and reinforcement learning for drug discovery. 2018, https://www.eurekalert.org/pub_releases/2018-05/imi-cga050918.php.

[23] Kazeminia S, Baur C, Kuijper A, Van Ginneken B, Navab N, Albarqouni A, Mukhopadhyay A. GANs for Medical Image Analysis. 2019, https://arxiv.org/abs/1809.06222.

[24] Sanfort V, Yan K, Pickhardt PJ, Summers RM. Data augmentation using generative adversarial networks (CycleGAN) to improve generalizability in CT segmentation tasks. Scientific Reports. 2019, 9, 16884. doi: 10.1038/s41598-019-52737-x.

[25] Pierre S. How GANs Can Improve Healthcare Analytics. 2020, https://towardsdatascience. com/how-gans-can-improve-healthcare-analytics-7d2379eff19e.

[26] Iqbal T, Ali H. Generative adversarial network for medical images (MI-GAN). Journal of Medical Systems. 2018, 42, 231. doi: https://doi.org/10.1007/s10916-018-1072-9.

[27] Armanious K, Jiang C, Fischer M, Küstner T, Hepp T, Nikolaou K, Gatidis S, Yang B. MedGAN: Medical image translation using GANs. Computerized Medical Imaging and Graphics. 2020, 79, 101684. doi: https://doi.org/10.1016/j.compmedimag.2019.101684.

[28] Cao J, Chen J, Du P, Ying J, Huang T-C, Chen L, Zhao B, Zhang X. Curved vertical-alignment liquid crystal display with uniform brightness: From pixel design to verification. Journal of Society for Information Display. 2020. doi: https://doi.org/10.1002/jsid.885.

[29] Iluchi I. Tools for Conviviality. 1973. https://arl.human.cornell.edu/linked%20docs/Illich_ Tools_for_Conviviality.pdf.

[30] Bharadwaj R. AI in Transportation – Current and Future Business-Use Applications. 2019. https://emerj.com/ai-sector-overviews/ai-in-transportation-current-and-future-business- use-applications/.

[31] AI In Transportation: Artificial Intelligence in the Transportation Industry. 2020. https://usm systems.com/ai-use-cases-in-transportation-sector/.

[32] Wilde GJS. The theory of risk homeostasis: implications for safety and health. Risk Analysis. 1982. doi: https://doi.org/10.1111/j.1539-6924.1982.tb01384.x.

[33] Rykiel EJ. Artificial intelligence and expert systems in ecology and natural resource management. Ecological Modelling. 1989, 46, 3–8. doi: 10.1016/0304-3800(89)90066-5.

[34] Ecological Artificial Intelligence. 2017, https://society-zero.com/icard/452549.

[35] Dao DQ. Everything You Need to Know About the AI Ecosystem. 2018. https://www.techfun nel.com/information-technology/everything-you-need-to-know-about-the-ai-ecosystem/.

[36] Jacquet F. Exploring the artificial intelligence ecosystem: AI, Machine Learning, and Deep Learning. 2017. https://dzone.com/articles/exploring-the-artificial-intelligence-ecosystem-fr.

[37] Satell G. How Artificial Intelligence Is Making The Shift From System To Ecosystem. 2019. https://www.digitaltonto.com/2019/how-artificial-intelligence-is-making-the-shift-from-sys tem-to-ecosystem/.

[38] Clune J, Norouzzadeh MS, Nguyen A, Kosmala M, Swanson A, Palmer M, Packer C. Ecology and AI. https://www.eurekalert.org/pub_releases/2018-07/hu-eaa071018.php.

[39] Hayashi K. AI network society and ecosystem. 2017. https://www.soumu.go.jp/main_con tent/000520421.pdf.

[40] Fernando J. Stakeholder capitalism. 2021. https://www.investopedia.com/terms/s/stake holder.asp.

[41] Christin S, Hervet É, Lecomte N. Applications for deep learning in ecology. 2019. https:// doi.org/10.1111/2041-210X.13256.

[42] Marturano A. Bioethics. 2009, 23, ii/iii. doi: 10.1111/j.1467-8519.2009.01764.x.

[43] Amer S. Ethical Concerns Regarding the use of Bioinformatics and Computational Genomics. Int'l Conf. e-Learning, e-Bus., EIS, and e-Gov. EEE'17. 2017, 58–62. https://csce.ucmss.com/ cr/books/2017/LFS/CSREA2017/EEE3568.pdf.

[44] Bayat A. Bioinformatics. BMJ. 2002, 324, 1018–22.

[45] Zhang S-Y, Liu S-L. Bioinformatics. Brenner's Encyclopedia of Genetics. 2nd Edition. 2013, Pages 338–340. doi: https://doi.org/10.1016/B978-0-12-374984-0.00155-8.

[46] Austin CP. Bioinformatics. NIH: National Human Genome Research Institute. https://www.ge nome.gov/genetics-glossary/Bioinformatics.

[47] Lan L, You L, Zhang Z, Fan Z, Zhao W, Zeng N, Chen Y, Zhou X. Generative adversarial networks and its applications in biomedical informatics. Front Public Health. 2020, 8, 164. doi: 10.3389/fpubh.2020.00164.

[48] Campbell AM, Heyer L. Genomics, proteomics and bioinformatics. The Yale Journal of Biology and Medicine. 2007, 80, 215, https://www.ncbi.nlm.nih.gov/pmc/articles/PMC2347366/.

[49] Hambuch TM, Mayfield J. Next generation sequencing. Pathobiology of Human Disease A Dynamic Encyclopedia of Disease Mechanisms. 2014, 4131–9. doi: https://doi.org/10.1016/ B978-0-12-386456-7.07717-0.

[50] Lengauer T, Hartmann C. Drug discovery technologies. Comprehensive Medicinal Chemistry II. 2007, 3, 315–47. doi: https://doi.org/10.1016/B0-08-045044-X/00088-2.

[51] Paszkiewicz KH, Van Dergiezen M. Omics, bioinformatics, and infectious disease research. Genetics and Evolution of Infectious Disease. 2011, 523–39. doi: https://doi.org/10.1016/ B978-0-12-384890-1.00018-2.

[52] Tsoka S, Ouzounis CA. Recent developments and future directions in computational genomics. FEBS Letters. 2000, 480, 42–8.

[53] Debouk C, Metcalf B. The impact of genomics on drug discovery. Annual Review of Pharmacology and Toxicology. 2000, 40, 193–208.

[54] Butler D. Are you ready for the revolution? Nature. 2001, 409, 758–60.

[55] Sinapayen L. Introduction to artificial life for people who like AI, The Gradient, 2019.

[56] Langton CG. Artificial Life: Proceedings of an Interdisciplinary Workshop on the Synthesis and Simulation of Living Systems. Addison-Wesley Longman Pub., Boston MA, 1989.

[57] Bedau MA. Artificial life: organization, adaptation and complexity from the bottom up. Trends in Cognitive Sciences. 2003, 7, 505–12.

[58] Abramovic V, Glynn R. Edge of chaos: artificial life based interactive art installation. ALIFE 2019: The 2019 Conference on Artificial Life. 2019, 493–4; https://www.mitpressjournals. org/doi/abs/10.1162/isal_a_00209.

[59] Volumetric Displays. https://www.gartner.com/en/information-technology/glossary/volumet ric-displays.

[60] Isono H. 3D display technology trends. Vision. 1996, 8, 106–18, http://www.visionsociety.jp/ vision/koumokuPDF/06kaisetu/E1996.08.02.02.pdf.

[61] Smalley D, Poon T-C, Gao H, Kvavle J, Qaderi K. Volumetric Displays: Turning 3-D Inside-Out. Optics & Photonics. 2018. https://www.osa-opn.org/home/articles/volume_29/june_2018/ features/volumetric_displays_turning_3-d_inside-out/.

[62] Hirayama R, Plasencia DM, Masuda N, Subramanian S. A volumetric display for visual, tactile and audio presentation using acoustic trapping. Nature, 2019, 575, 320–3.

[63] 3D Volumetric Display Technology. https://www.techbriefs.com/component/content/article/ tb/supplements/ptb/features/applications/21710.

[64] Held RT, Hui TT. A guide to stereoscopic 3D displays in medicine. Academic Radiology. 2011, 18, 1035–48.

[65] Van Beurden MHPH, Ijsselsteijin WA, Juola JF. Effectiveness of stereoscopic displays in medicine: A review. 3D Research. 2012, 3. doi: https://doi.org/10.1007/3DRes.01(2012)3.

[66] Liu J, Cui F, Li J, Shao W, Wang W, Li J, Liu M, He J. Development and clinical applications of glasses-free three-dimensional (3D) display technology for thoracoscopic surgery. Annals of Translational Medicine. 2018, 6, 214. doi: 10.21037/atm.2018.05.44.

[67] Freitag D. The Role Of 3D Displays In Medical Imaging Applications. 2015. https://www.med deviceonline.com/doc/the-role-of-d-displays-in-medical-imaging-applications-0001.

[68] Sun Z. 3D printing in medicine: current applications and future directions. Quantitative Imaging in Medicine and Surgery. 2018, 8. doi: 10.21037/qims.2018.12.06.

[69] About Parity Innovations. https://www.piq.co.jp/about_e.html.

[70] Parity mirror imaging. 2019; https://newswitch.jp/p/20409.

[71] "Parity Mirror®" moves images and touch panels in the air, just like in Science Fiction (SF) movies. 2019; https://shinkachi-portal.smrj.go.jp/en/webmagazine/qriii/.

[72] https://image.itmedia.co.jp/l/im/nl/articles/2102/14/l_miya_2102fptsystem02.jpg.

[73] Thaler SL. The emerging intelligence and its critical look at us. Journal of Near-Death Studies. 1998, 17, 21–9. doi: https://doi.org/10.1023/A: 1022990118714.

[74] Graziano M. Consciousness and the Social Brain. 2013, Oxford University Press.

[75] Artificial Intelligence: A Modern Approach. 2020; https://en.wikipedia.org/wiki/Artificial_In telligence:_A_Modern_Approach.

[76] Takata A. What is consciousness in the first place? Can artificial consciousness be created with artificial intelligence? https://robomind.co.jp/isikittenani/.

[77] Bongard J, Zykov V, Lipson H. Resilient machines through continuous self-modeling. Science. 2006, 314, 1118–21.

[78] Kwiatkowski R, Lipson H. Task-agnostic self-modeling machines. Science Robotics. 2019, 4, eaau9354. doi: 10.1126/scirobotics.aau9354.

[79] Tulving E. Memory and consciousness. Canadian Psychology. 1985, 26, 1–12. doi: https://doi.org/10.1037/h0080017.

[80] Mental time travel. https://en.wikipedia.org/wiki/Mental_time_travel.

[81] Kanai R. Answering the super-conundrum of "Can artificial intelligence be conscious" 2017; https://gendai.ismedia.jp/articles/-/52863.

[82] Haladjian HH, Montemayor C. Artificial consciousness and the consciousness-attention dissociation. Consciousness and Cognition. 2016, 45, 210–25.

[83] Hildt E. Artificial intelligence: does consciousness matter? Frontiers in Psychology. 2019, 10, 1535. doi: 10.3389/fpsyg.2019.01535.

[84] Solaiman SM. Legal personality of robots, corporations, idols and chimpanzees: a quest for legitimacy. Artificial Intelligence and Law. 2017, 25, 155–79. 19.2019 07:00 AM.

[85] Dennet DC. Will AI Achieve Consciousness? Wrong Question. 2019; https://www.wired.com/story/will-ai-achieve-consciousness-wrong-question/.

[86] Brain-Computer Interface. https://en.wikipedia.org/wiki/Brain%E2%80%93computer_interface

[87] Levine SP and 9 co-authors. A direct brain interface based on event-related potentials. IEEE Transactions on Rehabilitation Engineering. 2000, 8, 180–5.

[88] Zhang X, Ma Z, Zheng H, Li T, Chen K, Wang X, Liu C, Xu L, Wu X, Lin D, Lin H. The combination of brain-computer interfaces and artificial intelligence: applications and challenges. Annals of Translational Medicine. 2020, 8, 712. doi: 10.21037/atm.2019.11.109.

[89] Ienca M. Brain Machine Interfaces, Artificial Intelligence and Neurorights. 2017; https://brain.ieee.org/newsletter/2017-issue-3/brain-machine-interfaces-artificial-intelligence-neu rorights/.

[90] Smithson A. Artificial Intelligence + Brain-Computer Interface = The Ultimate Human Hack. 2019; https://alan-smithson.medium.com/artificial-intelligence-brain-computer-interface-the-ultimate-human-hack-3cbdd3395fc0.

[91] Schroeder R. We Are Underestimating Artificial Intelligence and BCI. 2020; https://www.insi dehighered.com/digital-learning/blogs/online-trending-now/we-are-underestimating-artifi cial-intelligence-and-bci.

[92]   7 Leading Brain-Computer Interface Companies and their Current and Prospective Products. https://rossdawson.com/futurist/companies-creating-future/leading-brain-computer-interface-companies-bci/.

[93]   Gonfalonieri A. What Brain-Computer Interfaces Could Mean for the Future of Work. Harvard Business Review; https://hbr.org/2020/10/what-brain-computer-interfaces-could-mean-for-the-future-of-work.

[94]   Pelagia M. AI in BCI: The new era of human factor design and research – InnoBrain. 2020; https://innobrain.se/2020/05/11/ai-in-bci-the-new-era-of-human-factor-design-and-research/.

[95]   Darling K. Extending legal protection to social robots: the effects of anthropomorphism, empathy, and violent behavior towards robotic objects in We Robot Conference 2012. 2012; https://papers.ssrn.com/sol3/papers.cfm?abstract_id=2044797.

[96]   Aguis C. Evolution of AI: Past, Present, Future. 2019. https://medium.com/datadriveninvestor/evolution-of-ai-past-present-future-6f995d5f964a.

[97]   Ruuse L. Artificial Intelligence: Mind-Boggling Future Predictions in 2019. https://www.scoro.com/blog/artificial-intelligence-predictions/.

[98]   Press G. 120 AI Predictions For 2020. Enterprise & Cloud. https://www.forbes.com/sites/gilpress/2019/12/09/120-ai-predictions-for-2020/#14f9439750cf.

[99]   Marr B. 5 Important Artificial Intelligence Predictions (For 2019) Everyone Should Read. https://www.forbes.com/sites/bernardmarr/2018/12/03/5-important-artificial-intelligence-predictions-for-2019-everyone-should-read/#5c84a82c319f.

[100] Marr B. 5 Key Artificial Intelligence Predictions: How Machine Learning Will Change Everything. 2018. https://bernardmarr.com/default.asp?contentID=1244.

[101] Top 5 Artificial Intelligence (AI) Predictions in 2020. 2020. https://www.geeksforgeeks.org/top-5-artificial-intelligenceai-predictions-in-2020/.

[102] Talwar R, Wells S, Whittington A, Koury A, Calle H. The evolution of AI: Seven stages leading to a smarter world. https://www.gigabitmagazine.com/ai/evolution-ai-seven-stages-leading-smarter-world.

[103] Kurzweil R. AI Will Not Displace Humans, It's Going to Enhance Us. 2017. https://futurism.com/ray-kurzweil-ai-displace-humans-going-enhance.

[104] Jajal TD. Distinguishing between Narrow AI, General AI and Super AI. 2018; https://medium.com/@tjajal/distinguishing-between-narrow-ai-general-ai-and-super-ai-a4bc44172e22.

[105] The technological singularity. https://en.wikipedia.org/wiki/Technological_singularity#cite_note-Singularity_hypotheses-3

[106] Liu F, Shi Y, Liu Y. Intelligence Quotient. Intelligence grade of artificial intelligence. Annals of Data Science. 2017, 4, 179–91.

[107] IQ of AI is now like 6-years-old, but exceeds human brain cells after few years. 2017. https://www.j cust.com/2017/10/12310927.html?p=all.

[108] Ito J. Extended Intelligence. 2021. https://www.researchgate.net/publication/322810327_Extended_Intelligence.

[109] Quantum computing. https://en.wikipedia.org/wiki/Quantum_computing.

[110] Lu D. What is a quantum computer? https://www.newscientist.com/question/what-is-a-quantum-computer/.

[111] Yano Y. What is a quantum computer? 2020. https://ledge.ai/quantumcomputer/.

[112] Quantum Computing: How it differs from classical computing? https://www.bbva.com/en/quantum-computing-how-it-differs-from-classical-computing/.

[113] Gil D. The Quantum Era of Accelerated Discovery. https://www.youtube.com/watch?v=zOGNoDO7mcU.

[114] Immunizing quantum computers against errors. 2019. https://www.sciencedaily.com/re leases/2019/02/190227131845.htm.

[115] Achieving scalability in quantum computing. 2018. https://cloudblogs.microsoft.com/quan tum/2018/05/16/achieving-scalability-in-quantum-computing/.

[116] Biamonte J, Wittek P, Pancotti N, Rebentrost P, Wiebe N, Lloyd S. Quantum machine learning. Nature. 2017, 549, 195–202.

[117] What is NISQ computing? https://q-ctrl.com/foundations/nisq/.

[118] Suchara M, Alexeev Y, Chong F, Finkel H, Hoffmann H, Larson J, Osborn J, Smith G. Hybrid Quantum-Classical Computing Architectures. 3rd Intl Workshop On Post-MOORE'S Era Supercomputing (PMES). 2018, https://cpb-us-w2.wpmucdn.com/voices.uchicago.edu/dist/ 0/2327/files/2019/12/HybridQuantumPMES.pdf.

[119] Karalekas PJ, Tezak NA, Peterson EC, Ryan CA, Da Silva MP, Smith RS. A quantum-classical cloud platform optimized for variational hybrid algorithms. Quantum Science and Technology. 2020, 5, 024003. doi: 10.1088/2058-9565/ab7559.

[120] Xia R, Kais S. Quantum machine learning for electronic structure calculations. Nature Communications. 2018, 9, 4195. doi: 10.1038/s41467-018-06598-z.

[121] How Do Quantum Computers Work? https://www.sciencealert.com/quantum-computers.

Closing remarks

The term "artificial intelligence" is immanent in a duo problem. It is a question whether intelligence can be reconstructed artificially and what is intelligence in the first place? Even though the full extent of human intelligence and wisdom has not yet been elucidated completely, the process of attempting to reconstruct human intelligence has highlighted what human intelligence is. There are two types of personality: personality of human in AI era and personality of personified AI (or artificial personality, AI personality, or robot personality). Both personalities are always changing, depending on surrounding environments. Furthermore, although there is another relationship between human's personality and AI (namely, research and assessment of human personality by use of AI/ML methods [1, 2]), it is out of scope what we are discussing here. (1) As to human personality, although there is no single definition of personality in human, it generally refers to a set of traits that predict a person's behavior. It can also refer to sets of behaviors or emotional patterns that derive from both biological and environmental factors. With the introduction of digital voice assistants on smartphones, tablets, computers, and smart home speakers and many other home appliances, humans are interacting with technology more than ever before. We are now at the dawn of a new era in AI technology, namely the so-called fourth industrial revolution that will reshape every industry. It is believed that AI cannot comprehend the context of what was input as well as outcomes that it produces. Our next generation in AI era is called as "AI native generation." Technology is already there ahead of them, so they do not even have the opportunity to overcome that high barrier. This AI native generation will get used to AI suggesting videos they want to watch, temperatures that should be comfortable, and deliveries before they are short in the fridge before they search or feel the need. Looking at the timelines on Twitter and Facebook, people's brains are changing plastically so that they do not think about a single thing. (2) An artificial personality or robotic personality is an advanced aspect of AI in which a device displays characteristic human behavior. Particularly, personality refers to the ability of a robot or software system to interact with people emotionally as well as on a logical level. The notion of robotic personality is based on anthropomorphism, a tendency for people to think of certain objects or machines as having human-like characteristics [3–7]. Ideally, these two types personality should be coexisting by equally influencing to each other; however, as mentioned above, AI technology appears to be overwhelming, so that these two are not in well balance any more.

The rise of the DL has caused a lot of excitement around the revolutionary capabilities of the artificially intelligent agents. But it has also raised fear and suspicion about what exactly is going on inside each algorithm. With DL, regulators cannot review the rationale or rules behind the algorithms, as neither is actually intelligible to humans – there is not a defined body of knowledge or rules of reasoning humans can comprehend. Essentially, DL agents use pre-existing data to find patterns and

https://doi.org/10.1515/9783110717853-015

predict future outcomes based on those patterns. Precisely how they arrive at a prediction is not truly known. So, if we are handing over important societal decisions to an AI agent, we need to know if we can trust the technology [8]. The great strength of DL lies in its ability to digest vast amounts of data that are used to identify patterns and make predictions. However, DL's extraordinary capacity has also raised fears and concerns. DL agents use pre-existing data to find patterns and predict future outcomes.

As mentioned earlier, and Latour [9] pointed out in the relationship between human and guns, human and AI are interpenetrative relationships in which one does not have the essence of either a subject or an object. We should look to the combined complex of these two and beyond. The focus on the human-technological complex and its consequences has been becoming even more important today, when the development of science and technology typified by an AI is directly and indirectly affecting our perspectives and the foundation of the way society should be. There has been a conflict between me personally as a human being and AI, which is an ever-advancing technology, while writing this book. This will continue now and in the future. The title of this book I imposed on me will forever be my lifetime homework theme.

In closing, I admit having a list of individual names to be mentioned with my memories and appreciation – my mentors who inspired me, friends and colleagues who challenged me, and most importantly, former students (some of them are now my colleagues) who continue to both inspire and challenge me. To all of them, I owe my knowledge and capacity to comprehend publications cited in this book. Those who were and are valuable to me, and this book should include, are V. Weiss, N. Abe, the late N. Iguchi, Y. Miwa, S. Wesugi, F. Farzin-Nia, G. Stookey, T. Miyazaki, T. Tominaga, Y. Ono, E. Tada, and D. Neugebauer. Special thanks to H. Yasukawa for comments and editing. Many thanks should also go to individual author(s) of articles cited in this book. Last but not least important, my sincere gratitude goes to the excellent editorial team of De Gruyter Publishing. Thank you all!

References

[1] Stachl C, Pargent F, Hilbert S, Harari GM, Schoedel R, Vaid S, Gosling SD, Bühner M. Personality Research and Assessment in the Era of Machine Learning. 2019. doi: 10.31234/osf.io/efnj8.

[2] Lim J. Personality in AI: Why Someday Soon Your Toaster Could be an Extrovert. Hanson Robotics. 2019; https://www.hansonrobotics.com/personality-in-ai-why-someday-soon-your-toaster-could-be-an-extrovert/.

[3] Robotic Personality. https://whatis.techtarget.com/definition/robotic-personality#:~:text=Ro botic%20personality%20is%20an%20advanced,as%20on%20a%20logical%20level.

[4] Mileounis A, Cuijpers RH, Barakova EI. (2015) Creating Robots with Personality: The Effect of Personality on Social Intelligence. In: Ferrández Vicente J. et al., ed., Artificial Computation in

Biology and Medicine. IWINAC 2015. Springer, Cham. https://doi.org/10.1007/978-3-319-18914-7_13.

[5]   Bremner PA, Celiktutan O, Gunes H. Personality Perception of Robot Avatar Teleoperators in Solo and Dyadic Tasks. Front Robot. AI. 2017, 23, https://doi.org/10.3389/frobt.2017.00016.

[6]   Mou Y, Shi C, Shen T, Xu K. A Systematic Review of the Personality of Robot: Mapping Its Conceptualization, Operationalization, Contextualization and Effects. International Journal of Human–Computer Interaction. 2020, 36; https://doi.org/10.1080/10447318.2019.1663008.

[7]   Endres LS. Personality engineering: Applying human personality theory to the design of artificial personalities. Advances in Human Factors/Ergonomics. 1995, 20, 477–82.

[8]   Chia H. The "Personality" in Artificial Intelligence. https://pursuit.unimelb.edu.au/articles/the-personality-in-artificial-intelligence.

[9]   Latour B. Pandora's Hope: Essays on the Reality of Science Studies. Harvard University Press, 1999.

Index

https://doi.org/10.1515/9783110717853-016